P9-ASI-372

Invasive Plants and Forest Ecosystems

Invasive Plants and Forest Ecosystems

Edited by
Ravinder Kumar Kohli
Shibu Jose
Harminder Pal Singh
Daizy Rani Batish

CRC Press is an imprint of the
Taylor & Francis Group, an informa business

CRC Press
Taylor & Francis Group
6000 Broken Sound Parkway NW, Suite 300
Boca Raton, FL 33487-2742

Printed in the United States of America on acid-free paper
10 9 8 7 6 5 4 3 2 1

International Standard Book Number-13: 978-1-4200-4337-2 (Hardcover)

Library of Congress Cataloging-in-Publication Data

Invasive plants and forest ecosystems / edited by Ravinder K. Kohli ... [et al.].
p. cm.
Includes bibliographical references and index.
ISBN-13: 978-1-4200-4337-2
ISBN-10: 1-4200-4337-4
1. Invasive plants. 2. Invasive plants--Ecology. 3. Forest ecology. I. Kohli, R. K. II. Title.

SB613.5.I57 2009
634.9'65--dc22 2008008314

Visit the Taylor & Francis Web site at
http://www.taylorandfrancis.com

and the CRC Press Web site at
http://www.crcpress.com

Contents

SECTION I Invasion Ecology

SECTION II Ecological Impacts

SECTION III Management of Invasive Plants

Preface

Plant species that invade an alien area and outgrow the native vegetation, establishing and increasing their own territory, often lead to negative economic, environmental, and social impacts. Even native species can behave like invasive species by their exponential spread. Similarly, not all non-native species are invasive. Many alien invasive species, however, do threaten the health and integrity of our terrestrial and aquatic ecosystems. As the human population explodes and trade becomes increasingly globalized, the transboundary movement of species from their places of origin to alien regions is escalating and is expected to continue so in the coming decades.

The ecological impacts of invasive species, in particular alien plants, on forest ecosystems have attracted the attention of researchers, managers, and policy makers the world over. Alien invaders are known to change the ecosystem structure and function, thereby minimizing the support for native flora, fauna and the provision of ecosystem services that directly benefit humans. There exists a large amount of scattered information on the ecology and management of alien invaders in forest ecosystems. The scientists and practitioners who attended the technical session, "Impact of exotic invasive plant species on forest ecosystems," during the XXII-IUFRO World Congress, Brisbane, 2005, (Ravinder K. Kohli, organizer; Shibu Jose, rapporteur) resolved that an edited volume that captures the current state of knowledge would benefit the global forestry and natural resources community. Invited papers and voluntary posters were presented at this session, which represented a cross section of the current global research being conducted in a variety of forest ecosystems. The editors of this volume accepted the task enthusiastically and immediately started working on the project. Selected authors (who presented their work at the IUFRO Congress) and other prominent researchers were invited to submit manuscripts. After a peer review process, we selected 21 chapters for the current volume.

The chapters are grouped into four sections. The first section examines invasion ecology through both synthesis and original research articles. Seven chapters are included in this section. The second section also has seven chapters. These give readers a flavor of the ecological impacts of alien invaders and include examples from both tropical and temperate regions of the world. The third section is an exploration into the adaptive collaborative management strategies that are central to successful control of invasive alien plants. The first chapter in this section describes the concept of adaptive collaborative restoration in great detail. The second chapter discusses the ecology and management of the world's number one invasive plant species of natural ecosystems. The remaining three chapters of the section give specific examples of invasive plant management, one each from a local, state, and country perspective. The fourth (last) section includes two chapters that discuss the socioeconomic and policy aspects of invasive species management using two examples, one from the tropics (Namibia) and another from the temperate region (the United States).

The book is intended as a reference book for students, scientists, professionals, and policy makers who are involved in the study and management of invasive alien plants in forested ecosystems of the world. We are grateful to a large number of individuals for their assistance in accomplishing this task, particularly the authors, for their commitment to the project and their original research or synthesis of the current knowledge. Moreover, the invaluable comments and suggestions made by the referees significantly improved the clarity and content of the chapters. We also extend our sincere thanks to John Sulzycki and Pat Roberson of CRC Press for their timely efforts in publishing this book.

Ravinder K. Kohli
Shibu Jose
Harminder P. Singh
Daizy R. Batish

Editors

Ravinder K. Kohli, PhD, a certified senior ecologist of the Ecological Society of America, is a senior professor of plant sciences and coordinator of the Centre for Environment and Vocational Studies at Panjab University, Chandigarh, India. He is also an adjunct professor of the Xishuangbanna Tropical Botanical Garden, Chinese Academy of Sciences, and holds the SAARC (South Asian Association for Regional Cooperation) chair at the University of Chittagong, Bangladesh. He is a fellow of some important academies, such as the National Academy of Agricultural Sciences (India), the National Academy of Sciences (India), the Indian Botanical Society, and the National Environment Science Academy (India). He is also the coordinator of the IUFRO (International Union of Forest Research Organization) Unit 8.02.04—Ecology of Alien Invasives—and IUFRO 4.02.02—Multipurpose Inventories. His research career of 28 years has been focused on ecological implications of introduced trees and invasive alien plants in India. He is on the editorial boards of several journals devoted to crop production, ecology, and environment. The honors and awards he has received include the B.P. Pal National Environment Fellowship Award (2001); the highest paid fellowship award of the Government of India, for work on Biodiversity; state honor from the government of Chandigarh for teaching and research in environment; and the Nanda Memorial National Young Scientist Award (1988) for research and forestry. He has to his credit about 100 research papers in international refereed journals, 70 research articles, 13 book chapters 7 reviews, 3 authored and 8 edited books apart from guiding 26 PhD theses and 3 patent applications.

Shibu Jose, PhD, is an associate professor of forest ecology at the School of Forest Resources and Conservation (SFRC), University of Florida, Gainesville. He is also co-director of the Cooperative for Conserved Forest Ecosystems at SFRC. He received his BS (Forestry) from India and MS and PhD (Forest Science) from Purdue University. He holds an affiliate faculty status at the School of Natural Resources and the Environment and the Soil and Water Science Department. His current research efforts focus on production ecology and ecophysiology of intensively managed pine and hardwood forests, restoration ecology of the longleaf pine ecosystem, invasive plant ecology and management, and ecological interactions in agroforestry systems and mixed species forest plantations. He serves on the editorial boards of *Journal of Forestry* (editor), *Agroforestry Systems* (editor-in-chief), *Forest Science* (book review editor), and *Research Letters in Ecology* (associate editor). He teaches forest ecology, applied forest ecology, advanced forest ecology, and ecology and restoration of the longleaf pine ecosystem. The honors and awards he has received include the Aga Khan International Fellowship (Switzerland), the Nehru Memorial Award for Scholastic Excellence (India), the UF CALS Junior Faculty Award of Merit, Award of Excellence in Research by the Southeastern Society of American Foresters (SAF), the Stephen Spurr Award by the Florida Division SAF, and the Young Leadership Award by the National SAF. He has over 150 publications to his credit, including 5 books.

Harminder P. Singh, PhD, an ecologist, is a lecturer of biotic environment at the Centre for Environment and Vocational Studies, Panjab University, Chandigarh, India. He received his PhD in botany from Panjab University. He guides researchers and teaches ecological principles and conservation of life support systems to students of environment and solid waste management at the masters level. His research interests include chemical ecology of plant interactions and impact of exotic invasive plants on native ecosystems. He has to his credit over 55 research papers in international refereed journals, 5 books, and 27 research articles. He is a recipient of the Young Scientist Award from Punjab Academy of Sciences, Indian Science Congress Association, and Dalela Educational Foundation (India) as well as the Junior Environmentalist Award from the National Environment Science Academy (India).

Daizy R. Batish, PhD, is a reader at the Department of Botany, Panjab University, Chandigarh, India. Her research interests include ecophysiology of interplant interactions, biology and ecology of invasive weeds, and ecological weed management. Apart from guiding doctoral research, she teaches ecology, environment botany, and forestry to undergraduate and postgraduate students. She has to her credit over 70 research papers in refereed journals, 5 books, and 36 research articles. She is a fellow of the National Environment Science Academy (India), a recipient of the Rajib Goyal Young Scientist Award in environmental sciences and a research award of the University Grants Commission (New Delhi, India). She is on the review committee of several international journals in plant sciences.

Contributors

Janaki R.R. Alavalapati
School of Forest Resources
and Conservation
University of Florida
Gainesville, Florida

Rachel Albritton
School of Forest Resources
and Conservation
University of Florida
Gainesville, Florida

Mary M. Apetorgbor
Forestry Research Institute of Ghana
Council for Scientific and Industrial
Research
Kumasi, Ghana

Daizy R. Batish
Department of Botany
Panjab University
Chandigarh, India

K. George Beck
Department of Bioagricultural Sciences
and Pest Management
Colorado State University
Fort Collins, Colorado

Paul P. Bosu
Forestry Research Institute of Ghana
Council for Scientific and Industrial
Research
Kumasi, Ghana

Daniel Bowker
Department of Forestry
University of Kentucky
Lexington, Kentucky

Mila Bristow
Walkamin Research Station
Walkamin, Australia

Alexandra R. Collins
School of Forest Resources
and Conservation
University of Florida
Gainesville, Florida

C. Mark Cowell
Department of Geography
University of Missouri–Columbia
Columbia, Missouri

Pedram Daneshgar
School of Forest Resources
and Conservation
University of Florida
Gainesville, Florida

Kuldeep Singh Dogra
Department of Botany
Panjab University
Chandigarh, India

James M. Dyer
Department of Geography
Ohio University
Athens, Ohio

Jichao Fang
Institute of Plant Protection
Jiangsu Academy of Agricultural
Sciences
Nanjing, China

Songlin Fei
Department of Forestry
University of Kentucky
Lexington, Kentucky

Johanna E. Freeman
School of Forest Resources
and Conservation
University of Florida
Gainesville, Florida

Andrew Gray
Forest Inventory and Analysis
USDA Forest Service
Pacific Northwest Research Station
Corvallis, Oregon

Mark H. Hansen
USDA Forest Service
Northern Research Station
Forest Inventory and Analysis
St. Paul, Minnesota

M.K. Hossain
Institute of Forestry and Environmental
Sciences
Chittagong University
Chittagong, Bangladesh

Michael A. Jenkins
Inventory and Monitoring Program
Great Smoky Mountains National Park
Gatlinburg, Tennessee

Kristine D. Johnson
Vegetation Management Program
Great Smoky Mountains National Park
Gatlinburg, Tennessee

Shibu Jose
School of Forest Resources
and Conservation
University of Florida
Gainesville, Florida

Dave F. Joubert
School of Natural Resources and Tourism
Polytechnic of Namibia
Windhoek, Namibia

Ravinder K. Kohli
Centre for Environment and Vocational
Studies
Panjab University
Chandigarh, India

Ningning Kong
Department of Forestry
University of Kentucky
Lexington, Kentucky

Robert E. Loeb
Departments of Biology and Forestry
The Pennsylvania State University
DuBois Campus
DuBois, Pennsylvania

Gregory E. MacDonald
Department of Agronomy
Institute of Food and Agricultural
Sciences
University of Florida
Gainesville, Florida

William H. McWilliams
USDA Forest Service
Northern Research Station
Forest Inventory and Analysis
Newtown Square, Pennsylvania

James H. Miller
Southern Research Station
USDA Forest Service
Auburn, Alabama

W. Keith Moser
USDA Forest Service
Northern Research Station
Forest Inventory and Analysis
St. Paul, Minnesota

Mark D. Nelson
USDA Forest Service
Northern Research Station
Forest Inventory and Analysis
St. Paul, Minnesota

J. Doland Nichols
School of Environmental Science and Management
Southern Cross University
Lismore, New South Wales, Australia

E. Corrie Pieterson
School of Natural Resources and the Environment
University of Florida
Gainesville, Florida

Alemayehu Refera
Ethiopian Institute of Agricultural Research
Forestry Research Centre
Addis Ababa, Ethiopia

John Schelhas
Southern Research Station
USDA Forest Service
Tuskegee University
Tuskegee, Alabama

Gritta Schrader
Institute of National and International Plant Health
Julius Kuehn Institute
Braunschweig, Germany

Harminder P. Singh
Centre for Environment and Vocational Studies
Department of Botany
Panjab University
Chandigarh, India

Uwe Starfinger
Institute of National and International Plant Health
Julius Kuehn Institute
Braunschweig, Germany

Jeffrey Stringer
Department of Forestry
University of Kentucky
Lexington, Kentucky

Fanghao Wan
Institute of Plant Protection
Chinese Academy of Agricultural Sciences
Beijing, China

Steven R. Wangen
Bioprotection and Ecology Division
Lincoln University
Canterbury, New Zealand

Christopher R. Webster
School of Forest Resources and Environmental Science
Michigan Technological University
Houghton, Michigan

Section I

Invasion Ecology

1 Invasive Plants: A Threat to the Integrity and Sustainability of Forest Ecosystems

Shibu Jose, Ravinder K. Kohli, Harminder P. Singh, Daizy R. Batish, and E. Corrie Pieterson

CONTENTS

1.1 INTRODUCTION

Charles Elton, a pioneer in population ecology, wrote of how ecological explosions were threatening the world (Elton 1958). Nearly half a century later, his early warning has become one of the most important environmental crises of our time. Biological invasions have caused more species extinctions than did human-induced climate change (D'Antonio and Vitousek 1992), and are the second leading cause of species extinctions after habitat loss. Biological invasion is one of the major reasons of biodiversity depletion. Invasive plants, in particular, are to blame for much of native species decline and ecosystem degradation (Wilcove et al. 1998). The invasion of native ecosystems by alien plants can lead to alterations in nutrient cycling, fire regime, hydrology, energy budgets, and native species abundance and survival (Mack et al. 2000).

Biological invasion occurs when species move from one geographical region to another, where they establish, proliferate, and persist (Mack et al. 2000). Several terms such as *introduced, nonnative* or *nonindigenous, exotic, alien, foreign*, or *invasive alien* have been applied to these species. In a community, a native species is the one that is naturally found in a given area, whereas introduced or exotic species result from human-induced introductions or accidental entries. In addition, there are some species whose origin is not clearly known and such species are known as *cryptogenic* (Carlton 1996).

In fact, there is a stalemate regarding much of the terminology in the discipline of invasion ecology. Richardson et al. (2000), on the basis of their critical analysis and extensive surveys, have tried to provide a clear insight into this problem. According to them, an invasive species is a naturalized alien that produces a large number of reproductive offsprings at considerable distances from the parent plant, and thus spread over large area (Richardson et al. 2000). This is in contrast to casual alien species, which do not form self-replacing populations, and hence cannot perpetuate for a long time. However, the World Conservation Union (formerly International Union for Conservation of Nature and Natural Resources), the Convention on Biological Diversity (CBD), and the Global Invasive Species Programme (GISP) prefer to use the term *invasive alien species* (IAS). As per GISP, an IAS is that alien species which proliferates and spreads in the new environment in ways that are destructive to human interests (McNeely et al. 2001). IAS has been addressed under Article 8(h) of the CBD.

In a survey of peer-reviewed literature regarding invasions and alien species, Pyšek et al. (2006) found 329 papers and over 27,000 citations in the period 1981–2003, reflecting the rapid growth of the field. The majority of the most commonly cited papers are related to plant invasions. Although the economic and ecological damage caused by alien animal and microbe species is astounding, the scope of this book is limited to alien plant invasions of forested ecosystems in various parts of the world. Several recently published books have focused on biological invasions (e.g., Mooney and Hobbs 2000; Pimentel 2002; Myers and Bazley 2003; Ruiz and Carlton 2003; Mooney et al. 2005; Sax et al. 2005; Cadotte et al. 2006; Lockwood et al. 2006), which is a clear indication of the growing body of literature on the topic. It also shows the seriousness with which the issue is being addressed by the scientific community. Since forests are one of the ecosystems most seriously affected by biological invasions, and in particular by invasive plants, this book is dedicated to the treatment of invasive plants in forested ecosystems. Authors from around the world contributed to this volume, and chapters have been organized into four sections: Invasion Ecology, Ecological Impacts, Management of Invasive Plants, and Socio-economic Policy Aspects. We summarize the scope of each section here.

1.2 INVASION ECOLOGY

This section begins with a chapter (Chapter 2, by Daneshgar and Jose) on the proposed mechanisms of invasion. Despite the number of studies on IAS, no grand unifying theory of invasions has emerged. Rather, researchers have proposed a suite of theories and hypotheses relating to the invading species, the habitats invaded, and their interactions. Inconsistent and varying terminology and difficulty in predicting which species will or will not become invasive contribute to the challenge of developing an overarching theory of biological invasions (Shrader-Frechette 2001). Nevertheless, the combination of ecological effects, societal concern, and economic costs have created the need for explanations of how to predict and prevent invasions and how best to treat them when they do occur.

The role of disturbance in the establishment and spread of invasive species is examined in Chapter 3 by Moser et al. These authors used the USDA Forest Service Northern Research Station's Forest Inventory and Analysis data for 2005 and 2006 and sampled for the presence and percent cover class of 25 selected nonnative invasive plant species in all the forested plots in the midwestern states (Indiana, Illinois, Missouri, Iowa, Minnesota, Wisconsin, and Michigan). Iowa, Indiana, and Illinois had relatively higher rates of invasive species presence, while Minnesota had the lowest. They also observed a strong latitudinal separation, particularly for woody invasives. Most subboreal forest types had lower percentages of invasive species. Similarly, Lake States of Minnesota, Wisconsin, and Michigan had lower invasive species presence than that in the southern-tier states. Grasses were particularly prominent in low-density or fragmented forestland. Metrics of disturbance and fragmentation, such as distance to road, county percent forest, or the forest intactness index, were significantly related to the presence and coverage of invasive species. They concluded that even the disturbance measures, lower basal area, and high road density could easily reflect the lingering influence of historic human disturbance as the microsite attributes that allowed invasive species to establish and expand.

Fei et al. argue in Chapter 4 that while disturbance and other landscape features play a major role in the establishment and spread of invasive plants, it is essential to determine how these factors influence the alien plant invasions. They used a geographic information system, high-accuracy global positioning system receivers, and high-resolution aerial photos to study the invasion patterns of alien plant species in eastern Kentucky. They have shown that invasive species occurrence was higher on or near roads than in the interior of the forest, although some species could establish in the forest interiors, where signs of anthropogenic disturbance were absent. Ground disturbance from harvesting, specifically skidding and road construction, could promote an increase in colonization and spread of invasive species.

Webster and Wangen (Chapter 5) use a case study from Michigan to explore the spatial and temporal dynamics of alien tree invasions. They used the example of *Acer platanoides*, a shade-tolerant invasive tree, and showed that because of the long generation times of trees relative to other organisms their invasion potential might not be easily recognized until they become a serious pest. With respect to the spatial spread patterns, they observed both thread-like patterns (e.g., roads and trails) and satellite populations that originated from human-mediated seed dispersal. The authors suggested that monitoring for invasive trees, therefore, should be proactive and where possible use risk assessment techniques to identify likely establishment sites. Using another case study, Dyer and Cowell (Chapter 6) demonstrated that habitat changes caused by the establishment of alien species through changes in natural disturbance regimes could steer communities to a new stable state, posing serious management concerns. Disruption of a natural disturbance—flooding, by the construction of a dam—caused invasive plants such as Japanese knotweed (*Polygonum cuspidatum*) and Reed canary grass (*Phalaris arundinacea*) to thrive at the expense of many resident overstory and understory plant species. These authors have shown that changes to disturbance regimes could alter both resource availability and competitive interactions in favor of alien species, with adaptive traits not present within the native community.

In Chapter 7, Loeb examines the biogeography of invasive plant species in urban forests and parks. Pre-twentieth century plantings comprised one-third of the invasive tree, shrub, and herbaceous species found in the urban forests of the mid-Atlantic region of the United States. Many of these invasive plants were available from nurseries before 1900. He concluded that urban forests are disturbance communities and could serve as centers for spread of invasive plant species.

1.3 ECOLOGICAL IMPACTS

Invasive species have contributed to the decline of 42% of the U.S. endangered and threatened species (Wilcove et al. 1998). In other parts of the world, as many as 80% of the endangered species are threatened because of the pressures of nonnative species (Pimentel et al. 2005). IAS may be intentionally or unintentionally introduced to new ecosystems. When a species is introduced, it may have both positive and negative consequences. This complicates the matter of how and whether to attempt to control or eradicate such species. In plantation forest ecosystems, many planted tree species are exotic (Ewel et al. 1999), which may also be invasive outside of the plantations. Research focusing on invasive plants in forest ecosystems contributes to our understanding of biological invasions in general and vice versa. This section focuses on the ecological impacts of alien invasive plants on forest ecosystems the world over.

The first three chapters in this section are from Asia. In Chapter 8, Hossain examines the effects of alien invasive plants on hill forest ecosystems of Bangladesh. According to Hossain, in addition to invasive understory species, the introduction of alien tree species for plantation forestry has also threatened the integrity of natural forest ecosystems. The ecological status of some of the invasive plants in one of the biodiversity hotspots in the world, the Himalayas, is discussed by Kohli et al. (Chapter 9). They concluded that establishment and spread of invasive plants such as *Ageratum conyzoides, Lantana camara*, and *Parthenium hysterophorus* have displaced native plant species and deteriorated the quality of native forest ecosystems in the region. In Chapter 10, Fang and Wan describe ecological impacts and management considerations of the major invasive pests with several examples. The invasive pests, including plants, cost China more than US$7 billion annually. These authors attributed the successful invasion and spread of the invasive species to anthropogenic factors, including intentional introduction and lack of rapid response mechanisms for eradicating potential invasive pests.

In Chapter 11, Nichols and Bristow discuss the invasive plants of subtropical and tropical Australian forests. Weeds cost Australia at least AU$3 billion per year, and this estimate is only for the cost of control and of losses due to agricultural and pastoral weeds. According to the authors, in the subtropical rainforest area, *L. camara* and *Cinnamomum camphora* (camphor laurel) often dominate much of the landscape. Although *C. camphora* aggressively colonizes areas that could support rainforest or mesic eucalypt forests, forming multistemmed thickets of trees with little commercial value, it also colonizes bare land and degraded pastures, creating conditions amenable for rainforest regeneration. Lantana has been threatening moist eucalypt forests and rainforests, and an estimated 80 plant species are threatened by the extensive coverage of the landscape by lantana. However, at the same time lantana has also become a keystone species for many species of animals. These authors also discuss the pine plantations in Australia, which cover ~1 million ha, and ask the question if these species are invasive as well.

In Chapter 12, Schrader and Starfinger focus on intentional introductions and the procedures of pest risk analysis for alien plants in European forests. They describe an approach to evaluate the probability of establishment and spread of invasive alien plants and the magnitude of the associated potential economic and environmental consequences (risk assessment), and how to deal with the identified risk using the case study of *Prunus serotina*. They demonstrate that careless planting of alien trees such as *P. serotina* could lead to a variety of negative impacts on biodiversity and economic values. However, the forestry sector in Europe has long benefited from the use of nonnative tree species. According to these authors, reference to the ecological and economic damage posed by alien trees would not be sufficient to influence policy decisions and forest owners. They identify risk analysis as an effective and important step in dealing with the use of alien trees in forests.

A wide range of approaches and types of data have been used to inventorise and monitor invasive plants. Gray (Chapter 13), using the USDA Forest Service's Forest Inventory and Analysis data from forestlands in California, Oregon, and Washington, has shown that these data could be used not only to monitor the presence and spread of invasive plants, but also to assess relationships between the occurrence of invasive alien species and climatic, topographic, and stand variables. The results showed a high percentage of plots with alien species. According to the author, these results could be quite surprising to policy makers and the public, as many of these stakeholders regard most of the regions' forestlands as rather pristine and consider invasive species to still be an emerging threat.

The last chapter in this section (Chapter 14) by Collins and Jose addresses the changes in soil chemical properties as a result of invasion by cogongrass (*Imperata cylindrica*), one of the most notorious invasive plants of forest ecosystems of the southeastern United States. The ability of invasive species to alter soil biochemistry, both through nutrient acquisition and allelopathy, remains a relatively unanswered question in invasion ecology and could offer important insights to potential mechanisms for invasion success. Field studies were conducted at two sites, a logged site and an unlogged site in Florida, to test the effects of *I. cylindrica* invasion on soil chemical properties. Analysis of soil samples, taken pairwise (*I. cylindrica* invaded and noninvaded areas) at both sites, showed significant differences in soil NO_3–N, K^+, and pH. Significantly lower levels of NO_3–N and K^+ were observed in *I. cylindrica* patches than in the surrounding native vegetation. The authors attributed the lower levels of these nutrients to the extreme ability of *I. cylindrica* to extract available resources from the area it invaded. The soil of the *I. cylindrica* patch was more acidic than that of the surrounding native vegetation. Although no direct evidence of any mechanisms responsible for lowering soil pH in *I. cylindrica* invaded patches was given, the authors suspected allelopathy or the preferential uptake of ammonium as a plausible mechanism.

1.4 MANAGEMENT OF INVASIVE PLANTS

This section begins with a discussion of the adaptive collaborative restoration concept of invasive plant management (Chapter 15). Miller and Schelhas reiterate the need for a concerted holistic effort that integrates science with management to predict, manage, and mitigate the spread of invasive species. The adaptive collaborative restoration approach incorporates elements from three key ecosystem management trends from the 1990s: adaptive management, collaborative management, and restoration management. Collaborating across institutional and property boundaries and across local and national levels to carry out the complex tasks of detection, prevention, eradication, and restoration of invasive species through a science-based adaptive learning process could be the key to effective invasive plant management. While it could always be a work in progress, this approach provides a framework with the potential to successfully combat invasive species.

The ecology and management of *I. cylindrica*, one of the 10 most troublesome weeds in the world and perhaps the worst weed of natural ecosystems, are discussed in Chapter 16 by MacDonald. This grass is considered a weedy pest in over 73 countries and is observed in every continent except Antarctica. Following an account of the biology and impacts of *I. cylindrica*, management strategies are discussed in detail. Although preventive, cultural, mechanical, biological, and chemical measures are identified and described, an integrated approach using multiple methods is recommended as the most effective way to manage *I. cylindrica* infestations. In Chapter 17, Jenkins and Johnson describe the alien invasive plant management program at the Great Smoky Mountains National Park in Tennessee as a case study of alien plant management in the U.S. National Park System. Since 2000, the National Park Service (NPS) has created 16 exotic plant management teams to assist 209 national parks with exotic plant control. Between 1999 and 2004, the NPS controlled alien plants on over 76,000 ha. According to the authors, $\sim$1 million ha still require control across the National Park System.

In Chapter 18, Beck focuses on the invasive weeds of Colorado forests and rangeland. Following a description of the forest types and rangelands and their current status, he discusses the major invasive plants and the current state of weed management efforts. Beck concludes that there has been progress in engaging more private and public land managers and landowners in the battle against invasive plants, albeit at a slow pace. He identified insufficient financial resources as one of the major limitations in Colorado and elsewhere in the United States in the battle against invasive species. The ecology and management of tropical Africa's forest invaders are discussed by Bosu et al. in Chapter 19. The introduction of alien plant species to tropical Africa dates as far back as the fifteenth century when the first Europeans arrived on the shores of the continent, but active and passive introductions of new species have continued throughout the centuries. Many of these introduced species now comprise a major proportion of the food, fiber, and wood resources on the continent. These authors describe some of the most troublesome invasive plants of tropical Africa and explain their management strategies. According to them, African scientists have realized the need to enhance the capacity and readiness to combat the spread of forest invasives on the continent, in particular with the formation of the Forest Invasive Species Network for Africa (FISNA). FISNA seeks to bring all forest health experts on the continent to work toward achieving the common objective of invasive species management.

1.5 SOCIOECONOMIC AND POLICY ASPECTS

The two chapters in this section address the socioeconomic and policy aspects of invasive species management. Joubert (Chapter 20) opens the section with a description of forest ecosystems and the current status of invasive plant problems in Namibia, one of the driest African countries south of the Sahara. According to the author, aridity and associated low population density have resulted in a relatively modest invasive plant problem in Namibia; however, a combination of propagule pressure and increasing invasibility of forest ecosystems in the north could escalate the problem. Joubert reviews the current policies and legislation regarding invasive alien species and their control and concludes that effective policies and programs need to be in place to prevent the introduction of likely invaders and the eradication of populations that are still in their establishment phase, so that potential economic and ecological losses can be avoided.

In their chapter, Freeman et al. (Chapter 21) explore the economics, law, and policy of invasive species management in the United States. The threats posed by invasive species to natural resources and to the economy are discussed first, followed by an exploration into the historical evolution of U.S. invasive species policy. The policy mechanisms in place to prevent new invasions and to manage existing ones are also discussed. They conclude the chapter by commenting on future directions for invasive species management in the United States.

1.6 CONCLUSION

It is clear that both deliberate and inadvertent introductions of alien plants have caused significant changes in structure and the function of forest ecosystems around the world. Many ecological functions are supported by a suite of species that are characteristic of a particular ecosystem. Alien invaders tend to alter the characteristic species composition of ecosystems, often by forming monospecific stands. The chapters in this volume make one appreciate the gravity of the situation, in particular, in forested ecosystems. Although commonalities exist in the mode of introduction and the spreading of alien invaders in new habitats, the extent of damage by the invaders varies depending on the socioeconomic and ecological realities of the affected regions. Although a vast majority of the alien invaders are considered harmful, some of them have formed alternate steady-state communities and have become keystone species in certain ecosystems. However, all authors agree that invasive plant management should be a priority in the management of forested ecosystems so that their health and integrity can be sustained. Although public awareness is increasing, policy-guided action plans are necessary to address the invasive species problem so that serious economic and ecological threats can be alleviated.

REFERENCES

Cadotte, M.W., McMahon, S.M., and Fukami, T., *Conceptual Ecology and Invasion Biology: Reciprocal Approaches to Nature*, Springer, Dordrecht, Netherlands, 505, 2006.

Carlton, J.T., Biological invasions and cryptogenic species, *Ecology*, 77, 6, 1653, 1996.

D'Antonio, C.M. and Vitousek, P.M., Biological invasions by exotic grasses, the grass/fire cycle, and global change, *Annu. Rev. Ecol. Syst.*, 23, 63, 1992.

Elton, C.S., *The Ecology of Invasions by Animals and Plants*, The University of Chicago Press, Chicago, IL, 181, 1958.

Ewel, J.J., O'Dowd, D.J., Bergelson, J., Daehler, C.C., D'Antonio, C.M., Gomez, L.D., Gordon, D.R., Hobbs, R.J., Holt, A., Hopper, K.R., Hughes, C.E., LaHart, M., Leakey, R.R.B., Lee, W.G., Loope, L.L., Lorence, D.H., Louda, S.M., Lugo, A.E., McEvoy, P.B., Richardson, D.M., and Vitousek, P.M., Deliberate introductions of species: Research needs, *BioScience*, 49, 8, 619, 1999.

Lockwood, J., Hoopes, M., and Marchetti, M., *Invasion Ecology*, Wiley-Blackwell, New York, 312, 2006.

Mack, R.N., Simberloff, D., Lonsdale, W.M., Evans, H., Clout, M., and Bazzaz, F.A., Biotic invasions: Causes, epidemiology, global consequences, and control, *Ecol. Appl.*, 10, 3, 689, 2000.

McNeely, J.A., Mooney, H.A., Neville, L.E., Schei, P., and Waage, J.K., Eds., *A Global Strategy on Invasive Alien Species, IUCN*, Gland, Switzerland, UK, 2001.

Mooney, H.A. and Hobbs, R.A., *Invasive Species in a Changing World*, Island Press, Washington, DC, 384, 2000.

Mooney, H.A., Mack, R.N., McNeely, J.A., Neville, L.E., Schei, P.J., and Waage, J.A., *Invasive Alien Species: A New Synthesis*, Island Press, Washington, DC, 368, 2005.

Myers, J.H. and Bazley, D., *Ecology and Control of Introduced Plants*, Cambridge University Press, New York, 328, 2003.

Pimentel, D., *Biological Invasions: Ecological and Environmental Cost of Alien Plant Animal and Microbe Species*, CRC Press, Boca Raton, FL, 369, 2002.

Pimentel, D., Zuniga, R., and Morrison, D., Update on the environmental and economic costs associated with alien-invasive species in the United States, *Ecol. Econ.*, 52, 3, 273, 2005.

Pyšek, P., Richardson, D., and Jarošík, V., Who cites who in the invasion zoo: Insights from an analysis of the most highly cited papers in invasion ecology, *Preslia*, 78, 437, 2006.

Richardson, D.M., Pyšek, P., Rejmánek, M., Barbour, M.G., Panetta, F.D., and West, C.J., Naturalization and invasion of alien plants: Concepts and definitions, *Divers. Distrib.*, 6, 93, 2000.

Ruiz, G.M. and Carlton, J., *Invasive Species: Vectors and Management Strategies*, Island Press, Washington, DC, 484, 2003.

Sax, D.F., Stachowicz, J.J., and Gaines, S.D., *Species Invasions: Insights into Ecology, Evolution and Biogeography*, Sinauer, Sunderland, MA, 495, 2005.

Shrader-Frechette, K., Non-indigenous species and ecological explanation, *Biol. Philos.*, 16, 507, 2001.

Wilcove, D.S., Rothstein, D., Dubow, J., Phillips, A., and Losos, E., Quantifying threats to imperiled species in the United States, *BioScience*, 48, 8, 607, 1998.

2 Mechanisms of Plant Invasion: A Review

Pedram Daneshgar and Shibu Jose

CONTENTS

2.1 INTRODUCTION

With increases in transport and commerce over the last thousand years, humans have been accidentally and deliberately dispersing and introducing plants to ecosystems beyond their native range (Mack et al. 2000). Plants making the transition to a new habitat must undergo a series of filters to become established: a historical filter, which asks whether or not the species arrives; a physiological filter, which asks whether or not the species can germinate, grow, survive, and reproduce; and lastly, a biotic filter, which addresses whether or not the species can compete and defend itself successfully (Lambers et al. 1998).

Not every introduction results in naturalization, and only a few of those that become naturalized become invasive. As a statistical generalization, Williamson and Fitter (1996) proposed the *tens rule* on the success of plants and animals as invaders when introduced to new ranges. This rule suggests that 1 in 10 of the biota brought into a region will escape and appear in the wild, 1 in 10 of those will become naturalized as a self-sustaining population, and 1 in 10 of those populations will become invasive. Although the percentage of plants crossing borders and becoming invasive seems low, the few that eventually do have radical effects on native species populations, communities, and ecosystem processes.

Since 1958, Charles Elton and other ecologists have made attempts to understand how introduced species become invasive in order to predict where and when invasions could occur. Dispersal, establishment, and survival are necessary for successful invasion of natural communities (Hobbs 1989), but what are the underlying mechanisms for invasion? There is a wide array of reasons as to why plant invaders may have rapid growth and spread in their new environments. Disturbance may reduce competition, allowing for the establishment of invaders. Exotic plants may escape herbivores or parasites, which keep their populations low in their native lands. The invaders may alter their new environment in order to promote their own growth while suppressing the growth of others. Empty niches may occur in a community that can be filled by an introduced plant. There are several plausible explanations and several mechanisms for invasion have been proposed.

In this chapter, we review many of the foremost theories of plant invasion of new communities. Several theories have been proposed in recent years, and some of the more prominent ones regarding plant invasion are addressed. The discussion begins with some of the theories of ecosystem susceptibility to invasion, and the factors that may determine whether a community is invaded. Some of the theories on how invasion is facilitated are then portrayed, followed by some of the suspected attributes of invading plants. It should be noted that some of these theories have been heavily researched and supported or refuted, while some of the more recent ones lack experimental proof. Some of the ideas discussed here are overlapping in concept and theory, while others are quite distinct.

2.2 BIOTIC RESISTANCE HYPOTHESIS

While some theories suggest that some plant species are able to easily invade new habitats because they do not encounter any herbivores that threaten their establishment and spread (the *natural enemies hypothesis*, discussed later in this chapter), the biotic resistance hypothesis says that exotics fail to establish and spread because of negative interactions between the introduced species and the native biota (Maron and Vila 2001). Enemies of the intruders occur in their introduced habitat, which can suppress their spread and establishment. The native communities therefore are able to resist invasion. In a common garden experiment conducted in southern Ontario, Canada, the impacts of herbivory were tested on 30 old-field plant species. It was observed that nonnative species experienced equal or greater herbivory than do natives (Agrawal and Kotanen 2003), suggesting some evidence of biotic resistance.

The hypothesis holds true as long as there are generalist herbivores in the community, which can attack the invaders, and the abundance of the invaders does not exceed the amount the herbivores can consume. Maron and Vila (2001) suggest that there is a threshold of exotic species abundance that generalist native herbivores can successfully limit. Beyond this threshold, the biotic resistance no longer exists. A species could rise above the threshold by means of propagule pressure. If a species is contributing large amounts of seed to a community, there is greater insurance of its establishment, survival, and spread (Hierro et al. 2005). This concept has lead to the *propagule pressure hypothesis*. Many acknowledge that the difference in the number of propagules arriving in a community plays a role in the level of invasion (Williamson 1996; Lonsdale 1999; Mack et al. 2000).

2.3 FLUCTUATING RESOURCE AVAILABILITY THEORY OF INVASIBILITY

Resource availability is one of the driving factors determining which species occur within a community. When resources are limited, fewer species are able to establish themselves within a community, and when resources increase either because of disturbance, heavy herbivory, or even fertilization, the window opens up for the establishment of new species. It was this concept that led Mark Davis and his colleagues to develop the fluctuating resource availability theory of invasibility (Davis et al. 2000). This states that an increase in the quantity of unused resources will allow a plant community to be susceptible to invasion. Their theory relies on the assumptions that available resources such as light, nutrients, and water are accessible to invading species and that as long as there is no severe competition from resident species for those resources, the species should successfully invade the community.

The theory rests on the concept that a community's susceptibility to invasion is not fixed (Davis et al. 2000). Fluctuations in resource availability will determine how prone a community is to invasion. Increases in resource availability can be driven by two means. First, resident use of resources could decline. Damage and mortality to resident species could occur as the result of a disturbance, thus reducing the uptake of resources. The other way resource availability can increase is by increasing the supply of resources at a rate faster than the uptake of the resident species. Examples of this include the following: higher precipitation than normal, which will increase water supply; eutrophication of soils, increasing nutrient availability; or loss of upper canopy trees, allowing for greater light availability. A community can maintain its resistance to invasion with increases in resource availability as long as the species in that community increase their uptake. Decreases in resource availability will increase competition between resident species in a community and make that community even harder to invade. According to Davis et al. (2000), the invasibility of a plant community is based upon a balance between resource uptake and gross resource supply (Figure 2.1). As long as these two are equivalent, the community should be resistant to invasion. Fluctuations away from this balance either increase or decrease the community's invasibility.

The literature has reports of several scenarios in which fluctuations in resource availability have affected an ecosystem's invasibility. It was demonstrated in Gros Morne National Park (Canada), a boreal ecosystem, that resources essential for alien plants were either not limiting to the resident species or were supplied by recent disturbances (Rose and Hermanutz 2004). The light availability and percent of bare ground, partially produced by moose trampling, were significantly higher than those in the undisturbed sites in this boreal ecosystem, which suggests that the increases in light availability allowed for invasion of aliens. Fluctuating light availability in a

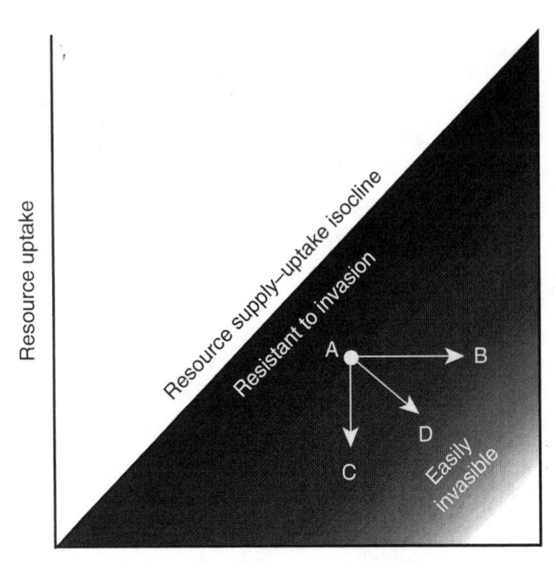

FIGURE 2.1 Balance between gross resource supply and resource uptake (as denoted by the isocline, where gross resource supply = resource uptake) represents a community's barrier to invasion. A quick increase in resource supply (A→B), a decline in resource uptake (A→C), or a combination of both (A→D) leads to an increase in a community's invasibility, because the resource supply is not matched by the uptake from the community. (Reproduced from Davis, M.A., Grime, J.P., and Thompson, K., *J. Ecol.*, 88, 528, 2000. With permission.)

podocarp/broad-leaved forest in New Zealand was also shown to be a driving factor for the invasion of *Tradescantia fluminensis*, which would reach its maximum biomass at 10%–15% full light (Standish et al. 2001).

2.4 EMPTY NICHE HYPOTHESIS

Davis et al. (2000) proposed that the fluctuation of resources leads to the invasion of plant communities by exotics. Disturbances often cause these fluctuations and the imbalance of resource supply and uptake. However, what if a relatively stable community just happened to have resources not utilized by the native community, and those resources are accessible to newcomers? The empty niche hypothesis states that exotic plants can be successful in a new community by accessing resources not utilized by the local species (Elton 1958; Levine and D'Antonio 1999; Mack et al. 2000).

To test the viability of the empty niche hypothesis, Hierro et al. (2005) suggested that parallel studies should be conducted of the invasive species in its native and introduced range in order to show that the invasive species is utilizing the unused resources in the new community while also showing that these resources were utilized by other plants in the native community. No such studies have been documented, but several studies of invasives in their introduced communities have alluded to the uptake of unused resources by invading plants. *Centaurea solstitialis* L.

is believed to dominate California grasslands with its extensive, deep roots, which access unused water below the shallow roots of the other vegetation (Roche' et al. 1994; Hierro et al. 2005 and the sources therein).

Studies involving cover crops that prevent the spread of invasives seem to be utilizing the empty niche hypothesis. If cover crops are planted to prevent spread, they are essentially used to occupy an unused niche and take up otherwise unused resources that could be utilized by exotic invaders. In field experiments conducted in Nigeria, it was observed that after 12 months of planting there was up to a 71% reduction in *Imperata cylindrica* (an invasive grass invading agricultural fields) biomass when cover crops such as velvet bean (*Mucuna pruriens* L.) were planted (Chikoye and Ekeleme 2001).

2.5 DIVERSITY–INVASIBILITY HYPOTHESIS

Species-rich communities are considered to utilize more resources and thus there are fewer empty niches to occupy. With fewer resources to tap and fewer niches to invade, species-rich communities may be less prone to invasion. This is one of the main concepts that led to the development of Charles Elton's diversity–invasibility hypothesis, which states that communities that are more diverse are less vulnerable to invasion. Several recent studies have been conducted to determine whether or not Elton's theory holds true. In most cases, species richness was used as a surrogate for species diversity. Stohlgren and his colleagues have led the argument that the hypothesis does not hold true and instead, diversity and invasibility are positively related (Stohlgren et al. 1999), whereas Tilman (2004) disagrees and supports Elton's theory.

Using two data sets of native and nonnative plant distributions from throughout the United States, Stohlgren et al. (2003) ran correlations of native and nonnative plant species richness on multiple scales to determine if there was any relationship between diversity of native species and nonnative invasive species. The results of this study demonstrated that native species richness was, in fact, positively correlated with invasive species distributions, and as spatial scales increased, the correlations grew stronger between native and nonnative distributions. They found that areas high in native species richness supported large numbers of nonnative species and proposed rapid turnover as the primary mechanism by which diverse habitats are able to support nonnative plants. Increases in richness lead to increases in turnover within the habitat, resulting in pulses of available resources, which promote the growth of both natives and nonnatives. This concept supports the fluctuating resource hypothesis. Stohlgren et al. (1999) proposed another mechanism in which the success of exotic colonization is driven by the same factors that contribute to high native richness, including high levels of propagule supply and resource availability as well as favorable environmental conditions.

Several other studies have produced results in support of Stohlgren (Wiser et al. 1998; Higgins et al. 1999; Lonsdale 1999; Smith and Knapp 1999). Stachowicz et al. (2002) suggest that there are more likely to be facilitators or important habitat-forming species that make conditions ideal for an invasive species in diverse communities. In a Kansas grassland, it was observed that a reduction in the dominance of the C4 grasses resulted in the reduction of the invasive species *Melilotus officinalis* (Smith et al. 2004). Removal of the dominants resulted in higher light availability of up to 35%, which negatively affected the establishment of *Melilotus* (Smith et al. 2004).

Tilman (2004) contends that as diversity increases, invasibility decreases. Recently, he justified this theory by citing stochastic theory for community assembly (Tilman 2004). In this theory, every new species entering a community is treated as an invader. There are three requirements for the establishment of an individual in a community. First, community assembly results from the successes and failures of propagules of invaders. Second, an invading propagule will survive and reproduce by only utilizing unconsumed resources. Third, successful establishment of an invader depends on the resource requirement of an invader relative to other species in its community. During assembly of a community, more and more invaders utilize unused resources: as a result, when the number of invaders increases, the amount of available resources decreases, making it harder for new invaders to establish themselves (Tilman 2004). With these assumptions, the number of invaders in a community will plateau as it becomes more diverse. This theory imitates the logistic theory and the idea of carrying capacity. As species numbers increase, the maximum number of species a system can support is approached. As species number increases, the probability of invasion by a new invader decreases (Tilman 2004). Several mechanisms have been proposed that explain how diverse systems are resistant to invasion, including the crowding effect, the complimentary effect, and the sampling effect.

The crowding effect is one mechanism by which diverse systems reduce their susceptibility to invasion. In a crowded community, there is little room for establishment of invasive seedlings. Kennedy et al. (2002) tested the relationship between species diversity and invasion in 147 experimental grassland plots of varying diversity, from 1 to 24 species. The success of 13 species of exotic plants was assessed. There was a 98% reduction in invader cover in the most diverse plots compared with monocultures, which was attributed to crowding (Kennedy et al. 2002).

The complementary effect refers to the ability of multiple species to utilize different resources or different sources of resources, in such a way that they can coexist in the same area. Plants that complement each other within a community efficiently utilize the various resources, allowing for the community to be resistant to invasion. Plants complement each other by occupying different niches (empty niche hypothesis). In Belgium, the effects of three invader grasses on European grasslands were assessed for varying levels of diversity, and it was observed that with increasing neighborhood richness, complementarity was enhanced, which negatively affected invader leaf length (Milabau et al. 2005). By complementing each other, plants in a community can discourage invasion.

Resources may be used more efficiently in diverse plots because they are more likely to have a species that is highly effective in capturing resources. This is referred to as the sampling effect, and it can play a role in a community's susceptibility to invasion. A highly diverse community is more likely to include a species that is capable of outcompeting an invader. The sampling effect represents a possible mechanism for why communities that are more diverse are less susceptible to invasion. One way to determine whether or not the sampling effect is playing a role in resistance is to test an invader in monoculture of a wide variety of species.

A single species may be tolerant of an invader, and this may be further demonstrated in communities of varying diversity containing that species. If the invader consistently fails to be successful in growth or establishment each time, it is paired with that particular species along different diversity gradients; it is then evident that there may be a sampling effect.

Fargione and Tilman (2005) demonstrated evidence of a sampling effect with prairie–savannah communities at varying levels of diversity (1, 2, 4, 8, and 16 species). It was observed that invader biomass was inhibited in plots in the presence of strongly competitive C_4 bunchgrasses (Fargione and Tilman 2005). Soil nitrate concentrations decreased, and root biomass of resident species increased with the presence of C_4 grasses across a diversity gradient, leaving the researchers to believe that communities are more resistant to invasion when they contain C_4 grasses.

Elton's theory of invasibility has brought on much debate since 1958 and currently the discrepancy is still unresolved. The diversity–invasibility hypothesis may hold only for certain types of systems or only on certain spatial scales. The debate will probably continue, and research will continue to test it with a variety of exotic species and invaded communities.

2.6 FACILITATION BY SOIL BIOTA

Soil-borne mutualists could facilitate the invasion of exotic plants. Soil biota can alter the soil conditions enough to favor the spread of an exotic species over a native species. Reinhart and Callaway (2006) proposed the enhanced mutualisms hypothesis, which acknowledges the possibility that there may be stronger facilitation of growth of invasives by soil microbes in new habitats than what the plants experienced in their native range. The mutualisms this hypothesis refers to are those formed with mycorrhizal fungi and nitrogen fixers.

The proponents of the enhanced mutualisms hypothesis suggest that the mutualism of invasive plants with mycorrhizae may not be just a two-way association (Reinhart and Callaway 2006). Instead, nonnative plants may use the associations of mycorrhizae and multiple plants and gain advantage by tapping into the mycelial network without providing the essentials for maintaining such a symbiosis. In the presence of arbuscular mycorrhizae and the North American native grass *Festuca idahoensis*, the invasive *Centaurea maculosa* showed a 66% increase in growth in comparison with the same species grown in the absence of the fungi (Marler et al. 1999).

The invasion of new habitats by plants, particularly legumes, may be facilitated by nitrogen-fixing bacteria. Nodule production by invading legumes requires a certain threshold of nitrogen-fixing bacteria (Parker 2001). Some legumes are able to invade with the aid of native bacteria and some succeed by bringing the nitrogen-fixing bacteria with them. For example, the nitrogen-fixing actinomycetes *Frankia* maintains a symbiosis with *Myrica faya* (both from the same habitat), allowing it to invade and alter the nitrogen cycle in ecosystems in Hawaii (Vitousek et al. 1987).

The presence of certain soil biota in the exotic habitat may not facilitate greater growth and spread of plant species, but rather provide less restraint than do biota in their native habitat. The activity of soil microbes in the native range may limit plant growth not only by limiting available nutrients, but also by providing negative feedback, which keeps the species under control. Thus, when the species is introduced elsewhere, it may not be constrained by the same mechanisms that disallow the plant growth and spread in its native range. Callaway et al. (2004) reported that *C. maculosa* experienced greater inhibitory effects by soil microbes in its native European soils than in North America. They attribute the differences in performance by the invasive species in the two soils to different feedback mechanisms. This study may be a demonstration of the plant escaping the inhibitory effects of its native soil biota, a theory that is discussed later in this chapter.

2.7 INVASION MELTDOWN HYPOTHESIS

Facilitation of invasion by plants may not only be facilitated by soil microbes and mycorrhizal fungi, but also by a variety of flora and fauna. The phenomenon of already invading exotic biota opening the door for the invasion by other aliens by altering site conditions and by providing a positive feedback has often been observed. The invasion meltdown hypothesis states that increasing numbers of exotic species facilitate additional invasions (Colautti et al. 2004). A meltdown of an ecosystem occurs as the number of invasive species increases.

Invasion meltdowns could be facilitated by plants or animals. Plants that alter the soil characteristics may also facilitate invasion of other species. In Hawaii, several studies have demonstrated how *M. faya*, a nitrogen-fixing shrub, has invaded volcanic nitrogen-poor soils and altered soil properties (Vitousek 1986; Vitousek and Walker 1989). Vitousek (1986) suggested that *M. faya* could further facilitate additional plant invasion. Hughes et al. (1991) experimentally showed that there was significant increase in biomass accumulation of the invader *Psidium cattleianum* in *M. faya* infested communities. Plants may also alter the soil characteristics by introducing chemicals that hinder the growth of native species (allelopathy, discussed later), which may allow for the establishment of exotics.

The introduction of pollinating wasps to south Florida has allowed for the establishment of *Ficus* species that depend on the pollinators for reproduction (Simberloff and Von Holle 1999). Several exotic plants invade ecosystems after herbivores from their native land have feasted on the plants in the new range. Exotic herbivores can reduce competition by native plants with the exotic plants. In a meta-analysis of 63 manipulative field studies, Parker et al. (2006) observed that grazing by exotic herbivores allowed for 52% greater abundance of exotic plants in native communities. They also observed that grazing by exotic herbivores led to an increase in exotic plant species richness, which they attributed to a reduction in the abundance of native species (Parker et al. 2006). An invasion meltdown of this sort requires that the invasive plants be preceded by generalist herbivores. Specialist herbivores are unlikely to affect the plants in a new habitat, enough to allow for the introduction and success of new plant species.

2.8 NATURAL ENEMIES HYPOTHESIS

According to the invasion meltdown hypothesis, some plant species are successful invaders because they followed exotic generalist herbivores into new habitats. Some invasive plants may be successful because, instead of following generalist herbivores, they escape from specialist herbivores that keep them from spreading in their native habitats. The natural enemies hypothesis assumes that natural enemies suppress plants in their native range, and it is the escape from these enemies that allows exotic populations to explode in their new habitat (Maron and Vila 2001). These natural enemies are not limited to herbivores; fungal pathogens and destructive soil biota may also be considered enemies. The natural enemies hypothesis has also been referred to as the enemy release, enemy escape, herbivore escape, predator escape, or ecological release hypotheses. Three points drive the basis of this hypothesis: (1) plant populations are regulated by natural enemies, (2) natives are affected more by enemies than are exotic species, and (3) reduction in enemy regulation should lead to increased plant population growth (Keane and Crawley 2002).

To demonstrate the natural enemies hypothesis, a study would have to show that native herbivores or pathogens reduce plant population sizes and growth rates, and at the same time show that those same plant species suffer little herbivory or diseases and have increased population size and growth rate in their introduced habitat. *Clidemia hirta*, a neotropical shrub native to Costa Rica, which is currently invading Hawaii, was used to test the natural enemies hypothesis with the use of both insecticides and fungicides. It was observed that *Clidemia* survival in Costa Rica increased 41% when both treatments were used (Figure 2.2) (DeWalt et al. 2004). Plant growth in Hawaii was unaffected by the fungicide, suggesting that fungal pathogens limit plant growth only in its native land (DeWalt et al. 2004). In this case, *Clidemia* escaped the suppression by fungal pathogens, which is why it was invasive in Hawaii.

Escape from natural enemies provides for the logic behind the use of biological control. If a plant is released from suppression by some sort of specialist herbivore or

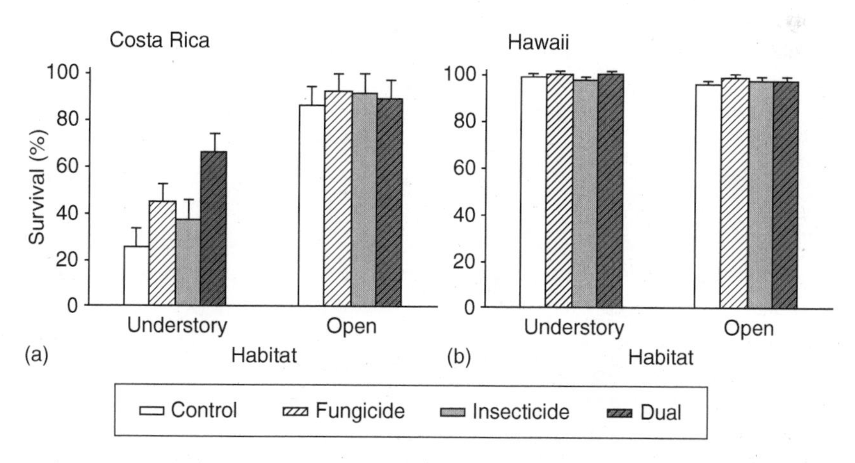

FIGURE 2.2 Survival of *Clidemia hirta* in (a) Costa Rica (native range) and (b) Hawaii (introduced range) in four natural enemy escape treatments in understory and open habitats. Survival was much higher in Hawaii where *Clidemia* has escaped herbivores and fungal pathogens, which have reduced its survival in its native land. (Reproduced from DeWalt, S.J., Denslow, J.S., and Ickles, K., *Ecology*, 85, 471, 2004. With permission.)

pathogen, then that specialist enemy could be used to regulate plant populations in habitats where that plant was introduced. A field study was conducted to test the biological control of saltcedars (*Tamarix*), an Asian tree species invading riparian areas of the United States with no real insect threat, by the Asian leaf beetle *Diorhabda elongata deserticola*. In this study, the saltcedars were caged in with the beetles (Lewis et al. 2003). A 60%–99% defoliation of the saltcedars was observed by the beetles as well as substantial dieback, mortality of young plants, and limited regrowth in the following growing season (Lewis et al. 2003). Many insects were observed to feed on the plant species in Asia, where it grows in isolated patches, leading to the coevolution of specialized insects that feed solely on it (Lewis et al. 2003 and sources therein). The successful invasion by *Tamarix* species in the United States is an example of natural enemies hypothesis because it has escaped the specialist herbivores from its native range.

2.9 EVOLUTION OF INCREASED COMPETITIVE ABILITY

Plants in the presence of specialist herbivores and pathogens develop defenses against and tolerances of these enemies in order to survive and proliferate in their native lands. Their relationships with specialist enemies are often coevolved in that, over time, the plant species have evolved to designate their resources toward surviving in the presence of enemies. When these plant species escape their native enemies and are introduced to new habitats, the resources they have been using for defense can be allocated to growth and reproduction (Blossey and Notzold 1995; Hanfling and Kollmann 2002). This may be what makes these introduced plants invasive. The evolution of increased competitive ability (EICA) hypothesis takes the natural enemies hypothesis a step further. It states that only when plants escape from coevolved specialist enemies are they able to gain advantage over other plants in their introduced community: they do so by using for growth and reproduction the resources that were previously used for defense (Blossey and Notzold 1995). It has been suggested that the liberation from herbivores allows for the selection of genotypes in the new community with increased competitive abilities (Blossey and Notzold 1995).

An efficient method of testing the EICA is to grow the particular invasive species from seed from both the native and introduced ranges in a common garden or in identical conditions while excluding pests. Support for the hypothesis comes from observing that the plants from the introduced habitat perform better than those from the native habitat, since they experience little herbivore pressure and have adapted to allocating their resources toward growth. Purple loosestrife plants (*Lythrum salicaria* L.) from two locations, one with herbivory (Leselle, Switzerland, where it is native) and one without (Ithaca, New York, where it was introduced) were grown in identical conditions, and it was observed that plants from the region that experienced little herbivore pressure had greater vegetative growth (Figure 2.3) (Blossey and Notzold 1995). The results could be explained by the fact that the introduced plants had escaped pressures of herbivory and were able to allocate resources for growth.

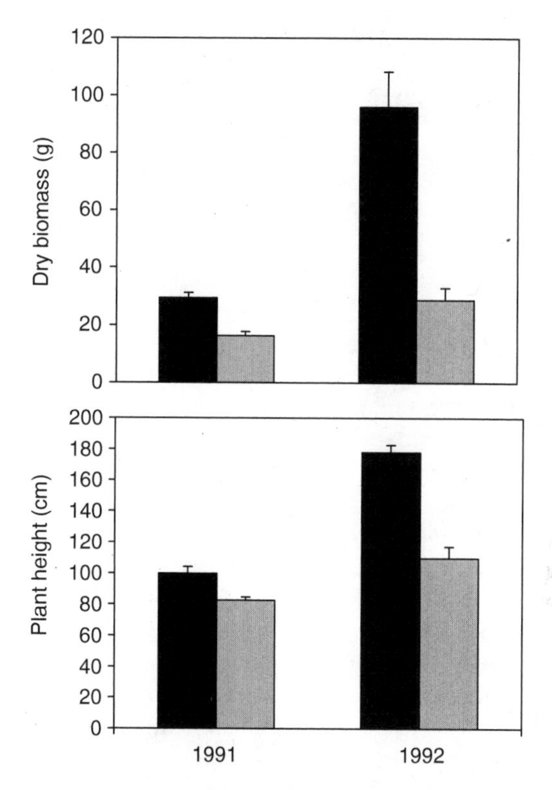

FIGURE 2.3 Mean (±SE) dry biomass and height of purple loosestrife from Ithaca, NY (*black*) and Lucelle, Switzerland (*grey*) after the growing season, grown in common gardens under identical conditions. (From Blossey, B. and Notzold, R., *J. Ecol.*, 83, 887, 1995.)

2.10 REPRODUCTIVE TRAITS

Some invasion success could be explained by the reallocation of resources from defense to reproduction; alternatively, some species may be good invaders simply because they have the ability to reproduce quickly and in great numbers. Several authors have pointed out that invasive plants tend to be r-strategists, considering that they tend to invade disturbed habitats (Rejmanek 1989; Hobbs 1991). This seems to be the strategy used by several species of *Pinus* to invade regions outside their natural range throughout the world. Rejmanek and Richardson (1996) have identified the main reproductive characteristics that cause certain species of pine (particularly *Pinus radiata, P. contorta, P. halepensis, P. patula, P. pinaster*) to be more invasive. These include short juvenile period, short intervals between large seed crops and small seed mass. Some of the other characteristics they identified include the large number of seeds produced, better dispersal, shorter chilling period needed to overcome dormancy, high initial germinability, and higher relative growth rate (Rejmanek and Richardson 1996 and sources therein). It seems apparent that all these characteristics allow for quick and efficient spread of the pines.

2.11 SUPERIOR COMPETITOR

Specialized reproductive abilities allow for some species to establish earlier, faster, and in greater numbers, and these abilities put invasives at a competitive advantage. After establishment, some plant species may compete better for resources than does native vegetation, which may make these species more successful. Bakker and Wilson (2001) proposed that differences in competitive ability may determine what species invade new areas. In their field study, they demonstrated that an introduced C_3 grass, *Agropyron cristatum* (L.) Gaertn, had a stronger ability to resist competition than did the native C_4 grass *Bouteloua gracilis* (HBK). In this case, *Agropyron* was a superior competitor than the native vegetation.

Several other studies have demonstrated an invasive's ability to outcompete native species for resources. In a study examining the invasion of a longleaf pine (*P. palustris*) savanna in southeastern United States by the exotic grass, *I. cylindrica*, it was observed that the clonal expansion of the grass was reduced when plots were fertilized with phosphorus, suggesting that the invasive species was a better competitor for the resource (Brewer and Cralle 2003). When photosynthetic measurements were taken of two exotic invasive species of *Rubus* and compared with those of two native species, it was observed that the exotic species had higher rates of photosynthesis than their native cousins, implicating why the exotics were better competitors.

In addition to being a superior competitor in an introduced habitat under normal conditions, some invasives may be better competitors in stressful situations, such as those induced by disturbance. Ruderals, highly competitive natives and stress-tolerant species, are expected in communities immediately after a disturbance event (Grime 1979). However, it has been proposed that certain invasives are much more adapted to disturbance and thus are better competitors in the presence of disturbance than natives simply because the natives have not experienced as much disturbance over time (Gray 1879; Mack et al. 2000). This concept, known as the disturbance hypothesis, has received little attention since it was proposed by Gray in 1879, but would be fairly easy to test and should receive attention.

2.12 NOVEL WEAPONS HYPOTHESIS

Over time, plants may have persisted in their native range by exuding chemicals that help them deal with competition by inhibiting the activity of neighbors (allelopathy). Because of repeated exposure to these chemical exudates over time, the neighboring plants may have evolved resistance and thus are not impeded by them. However, when these allelopathic plants are introduced to new communities, they have "weapons" that the plants in the new community have never experienced before and as a result may suffer the inhibitory effects. The inhibition of plant growth allows the introduced species to be at a competitive advantage over its new neighbors, allowing it to become invasive. The novel weapons hypothesis states that biochemicals released by a plant, which are ineffective against native neighbors, are inhibitory to plants or soil microbes in a new community, contributing to the plant's invasiveness (Callaway and Ridenour 2004). The introduced species are invasive because they present their new competitors with weapons to which they have not been exposed previously.

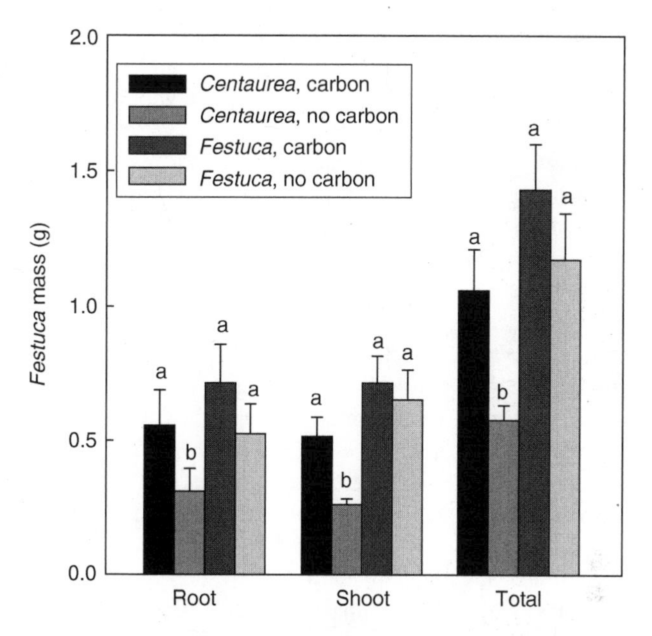

FIGURE 2.4 Biomass of roots, shoots, and total of *Centaurea* and *Festuca* grown together in pots with and without activated carbon, which has a high affinity for organic chemicals. Shared letters designating means that are not significantly different. (Reproduced from Ridenour, W.M. and Callaway, R.M., *Oecologia*, 126, 444, 2001. With permission.)

Because of its high affinity for adsorbing organic compounds, activated carbon is often used to test allelopathy. Ridenour and Callaway (2001) used activated carbon to test whether the noxious weed *C. maculosa* used novel weapons (allelopathy) to hinder the growth of the native bunchgrass, *F. idahoensis*. They observed that *Festuca* had reduced root and shoot growth in the presence of *Centaurea* in pure sand compared with sand mixed with activated carbon (Figure 2.4), implying that *Centaurea* uses biochemicals to gain competitive advantage. Another activated carbon study demonstrated the use of allelopathy by *C. diffusa*, which had stronger negative effects on North American species than Eurasian species (Callaway and Aschehoug 2000).

As allelopathy continues to work for a species, natural selection will favor its reproduction and growth over other intruding species, which are able to compete. This has been referred to as the allelopathic advantage against resident species (AARS) (Callaway and Ridenour 2004). Just as the species that reallocated resources toward growth and reproduction (the EICA), species having success with biochemical weapons may allocate more to the production of these chemicals, thus increasing their success. Support for the AARS could then be observed by the introduced species being even more allelopathic than source populations (Callaway and Ridenour 2004).

2.13 INTEGRATED MECHANISMS

Although several mechanisms for invasion have been proposed and demonstrated through examples, it is clear that many species use more than one mechanism to gain competitive dominance in their new habitats. Blumenthal (2005) suggested that there might be an interaction between the natural enemies hypothesis and the fluctuating resource theory because fast growing species, which require a high amount of resources, tend to be susceptible to enemies. This is because high-resource species tend to be nutritious and lack both structural material and defensive chemicals, tendencies that encourage herbivory. In a new range, the invader escapes its enemies and encounters a flush of resources, which may result from a disturbance. The mechanisms for invasion, therefore, may be integrated.

We propose a new hypothesis called the rhizochemical dominance hypothesis that integrates several mechanisms in explaining the success of invasive plants. This hypothesis attributes invasive species success to allelopathy (novel weapons) and alteration of soil chemical properties by the rhizosphere exudates of the invader (see Collins and Jose, Chapter 14), which in turn favors its own growth while inhibiting the growth of competing vegetation. These chemical alterations may include changes in soil pH, and nutrient levels and availability. *I. cylindrica* was shown to alter soil pH and decrease soil nitrate and potassium levels in invaded areas rather than in noninvaded areas (Collins and Jose, Chapter 14). *Imperata* has also been shown to be allelopathic, suppressing the growth of crops such as tomato and cucumber (Eussen 1979) and having greater impacts at lower pH (Eussen and Wirjahardia 1973). In this case, it seems that *Imperata* increases the potency of its weapons by altering soil chemical properties. The genus *Centaurea* has been proposed to have invasive success by suppressing the growth of native species by phytotoxicity or by altering soil microbial activity, leading to restrictions of nutrient availability (LeJeune and Seastedt 2001). By root-induced mechanisms, the genus may be gaining dominance.

2.14 CONCLUSIONS

The fact that several mechanisms for invasion have been proposed in recent years (many of which were discussed in this chapter), and that basically no generalizations can be made about the nature of invasive plants, indicates that research in this area is still fairly new and needs much attention. No one has yet explained invasion patterns across a large range of systems, and this may be simply due to the fact that each invasive species is unique and that invasions are unpredictable (Williamson 1999; Dietz and Edwards 2006 and references therein). It has recently been proposed, however, that the conflicts in invasion theory result from the examination of different parts of the invasion process, and it should be recognized that the processes enabling a species to invade change over the course of the invasion (Dietz and Edwards 2006). Consideration of these different phases may allow for some generalizations to be made.

REFERENCES

Agrawal, A.A. and Kotanen, P.M., Herbivores and the success of exotic plants: A phylogenetically controlled experiment, *Ecol. Lett.*, 6, 712, 2003.

Bakker, J. and Wilson, S., Competitive abilities of introduced and native grasses, *Plant Ecol.*, 157, 117, 2001.

Blossey, B. and Notzold, R., Evolution of increased competitive ability in invasive nonindigenous plants: A hypothesis, *J. Ecol.*, 83, 887, 1995.
Blumenthal, D.M., Interrelated causes of plant invasion: Resources increase enemy release, *Science*, 310, 243, 2005.
Brewer, J.S. and Cralle, S.P., Phosphorus addition reduces invasion of a longleaf pine savanna (Southeastern USA) by a non-indigenous grass (*Imperata cylindrica), Plant Ecol.*, 167, 237, 2003.
Callaway, R.M. and Aschehoug, E.T., Invasive plants versus their new and old neighbors: A mechanism for exotic invasion, *Science*, 290, 521, 2000.
Callaway, R.M. and Ridenour, W.M., Novel weapons: Invasive success and the evolution of increased competitive ability, *Front. Ecol. Environ.*, 2, 8, 436, 2004.
Callaway, R.M., Thelen, G.C., Rodriguez, A., and Holben, W.E., Soil biota and exotic plant invasion, *Nature*, 427, 731, 2004.
Chikoye, D. and Ekeleme, F., Cogongrass suppression by intercropping cover crops on corn/cassava systems, *Weed Sci.*, 49, 658, 2001.
Colautti, R.I., Ricciardi, A., Grigorovich, I.A., and MacIsaac, H.J., Does the enemy release hypothesis predict invasion success? *Ecol. Lett.*, 7, 721, 2004.
Davis, M.A., Grime, J.P., and Thompson, K., Fluctuating resources in plant communities: A general theory of invasibility, *J. Ecol.*, 88, 528, 2000.
DeWalt, S.J., Denslow, J.S., and Ickles, K., Natural-enemies release facilitates habitat expansion of invasive tropical shrub *Clidemia hirta, Ecology*, 85, 2, 471, 2004.
Dietz, H. and Edwards, P.J., Recognition that causal processes change during plant invasion helps explains conflicts in evidence, *Ecology*, 87, 6, 1359, 2006.
Elton, C., *The Ecology of Invasions by Animals and Plants*, University of Chicago Press, Chicago, IL, 1958.
Eussen, J.H.H., Some competition experiments with alang-alang [*Imperata cylindrica* (L.) Beauv.] in replacement series, *Oecologia*, 40, 351, 1979.
Eussen, J.H.H. and Wirjahardia, S., Studies of an alang-alang, *Imperata cylindrica* (L.) Beauv. vegetation, *Biotrop. Bull.*, 6, 1, 1973.
Fargione, J. and Tilman, D., Diversity decreases invasion via both sampling and complementarity effects, *Ecol. Lett.*, 8, 604, 2005.
Gray, A., The predominance of pertinacity of weeds, *Am. J. Sci. Arts*, 118, 161, 1879.
Grime, J.P., *Plant Strategies and Vegetation Processes*, Wiley, New York, 1979.
Hanfling, B. and Kollmann, J., An evolutionary perspective of biological invasions, *Trends Ecol. Evol.*, 17, 12, 545, 2002.
Hierro, J.L., Maron, J.L., and Callaway, R.M., A biogeographical approach to plant invasions: The importance of studying exotics in their introduced and native range, *J. Ecol.*, 93, 5, 2005.
Higgins, S.I., Richardson, D.M., Cowling, R.M., and Trinder-Smith, T.H., Predicting the landscape-scale distribution of alien plants and their threat to plant diversity, *Conserv. Biol.*, 13, 303, 1999.
Hobbs, R.J., The nature and effects of disturbance relative to invasions, in *Biological Invasions: A Global Perspective*, Drake, J.A., Mooney, H.A., di Castri, F., Groves, R.H., Kruger, F.J., Rejmanek, M., and Williamson, M., Eds., Wiley, Chichester, UK, 389, 1989.
Hobbs, R.J., Disturbance as a precursor to weed invasion in native vegetation, *Plant Prot. Q.*, 6, 99, 1991.
Hughes, R., Vitousek, P.M., and Tunison, T., Alien grass invasion and fire in the seasonal submontane zone of Hawaii, *Ecology*, 72, 743, 1991.
Keane, R.M. and Crawley, M.J., Exotic plant invasions and the enemy release hypothesis, *Trends Ecol. Evol.*, 17, 164, 2002.
Kennedy, T., Naeem, S., Howe, K., Knops, J., Tilman, D., and Reich, P., Biodiversity as a barrier to ecological invasion, *Nature*, 417, 636, 2002.

Lambers, H., Chapin, F.S., III, and Pons, T.L., Eds., *Plant Physiological Ecology*, Springer-Verlag, New York, 2, 1998.

LeJeune, K.D. and Seastedt, T.R., *Centaurea* species: The forb that won the west, *Conserv. Biol.*, 15, 6, 1568, 2001.

Levine, J. and D'Antonio, C.M., Elton revisited: A review of evidence linking diversity to invasibility, *Oikos*, 87, 15, 1999.

Lewis, P.A., DeLoach, C.J., Knutson, A.E., Tracy, J.L., and Robbins, T.O., Biology of *Diorhaba elongata deserticola* (Coleoptera: Chrysomelidae), an Asian leaf beetle for biological control of saltcedars (*Tamarix* spp.) in the United States, *Biol. Control*, 27, 101, 2003.

Lonsdale, W.M., Global patterns of plant invasions and the concept of invasibility, *Ecology*, 80, 5, 1522, 1999.

Mack, R.N., Simberloff, D., Lonsdale, W.M., Evans, H., Clout, M., and Bazzaz, F.A., Biotic invasions: Causes, epidemiology, global consequences, and control, *Ecol. Appl.*, 10, 3, 689, 2000.

Marler, M.J., Zabinksi, C.A., and Callaway, R.M., Mycorrhizae indirectly enhance competitive effects of an invasive forb on a native bunchgrass, *Ecology*, 80, 1180, 1999.

Maron, J.L. and Vila, M., When do herbivores affect plant invasion? Evidence for the natural enemies and biotic resistance hypotheses, *Oikos*, 95, 361, 2001.

Milabau, A., Nijs, I., De Raedemaecker, F., Reheul, D., and De Cauwer, B., Invasion in grassland gaps: The role of neighbourhood richness, light availability and species complementarity during two successive years, *Funct. Ecol.*, 19, 27, 2005.

Parker, J.D., Burkepile, D.E., and Hay, M.E., Opposing effects of native and exotic herbivores on plant invasions, *Science*, 311, 1459, 2006.

Parker, M.A., Mutualism as a constraint on invasion success for legumes and rhizobia, *Divers. Distrib.*, 7, 125, 2001.

Reinhart, K.O. and Callaway, R.M., Soil biota and invasive plants, *New Phytol.*, 170, 445, 2006.

Rejmanek, M., Invasibility of plant communities, in *Biological Invasions. A Global Perspective*, Drake, J.A., Mooney, H.A., di Castri, F., Groves, R.H., Kruger, F.J., Rejamnek, M., and Wiliamson, M., Eds., Wiley, Chichester, UK, 369, 1989.

Rejmanek, M. and Richardson, D.M., What attributes make some plant species more invasive, *Ecology*, 77, 6, 1655, 1996.

Ridenour, W.M. and Callaway, R.M., The relative importance of allelopathy in interference: The effect of an invasive weed on a native bunchgrass, *Oecologia*, 126, 444, 2001.

Roche', B.F., Jr., Roche', C.T., and Chapman, R.C., Impacts of grassland habitat on yellow starthistle (*Centaurea solstitialis* L.) invasion, *Northwest Sci.*, 68, 86, 1994.

Rose, M. and Hermanutz, L., Are boreal ecosystems susceptible to alien plant invasion? Evidence from protected areas, *Oecologia*, 139, 467, 2004.

Simberloff, D. and Von Holle, B., Positive interactions of nonindigenous species: Invasional meltdown? *Biol. Invasions*, 1, 21, 1999.

Smith, M., Wilcox, J., Kelly, T., and Knapp, A.K., Dominance not richness determines invasibility of tallgrass prairie, *Oikos*, 106, 253, 2004.

Smith, M.D. and Knapp, A.K., Exotic plant species in a C4-dominated grassland: Invasibility, disturbance and community structure, *Oecologia*, 120, 605, 1999.

Stachowicz, J.J., Fried, H., Osman, R.W., and Whitlach, R.B., Reconciling pattern and process in marine bioinvasions: How important is diversity in determining community invasibility? *Ecology*, 83, 2575, 2002.

Standish, R.J., Robertson, A.W., and Williams, P.A., The impact of an invasive weed *Tradescantia fluminensis* on native forest regeneration, *J. Appl. Ecol.*, 38, 1253, 2001.

Stohlgren, T.J., Barrett, D.T., and Kartesz, J.T., The rich get richer: Patterns of plant invasions in the United States, *Front. Ecol. Environ.*, 1, 1, 11, 2003.

Stohlgren, T.J., Binkley, D., Chong, G.W., Kalkhan, M.A., Schell, L.D., Bull, K.A., Otsuki, Y., Newman, G., Bashkin, M., and Son, Y., Exotic plant species invade hot spots of native plant diversity, *Ecol. Monogr.*, 69, 25, 1999.

Tilman, D., Niche tradeoffs, neutrality, and community structure: A stochastic theory of resource competition, invasion, and community assembly, *Proc. Natl. Acad. Sci.*, 101, 30, 2004.

Vitousek, P.M., Biological invasions and ecosystem properties: Can species make a difference? in *Ecology of Biological Invasions of North America and Hawaii*, Mooney, H.A. and Drake, J.A., Eds., Springer-Verlag, New York, 163, 1986.

Vitousek, P.M. and Walker, L.R., Biological invasion by Myrica faya in Hawaii: Plant demography, nitrogen fixation, ecosystem effects, *Ecol. Monogr.*, 59, 247, 1989.

Vitousek, P.M., Walker, L.R., Whiteaker, L.D., Mueller-Dombois, D., and Matson, P.A., Biological invasion by *Myrica faya* alters ecosystem development in Hawaii, *Science*, 238, 802, 1987.

Williamson, M., *Biological Invasions*, Chapman & Hall, London, 1996.

Williamson, M., Invasions, *Ecography*, 22, 5, 1999.

Williamson, M. and Fitter, A., The varying success of invaders, *Ecology*, 77, 6, 1661, 1996.

Wiser, S.K., Allen, R.B., Clinton, P.W., and Platt, K.H., Community structure and forest invasion by an exotic herb over 23 years, *Ecology*, 79, 2071, 1998.

3 Relationship of Invasive Groundcover Plant Presence to Evidence of Disturbance in the Forests of the Upper Midwest of the United States

W. Keith Moser, Mark H. Hansen, Mark D. Nelson, and William H. McWilliams

CONTENTS

3.1 INTRODUCTION

Nonnative invasive plants (NNIPs) have been introduced to North America by humans since European settlement. Much like other exotic-invasive organisms, NNIPs typically have some advantage over native plants, such as prolific seed production and dispersal. Native forest ecosystems that developed over centuries are limited in their ability to compete against these invaders. Some species, such as kudzu, were deliberately introduced (Mitich 2000), while others were introduced inadvertently, such as in contaminated crop seed. Introduction, however, does not necessarily mean establishment. Although a particular NNIP may have a competitive advantage over native species, timing of emergence and seed dispersal, site quality, and other factors determine whether an NNIP will take hold in an ecosystem. Once established, NNIP threatens the sustainability of native forest composition, structure, function, and resource productivity (Webster et al. 2006).

NNIPs occur in all the major life forms found in forest ecosystems: trees, shrubs, vines, forbs/grasses, and other herbs. Although there is scant knowledge of other life forms, such as lichens and mosses, they are likely to be impacted as well. Examples of invasive trees include Norway maple (*Acer platanoides*) and tree-of-heaven (*Ailanthus altissima*). Many shrub species currently influence North American forests. Some very important examples that have spread nationwide include multiflora rose, bush honeysuckles, Russian/autumn olive, and privet.* Examples of invasive vines include kudzu, Japanese honeysuckle, oriental bittersweet, and mile-a-minute vine. Forbs/grasses and other herbs include a host of species, such as garlic mustard, Japanese and giant knotweed, and Japanese stiltgrass.

NNIPs pose significant challenges for decision makers attempting to develop policies for control and amelioration. There is an underlying need for improvement of inventory and monitoring efforts nationwide. New methods of control and restoration of impacted systems are also needed, and these will require novel approaches because of a dearth of relevant literature. Information on the invasion process indicates that efforts to control invasive plants should focus on the establishment phase (Webster et al. 2006). This introduces further difficulties for monitoring and planning efforts

* Scientific names of the species in this study are listed in Table 3.1.

because populations are sparse in the establishment phase and the subsequent expansion and saturation phases occur rapidly. Once saturation has occurred, the challenges of control and restoration become immense. The high cost of managing impacted forest ecosystems is prohibitive, especially when costly mechanical and chemical activities are required. In some cases, there are no known biological controls for invasive plants. More research on the impacts of invasives and related science is sorely needed.

This issue has become a hot topic not only in local communities, but also in Washington, DC. Dale Bosworth, former Chief of the U.S. Department of Agriculture Forest Service, has listed the threat from invasives among the top four threats facing our forests today:

> Another threat is from the spread of invasive species. These are species that evolved in one place and wound up in another, where the ecological controls they evolved with are missing. They take advantage of their new surroundings to crowd out or kill off native species, destroying habitat for native wildlife. ...—at a cost that is in the billions [of dollars]
>
> **Four Threats to the Nation's Forests and Grasslands, U.S. Forest Service Chief Dale Bosworth at the Idaho Environmental Forum, Boise, Idaho—January 16, 2004. http://fsweb.wo.fs.fed.us/pao/four-threats/.**

3.1.1 Definition of NNIPs

We define NNIPs as those plants that (1) are not indigenous to the ecosystem and (2) have a competitive advantage that causes deleterious impacts on structure, composition, and growth in forested ecosystems.

3.1.2 Study Region

The Upper Midwest region of the United States is at the nexus of several ecoregions. Historically, the Upper Midwest region encompassed many different forest compositions and structures, ranging from closed-canopy forest in the Lake States and the Ozarks to woodland ecosystems in southern Wisconsin to savannas and prairies in Iowa, Illinois, and Indiana. The fertile soils of this region were ideal for farming, and settlers proceeded to clear the land for agriculture. In the heavily timbered areas of northern Minnesota, Wisconsin, and Michigan and in southern Missouri, large-scale commercial harvesting exploited the magnificent stands of eastern white pine, shortleaf pine, and other species. Subsequent fires and lack of scientific management resulted in a radically altered forested landscape. The combination of settlement or clearing and timber harvesting created a highly fragmented landscape, offering many opportunities for NNIPs to establish in its forests.

3.1.3 Factors Influencing Invasive Establishment

Studies have identified elements of four factors that influence invasion success: disturbance, competitive release, resource availability, and propagule pressure (Richardson and Pyšek 2006). To gauge if a plant community or habitat is more invasible, investigators must ask not only if there are more potential invaders present but also whether the habitat is more susceptible to invasions (Lonsdale 1999; Richardson and Pyšek 2006). We will examine how these factors might influence invasibility in the Midwest United States.

3.1.4 Site

Site quality is an important factor influencing invasion success. Gelbard and Belnap (2003) found that plant communities with high resource availability (i.e., deep, silty, or otherwise fertile soils) were particularly susceptible to disturbance and invasion. They add that disturbance, when combined with high site conditions, maximized a plant community's vulnerability to invasives. Richardson and Pyšek (2006) postulated that resource availability was a facilitator of invasiveness at larger spatial scales. Much has been made of the role of diversity in defending against nonnative invasives (Elton 1958). Yet, Huston and DeAngelis (1994) pointed out that species-rich communities occur in habitats with high levels of heterogeneity in terms of climate, soil, and topography and that alien species are more likely to gain a toehold in such sites than in those habitats that are less heterogeneous. Climate also plays a role. NNIPs often successfully establish in habitats with climate similar to that of their native ecosystems (Richardson and Pyšek 2006). According to Sax (2001), Rapoport's rule states that the number of naturalized species is negatively correlated, and geographic range size is positively correlated, with latitude. We will see evidence of this characteristic later in this chapter.

3.1.5 Disturbance

Disturbance increases the resource availability for plants, including invasive species. Disturbance can upset the competitive balance and site occupancy of prior plant communities, making abiotic factors more important as determinants of invasion success than biotic factors (Richardson and Bond 1991; Hood and Naiman 2000). The larger the difference between gross resource supply and resource uptake, the more vulnerable a plant community becomes to invasive species. Even intermittent or short fluctuations in resource availability have long-term impacts on the outcome of an invasion, particularly if these fluctuations coincide with the availability and arrival of suitable propagules (Richardson and Pyšek 2006).

One prominent indicator of disturbance and a correlate of other measures of disturbance is the density of roads. A study in Utah found that the activities of expanding roads in interior forest areas (road construction, maintenance, and vehicle traffic) "corresponded with greater cover and richness of exotic species and lower richness of native species" (Gelbard and Belnap 2003). An inverse relationship exists between distance to road and prevalence of exotic species (e.g., Watkins et al. 2003), although the influence is most pronounced within 15–30 m of a road. Forman and Alexander (1998) could not document many cases where species spread more than 1 km from a road. For these reasons, this study assumed that distance to the nearest road is a surrogate for human activity, rather than a direct conduit for invasive exotics.*

* In examining the relationship between density of roads and the presence and coverage of invasive species, the assumption was that road density was correlated with any point's distance to the nearest road.

3.1.6 Competition

As we stated earlier, Elton (1958) suggested that there is a negative relationship between native species diversity and community invasibility. He apparently based his hypothesis on the idea that, with less diverse assemblages of species, interspecific competition is less robust because there are empty niches available. Richardson and Pyšek (2006) found numerous studies that supported Elton's hypothesis, but also reported that others found that areas with a high species diversity harbored more alien species. They noted Levine and D'Antonio's (1999) conclusion that species richness may be too broad a factor to explain observed differences in community invasibility, given that other factors (disturbance, nutrient availability, climate, and propagule pressure) are frequently covariates with species richness.

3.1.7 Spread

We can sometimes determine the date an invasive plant is introduced into the country or region. The likelihood of invasion increases with the time since the original introduction. "Minimum residence time" (MRT) is often used when the initial introduction of an inoculum is unknown. MRT integrates the time of potential establishment opportunity, the size of the propagule bank (seeds and shoots), and (with expanding populations) the area from which the propagules originate (Richardson and Pyšek 2006). Such knowledge does not always help us to determine the rate of spread. Plant invasions do not move across the landscape in a continuous wave or front; both local and long-distance transport will determine the spatial pattern (Pyšek and Hulme 2005). These authors summarized data about 100 taxa worldwide and found an average local spread rate <400 m per year, but a long-distance dispersal rate that was 2–3 orders of magnitude greater. For this reason, Richardson and Pyšek (2006) concluded that invasions are often faster than most natural migrations. Given the serendipitous nature of inadvertent human transport, the most significant driver of postinvasion spread (Hodkinson and Thompson 1997), it is hard to predict the source and final destination of many invasive species. Those species that are widespread (1) have shown that they are adapted to a wide range of conditions and (2) have a greater source of propagules to continue the spread (Booth et al. 2003).

3.1.8 Adaptation

There are several interesting precepts of NNIPs that this study had hoped to examine. Daehler (2001) postulated that exotic species in an area with native species of the same genus have a better chance of naturalizing because they share a certain amount of preadaptation to the conditions of the region. Yet Daehler (2003) also concluded that invasive species have greater phenotypic plasticity than co-occurring native species, suggesting that the shared features are less important than the individually unique ones. Some invaders benefit from release of fitness constraints present in their original habitat while others evolve after arriving in this country (Ellstrand and Schierenbeck 2000). Crawley et al. (1999) speculated that NNIPs may occupy vacant niches at either end of the plant performance spectrum perhaps by growing either very small or very big in size, by flowering very early in the season or very late, or by foregoing dormancy, or by exhibiting a very long dormant period. However, the interaction of ecological and evolutionary forces is context-dependent (Daehler 2003) and unique to each invasive episode (Richardson and Pyšek 2006).

3.2 STUDY OBJECTIVES

Drake et al. (1989) summarized the then-current knowledge around three fundamental topics: which species invade, which habitats are invaded, and how these invaded ecosystems can be managed. Research to understand the factors influencing the invasibility of a site focuses on the availability of resources, disturbance effects, competition, and the availability of invasive material ("propagule pressure") (Richardson and Pyšek 2006). We examined characteristics such as overstory basal area, basal area of oak species, stand age, and stand density for clues of their influence over invasive species' presence and coverage. Variations in overstory diversity, whether compositional (species) or structural (height or diameter), influence available growing space and present opportunities for ground flora. We examined characteristics such as overstory species (compositional) diversity and diameter and height (structural) diversity. Finally, we examined the impact of human influences, such as forest fragmentation and roads.

3.3 METHODOLOGY

Meaningful NNIP inventory requires a large network of sample plots measured consistently over time. Over the past decade, the U.S. Forest Service, Northern Research Station, Forest Inventory and Analysis (NRS-FIA) unit has implemented a new inventory system that embodies the challenges of developing national and international consistency. Complete documentation of the plot design and all measurements can be found at http://socrates.lv-hrc.nevada.edu/fia/dab/databandindex.html and North Central Research Station, Forest Inventory and Analysis (NCRS-FIA 2005).

The FIA program utilizes three phases of inventory designed to produce estimates of forest extent, composition, structure, health, and sustainability. Phase 1 uses remote sensing (currently LandSat TM imagery) to identify accessible forestland and to develop stratification layers for improving precision of postsampling stratified estimates. Phase 2 consists of a systematic grid of ground samples where detailed measurements of tree and forest attributes are taken on a 5 year remeasurement cycle with one-fifth of the grid measured every year. Each Phase 2 sample consists of 4–7.3 m (24 ft) radius subplots at an intensity of 1 per 2400 ha (5960 acres) (McRoberts 1999). We included data from the 2005 and 2006 inventory of forest resources on Phase 2 plots in Indiana, Illinois, Iowa, Missouri, Michigan, Wisconsin, and Minnesota. Phase 3 ground samples include more detailed forest health protocols, including a complete ground vegetation sample, that are measured during the summer growing season months on a subset of the Phase 2 samples. Each Phase 3 sample represents about 39,000 ha (96,000 acres). In lieu of national protocols for monitoring all vegetation on Phase 2 samples, some regional FIA programs, including the NRS, have implemented exotic-invasive plant surveys to address the burgeoning need for this information (Rudis et al. 2006).

TABLE 3.1
NNIPs Surveyed on FIA Plots in the Upper Midwest of the United States in 2005–2006

Common Name	Scientific Name
Woody species	
Multiflora rose	*Rosa multiflora*
Japanese barberry	*Berberis thunbergii*
Common buckthorn	*Rhamnus cathartica*
Glossy buckthorn	*Frangula alnus*
Autumn olive	*Elaeagnus umbellata*
Nonnative bush honeysuckles	*Lonicera* spp.
European privet	*Ligustrum vulgare*
Vines	
Kudzu	*Pueraria montana*
Porcelain berry	*Ampelopsis brevipedunculata*
Asian bittersweet	*Celastrus orbiculatus*
Japanese honeysuckle	*Lonicera japonica*
Chinese yam	*Dioscorea oppositifolia*
Black swallowwort	*Cynanchum louiseae*
Wintercreeper	*Euonymus fortunei*
Grasses	
Reed canary grass	*Phalaris arundiacea*
Phragmites, Common reed	*Phragmites australis*
Nepalese browntop, Japanese stiltgrass	*Microstegium vimineum*
Herbaceous	
Garlic mustard	*Alliaria petiolata*
Leafy spurge	*Euphorbia esula*
Spotted knapweed	*Centaurea biebersteinii*
Dame's rocket	*Hesperis matronalis*
Mile-a-minute weed, Asiatic tearthumb	*Polygonum perfoliatum*
Common burdock	*Arctium minus*
Japanese knotweed	*P. cuspidatum*
Marsh thistle	*Cirsium palustre*

During 2005–2006, 8663 Phase 2 forested plots were surveyed for presence and cover of any of 25 noninvasive plant species (Table 3.1) (Olson and Cholewa 2005; NCRS-FIA 2005). If one or more of these species was observed, the percent cover was estimated on each subplot and placed into one of seven ordinal categories (Table 3.2). Where an NNIP was found on a plot but had not been previously documented to exist in that state, a specimen was collected and sent to NRS-FIA staff in St. Paul, MN, for positive identification. In winter, the crews treated the plants as if they were in a leaf-on condition for purposes of cover calculation.

Measures of individual trees resulted in summaries of species, diameters, and heights and estimates of density using overstory basal area and the stand density index (SDI) (Reineke 1933; Woodall and Miles 2006) and diversity, using the

TABLE 3.2
Cover Codes and Ranges of Percent Cover of NNIPs Used in Recording Invasive Species' Presence, FIA Plots in 2005–2006

Cover Code	Range of Percent Cover
1	<1%, trace
2	1%–5%
3	6%–10%
4	11%–25%
5	26%–50%
6	51%–75%
7	76%–100%

Shannon index (H′) (Shannon 1948; Magurran 1988) for species, heights, and diameters. To convert continuous variables like height or diameter into categorical ones, we assigned heights to 0.9 m (3 ft) classes and diameters to 5 cm (2 in.) classes.

3.3.1 Measurement of Distance from Plots to Roads

Distances (km) from NRS-FIA plots to roads were calculated with a geographic information system (GIS), for each of five categories of roads within the ESRI StreetMaps dataset, version 2005, with 2006 updates. Distances were calculated simultaneously from all plots across the seven states to National Freeway, State Freeway, and Major Highway features. Because a large number of local roads within Minor Highway and Local Street features resulted in unwieldy processing, these features were subset and analyzed on a state-by-state basis, including a 20 km buffer around each state to allow for more representative calculations for plots near state boundaries. In addition, the "NEAR" command was constrained to search for the nearest Minor Highway and Local Street within a 10–25 km buffer radius of each plot, with buffer radius varying by state.

3.3.2 Fragmentation Data

To evaluate fragmentation, we divided total forestland area by total land area in each county. This metric is useful for establishing the relative amount of forested growing space and, by inference, the amount of edge-to-area ratio that would provide the entry for forest invasives. On a smaller scale, we looked at the percent of each plot that was forested. While most of the plots were 100% forested, some of the plots intersected a forest–nonforest edge, giving us a sample of local fragmentation to compare with the county-level measures.

One measure of the effects of disturbance on forests is "forest intactness," a composite of several metrics of class- and landscape-level fragmentation. We used a sum of ordinal measures prepared by Heilman et al. (2001) in their analysis of fragmentation in 39 forested ecoregions in the lower 48 states. Using

TABLE 3.3
List of Candidate Stand, Site, and Disturbance Variables for Our Models of NNIP Invasiveness

Stand Variables	Site Variables
Total basal area	Site index
Oak basal area	Aspect (cosine)
Stand age	Physiographic class code
SDI—All	
SDI—5″ plus	**Disturbance Variables**
H′ species	
H′ diameter	County percent forest
H′ height	Plot percent forest
	Minimum distance to the nearest road
	Index of forest intactness (sum of all expanded ordinal scores)

FRAGSTATS 2.0, they combined ordinal scores of road density, class area (the amount of the landscape comprised of a particular patch type), core area (the area within a patch beyond a specified buffer), percentage of landscape (the percentage of the total landscape occupied by a particular class), and nearest neighbor (the edge-to-edge distance to the nearest patch of the same type) (McGarigal and Marks 1995). This summary index combines most measures of fragmentation and provides a representative measure of the entry and establishment opportunities of NNIPs. Because of the way Heilman et al. (2001) defined forests, this index did not apply to all of our plots, omitting many plots in the central part of our region.

3.3.3 Variable Reduction

On the basis of the considerations outlined earlier, we started out with 15 candidate variables (Table 3.3). We then employed bidirectional stepwise regression of a linear model of these variables. The stepwise regression was checked by manually removing variables that would reduce the AIC level (Akaike Information Criterion; Akaike 1974) the most. The model with the reduced dataset was rerun to obtain the coefficient and Pr(t) values, which were used to determine the level of significance.

3.4 RESULTS

3.4.1 Presence and Location

Present-day forested areas are concentrated in the northern and southern edges of the region and include a mix of private and public forest (Figure 3.1). NNIPs are present on many FIA plots throughout the Upper Midwest. Figure 3.2 displays those plots with at least one NNIP present. When comparing this figure to Figure 3.1, we found that invasive species are highly associated with fragmented forests. The juxtaposition of developed areas, agricultural land, and forest has created a vehicle for the establishment and spread of NNIPs in forested landscapes.

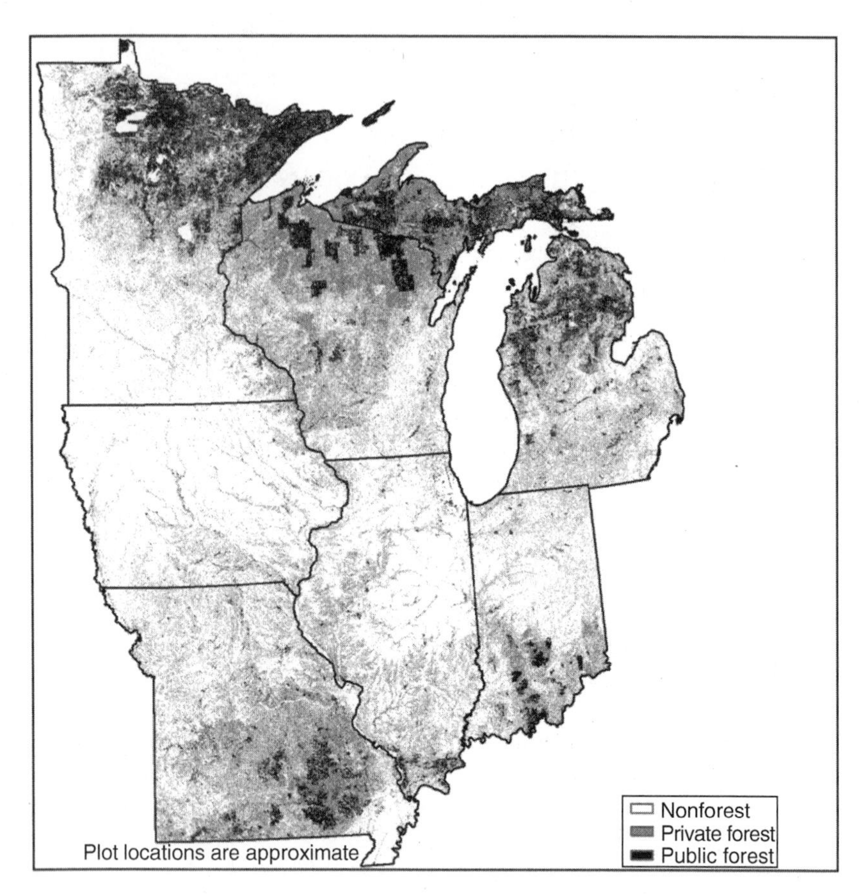

FIGURE 3.1 Distribution of forestland by owner group, Upper Midwest, United States. (From States—ESRI Data & Maps, 2002 ESRI 2002; Ownership—Protected Areas Database, Della Salla et al. 2001.)

Approximately 25% of the forested plots visited in 2005 and 2006 had at least one occurrence of an NNIP (Table 3.4). The top five occurrences included three woody species (multiflora rose, nonnative bush honeysuckles, and common buckthorn), one herbaceous species (garlic mustard), and one vine (Japanese honeysuckle). The second five included common burdock, autumn olive, Japanese barberry, reed canary grass, and marsh thistle.

Figure 3.3 displays the dramatic differences in proportion of plots that have at least one invasive species. In Iowa, Indiana, and Illinois, over 70% of the plots had at least one NNIP on them, while in Minnesota less than 10% of the plots had at least one NNIP.

3.4.2 Relationship between NNIP and Overstory Forest Type

Of the 61 total forest types within the Upper Midwest region, 21 were present on the bulk of the plots and three types—aspen, white oak-red oak-hickory, and sugar

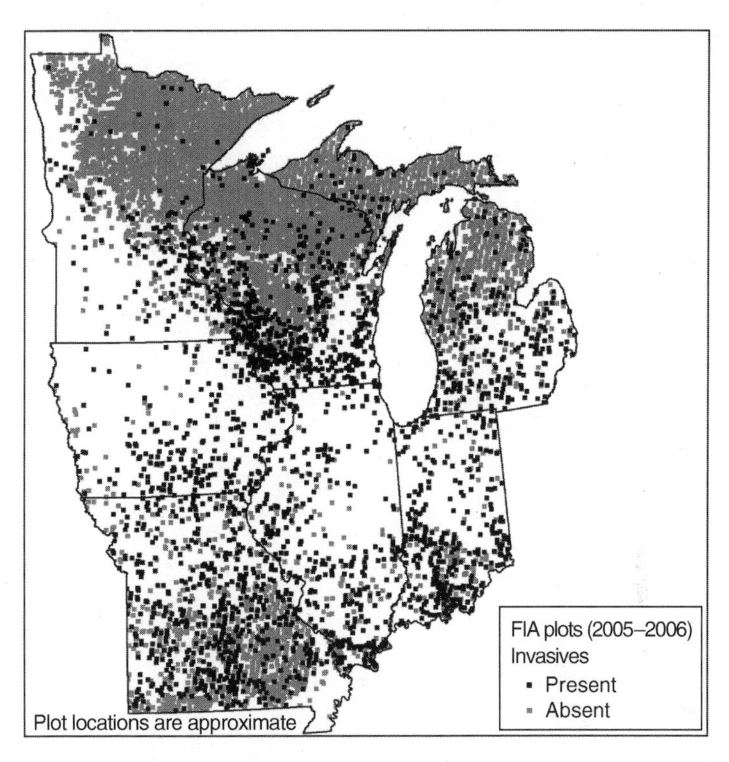

FIGURE 3.2 Distribution of plots with and without invasives of any type, Upper Midwest, 2005–2006.

maple-beech-yellow birch—comprised almost half of the plots (Figure 3.4). Several forest types, including white oak-red oak-hickory, mixed upland hardwoods, and sugarberry-hackberry-elm-green ash had proportions of plots with NNIPs near 50% or greater. To provide an indication of overstory tree density relative to NNIP presence, our NNIP distribution maps overlay the volume per acre of the top 12 species (by volume) in the Upper Midwest.

3.4.3 Woody Species

Woody NNIPs were the predominant life forms among the invasive plants sampled in this study of the Midwest's forests (Figure 3.5). Because of their perennial nature, woody NNIPs often were planted for aesthetic and wildlife purposes. Examination of the distribution of woody NNIPs by percent cover reveals strong geographical trends (Figure 3.6). Multiflora rose and nonnative bush honeysuckles were the most prominent species in the region, particularly in Illinois, Indiana, Iowa, and Missouri. Common buckthorn was prominent in most states, particularly Minnesota and Wisconsin, while Japanese barberry was evident in Illinois and Michigan.

TABLE 3.4
Number of Occurrences (Plots) for Each of the 25 Nonnative Invasive Plant Species Sampled during the 2005 and 2006 Panels, Ranked by Number of Occurrences

Seven States Total	Cover Class							
Invasive species	0	1	2	3	4	5	6	7
Multiflora rose	7343	351	473	208	136	106	31	15
Nonnative bush honeysuckles	7875	169	231	143	99	75	45	26
Common buckthorn	8248	120	108	56	60	27	28	16
Garlic mustard	8397	57	57	51	36	33	21	11
Japanese honeysuckle	8441	46	54	32	33	31	18	8
Common burdock	8465	93	62	27	9	5	2	0
Autumn olive	8491	39	61	21	21	18	11	1
Japanese barberry	8549	41	49	9	10	3	1	1
Reed canary grass	8574	13	21	10	7	11	12	15
Marsh thistle	8628	18	13	2	0	2	0	0
Spotted knapweed	8634	9	9	3	1	3	4	0
Glossy buckthorn	8637	5	5	3	9	3	1	0
Nepalese browntop	8651	3	3	3	0	1	2	0
Wintercreeper	8652	5	1	1	2	0	1	1
Asian bittersweet	8654	3	4	1	0	1	0	0
Chinese yam	8657	4	1	1	0	0	0	0
European privet	8658	1	0	1	2	1	0	0
Dames rocket	8658	0	1	1	1	1	0	1
Phragmites	8658	1	0	1	1	0	2	0
Japanese knotweed	8660	2	1	0	0	0	0	0
Kudzu	8660	1	1	1	0	0	0	0
Leafy spurge	8662	1	0	0	0	0	0	0
Black swallowwort	8662	1	0	0	0	0	0	0
Mile-a-minute weed	8663	0	0	0	0	0	0	0
Porcelain berry	8663	0	0	0	0	0	0	0

Note: Cover class categories are as follows: 0, none found; 1, <1%, trace; 2, 1%–5%; 3, 6%–10%; 4, 11%–25%; 5, 26%–50%; 6, 51%–75%; 7, 76%–100%.

3.4.3.1 Multiflora Rose

Multiflora rose is a widespread shrub introduced as rootstock for ornamental roses in 1866 (Plant Conservation Alliance 2006). The species was distributed and planted widely for erosion control, "living fences" for livestock, and cover for wildlife. Multiflora rose spreads quickly and establishes dense cover that shades out other plants. Its seeds are dispersed by birds and remain viable in soils for many years. It is currently found across the United States and is classified as "noxious" in several states. Control methods include mechanical and chemical methods that require repeated application for success, making control very expensive (Evans 1983).

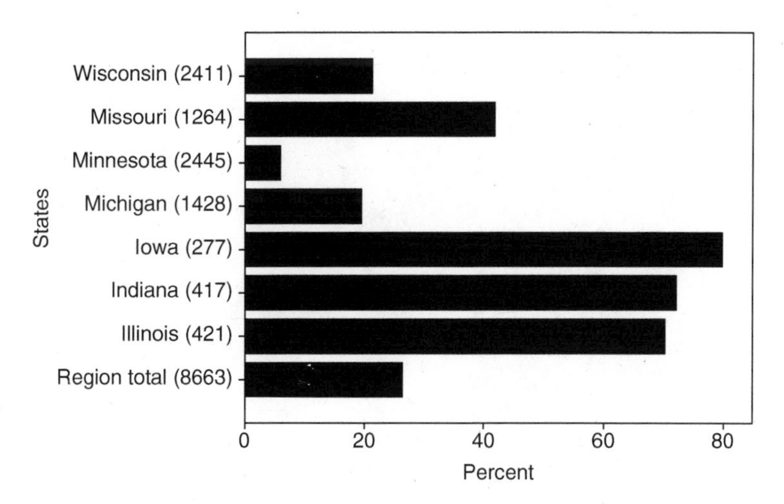

FIGURE 3.3 Percent of all plots sampled that have at least one NNIP present, by state and region total, 2005 and 2006 panels. Numbers next to state names are the number of forested plots in the state.

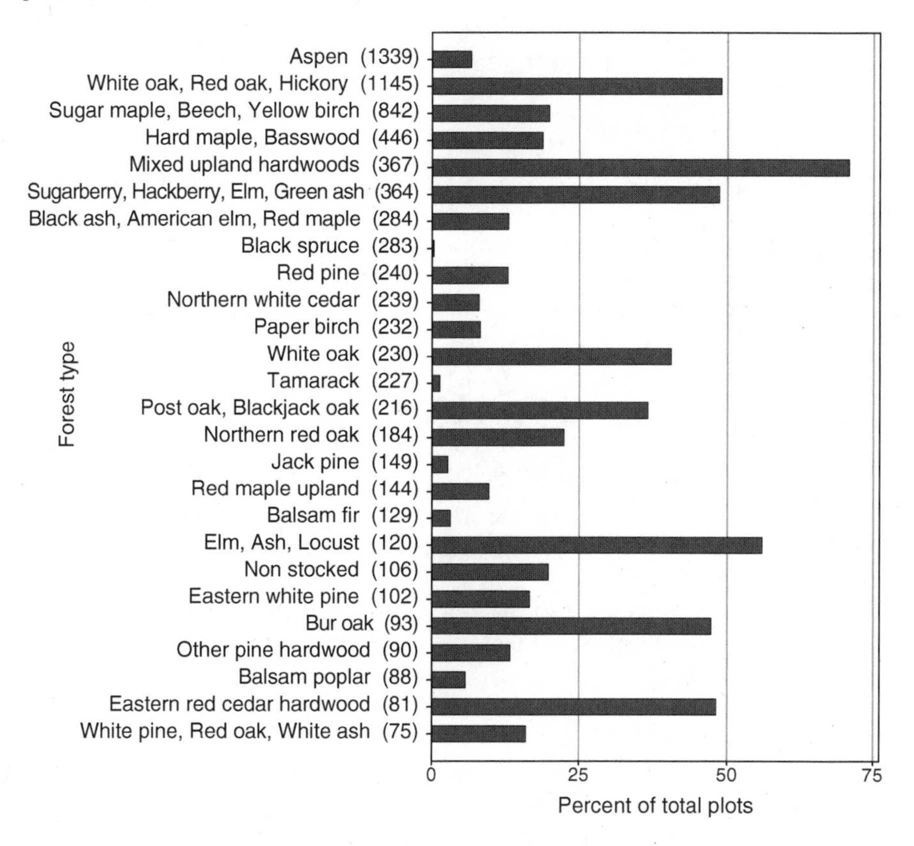

FIGURE 3.4 Percentage of plots in each forest type with 75 or more plots with at least one NNIP, 2005–2006.

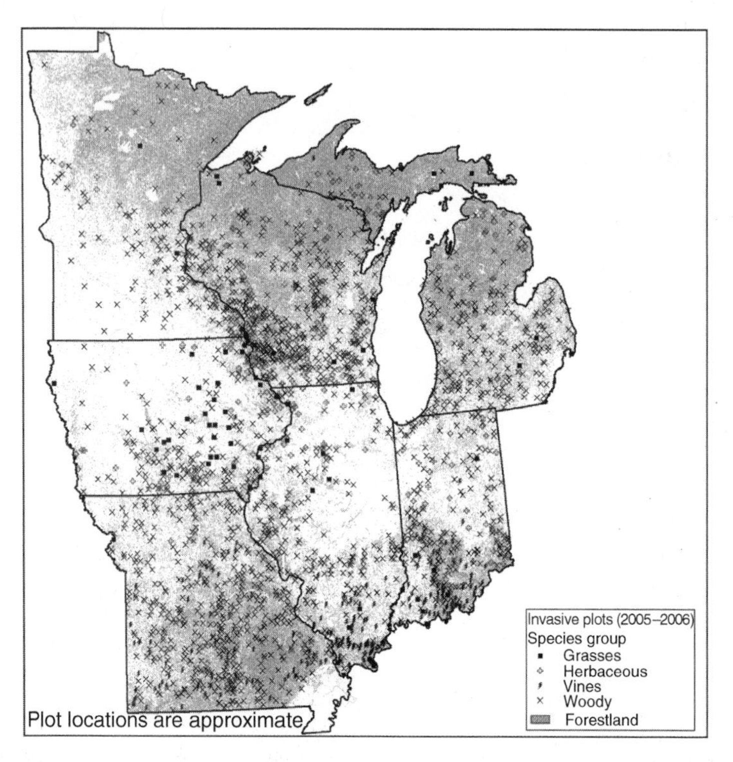

FIGURE 3.5 Distribution of plots with invasives in the Upper Midwest, by life form in 2005–2006.

In this study, multiflora rose was the most frequently found NNIP in the Upper Midwest (Table 3.4). It was detected on over 14% of all plots, with cover classes greater than 10% occurring on over 3% of all plots in the seven-state region (Figure 3.7).

3.4.3.2 Nonnative Bush Honeysuckles

Nonnative bush honeysuckles were recorded on 9% of plots sampled in 2005 and 2006 and were distributed over most of the forested areas in the region except for the extreme north (Figure 3.8).

These honeysuckles are natives of eastern Asia and were imported to the United States for use as ornamentals and for wildlife habitat. Fragmented forest remnants are vulnerable to honeysuckle invasion and establishment, particularly sites with limestone geology, which is prominent in the southern part of our region. Bush honeysuckles frequently become well established on the forest edge (Luken and Goessling 1995). They not only outcompete native shrubs, but also reduce understory diversity by shading the forest floor. The bush honeysuckles produce small juicy berries that are eaten and distributed by many species of small mammals

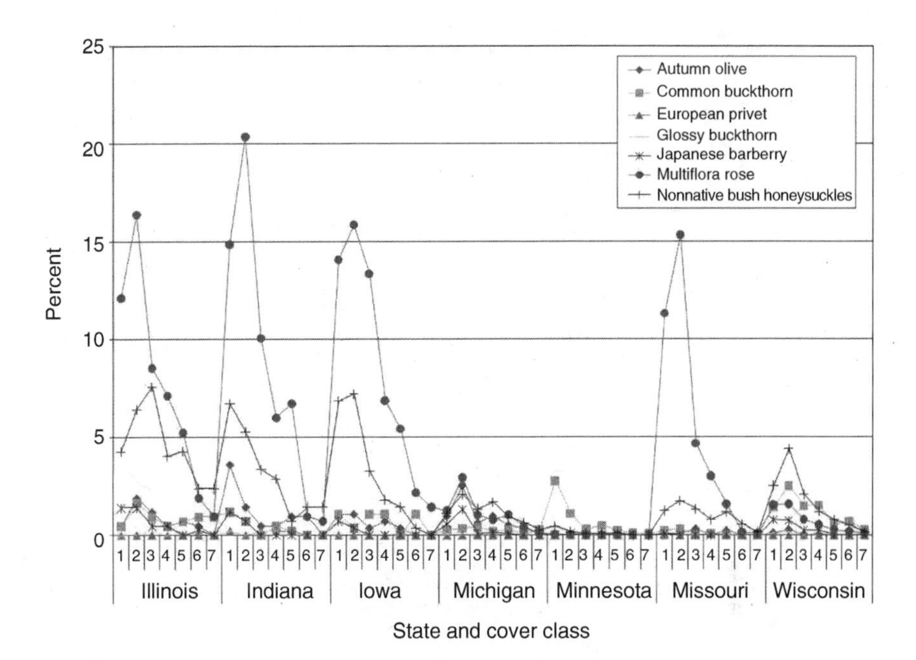

FIGURE 3.6 Presence of nonnative woody species in the seven states of the Upper Midwest, as measured by percent of all forested inventory plots sampled in 2005 and 2006, by state and cover class category. Cover class categories: 1 = <1%, trace; 2 = 1%–5%; 3 = 6%–10%; 4 = 11%–25%; 5 = 26%–50%; 6 = 51%–75%; and 7 = 76%–100%. Cover class 0—no invasives found—is not shown so as to preserve graphic scale.

and birds. Honeysuckles are generally believed to have a minimal interval between dispersal and germination and a short-lived seed bank. The species relies on the heavy seed output and sprouting from buds at the base of the stems on large plants (Luken 1988).

3.4.3.3 Common Buckthorn

Both of the major species of buckthorns found in eastern United States (glossy and common buckthorn) were introduced from Europe. Now common to the Midwest and New England, the species have been utilized as ornamental plantings and for wildlife habitat (Webster et al. 2006). Both of the buckthorns exhibit classic hyper-competitor behavior: they leaf out earlier than their native competitors, resprout vigorously, and produce large amounts of seeds that are spread by birds (Harrington et al. 1989). Buckthorns can suppress seedling height and diameter growth, both by shading as well as belowground competition from their extensive root systems. In one study, tree seedling survival was found to be about half that of open-grown seedlings (Fagan and Peart 2004). Common buckthorn observations were most frequent along the prairie forest "tension zone"—a diagonal line extending from

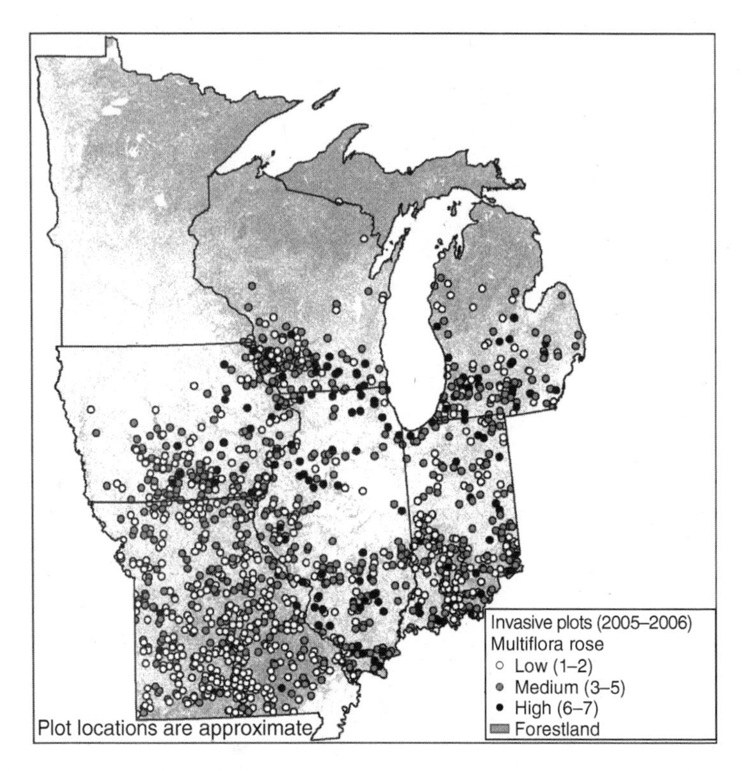

FIGURE 3.7 Distribution of plots with multiflora rose, Upper Midwest, 2005–2006.

central Minnesota through southeastern Wisconsin. Other states contained scattered observations (Figure 3.9).

In contrast with other NNIP species that have the highest frequency of observations in the 1%–5% cover class, buckthorn observations were most numerous in the cover category of <1% (trace) (Table 3.4).

3.4.4 Herbaceous Species

Garlic mustard and common burdock were the most prominent herbaceous NNIPs in the Upper Midwest (Figure 3.10). Garlic mustard had a greater extent in Illinois and Indiana, whereas the two species were similar in extent in Iowa and Wisconsin. Minnesota had the lowest overall percentage of plots with NNIP herbaceous species present.

Garlic mustard is an herb from Europe that was originally introduced to the United States in the mid-1800s (Meekins and McCarthy 1999), but is now present throughout the eastern United States (Nuzzo 1993). The species is very common in disturbed forests, which occur primarily in the central portion of our study region (Figure 3.11).

It has the capability, considered unusual for an invasive plant species, to invade mature second-growth forests (McCarthy 1997). The species reproduces

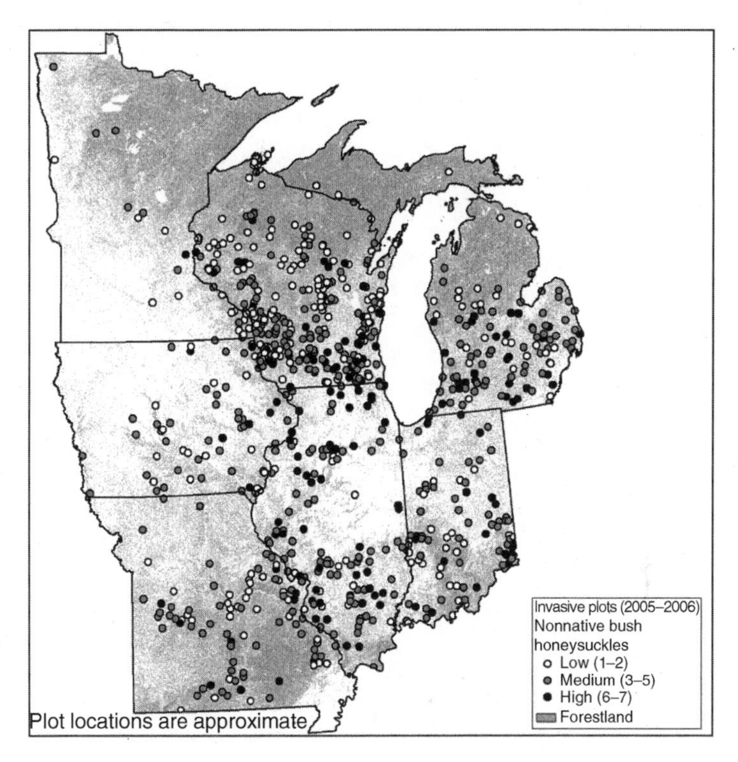

FIGURE 3.8 Distribution of plots with nonnative bush honeysuckles, Upper Midwest, 2005–2006.

sexually, after which the plants die (Nuzzo 1999). By its presence and superior competitive ability, species richness and growth of the ground-level flora and tree regeneration are suppressed. Some forest types, such as upland oak types with species such as *Quercus prinus* (chestnut oak), are particularly susceptible to being invaded and the tree regeneration outcompeted by garlic mustard (Meekins and McCarthy 1999). Garlic mustard was the most prominent herbaceous NNIP found in the Upper Midwest, occurring on 3% of the plots inventoried in 2005–2006 (Table 3.4).

3.4.5 Vines

Invasive vines were concentrated in the southern part of the region, particularly along the Ohio River watershed in Illinois and Indiana (Figure 3.5). Japanese honeysuckle was the most prominent species in this category (Figure 3.12).

Japanese honeysuckle is a persistent vine introduced as an ornamental and for erosion control and wildlife habitat in the mid-1800s (Plant Conservation Alliance 2006). The species thrives in a wide variety of habitats and quickly becomes established on disturbed sites (Rhoads and Block 2000). It is currently distributed in most states including Hawaii, but is limited by cold temperatures and low

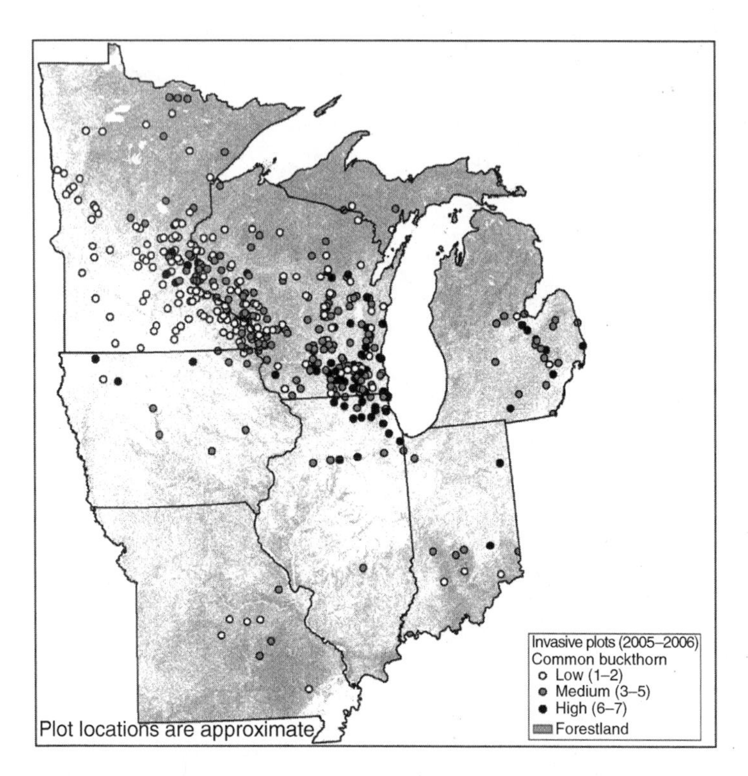

FIGURE 3.9 Distribution of plots with common buckthorn in the Upper Midwest, 2005–2006.

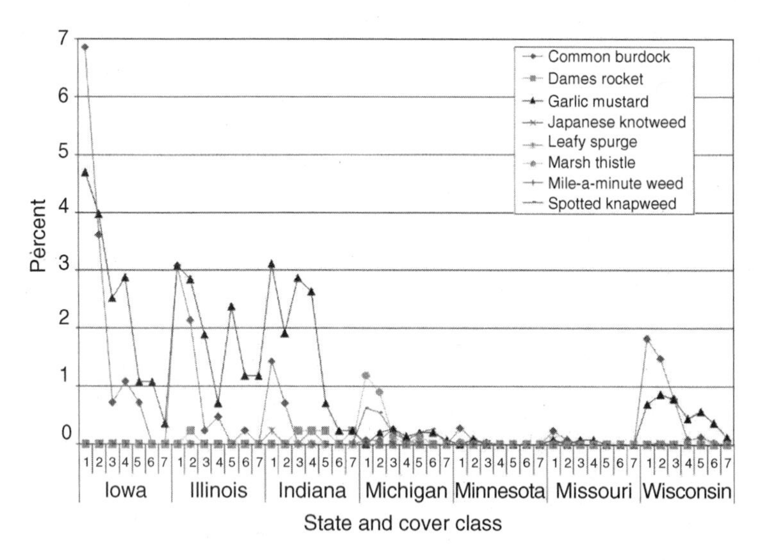

FIGURE 3.10 Presence of nonnative invasive herbaceous species in the seven states of the Upper Midwest, as measured by percent of all forested inventory plots sampled in 2005–2006, by state and cover class category. Cover class categories: 1 = <1%, trace; 2 = 1%–5%; 3 = 6%–10%; 4 = 11%–25%; 5 = 26%–50%; 6 = 51%–75%; and 7 = 76%–100%. Cover class 0—no invasives found—is not shown so as to preserve graphic scale.

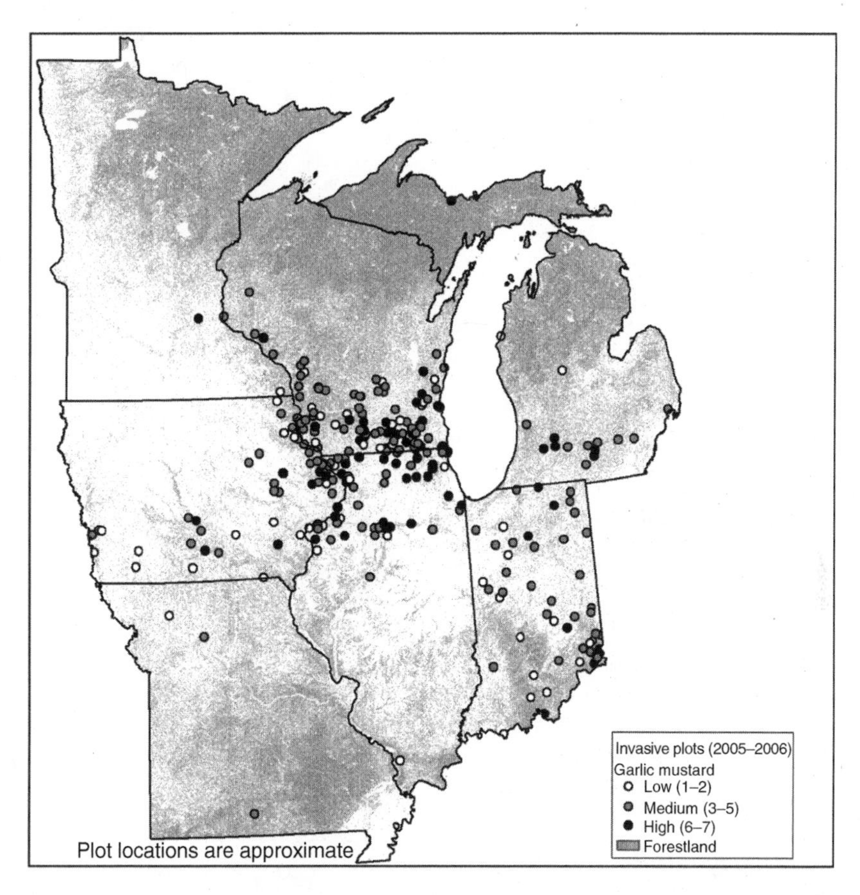

FIGURE 3.11 Distribution of plots with garlic mustard, Upper Midwest, 2005–2006.

precipitation (U.S. Department of Agriculture, NRCS 2007; Plant Conservation Alliance 2006). Japanese honeysuckle spreads by vegetative runners, underground rhizomes, and seed dispersal, particularly by birds. It quickly becomes established and crowds out native plants (Missouri Department of Conservation 1997). Among vines in this study, Japanese honeysuckle was the most prevalent and was found in 2.5% of all plots inventoried in 2005–2006 (Table 3.4). Plots with high cover classes (4–7) were more common for this species than for other NNIPs (Table 3.4). This NNIP species occurs primarily along the Ohio and Mississippi River basins in Indiana, Illinois, and Missouri (Figure 3.13).

3.4.6 Grasses

Invasive grasses were observed largely in forests near extensive agricultural land—eastern Iowa and northern Illinois—rather than areas with extensive closed-canopy forests like northern Minnesota or southern Missouri. Reed canary grass was the

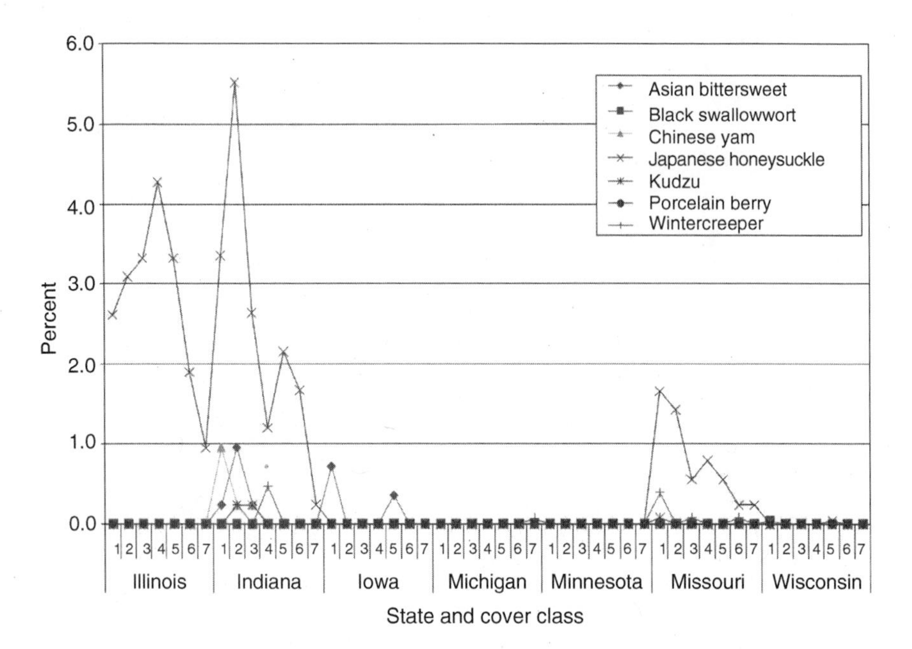

FIGURE 3.12 Presence of nonnative invasive vines in the seven states of the Upper Midwest, as measured by percent of all forested inventory plots sampled in 2005–2006, by state and cover class category. Cover class categories: 1 = <1%, trace; 2 = 1%–5%; 3 = 6%–10%; 4 = 11%–25%; 5 = 26%–50%; 6 = 51%–75%; and 7 = 76%–100%. Cover class 0—no invasives found—is not shown so as to preserve graphic scale.

most prominent NNIP grass on forested plots in the Upper Midwest, particularly Iowa (Figure 3.14).

Among the NNIP grasses, reed canary grass was the most prominent at 1% of forested plots inventoried in 2005–2006. It was the only NNIP where plots with cover classes greater than 10% exceeded the number of plots with cover classes 10% or less (Table 3.4). Although it is widespread in nonforested areas, the principal indication of reed canary grass presence on forested lands was in Iowa (Figure 3.15).

3.4.7 Regional and Climatic Limitations

In preparation for analysis of site factors that might relate to invasive presence, we examined regional limits of each species as an indicator of climatic influence. Initial attempts at variable reduction always found that latitude and longitude were the major influences on NNIP presence. Some of this distribution might be due to historical factors, such as the preference for a species by a state agency, and some of the cause may be due to climatic limitations, such as a species' intolerance of extreme cold. We used these apparent limits as sideboards to closely examine the influence of site- and stand-specific variables. We visually estimated the latitudinal and longitudinal ranges that contained ~90%–95% of the plots with each species,

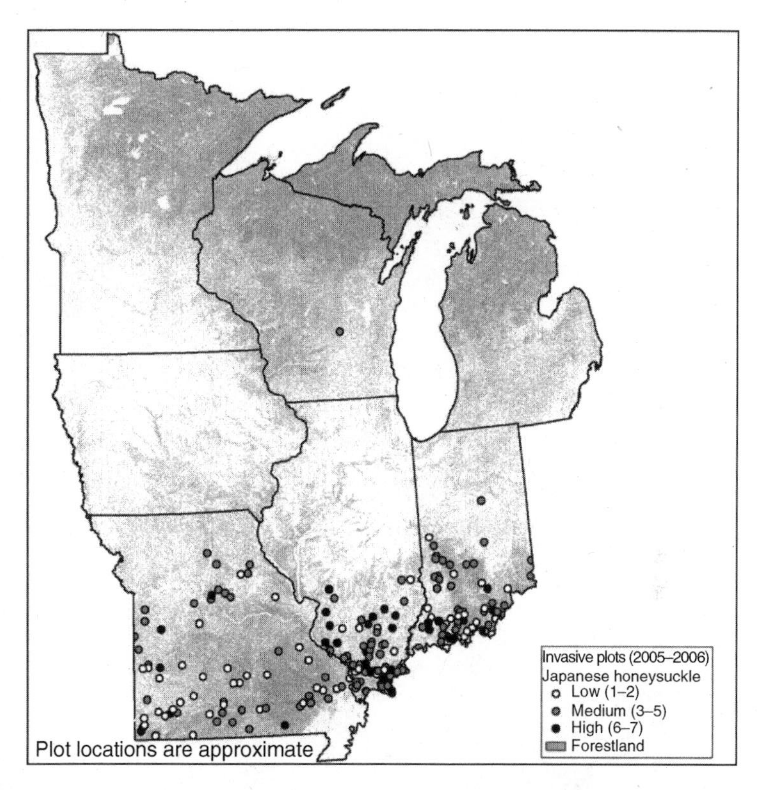

FIGURE 3.13 Distribution of plots with Japanese honeysuckle, Upper Midwest, 2005–2006.

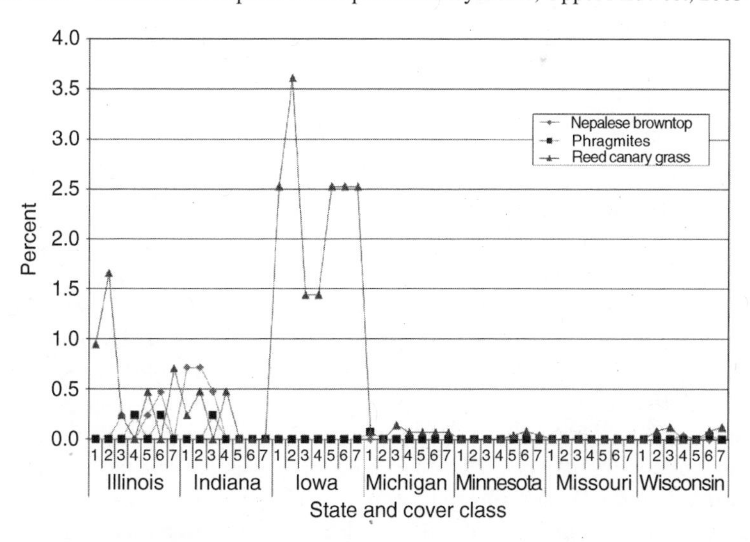

FIGURE 3.14 Presence of nonnative invasive grasses in the seven states of the Upper Midwest, as measured by percent of all forested inventory plots sampled in 2005–2006, by state and cover class category. Cover class categories: 1 = <1%, trace; 2 = 1%–5%; 3 = 6%–10%; 4 = 11%–25%; 5 = 26%–50%; 6 = 51%–75%; and 7 = 76%–100%. Cover class 0—no invasives found—is not shown so as to preserve graphic scale.

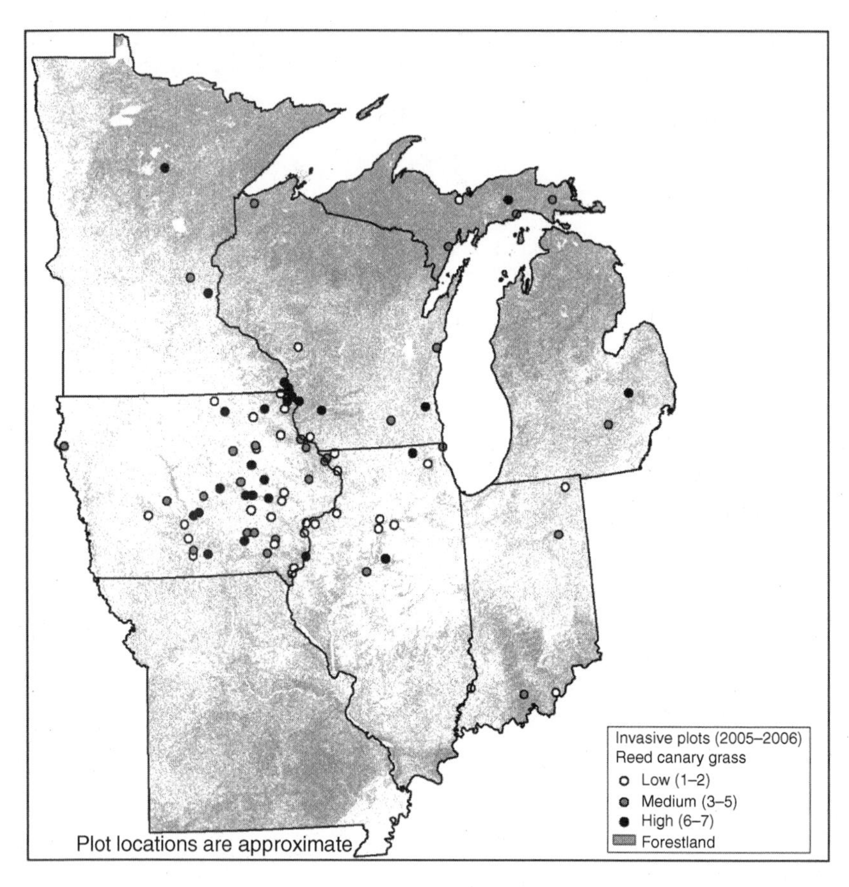

FIGURE 3.15 Distribution of plots with reed canary grass, Upper Midwest, 2005–2006.

then subset the dataset for each species, and examined site and disturbance relationships within this subset.

There were some pronounced regional influences on NNIP presence (Figure 3.16). Regarding multiflora rose, there was no noticeable longitudinal influence while there was a noticeable latitudinal cutoff north of 44° N (Figure 3.7). Other investigators have suggested that the species is not tolerant of cold weather (Amrine 2002; Munger 2002). Nonnative bush honeysuckles, on the other hand, were distributed widely throughout the region (Figure 3.8). Situated across the Midwest and along all but the northernmost latitudes, these species were present in most of the Midwest's major forest types. Common buckthorn was present in the longitudinal range centering on Wisconsin and Minnesota (Figure 3.9). Although there are occurrences farther south, buckthorn presence generally decreased south of 42° N latitude. Garlic mustard was widespread longitudinally, but exhibited a sharp decline above 44° N latitude (Figure 3.11). Research suggests that Japanese honeysuckle prefers warmer climates (Leatherman 1955). Our data support this contention;

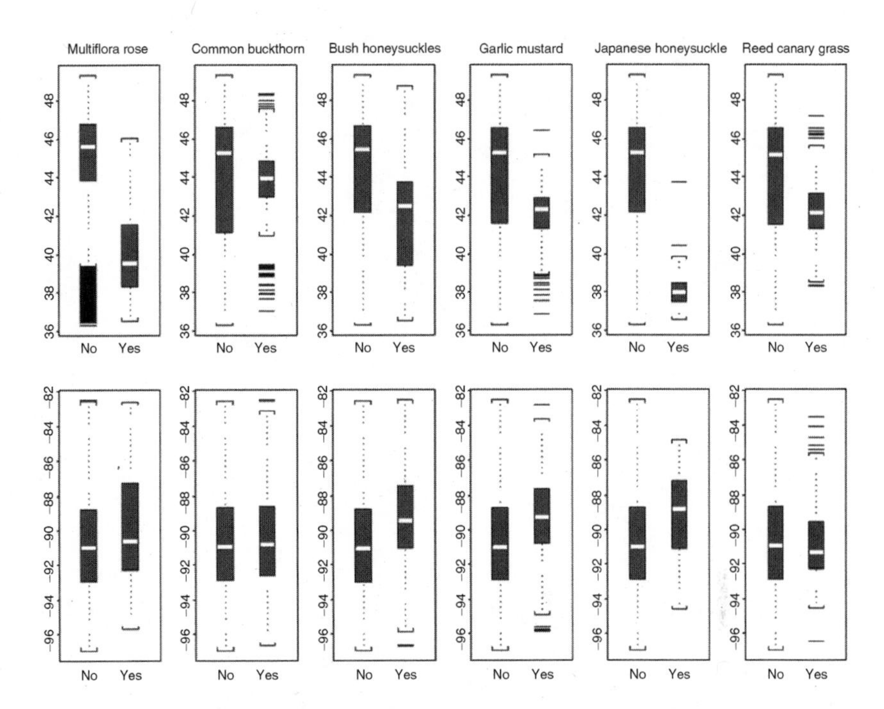

FIGURE 3.16 Latitudinal and longitudinal distribution of the six NNIPs of interest. Presence ("Yes") is on the right side of each plot; absence ("No") is on the left side. Latitudinal range is 36°–48° N; longitudinal range is 82°–96° W.

observations of the species centered on southern Illinois and southern Missouri (Figure 3.13). Reed canary grass was primarily found in the middle latitudes and longitudes of the region, with the North–South observations particularly clustered (Figure 3.15).

3.4.8 Stand, Site, and Disturbance Factors

We used a bidirectional stepwise regression to reduce the candidate variables to a subset considered to have a significant relationship with NNIP abundance. Table 3.5 displays the complete list of variables with a value and level of significance for those variables remaining in each model. Measurements of disturbance and fragmentation had a significant relationship with NNIP presence and cover across the board (Table 3.5). County percent forest had a highly significant relationship with almost every invasive species and 4 of the 6 individual species we tracked. This variable also displayed a significant relationship with Japanese honeysuckle. The three woody species and reed canary grass had a significant negative relationship with forest intactness. Common buckthorn and nonnative bush honeysuckles displayed a significant negative relationship with increasing distance from the road while, conversely, reed canary grass had a positive relationship.

TABLE 3.5
Site and Disturbance Variables Significantly Related to the Presence and Abundance of NNIPs in the Upper Midwest, United States in 2005–2006

Variable	Any Invasive Species	Multiflora Rose	Common Buckthorn	Nonnative Bush Honeysuckles	Garlic Mustard	Japanese Honeysuckle	Reed Canary Grass
Stand variables							
Total basal area							−0.2557***
Oak basal area	−0.08102***	−0.02725**				−0.0343**	−0.09407***
Stand age	−0.0798**			−0.03482***			−0.04406
SDI—All	−0.04684**	−0.008662		−0.01202*			−0.106**
SDI—5″ plus	0.06826**			0.01189		0.01637*	
H′ species				0.7033			−3.559**
H′ diameter				−0.9138	3.338*		
H′ height		2.211**			−3.009*		
Site variables							
Site index		0.05335**	−0.0326*				
Aspect (cosine)				−0.7693**		−0.701	
Physiographic class code			Mesic: 1.386, Xeric: −0.2375				
Disturbance variables							
County percent forest	−17.58***	−5.139***	−6.264***	−4.748***	−10.28***	−5.113*	
Plot percent forest		2.360					
Minimum distance to the nearest road			−0.001135	−0.001175*			0.009***
Index of forest intactness (sum of all expanded ordinal scores)		-3.448×10^{-5}*	-5.542×10^{-5}***	-1.898×10^{-5}	4.873×10^{-4}*	3.929×10^{-5}	-6.186×10^{-5}*
R^2	0.0595	0.0577	0.0562	0.0423	0.0249	0.0724	0.133

Note: The values in each cell represent the parameter estimates in the final linear model.

* Indicates a positive (negative) relationship at the 0.05 level.

** Indicates a positive (negative) relationship at the 0.01 level.

*** Indicates a positive (negative) relationship at the 0.001 level; blank predictor variables were dropped during stepwise regression.

The correlation with site index was less conclusive. Site index was positively significant for multiflora rose coverage and was negatively related to nonnative bush honeysuckles coverage. We found little correlation between aspect and NNIP coverage. Stand variables, surprisingly, did not display a consistent relationship with NNIP abundance. Oak basal area, stand age, and all-tree stand density indices were negatively significantly related to any NNIP abundance, but relationships with individual species were more variable. Stand age had a significant negative relationship with multiflora rose and nonnative bush honeysuckles, consistent with expectations of high basal area and time since the last disturbance, yet total basal area was not significant for these species.

Our results could have emanated from alternate yet co-occurring influences. For example, the negative relationship of oak basal area and NNIPs could be the result of *Quercus* species' frequent presence on drier sites, a difficult habitat for moisture-loving Japanese honeysuckle and reed canary grass. The apparent contradiction between SDI of all trees' (usually negative) relationship and SDI of trees 12 cm and greater (often positive) relationship points to the impact of low shade, more frequently found in stands with diameters ranging from 2.4 cm to sawtimber size (30 cm), than in stands where the bulk of the trees is much larger.

3.5 DISCUSSION AND CONCLUSIONS

This study examined patterns of distribution and relationships with selected forest and site characteristics for 25 exotic plant species/species groups of interest in the Upper Midwest of the United States. NRS-FIA recorded one or more of these 25 species on one-quarter of the forested plots inventoried in 2005 and 2006. In some portions of the region, plots had even higher rates of NNIP presence. Iowa, Indiana, and Illinois had the highest overall proportion of plots with invasives, while Minnesota had the lowest. Our data revealed a strong latitudinal separation, particularly for woody invasives. Common buckthorn was prominent in Wisconsin and Minnesota, while multiflora rose was more prevalent in Missouri, Illinois, and Indiana.

Most subboreal forest types had lower percentages of NNIP presence. Accordingly, we also saw lower occurrences in the Lake States of Minnesota, Wisconsin, and Michigan than those in the southern-tier states. Early successional forest types in the center of the region appear to have a higher percentage of plots with NNIPs, but it was difficult to separate any relationship from the sampling effect, as these early successional forest types were often the most predominant on the landscape. Grasses were particularly prominent in fragmented forestland in the center of our study area. Our results support Richardson and Pyšek's (2006) contention that agricultural or urban sites are the most invasible biomes. The nature of our sample set meant that we were only examining successful invaders on forested plots, not those species that failed to establish in the region. This likely skewed our examination of particular plant characteristics or affinities for a site or disturbance pattern. Although we could not conclusively tie the presence of NNIPs to particular forest types, disturbance likely played a role in the life history strategy of both overstory tree species and understory invasive plants. The predominant forest types in the southern two-thirds of the region—oaks—are mid-shade tolerants and rely upon disturbance to maintain their position in most parts of the genus' range (Johnson et al. 2002).*

We did not assess time between the start of an invasion and the typical phase of exponential increase. Our inability to determine the contribution of site and stand factors to invasiveness is likely because the NNIP patterns we observe today are largely the net result of introductions and prevailing conditions and processes from 50 to 100 years ago or more. This conclusion suggests a kind of built-in inertia, where the number of naturalized and invasive species will increase in the future even if no additional introductions occur (Kowarik 1995; Richardson and Pyšek 2006).

Metrics of disturbance and fragmentation, such as distance to road, county percent forest or the forest intactness index, were significantly related to NNIP presence and coverage. NRS-FIA treatment or disturbance codes and other measures such as the ratio of tree removals to current volume revealed no significant connection with NNIP presence. Disturbances that initiated an invasive plant's presence likely occurred several decades ago (Hulme 2003; Richardson and Pyšek 2006), which is why patterns of fragmentation and landscape-level forest proportions are better measures of disturbance history.

While some of our results may be extrapolated to other species or regions, we should remember Pyšek's (2001) words that "in predicting the success of potential invaders, it is easier to predict invaders than non-invaders among exotic species." A posteriori analysis of invasive species at one point in time is usually not sufficient to evaluate trends in regeneration, expansion, or growth (Rejmánek 1989). The NRS-FIA database tracks disturbance and silvicultural treatments, but only in the interval since the previous inventory. The anthropogenic activities that resulted in the establishment of these nonnative invasive species likely occurred many years ago. Repeated measures on a wide scale will be necessary to verify any trends.

Given the history of natural and human-caused disturbance and forest types whose shade tolerance results in understory growing space that is not completely occupied, we expected to find multiple relationships between NNIP and forest and site characteristics. When looking at disturbance, we observed that multiflora rose, Japanese honeysuckle, and reed canary grass significantly benefited from lower overstory basal areas, but this relationship did not apply to other species. Another measure of disturbance, distance to nearest road, had a significant negative relationship with the presence of nonnative bush honeysuckles and reed canary grass.

The percent of total land area in a county that is forest provided a striking indicator of historic disturbance. This metric displayed an almost universally significant negative relationship with NNIP abundance. These results are not surprising; invasive species are known to thrive on sites with more available resources (Richardson and Pyšek 2006). The challenge is separating the human influence from the ecological. One could easily argue that our results reflect the heavily disturbed nature of the Midwest's second- and third-generation forests, which either reestablished following the abandonment of farmland or pasture or were influenced by heavily disturbed adjacent land. The characteristics of the landscape that influenced invasive species

* In fact, the lack of disturbance is resulting in a shift in species composition of regeneration in oak forests throughout the genus' range (Moser et al. 2006).

presence may also be a significant relationship with homestead choice by settlers. Even our disturbance measures, lower basal area, and high road density could reflect the lingering influence of historic human disturbance as the microsite attributes that allowed them to thrive.

Disturbance events, coupled with anthropogenic establishment of individual species, displayed lingering effects on the Midwest forest ecosystem long after they occurred. Site conditions and stand structures cannot be relied upon to reverse these trends. As with most situations where ecological restoration is the goal, elimination of NNIPs in the Upper Midwest will demand both aggressive action to stop the spread of the species and significant investment in efforts to restore invaded ecosystems to their pre-NNIP state.

ACKNOWLEDGMENTS

The authors thank Barry T. Wilson for developing the map template used as the background for the NNIP distribution maps and Sonja Oswalt and Andy Gray for their reviews of earlier versions of this manuscript.

REFERENCES

Akaike, H., A new look at the statistical model identification, *IEEE Trans. Automat. Contr.*, 19, 6, 716, 1974.

Amrine, J.W., Multiflora rose, in *Biological Control of Invasive Plants in the Eastern United States*, Van Driesche, R., Blossey, B., Hoddle, M., Lyon, S., and Reardon, R., Eds., USDA Forest Service Publication FHTET-2002–04, Washington, DC, 265, 2002.

Booth, B.D., Murphy, S.D., and Swanton, C.J., *Weed Ecology in Natural and Agricultural Systems*, CABI, Wallingford, 2003.

Crawley, M.J., Brown, S.L., Heard, M.S., and Edwards, G.G., Invasion-resistance in experimental grassland communities: Species richness or species identity? *Ecol. Lett.*, 2, 140, 1999.

Daehler, C.C., Darwin's naturalization hypothesis revisited, *Am. Nat.*, 158, 324, 2001.

Daehler, C.C., Performance's comparisons of co-occurring native and alien invasive plants: Implications for conservation and restoration, *Annu. Rev. Ecol. Syst.*, 34, 183, 2003.

Della Sala, D.A., Stans, N.L., Stritthott, J.R., Hackman, A., and Lacobelli, A. An updated protected areas database for the United States and Canada, *Nat. Areas J.*, 21, 124, 2001.

Drake, J.A., Mooney, H.A., Di Castri, F., Groves, R.H., Kruger, F.J., Rejmánek, M., and Williamson, M., *Biological Invasions: A Global Perspective*, Wiley, Chichester, UK (published on behalf of the Scientific Committee on Problems of the Environment (SCOPE) of the International Council of Scientific Unions, Series SCOPE report, no. 37), 1989.

Ellstrand, N.C. and Schierenbeck, K.A., Hybridization as a stimulus for the evolution of invasiveness in plants? *Proc. Natl. Acad. Sci. USA*, 97, 7043, 2000.

Elton, C.S., *The Ecology of Invasions by Animals and Plants*, University of Chicago Press, Chicago, IL, 1958.

ESRI Data & Maps, 2002. CD-Rom, Environment Systems Research Institute, Inc. 2002.

Evans, J.E., A literature review of management practices for multiflora rose *(Rosa multiflora)*, *Nat. Areas J.*, 3, 1, 6, 1983.

Fagan, M.E. and Peart, D.R., Impact of the invasive shrub glossy buckthorn (*Rhamnus frangula* L.) on juvenile recruitment by canopy trees, *For. Ecol. Manage.*, 194, 1–2, 95, 2004.

Forman, R.T.T. and Alexander, L.E., Roads and their major ecological effects, *Ann. Rev. Ecol. Syst.*, 29, 207, 1998.

Gelbard, J.L. and Belnap, J., Roads as conduits for exotic plant invasions in a semiarid landscape, *Conserv. Biol.*, 17, 2, 420, 2003.

Harrington, R.A., Brown, B.J., Reich, P.B., and Fownes, J.H., Ecophysiology of exotic and native shrubs in Southern Wisconsin. II. Annual growth and carbon gain, *Oecologia*, 80, 368, 1989.

Heilman, G.E., Slosser, N.C., and Strittholt, J.R., *Forest intactness of the coterminous United States* (CD-ROM database prepared for the World Wildlife Fund and the World Resources Institute's Global Forest Watch), Conservation Biology Institute, Corvallis, OR, 2001.

Hodkinson, D.J. and Thompson, K., Plant dispersal: The role of man, *J. Appl. Ecol.*, 34, 1484, 1997.

Hood, W.G. and Naiman, R.J., Vulnerability of riparian zones to invasion by exotic plants, *Plant Ecol.*, 148, 105, 2000.

Hulme, P.E., Biological invasions: Winning the science battles but losing the conservation war? *Oryx*, 37, 178, 2003.

Huston, M.A. and DeAngelis, D.L., Competition and coexistence—The effects of resource transport and supply rates, *Am. Nat.*, 144, 954, 1994.

Johnson, P.S., Shifley, S.R., and Rogers, R., *The Ecology and Silviculture of Oaks*, CABI Publishing, New York, 501, 2002.

Kowarik, I., Time lags in biological invasions with regard to the success and failure of alien species, in *Plant Invasions: General Aspects and Special Problems*, Pyšek, P., Prach, K., Rejmanek, M., and Wade, M., Eds., SPB Academic, Amsterdam, 15, 1995.

Leatherman, A.D., Ecological life-history of *Lonicera japonica* Thunb., University of Tennessee, Knoxville, TN, 97, 1955 (unpublished dissertation).

Levine, J.M. and D'Antonio, C.M., Elton revisited: A review of evidence linking diversity and invasibility, *Oikos*, 87, 1, 15, 1999.

Lonsdale, W.M., Global patterns of plant invasions and the concept of invasibility, *Ecology*, 80, 1522, 1999.

Luken, J.O., Population structure and biomass allocation of the naturalized shrub *Lonicera maackii* (Rupr.) Maxim. in forest and open habitats, *Am. Midl. Nat.*, 119, 2, 258, 1988.

Luken, J.O. and Goessling, N., Seedling distribution and potential persistence of the exotic shrub *Lonicera maackii* in fragmented forests, *Am. Midl. Nat.*, 133, 1, 124, 1995.

Magurran, A.E., Ecological diversity and its measurement. Princeton University Press, Princeton, New Jersey, 192, 35, 1988.

McCarthy, B., Response of a forest understory community to experimental removal of an invasive nonindigenous plant (*Alliaria petiolata*, Brassicaceae), in *Assessment and Management of Plant Invasions*, Luken, J.O. and Thieret, J.W., Eds., Springer-Verlag, New York, 117, 1997.

McGarigal, K. and Marks, B.J., FRAGSTATS: Spatial pattern analysis program for quantifying landscape structure, USDA, Forest Service Pacific Northwest Research Station, Gen. Tech. Rep. 351, Portland, OR, 122, 1995.

McRoberts, R.E., Joint annual forest inventory and monitoring symposium: The North Central perspective, *J. For.*, 97, 12, 27, 1999.

Meekins, J.F. and McCarthy, B.C., Competitive ability of *Alliaria petiolata* (garlic mustard, Brassicacaeae), an invasive, nonindigenous forest herb, *Int. J. Plant Sci.*, 160, 4, 743, 1999.

Missouri Department of Conservation, Missouri vegetation management manual, Jefferson City, MO, 161, 1997.

Mitich, L.W., Intriguing world of weeds: Kudzu (*Pueraria lobata* (Willd.) Ohwi), *Weed Technol.*, 14, 231, 2000.

Moser, W.K., Hansen, M.H., McWilliams, W., and Sheffield, R., Oak composition and structure in the Eastern United States, in *Fire in Eastern Oak Forests: Delivering Science to Land Managers* (Proceedings of a conference, November 15–17, 2005, Columbus, OH), Dickinson, M.B., Ed., Gen. Tech. Rep. NRS-P-1, U.S. Department of Agriculture, Forest Service, Northern Research Station, Newtown Square, PA, 49, 2006.

Munger, G.T., Rosa multiflora, in Fire effects information system [Online], U.S. Department of Agriculture, Forest Service, Rocky Mountain Research Station, Fire Sciences Laboratory (Producer), Missoula, MT, 2002. Available at http://www.fs.fed.us/database/feis/, accessed November 15, 2007.

North Central Research Station, Forest Inventory and Analysis (NCRS-FIA), *Forest Inventory and Analysis National Core Field Guide, Volume 1: Field Data Collection Procedures for Phase 2 Plots, Ver. 2.0.*, USDA, Forest Service North Central Research Station, St. Paul, MN, 290, 2005.

Nuzzo, V.A., Distribution and spread of the invasive biennial *Alliaria petiolata* [(Bieb.) Cavara & Grande] in North America, in *Biological Pollution: Control and Impact of Invasive Exotic Species*, McKnight, B.L., Ed., Indiana Academy of Sciences, Indianapolis, IN, 115, 1993.

Nuzzo, V.A., Invasion pattern of the herb garlic mustard (*Alliaria petiolata*) in high quality forests, *Biol. Invasions*, 1, 1, 169, 1999.

Olson, C.L. and Cholewa, A.F., Nonnative invasive plant species of the North Central region. A guide for FIA field crews, USDA, Forest Service, St. Paul, MN, 120, 2005 (unpublished field guide).

Plant Conservation Alliance, Various species' websites, U.S. Department of the Interior, Bureau of Land Management, Plant Conservation Alliance, Alien Plant Working group, Washington, DC. Available at http://www.nps.gov/plants/alien/fact.htm, accessed September 24, 2007, 2006.

Pyšek, P., Past and future of predictions in plant invasions: A field test by time, *Divers. Distrib.*, 7, 145, 2001.

Pyšek, P. and Hulme, P.E., Spatio-temporal dynamics of plant invasions: Linking pattern to process, *Ecoscience*, 12, 345, 2005.

Reineke, L.H., Perfecting a stand-density index for even-aged stands, *J. Agric. Res.*, 46, 627, 1933.

Rejmánek, M., Invasibility of plant communities, in *Biological Invasions: A Global Perspective*, Drake, J.A., Mooney, H.A., di Castri, F., Groves, R.H., Kruger, F.J., Rejmanek, M., and Williamson, M., Eds., Wiley, Chichester, UK, 369, 1989.

Rhoads, A.F. and Block, T.H., *The Plants of Pennsylvania, An Illustrated Manual*, Morris Arboretum of the University of Pennsylvania, University of Pennsylvania Press, Philadelphia, PA, 1060, 2000.

Richardson, D.M. and Bond, W.J., Determinants of plant-distribution—Evidence from pine invasions, *Am. Nat.*, 137, 639, 1991.

Richardson, D.M. and Pyšek, P., Plant invasions: Merging the concepts of species invasiveness and community invasibility, *Prog. Phys. Geogr.*, 30, 3, 409, 2006.

Rudis, V.A., Gray, A., McWilliams, W., O'Brien, R., Olson, C., Oswalt, S., and Schulz, B., Regional monitoring of nonnative plant invasions with the Forest Inventory and Analysis program, in *Proceedings of the Sixth Annual FIA Symposium* (September 21–24, 2004, Denver, CO), McRoberts, R.E., Reams, G.A., Van Deusen, P.C., and McWilliams, W.H., Eds., Gen. Tech. Rep. WO-70, U.S. Department of Agriculture, Forest Service, Washington, DC, 49, 2006.

Sax, D.F., Latitudinal gradients and geographic ranges of exotic species: Implications for biogeography, *J. Biogeogr.*, 28, 139, 2001.

Shannon, C.E., A mathematical theory of communication, *Bell System Tech. J.*, 27, 379, 1948.

U.S. Department of Agriculture, The PLANTS database, Natural Resource Conservation Service, National Plant Data Center, Baton Rouge, LA, 2007. Available at http://plants.usda.gov, accessed September 24, 2007.

Watkins, R.Z., Chen, J., Pickens, J., and Brosofske, K.D., Effects of roads on understory plants in a managed hardwood landscape, *Conserv. Biol.*, 17, 2, 411, 2003.

Webster, C.R., Jenkins, M.A., and Jose, S., Woody invaders and the challenges they pose to forest ecosystems in the eastern United States, *J. For.*, 104, 7, 366, 2006.

Woodall, C.W. and Miles, P.D., New method for determining the relative stand density of forest inventory plots, in *Proceedings of the Sixth Annual Forest Inventory and Analysis Symposium* (September 21–24, 2004, Denver, CO), McRoberts, R.E., Reams, G.A., Van Deusen, P.C., and McWilliams, W.H., Eds., Gen. Tech. Rep. WO-70, U.S. Department of Agriculture Forest Service, Washington, DC, 105, 2006.

4 Invasion Pattern of Exotic Plants in Forest Ecosystems

Songlin Fei, Ningning Kong, Jeffrey Stringer, and Daniel Bowker

CONTENTS

4.1 INTRODUCTION

Invasion by exotic species is a significant problem, reaching epidemic proportions and costing the American public an estimated US$137 billion or about US$1300 per household each year as determined by loss of productivity, costs of herbicides, and other measures (Pimentel et al. 2000). The health and longevity of many forest ecosystems are also at risk because of significant invasion by exotic plants, resulting in loss of native species, deterioration of natural regenerative processes, decreases in forest productivity, and degradation of the environment. Moreover, the effects of invasion by exotic species have proved to be long lasting and cumulative (Chornesky et al. 2005).

Much attention has been given to invasive exotic insects and pathogens. Damages resulting from invasive exotic insects and pathogens are most often species or genus specific and can be devastating. The American chestnut (*Castanea dentata*), one of the dominant tree species of the eastern North American forest, was functionally wiped out because of the invasion of chestnut blight (*Cryphonectria parasitica*). The majority of American elm (*Ulmus americana*) trees, one of the favorite shade trees in the United States, was severely impacted by the invasion of

Dutch elm disease (*Ophiostoma* spp.) and was essentially eliminated as a viable shade tree. Eastern hemlock (*Tsuga canadensis*), one of the most important riparian species in the Appalachian region, is endangered due to the invasion of hemlock woolly adelgid (*Adelges tsugae*). There are many other invasive exotic insects and diseases, such as butternut canker (*Sirococcus clavigignenti-juglandacearum*), beech bark disease (*Nectria galligena*), emerald ash borer (*Agrilus planipennis*), sudden oak death (*Phytophthora ramorum*), and dogwood anthracnose (*Discula destructiva*), that are putting many tree species in great danger.

Invasions of exotic plants, on the other hand, often cause less prominent impacts than that of invasive exotic insects and pathogens. Invasion of exotic plants often does not result in the elimination of a native tree species. However, exotic plants can cause profound and devastating negative impacts on forest ecosystems. Many invasive exotic plants can act as ecosystem simplifiers by smothering an entire ecosystem and converting it into a monocultural system. A decrease in habitat complexity and heterogeneity associated with the proliferation of introduced species often leads to a decline in native species richness (Lodge 1993; Hanowski et al. 1997; Deckers et al. 2005). For example, invasive exotic plants such as bush honeysuckle (*Lonicera maackii*) and winter creeper (*Euonymus fortunei*) often form dense mid- and/or understories in the forests of central Kentucky, limiting the regeneration of native tree species (Figure 4.1). Unlike the invasion of exotic insects or pathogens that often target an individual native species or genus, the invasion of an exotic plant in a natural forest can cause the elimination of a range of unrelated but cohabitating tree species over time due to the impediment of tree regeneration.

FIGURE 4.1 Examples of invasive exotic plants dominating forest understory: (a) bush honeysuckle (*Lonicera maackii*) and (b) winter creeper (*Euonymus fortunei*).

The onset and aggressiveness of the invasion of an exotic plant in a forest ecosystem can be affected by many factors. Landscape structure may affect any or all of the stages of the invasion process, including introduction, colonization, establishment, and dispersal (With 2002). Disturbance, especially harvest-induced disturbance, is also an important factor that can increase the abundance of exotic plants in forest ecosystems (Kotanen 1997; Mack and D'Antonio 1998; Mooney and Hobbs 2000). Overstory harvest changes microclimate at the ground surface by increasing temperature maxima, diurnal temperature range, and surface soil moisture, and by decreasing relative humidity. Changes in these factors and the sudden availability of resources provide the opportunity for exotic plants to invade forest ecosystems.

To prevent and/or mitigate exotic plant invasion, the most critical questions are not *whether* the aforementioned factors promote establishment of nonnatives. Rather, it is essential to determine *how* these factors influence exotic plant invasion. Studying spatial patterns of invasion can offer insights into the role of a number of influential factors in invasion. However, relatively few studies have examined patterns of invasion or the factors contributing to the process because of the lack of fine-resolution spatial data (Kolb et al. 2002). Fortunately, with the recent availability of fine-resolution remotely sensed imagery, digital elevation models, and high-accuracy global positioning system (GPS) receivers, this obstacle can be overcome.

In this study, we used geographic information system (GIS), high-accuracy GPS receivers, and high-resolution aerial photos to study the invasion patterns of exotic plant species in eastern Kentucky. We analyzed the association of invasive exotic species occurrence in relation to adjacent land use, disturbance, and geomorphologic features. The results of this study can be used to improve forest management practices through adaptive management.

4.2 STUDY AREA

Robinson Forest, managed by the University of Kentucky, Department of Forestry, is one of the largest research and educational forests in the eastern United States. Robinson Forest is a collection of seven tracts totaling 6000 ha, and is located on the Cumberland Plateau in southeastern Kentucky. This study was conducted on the largest section of Robinson Forest—a contiguous area of ~4000 ha. The region has a deeply dissected topography and is occupied by a mixed mesophytic forest consisting mainly of second-growth hardwood species 80–90 years of age. The main block of Robinson Forest is surrounded by reclaimed surface-mined lands (Figure 4.2), which contain many invasive exotic plants. These exotic plants represent an invasive threat when a disturbance such as timber harvesting occurs.

4.3 MATERIALS AND METHODS

Two sampling schemes were employed to survey the distribution of invasive exotic plants in 10 selected watersheds in Robinson Forest (Figure 4.2). The first sampling scheme was designed to capture the overall distribution and spatial variation of invasive plants using the existing Continuous Forest Inventory (CFI) grid system. The CFI plot system is a permanent $384 \times 384\,m^2$ grid of forest inventory plots

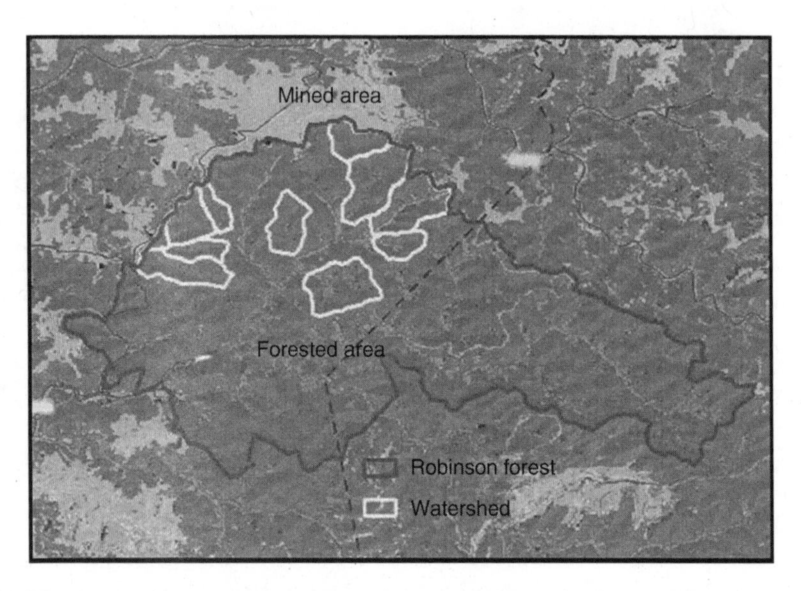

FIGURE 4.2 Landcover within and around the main block of Robinson Forest study area. (Based on Kentucky Land Cover Dataset—Anderson level III, 2001 (Kentucky Division of Geographic Information 2007).)

established in 1994. Invasive plant species were surveyed both in the $400\,m^2$ CFI plots and along 10 m wide transects between the sequentially spaced CFI plots. A total of 48 CFI plots and 18.1 km of transects were surveyed. The second sampling scheme was designed to capture the distribution of invasive plants on or near the forest road and trail systems. The survey was carried out on all hiking trails and both active and inactive roads in 10 watersheds occurring in the 4000 ha section of the forest. Invasive plants were inventoried along the road and a 5 m zone on each side of the road. A total of 42.5 km of roads were surveyed. Trimble Pathfinder GeoXM GPS receiver was used to georeference the location of invasive plants.

Invasive exotic plants were divided into three groups: herbaceous, shrubs/vines, and trees. For each location, patch size, average patch width, patch length, and percentage cover were recorded for invasive herbaceous species; average patch size, number of clumps, and average crown width were recorded for invasive shrub species; while stem diameter, number of stems, and average crown width were recorded for invasive tree species. Descriptive statistics of invasive species were summarized for the study area, and overall spatial patterns of invasive exotic plants were analyzed in ArcGIS V9.2 (ESRI Inc., Redlands, CA).

Association between the measured invasive species metrics and road surface type and road traffic intensity were analyzed. Road surface type and road traffic intensity were obtained from a previously surveyed GIS data layer. A mixed model was used to test the statistical association in SAS V8.0 (SAS Institute Inc., Cary, NC). The most abundant herbaceous species (Japanese stiltgrass, *Microstegium vimineum*) and the top three shrub species (autumn olive, *Elaeagnus umbellata*; bush honeysuckle; and multiflora rose, *Rosa multiflora*) were included in this analysis.

Logistic regression was employed to analyze how potential factors influence the distribution of invasive species. Factors including elevation, site wetness, slope, aspect, road density, and distance to adjoining reclaimed surface mine were used in the analysis. Elevation, site wetness, slope, and aspect were derived from 10 m resolution digital elevation models from the U.S. Geological Survey (USGS 2007), using Spatial Analyst in ArcGIS. Road density was calculated using the line density function in Spatial Analyst. The reclaimed surface mine was digitized based on 1 m resolution aerial photos of 2004 (Kentucky Division of Geographic Information 2007), and Euclidean distance from the mine areas was calculated. Each of the aforementioned six factors was saved as a thematic layer in ArcGIS. Surveyed invasive species points were then superimposed on these thematic layers, and the corresponding values of each thematic layer were extracted for these points. Since invasive absence points were not recorded in our sampling procedure, random points (same number as the presence points) were generated outside a 5 m buffer zone of known presence locations for each species to represent the absent points to be used in logistic regression. Random points were then also superimposed on the six thematic layers and their corresponding values extracted.

4.4 RESULTS

A total of 11 invasive exotic plant species were identified in the 10 selected watersheds in Robinson forest, including 5 herbaceous species, 5 shrub/vine species, and 1 tree species (Table 4.1). The main invasive herbaceous species was Japanese stiltgrass (*Microstegium vimineum*). The total linear length of Japanese stiltgrass occurrence was 15.7 km, ~37% of the total road length surveyed. The main invasive shrub species were multiflora rose, autumn olive, and bush honeysuckle. The largest patch of invasive exotic shrub observed was multiflora rose (0.84 ha). Tree-of-heaven (*Ailanthus altissima*) was the only invasive tree species observed in the surveyed area with limited occurrence (10 times).

TABLE 4.1
Invasive Exotic Plants Observed in Robinson Forest

Scientific Name	Common Name	Observed Occasions
Microstegium vimineum	Japanese stiltgrass	518
Sorghum halepense	Johnsongrass	15
Festuca arundinacea	KY 31 fescue	12
Melilotus spp.	White or yellow sweet clover	6
Miscanthus sinensis	Chinese silver grass	3
Rosa multiflora	Multiflora rose	472
Elaeagnus umbellata	Autumn olive	191
Lonicera maackii	Bush honeysuckle	100
Lespedeza cuneata	Sericea lespedeza	35
Lonicera japonica	Japanese honeysuckle	1
Ailanthus altissima	Tree-of-heaven	10

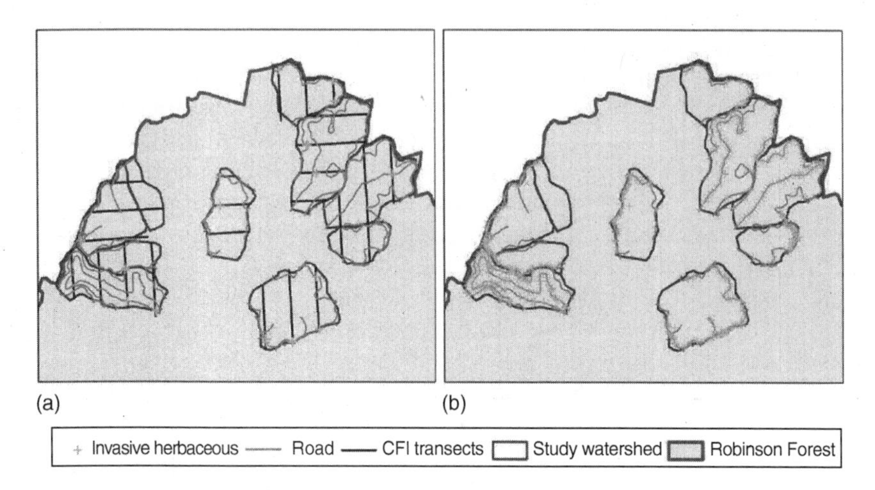

FIGURE 4.3 Distribution of invasive exotic herbaceous species along CFI plots and transects (a) and along roads (b).

4.4.1 Spatial Patterns of Invasive Herbaceous Species

Invasive exotic herbaceous species were all distributed on or near the road system (Figure 4.3). Of all the surveyed 48 CFI plots and 18.1 km transects, there were only 20 occasions where invasive herbaceous species were observed. All 20 observations were within 5 m distance from the nearest road (Figure 4.3a). For invasive herbaceous species recorded in the second sampling scheme, 84% of invasive herbaceous species were located on the road, 10% were located within 1 m buffer from the edge of the road, and 6% were located within 1–5 m buffer from the edge of the road (Figure 4.3b). All 10 watersheds had some level of invasive herbaceous plants, most often Japanese stiltgrass.

Analysis of the association between Japanese stiltgrass and environmental factors at the road level revealed that the occurrence and abundance of Japanese stiltgrass was significantly associated with road surface type and road traffic intensity. Japanese stiltgrass was more commonly found on roads with moderate levels of traffic, but had significantly lower occurrence on roads with no traffic (Figure 4.4). Average patch cover intensity of Japanese stiltgrass was only significantly associated with road surface type (Figure 4.5). Japanese stiltgrass cover intensity was significantly lower on roads with a surface type covered with sapling structures than on roads covered with grass or seedlings.

Logistic regression also indicated that the presence of Japanese stiltgrass was significantly associated with elevation, road density, and distance to invasive source (Table 4.2). The probability of finding Japanese stiltgrass was higher on sites with low elevation, high road density, and proximity to an invasive source.

4.4.2 Spatial Patterns of Invasive Shrub Species

Different spatial distribution patterns were observed for different invasive exotic shrub species (Figure 4.6). For all CFI plots and transects pooled, there were 91

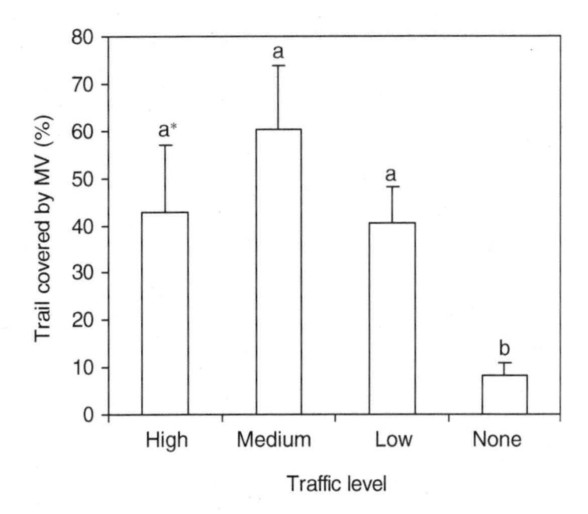

FIGURE 4.4 Association between traffic level intensity and mean and standard error for percentage of Japanese stiltgrass (MV) coverage in length on road system (total length of MV/total length of road). *Groups not sharing the same letter are statistically significant at $p < .05$.

observed invasive shrub locations, including 12 autumn olive, 3 sericea lespedeza (*Lespedeza cuneata*), 7 bush honeysuckle, and 69 multiflora rose locations. All three sericea lespedeza locations were found within 5 m of a road. Seventy-five percent of the autumn olives were located within 10 m of a road, and the furthest autumn olive was located ~200 m away from the nearest road (Figure 4.6a). All seven bush honeysuckles were located within 50 m of a road (Figure 4.6b). Multiflora rose was more widely distributed, where 28%, 36%, 20%, and 16% of the observed multiflora roses were located within 0–10, 10–50, 50–100, and over 100 m of a road, respectively (Figure 4.6c). Multiflora rose was observed as far as 308 m from the nearest road.

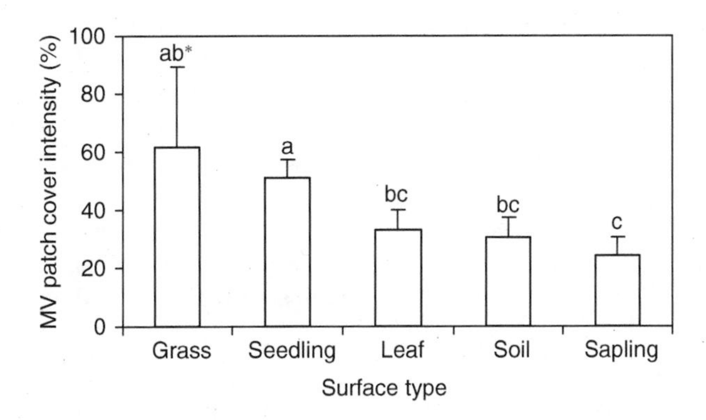

FIGURE 4.5 Association between road surface type and mean and standard error for patch cover intensity of Japanese stiltgrass (MV). *Groups not sharing the same letter are statistically significant at $p < .05$.

TABLE 4.2
Factors Associated with the Presence and Absence of the Four Most Abundant Exotic Invasive Species in Robinson Forest Based on Logistic Regressions

Factors	Japanese Stiltgrass	Autumn Olive	Multiflora Rose	Bush Honeysuckle
Elevation	−	−	+	0
Road density	+	+	+	+
Wetness	0	0	0	0
Slope	0	0	+	+
Aspect	0	0	0	0
Distance[a]	−	−	−	−

Note: 0, no significant association; +, positive association; −, negative association at $p < .05$.

[a] Euclidean distance between sampling point (presence or absence point) and the invasive source on the surface mine lands.

Invasive shrub data from the two sampling schemes indicated that invasive exotic shrubs were more abundant on or within 5 m of a road (Figure 4.6). With the consideration of total length of roads and transects surveyed, there were 6.4 times more autumn olives found on the road system than on the CFI transects, and 5.7 and 2.5 times more for bush honeysuckle and multiflora rose, respectively. Different spatial patterns were also observed for invasive shrubs that were established on or within 5 m of a road (Figure 4.6d–4.6f). Autumn olives were more abundant in watersheds that were close to the reclaimed surface mines, but less abundant in watersheds that were in the interior of the forest. Bush honeysuckles were only observed in two watersheds near the western border of Robinson Forest. Multiflora roses were widely distributed in 9 of the 10 watersheds. No statistically significant associations were found between invasive shrub abundance and road surface type and road traffic intensity.

The occurrence of the three most abundant invasive shrub species (autumn olive, multiflora rose, and bush honeysuckle) had the same associations with some environmental factors (Table 4.2). All the three species were favored by high road density and a short distance to the invasive source (reclaimed surface mines), and were mutually associated with site wetness and aspect. Species-specific associations with other environmental factors were also found. Autumn olive had a higher occurrence frequency on sites with lower elevation and had no significant association with slope steepness. Multiflora rose had a higher occurrence frequency on sites with higher elevation and steeper slopes. Bush honeysuckle had a higher occurrence frequency on sites with steeper slopes, but had no significant association with elevation.

4.4.3 Spatial Patterns of Invasive Tree Species

Tree-of-heaven, the only invasive tree species observed in the study area, had limited occurrence. All inventoried tree-of-heaven trees were located on or within 5 m of a road. Nine of the ten of the observed tree-of-heaven locations were

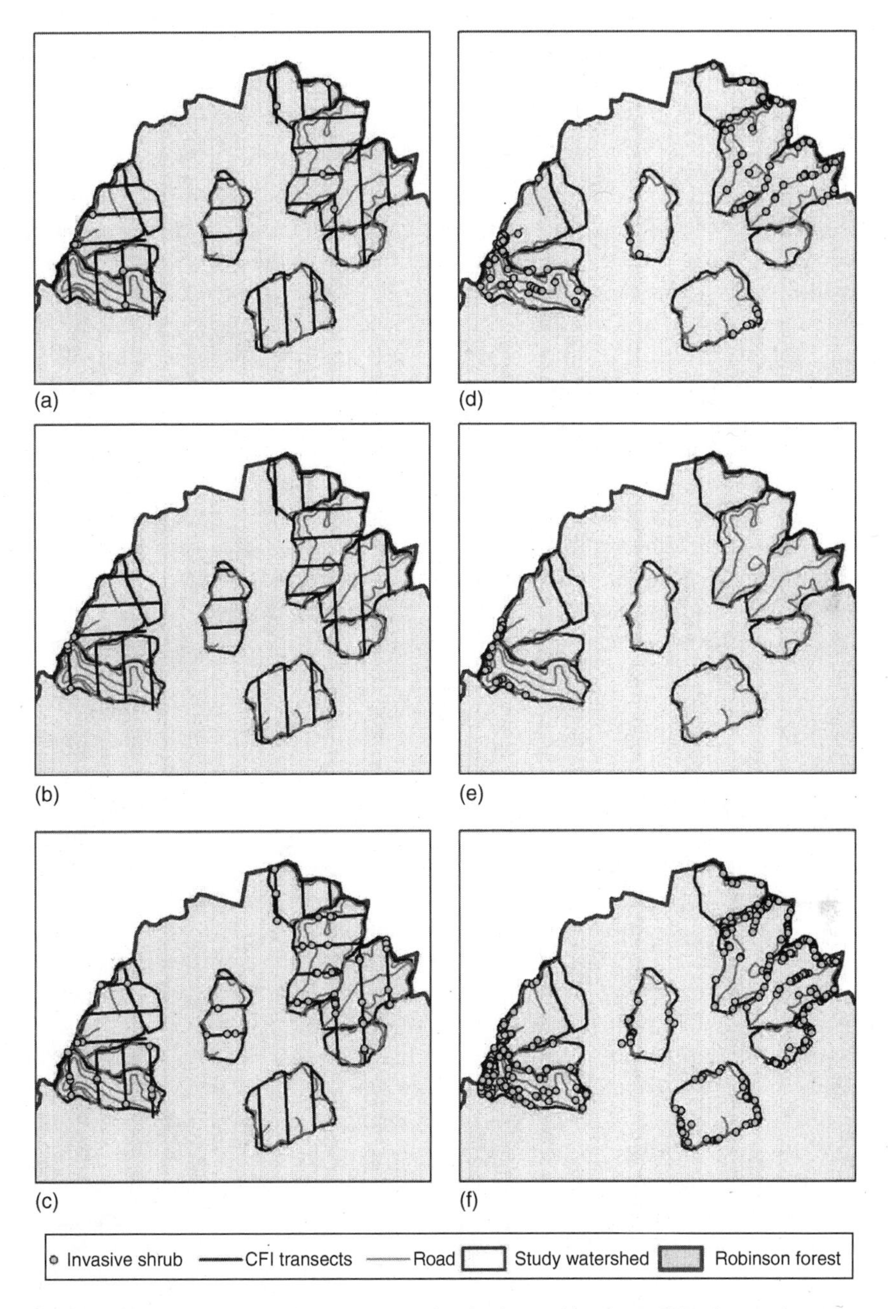

FIGURE 4.6 Distribution of invasive exotic shrub species along CFI plots and transects (*left* column, a–c) and along roads (*right* column, d–f) for autumn olive (a and d), bush honeysuckle (b and e), and multiflora rose (c and f).

distributed near the transition zone of Robinson Forest and the surrounding reclaimed surface mines.

4.5 DISCUSSION

One of the general trends observed in this study was that invasive exotic plants were more frequently distributed on or near roads than in the interior of the forest. Nearly all invasive herbaceous species were found close to roads, and the majority of invasive shrub species were also found close to roads. Numerous studies have documented that plant invasions in forests follow disturbance (Brothers and Spingarn 1992; Silveri et al. 2001), and our results confirmed this pattern. In addition, we found that the relationship between disturbance intensity and invasive abundance is not linear for herbaceous species. Invasive Japanese stiltgrass was most abundant on roads with a medium traffic level. High traffic levels may increase propagule pressure and available habitat, but it may also decrease survival probability. In contrast, medium traffic levels may introduce ample propagule pressure, create sufficient viable habitats, and allow good survival rate. Although invasive shrub species were also found more abundantly on or near roads, no statistically significant association between disturbance intensity and invasive shrub abundance was found. This may be due to our sampling design. As the results indicated, invasive shrub species can grow deeper in the forest interior than invasive herbaceous species. In these instances, a 5 m buffer zone may not be wide enough to sufficiently sample the majority of invasive shrub species. Wider sample zones need to be considered in future studies for shrubs.

As Jules et al. (2002) pointed out, understanding biological invasions requires information on the spatial spread of invasive species, as well as measures of environmental heterogeneity. With high-resolution GPS receivers, digital elevation models, and aerial photos, we are able to gather high-resolution measures of environmental heterogeneity. There were some common factors associated with all invasive plants in this study. Both invasive herbaceous and shrub species were favored by high road density and close distance to the source of invasive species. Other factors associated with invasive plants were more species-specific. For example, Japanese stiltgrass and autumn olive were more frequently found at lower elevations but multiflora rose was frequently found at higher elevation.

One interesting result that deserves further investigation is the association between road surface type and average patch cover percentage of Japanese stiltgrass. Average patch cover percentage for Japanese stiltgrass was significantly lower on roads that were covered with saplings than on roads covered with seedlings. Total percentage of roads covered by Japanese stiltgrass patches was also lower on roads that were covered with saplings than on roads covered with seedlings, although the difference was not statistically significant. The difference in Japanese stiltgrass abundance between these two surface types cannot be explained easily. There are two possible explanations for the observed difference. The first explanation is that saplings can outcompete Japanese stiltgrass better than seedlings. However, Japanese stiltgrass is a very shade-tolerating species and can grow in open to shady and moist to dry locations (Barden 1987). The likelihood for it to be outcompeted is very low after its establishment. The second explanation is that roads with seedling surface type had higher propagule pressure. Invasion of exotic plants into a forested area is a function of the germ plasm of invasive exotic species surrounding the forest and the degree of disturbance in the forest providing niches for invasive species colonization. Ground disturbance from harvesting, especially skidding and road construction, may promote an increase in invasive species colonization in disturbed areas. Roads with seedling surface are more likely to be younger than roads with sapling surface. Propagule pressure was much higher on the seedling surface type road when it was constructed, because Japanese stiltgrass had already established on the older sapling surface road. If this is the case, the probability of new sites becoming infested by Japanese stiltgrass will be much higher if a major disturbance occurs in the future.

In conclusion, invasive exotic plants had a higher occurrence frequency on or near roads compared with the forest interior at Robinson Forest. Invasive shrub species, such as multiflora rose and autumn olive, can distribute deep in the forest interior where anthropogenic disturbance is minimal. The occurrence of invasive tree species was relatively low. Factors such as high road density and short distance to invasive sources can increase the probability of exotic plant invasion, and other environmental factors can cause more species-specific influences on exotic plant invasion.

REFERENCES

Barden, L.S., Invasion of *microstegium vimineum* (Poaceae), an exotic, annual, shade-tolerant, C4 grass, into a North Carolina floodplain, *Am. Midl. Nat.*, 118, 40, 1987.

Brothers, T.S. and Spingarn, A., Forest fragmentation and alien plant invasion of central Indiana old-growth forests, *Conserv. Biol.*, 6, 91, 1992.

Chornesky, E.A., Bartuska, A.M., Aplet, G.H., et al., Science priorities for reducing the threat of invasive species to sustainable forestry, *Bioscience*, 55, 335, 2005.

Deckers, B., Verheyen, K., Hermy, M., and Muys, B., Effects of landscape structure on the invasive spread of black cherry in an agricultural landscape in Flanders, Belgium, *Ecography*, 28, 99, 2005.

Hanowski, J.M., Niemi, G.J., and Christian, D.C., Influence of within-plantation heterogeneity and surrounding landscape composition on avian communities in hybrid poplar populations, *Conserv. Biol.*, 11, 936, 1997.

Jules, E.S., Kaufmann, M.J., Ritts, W., and Carroll, A.L., Spread of an invasive pathogen over a variable landscape: A non-native root rot on Port Orford cedar, *Ecology*, 83, 3167, 2002.

Kentucky Division of Geographic Information, http://kymartian.ky.gov/, accessed October 2007.

Kolb, A., Alpert, P., Enters, D., and Holzapfel, C., Patterns of invasion within a grassland community, *J. Ecol.*, 90, 871, 2002.

Kotanen, P.M., Effects of experimental soil disturbance on revegetation by natives and exotics in coastal Californian meadows, *J. Appl. Ecol.*, 34, 631, 1997.

Lodge, D.M., Biological invasions: Lessons from ecology, *Trends Ecol. Evol.*, 13, 195, 1993.

Mack, M.C. and D'Antonio, C.M., Impacts of biological invasions on disturbance regimes, *Trends Ecol. Evol.*, 13, 195, 1998.

Mooney, H.A. and Hobbs, R.J., *Invasive Species in a Changing World*, Island Press, Washington, DC, 2000.

Pimentel, D., Lach, L., Zuniga, R., and Morrison, D., Environmental and economic costs of nonindigenous species in the United States, *Bioscience*, 50, 53, 2000.
Silveri, A., Dunwiddie, P.W., and Michaels, H.J., Logging and edaphic factors in the invasion of an Asian woody vine in a mesic North American forest, *Biol. Invasions*, 3, 379, 2001.
U.S. Geological Survey (USGS), http://seamless.usgs.gov/, accessed October 2007.
With, K.A., The landscape ecology of invasive spread, *Conserv. Biol.*, 16, 1192, 2002.

5 Spatial and Temporal Dynamics of Exotic Tree Invasions: Lessons from a Shade-Tolerant Invader, *Acer platanoides*

Christopher R. Webster and Steven R. Wangen

CONTENTS

5.1 INTRODUCTION

Persistent colonies composed of invasive woody perennials alter the structure and function of forest ecosystems by inhibiting the growth and development of native species (Webster et al. 2006). The ability of these species to inhibit the recruitment of native species raises the possibility that large-scale invasions may fundamentally change the successional trajectory of forest ecosystems. Such a change would have far-reaching implications for numerous plant and animal species that rely on native plant communities and their successional pathways. For example, research by Brown et al. (2006) in the Luquillo Mountains of Puerto Rico suggests that invasion by the invasive tree *Syzygium jambos* has produced a new vegetation assemblage, which is likely to foster long-term changes in community structure, composition, and successional trajectory. Similarly, in the eastern United States, the small shade-tolerant, invasive tree *Rhamnus frangula* aggressively invades forest understories, forming dense thickets that inhibit tree regeneration and reduce herbaceous layer cover and species diversity (Frappier et al. 2003a; Fagan and Peart 2004). Invasions by woody perennials into nonforested habitats have also had significant consequences for local plant and animal communities (Braithwaite et al. 1989; Knopf 1994; Bruce et al. 1995). More cryptic, but equally weighty, changes may arise from an invader's influence on nutrient cycling (Rice et al. 2004) even if it does not completely replace the native overstory.

Long-lived woody invaders may be especially insidious since they are capable of gradually displacing and competitively excluding dominant overstory species as forests turn over in response to natural or anthropogenic disturbance (Loehle 2003). In this chapter, we explore the spatial and temporal dynamics of exotic tree invasions, with emphasis on the shade-tolerant invader *Acer platanoides*.

5.1.1 Invasion Process

A number of empirical and theoretical models have been developed in an effort to describe and understand the invasion process (Andow et al. 1990; Andow 1993; Higgins and Richardson 1996). In its simplest form, the invasion process can be described by a logistic growth curve (Figure 5.1), with a lag phase following introduction before the

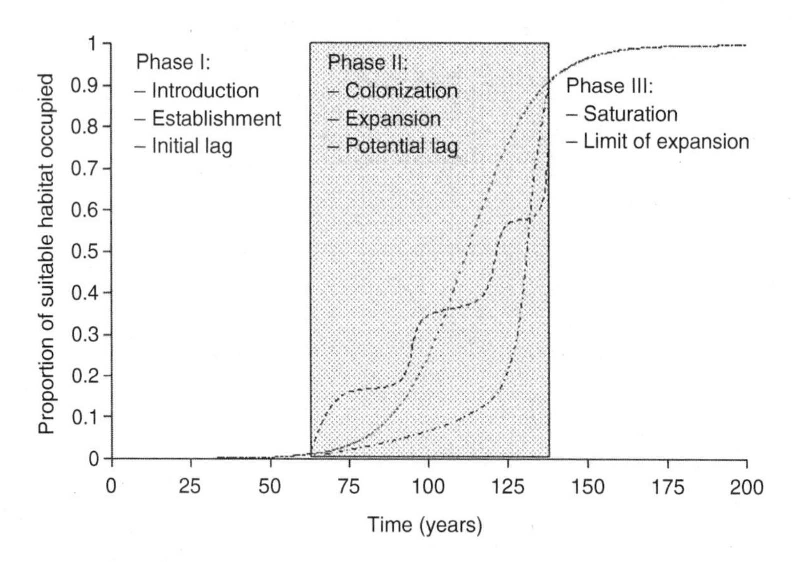

FIGURE 5.1 Conceptual diagram of the invasion process as it relates to invasive exotic trees. The length of phase I is highly variable, but comparatively long for exotic trees. The shape of the phase II curve will vary in response to the interaction between species life-history traits and the landscape context of the invasion. Additional lags may occur during this phase if there is a long delay between colonization of new sites and the trees becoming reproductively active. Phase III is reached once a species occupies all suitable habitats and/or reaches a geographic limit to its spread.

population begins to grow rapidly (Radosevich et al. 2003). The length of this lag depends on both the biology of the invading organism and the composition and structure of the habitat. Some species, however, would likely remain in this initial lag phase indefinitely if not for a change in extrinsic factors, such as land use or disturbance regime. Consequently, successful invasions often hinge on very rare recruitment events and chance fluctuations in extrinsic factors (Radosevich et al. 2003; Rilov et al. 2004). The next phase of invasion typically involves expansion of the population via the colonization of new habitats. This growth phase may occur at an asymptotic rate (diffusion, Andow et al. 1990) or proceed in a more biphasic manner (stratified diffusion, Shigesada et al. 1995). The shape of the invasion curve is influenced by life-history characteristics (intrinsic factors, Radosevich et al. 2003) such as seed dispersal mechanisms (Shigesada et al. 1995). For example, species with a high propensity for long-distance dispersal typically spread more quickly than those relying solely on short-distance dispersal (Moody and Mack 1988). In this idealized invasion, population growth continues until all suitable habitats have been colonized or the invasion reaches the geographic limit of spread (Shigesada et al. 1995). This "saturation" phase, which may take millennia to reach, is represented by the asymptote of the logistic curve (Shigesada et al. 1995).

In recent years, there has been a growing effort to link the temporal dynamics of invasion to landscape-level processes that modulate the movement of exotic organisms across heterogeneous landscapes (Bergelson et al. 1993; With 2002). Two ways by which landscapes influence invasions are through the variable susceptibility of certain community assemblages to invasion (Richardson et al. 1994; Foxcroft et al. 2004) and the efficacy of different landscape features, such as roads, to act as vectors or barriers for dispersal and establishment (Buckley et al. 2003; Spooner et al. 2004; Wangen and Webster 2006). Disturbance (natural or anthropogenic), which is a spatial process, may interact synergistically or antagonistically with the existing landscape to create novel invasion patterns (Rouget et al. 2004). These patterns may differ markedly from those that are formed solely due to dispersal ability of the invader (Bergelson et al. 1993). Nevertheless, human-mediated dispersal may trump natural factors governing invasion rate and pattern (Suarez et al. 2001; Foxcroft et al. 2004). For example, Suarez et al. (2001) found that human-mediated, long-distance "jump dispersal" led to dispersal distances three orders of magnitude greater than local population expansion in Argentine ants (*Linepithema humile*).

5.1.2 Trees as Invaders

The vast majority of exotic woody plant invasions around the world are the result of intentional introductions (82%, Reichard and Hamilton 1997). Surprisingly, many species, while recognized as problematic invaders, are still widely planted as ornamental or commercial timber species (Nowak and Roundtree 1990; Silander and Klepeis 1999; Rouget et al. 2002). In a recent review, Webster et al. (2006) found that ~30% of the land grant universities in the eastern United States still offer publications that promote the use of invasive exotic woody plants for plantations, wildlife plantings, and/or as ornamentals. The sale of even the most aggressive invaders is largely unregulated, and species that are aggressively controlled in natural areas are often readily available at local garden centers. This widespread availability and use, unfortunately, assures a continuous supply of future source populations. An especially troubling aspect of this trend is that these anthropogenic source populations are not subject to the same stresses encountered by other invading organisms. In fact, a great deal of time and expense is exerted to keep ornamental plantings healthy, which enhances their chances of survival and reproductive output.

While the current dilemma posed by invasive trees has a ring of absurdity to it, there may be a rational explanation for the apparent disconnect between risk and response. First and foremost, trees in comparison with other plants are long-lived and relatively slow growing. Trees also typically have longer generation times (greater ages to first reproduction) than other invasives, such as herbs and animals. Consequently, there may be a long lag between introduction, colonization, and rapid range expansion that masks their invasive potential (Frappier et al. 2003b; Wangen and Webster 2006). Second, exotic trees do not typically strangle existing vegetation or kill mature forests. Rather, they tend to colonize following disturbance (Call and Nilsen 2003; Paynter et al. 2003) or, even more subtly, gradually attain dominance through gap capture (Knapp and Canham 2000; Webb et al. 2000; Lichstein et al. 2004; Webster et al. 2005).

Since forests turn over slowly, the transition to exotic dominance may not be readily apparent for centuries (Loehle 2003). The turnover rate, however, may be accelerated by anthropogenic disturbance, native or exotic insect outbreaks, and less frequent catastrophic natural disturbances. For example, *Ailanthus altissima* can colonize canopy gaps and form source populations (Knapp and Canham 2000) that are capable of responding rapidly to more widespread disturbances (Orwig and Foster 1998). Conversely, species such as *A. platanoides*, which are tolerant of shade (Nowak and Roundtree 1990), may build up dense persistent sapling layers in the understory (Bertin et al. 2005). This dense growth of advance regeneration is then in a position to capture the site following overstory removal through a process known as "disturbance-mediated accelerated succession" (Abrams and Scott 1989; Webb and Scanga 2001). Several species, however, are flexible in their invasion strategy and can invade across a continuum of disturbance intensities. For example, *Ligustrum lucidum* (Lichstein et al. 2004) and *A. platanoides* (Webster et al. 2005), both shade-tolerant, are capable of readily invading highly disturbed habitats, canopy gaps, and intact forest understory.

5.1.3 *A. platanoides*

A. platanoides (Figure 5.2) is a shade-tolerant escapee from ornamental plantings in the eastern and central United States (Webb et al. 2000; Bertin et al. 2005; Webster et al. 2005), which has recently emerged as a potentially serious invader in some mountainous regions of the west as well (Reinhart et al. 2005). *A. platanoides* is native to Eurasia where it is generally found in small groups or as scattered individuals in mixed forests (Nowak and Rowntree 1990). In its introduced range, this species is capable of forming monotypic stands (Figure 5.2), which negatively affect native plant communities (Wyckoff and Webb 1996; Martin 1999; Fang 2005). *Acer platanoides* has been shown to utilize light, water, and nutrients more

FIGURE 5.2 Dense stand of *A. platanoides* saplings under an open grown *A. saccharum* in an area of wooded parkland in Houghton, MI. (Photo courtesy of C.R. Webster.)

efficiently than the native *Acer saccharum* (Kloeppel and Abrams 1995), and is capable of competitively displacing the aggressive native *Acer rubrum* (Fang 2005). Research by Sanford et al. (2003) suggests that *A. platanoides* is also subject to lower mortality in both sun and shade than its native congener, *A. saccharum*. Relative to native deciduous species, *A. platanoides* displays greater leaf longevity (191 days for *A. platanoides* vs. ~180 days for *A. saccharum*, Kloeppel and Abrams 1995), which can relay a competitive advantage for an invader and help provide access to additional resources, increasing the chance of successful establishment (Harrington et al. 1989). This longer leaf-on period in conjunction with high light interception may also reduce understory light levels sufficiently to depress native tree regeneration and herbaceous layer diversity proximate to established trees (Wyckoff and Webb 1996; Martin 1999; Reinhart et al. 2005).

While first introduced to the eastern United States around 1756, *A. platanoides* was an occasional ornamental until the devastation wrought by Dutch elm disease [disease complex; fungus *Ophiostoma ulmi* transmitted by two vectors: European elm bark beetle (*Scolytus multistriatus*) and the native elm bark beetle (*Hylurgopinus rufipes*)] prompted its widespread application as a replacement for *Ulmus americana* (Nowak and Rowntree 1990). *A. platanoides* has several desirable traits that made it well suited for the harsh conditions associated with the urban environment. These traits include the following: vigorous early growth, diversity in cultivar color and form, pollution tolerance, drought tolerance, shade tolerance, and the ability to grow well across a wide range of soil conditions (Nowak and Rowntree 1990). However, in arid regions of the western United States, *A. platanoides* may be restricted to mesic habitats (Reinhart et al. 2005). On Mackinac Island, MI, which has shallow calcareous soils, open grown *A. platanoides* averaged 5.9 mm per year in radial growth, while codominant and dominant crown class trees in the forest averaged 4.2 mm per year (Webster et al. 2005). These mean growth rates are similar to maximum growth rates reported for native *Acer* spp. in well-tended stands on fertile sites in the region (Burns and Honkala 1990).

A. platanoides is a prolific seeder that usually begins bearing seed at 25–30 years of age (Gordon and Rowe 1982). Large seed crops are produced every 1–3 years (Gordon and Rowe 1982). Seeds are wind-dispersed, and samaras can travel ~50 m from a parent tree, given a gentle wind (10 km/h; Matlack 1987). *A. platanoides* also sprouts prolifically from cut stumps and forms dense sprout clumps (Figure 5.3). This species is also an aggressive competitor in forest understory environments (Webb and Kauzinger 1993; Martin 1999; Webb et al. 2000) that can persist as advance regeneration for at least 30–40 years (Bertin et al. 2005). Meiners (2005) has suggested that the greater relative success of *A. platanoides* in the seedling bank compared with native species is at least partially attributable to its greater diaspore mass and relatively unpalatable seeds.

Given these traits and *A. platanoides*' relatively pest-free status in its introduced range (Nowak and Rowntree 1990), the fact that it is now invading native forest may come as little surprise. What is surprising is that it is still one of the most widely planted ornamental tree species in the United States, and that most people do not perceive the risk and consequences of its invasion into native forests. However, as

FIGURE 5.3 *A. platanoides* sprout clump (Mackinac Island, MI). (Photo courtesy of S.R. Wangen.)

discussed earlier and illustrated by the invasion of *A. platanoides* on Mackinac Island chronicled later, appreciating the risks posed by exotic tree invasions requires a long-term perspective.

5.2 CASE STUDY: PUNCTUATED INVASION OF *A. PLATANOIDES* ON MACKINAC ISLAND

5.2.1 Island Setting

Mackinac Island is located in Lake Huron, ~5 km east of St. Ignace, MI. Approximately 82% of the 1130 ha island is encompassed by the Mackinac Island State Park, which is administered by the Mackinac Island State Park Commission. No cars are allowed in the island; horses, bikes, and foot traffic are the primary modes of transportation. Most of the parkland is composed of second-growth forest resulting from extensive wood cutting for fuel and building materials during the mid- to late 1800s. Species composition is heavily influenced by the shallow bedrock of erosion-resistant limestone breccia (Milstein 1987). Typical soils include St. Ignace silt loams and Alpena gravelly loams, both of which are well-drained upland soils (Whitney 1997).

During the summer of 2003, we mapped the location of every *A. platanoides* ≥50 cm in height that had invaded the 728 ha park and noted the locations of street trees within developed areas. The diameter at breast height (1.37 m) of each mapped tree was measured and its position in the canopy was recorded. Heights, crown measurements, and increment cores were collected on randomly selected subsets of the population. Detailed results of this study are reported in Webster et al. (2005), Wangen and Webster (2006), and Wangen et al. (2006).

5.2.2 Lag Phases and Invasion Dynamics

A. platanoides was intentionally introduced as an ornamental to Mackinac Island sometime during the early 1900s, and the oldest extant individual found during a survey by Webster et al. (2005) established in 1938. The species was planted extensively as a street tree in the City of Mackinac Island, which is located on the Island's southern coast. Nevertheless, in spite of its abundance in town, its invasion into nearby forested parklands was nearly imperceptible until around 1965 when the population entered the expansion phase (Figure 5.4). This long initial lag phase was followed by a period of exponential population growth, and rapid range expansion (Wangen and Webster 2006). During the height of the expansion phase, the population was adding 254 individuals per year (that were still alive as of 2003) and increasing the area occupied by 5.6 ha per year (Wangen and Webster 2006).

The primary driver of the increase in population size has been wave-front movement away from source populations. This is a relatively slow process given a mean natural dispersal distance of about 50 m and the relatively long time required for *A. platanoides* to become reproductively active (25–30 years for open grown trees). On the basis of these values, we would predict that between

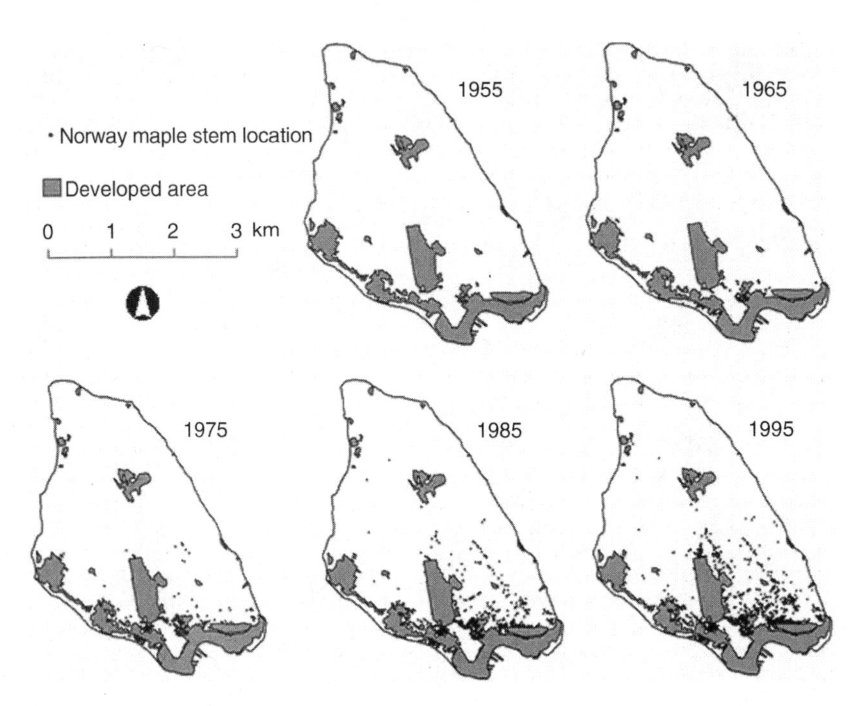

FIGURE 5.4 Locations of individual *A. platanoides* trees over the course of the invasion reconstructed by Wangen and Webster (2006). The presence of a tree during a given time period is based on an estimate of its age, which was derived from a regression of *A. platanoides* tree age as a function of stem diameter presented by Webster et al. (2005).

1938 and 1995 [the end date of the reconstruction by Wangen and Webster (2006)] *A. platanoides* could have easily invaded about 100 m away from source populations because of wind-dispersed seed. However, the range radius of the population actually increased by over 1 km, and the area occupied by the invasion (≥1 tree per ha) swelled to over 200 ha (Wangen and Webster 2006). Consequently, while population growth has primarily been determined by short-distance dispersal, long-distance dispersal has been the primary driver of population spread (see also Neubert and Caswell 2000).

Both population growth and spread began to slow during the early 1990s as the invasion entered a second lag phase (Wangen and Webster 2006). This second lag was somewhat unexpected, but is logical given the reproductive biology of the species and the stand dynamics of the native forest. The deceleration in population growth appears to have been prompted by the saturation of easily colonized sites near source populations as well as the fact that most of the satellite populations have not reached reproductive maturity. Tree-ring analysis of contemporary *A. platanoides* canopy trees has suggested that most of the early colonists (source populations within the forest reserve) were not subjected to protracted periods of suppression in the understory (Webster et al. 2005). Instead, these trees appear to have colonized relatively open sites and quickly reached reproductive maturity. Nevertheless, the bulk of the contemporary invasion is represented by suppressed saplings that are not yet reproductively active. Suppressed saplings of *A. platanoides*, even if they are sufficiently old (≥25 years), seldom produce viable seed, but can persist as prereproductive subcanopy trees for at least 30–40 years (Bertin et al. 2005). Consequently, most of the population spread that was observed was the result of the reproductive effort of a comparatively small number of trees. In terms of long-distance dispersal, there is likely an upper limit to the number of suitable establishment sites that standard and nonstandard (sensu Higgins et al. 2003) dispersal events can reach in a heterogeneous environment, especially given a small population of reproductive trees.

The next expansion phase will be strongly influenced by the stand dynamics of *A. platanoides* and the turnover rate of the native forest. A number of scenarios are possible depending on the intensity and severity of both natural and anthropogenic disturbances, and the ability of the park and city to implement a coordinated control program. At the low end of the disturbance continuum, we might expect a slow resumption of population growth and spread as *A. platanoides* is released from suppression and captures canopy gaps. This likely would result in an acceleration of the asymptotic rate of increase compared with the previous expansion phase once a critical mass of satellite populations mature and begin to expand. Alternately, a major stand-replacing disturbance could lead to a rapid transition to *A. platanoides* where it is present in the contemporary understory. Under favorable light conditions, trees would quickly form a canopy and reach reproductive maturity. This would initiate a period of rapid range expansion into formally unoccupied areas. In a large contiguous landscape, the invasion might again slow as the next generation of satellite populations awaits a gap in the canopy.

5.2.3 Stand Dynamics

Detailed tree-ring analyses of a randomly selected subset of the *A. platanoides* that had successfully invaded the forest reserve on Mackinac Island revealed that while residence times in the understory were variable, most of the contemporary *A. platanoides* overstory trees showed little evidence of past suppression in the understory (Webster et al. 2005). This lack of suppression in overstory trees suggests that the initial vanguard of the invasion probably established under open canopy conditions and/or in canopy gaps (Figure 5.5). Consequently, while disturbance may not be directly linked to invasion by *A. platanoides* (Webb et al. 2000), it may help to accelerate the invasion by facilitating more rapid canopy recruitment and seed production. Given that most of the *A. platanoides* invasion on Mackinac Island is still restricted to the understory, canopy gap dynamics are likely to dictate the rate at which the invasion proceeds.

Height growth comparisons with native species suggest that *A. platanoides* should be capable of outcompeting most of its native competitors in canopy gaps (Webster et al. 2005). In fact, even assuming optimal growing conditions for natives and mean conditions for *A. platanoides*, a gap capture simulation still predicted that *A. platanoides* would reach the overstory in half the time it would take for most of its competitors

FIGURE 5.5 *A. platanoides* sapling towering over the competition in a large canopy opening (Houghton, MI). (Photo courtesy of C.R. Webster.)

(Table 5.1). This potential competitive superiority, however, may not be realized if the understory of native shade-tolerant species has a substantial head start or is dense enough to discourage establishment (S.R. Wangen and C.R. Webster, unpublished data). Nevertheless, the combination of superior survival (Sanford et al. 2003; Bertin et al. 2005) and growth rate (Webster et al. 2005; Kloeppel and Abrams 1995) in

TABLE 5.1
Estimated Canopy Ascension Times for Native Species and Exotic *A. platanoides* in Mixed Conifer-Hardwood Stands on Mackinac Island Based on Height Growth Comparisons

Species	Years Required to Reach Canopy
A. platanoides	38
A. saccharum	70
Fagus grandifolia	77
Picea glauca	64

Note: Canopy ascension is assumed once trees reach crown shoulder height, which represents the height trees must reach to avoid overtopping and recruit into the overstory.

conjunction with its oft-observed numerical supremacy over native shade-tolerant species (Webb and Kaunzinger 1993; Wyckoff and Webb 1996; Martin 1999; Webb et al. 2000; Bertin et al. 2005; Fang 2005) suggests that *A. platanoides* is fully capable of eventually displacing native competitors in both disturbed and undisturbed environments.

5.2.4 Invasion Pattern

The invasion of *A. platanoides* on Mackinac Island displayed a high degree of aggregation across multiple spatial scales (Wangen et al. 2006). *A. platanoides* individuals were more closely associated with conspecifics than would be predicted, given a random distribution of neighbors (Wangen et al. 2006). This affinity was consistent even when higher order neighbors or larger search radii were examined. Several possible mechanisms probably contributed to this pattern. First, the obvious explanation is that the limited seed shadow of parent trees results in a high density of recruits proximate to parents. This dispersal pattern would result in a wave-front pattern of spread (Shigasada et al. 1995), which was observed close to town (Wangen and Webster 2006; Wangen et al. 2006). A similar pattern was recently reported by Fang (2005) for a mapped population of *A. platanoides* in New York State. Second, established individuals may modify the site and enhance the survival of new propagules, especially in harsh environments. This sort of facilitation was recently documented in a western population of *A. platanoides* by Reinhart et al. (2006) and warrants further investigation.

Mapping of the invasion on Mackinac Island also revealed that the invasion had stream-like tendencies as well as numerous satellite populations (Figure 5.4). These stream- or thread-like patterns suggest that either dispersal and/or establishment are influenced by vectors that accelerate or channel the invasion along certain corridors. In the case of Mackinac Island, the invasion showed a strong affinity with roads and trails (Wangen et al. 2006). This affinity was probably the result of nonstandard mechanisms of propagule dispersal along roadways, which appear to have been an important factor in the initial movement of seeds beyond the wildland–urban interface (Wangen and Webster 2006). For example, the discontinued practice of transporting horse manure and other organic street sweepings into the forest for composting likely resulted in the establishment of some of the older satellite populations. Similarly, samaras from the numerous *A. platanoides* that line the streets of the city may hitchhike on passing vehicles (e.g., open horse-drawn wagons or drays). These modes of human-mediated dispersal are likely to carry propagules varying distances, but would tend to reinforce the association between *A. platanoides* and road corridors on the Island. Research by Anderson (1999) also suggests that disturbed environments found along roads may also help facilitate invasion.

Collectively, these results indicate that while the central invasion tendency of *A. platanoides* is probably a slow-radiating wave front, its colonization of new sites is very opportunistic, resulting in a flexible invasion pattern. Consequently, the realized pattern is derived from a combination of influences. As discussed earlier, although the advancing front may contribute the most new recruits to the populations (especially since it may take satellite populations of exotic trees decades to mature), it is the chance long-distance dispersals that ultimately govern the rate of spread. Work by Moody and Mack (1988), Shigesada et al. (1995), and Neubert and Caswell (2000) suggest that while the details may vary from organism to organism, this paradox underlies the initial stages of most biological invasions.

5.2.5 Potential for Effective Control

The earlier case study illustrates that the potential for control is high during early stages of invasion, but once the population is allowed to progress out of its initial lag phase the prospects for effective control diminish rapidly. Nevertheless, additional lag phases, should they occur, may provide critical windows of opportunity for control. The most effective control techniques usually begin with outlying populations and work back toward the invasion front (Moody and Mack 1988). However, since the seedling bank is very resilient to control because it replenishes quickly (Webb et al. 2001), a prudent approach may be to target seed trees first. In general, control activities should begin before a species demonstrates that it is likely to become a problematic invader at a particular site. Knowledge of species life-history traits, the disturbance regime of the native forest, and invasion histories elsewhere should inform when rapid response is warranted, rather than a wait-and-see approach.

Expansion phases driven by chance, long-distance dispersal events may be particularly recalcitrant to control efforts. This is because, in addition to the rapid increase in the number of individuals/satellite populations requiring control, the fact that these populations are disjunct from the parent population makes them inherently difficult to find. Consequently, monitoring for invasive trees should be proactive, and where possible utilize risk assessment techniques to identify likely establishment sites.

5.3 CONCLUSION

The greatest risk associated with invasive trees is that their potential as invaders may not be easily recognized until they have become a serious pest. This risk is exacerbated by the long generation times of trees relative to other organisms. For example, during the initial lag phase on Mackinac Island, even a trained observer would have been hard pressed to find worrisome numbers of *A. platanoides* in the forest much less perceive the rapid population expansion that would occur in the coming decades. Additionally, the slow turnover rates of many native forests in the absence of major disturbance events may foster a slow transition to invasive dominance that is difficult to perceive until it is well under way. This may be especially true for exotics that superficially resemble native species (e.g., *A. platanoides* which closely resembles the native *A. saccharum*). For example, the extent of the invasion of *A. platanoides* on Mackinac Island was not readily apparent to the public until an epidemic of tar spot (*Rhytisma* spp.) preferentially infested the Island's host of *A. platanoides*. The copious black spots on the leaves of *A. platanoides* awakened the public to an invasion that had literally crept around them as they slept.

ACKNOWLEDGMENTS

We gratefully acknowledge the Mackinac Island Community Foundation, Mackinac Island State Park Commission, and the city of Mackinac Island for their support in this endeavor. The original funding for our investigation of *A. platanoides* on Mackinac Island was provided by The School of Forest Resources and Environmental Science, Michigan Technological University, and the McIntire-Stennis Cooperative Forestry Program. Invaluable logistical support was provided by The Mackinac Island Community Foundation, Mackinac Island State Park Commission, and local residents.

REFERENCES

Abrams, M.D. and Scott, M.L., Disturbance-mediated accelerated succession in two Michigan forest types, *For. Sci.*, 35, 1, 1989.

Anderson, R., Disturbance as a factor in the distribution of sugar maple and the invasion of Norway maple into a modified woodland, *Rhodora*, 101, 264, 1999.

Andow, D., Spread of invading organisms: Patterns of spread, in *Evolution of Insect Pests: The Pattern of Variations*, Kim, K.C., Ed., Wiley, New York, 219, 1993.

Andow, D.A., Kareiva, P.M., Levin, S.A., and Okubo, A., Spread of invading organisms, *Landscape Ecol.*, 4, 177, 1990.

Bergelson, J., Newman, J.A., and Floresroux, E.M., Rates of weed spread in spatially heterogeneous environments, *Ecology*, 74, 999, 1993.

Bertin, R.I., Manner, M.E., Larrow, B.F., Cantwell, T.W., and Berstene, E.M., Norway maple (*Acer platanoides*) and other non-native trees in urban woodlands of central Massachusetts, *J. Torrey Bot. Soc.*, 132, 225, 2005.

Braithwaite, R.W., Lonsdale, W.M., and Estbergs, J.A., Alien vegetation and native biota in tropical Australia: The impact of *Mimosa pigra, Biol. Conserv.*, 48, 189, 1989.

Brown, K.A., Scatena, F.N., and Gurevitch, J., Effects of an invasive tree on community structure and diversity in a tropical forest in Puerto Rico, *Forest Ecol. Manage.*, 226, 145, 2006.

Bruce, K.A., Cameron, G.N., and Harcombe, P.A., Initiation of a new woodland type on the Texas coastal prairie by the Chinese tallow tree (*Sapium sebiferum* (L.) Roxb.), *Bull. Torrey Bot. Club*, 122, 215, 1995.

Buckley, D.S., Crow, T.R., Nauertz, E.A., and Schulz, K.E., Influence of skid trails and haul roads on understory plant richness and composition in managed forest landscapes in Upper Michigan, USA, *Forest Ecol. Manage.*, 175, 509, 2003.

Burns, R.M. and Honkala, B.H., *Silvics of North America: 1. Conifers; 2. Hardwoods*, Agriculture Handbook 654, U.S. Department of Agriculture, Forest Service, Washington, DC, 2, 1990.

Call, L.J. and Nilsen, E.T., Analysis of spatial patterns and spatial association between the invasive tree-of-heaven (*Ailanthus altissima*) and the native black locust *(Robina pseudoacacia), Am. Midl. Nat.*, 150, 1, 2003.

Fagan, M.E. and Peart, D.R., Impact of the invasive shrub glossy buckthorn (*Rhamnus frangula* L.) on juvenile recruitment by canopy trees, *Forest Ecol. Manage.*, 194, 95, 2004.

Fang, W., Spatial analysis of an invasion front of *Acer platanoides*: Dynamic interferences from static data, *Ecography*, 28, 283, 2005.

Foxcroft, L.C., Rouget, M., Richardson, D.M., and Fadyen, S.M., Reconstructing 50 years of *Opuntia stricta* invasion in the Kruger national park, South Africa: Environmental determinants and propagule pressure, *Divers. Distrib.*, 10, 427, 2004.

Frappier, B., Eckert, R.T., and Lee, T.D., Potential impacts of the invasive exotic shrub *Rhamnus frangula* L. (glossy buckthorn) on forests of southern New Hampshire, *Northeast. Nat.*, 10, 277, 2003a.

Frappier, B., Lee, T.D., Olson, K.F., and Eckert, R.T., Small-scale invasion pattern, spread rate, and lag-phase behavior of *Rhamnus frangula* L., *Forest Ecol. Manage.*, 186, 1, 2003b.

Gordon, A.G. and Rowe, D.C.F., *Seed Manual for Ornamental Trees and Shrubs*, Her Majesty's Stationary Office, London, 80, 1982.

Harrington, R.A., Brown, B.J., and Reich, P.B., Ecophysiology of exotic and native shrubs in Southern Wisconsin. I. Relationship of leaf characteristics, resource availability, and phenology to seasonal patterns of carbon gain, *Oecologia*, 80, 356, 1989.

Higgins, S.I., Nathan, R., and Cain, M.L., Are long-distance dispersal events in plants usually caused by nonstandard means of dispersal? *Ecology*, 84, 1945, 2003.

Higgins, S.I. and Richardson, D.M., A review of models of alien plant spread, *Ecol. Model.*, 87, 249, 1996.

Kloeppel, B.D. and Abrams, M.D., Ecophysiological attributes of the native *Acer saccharum* and the exotic *Acer platanoides* in urban oak forest in Pennsylvania, USA, *Tree Physiol.*, 15, 739, 1995.

Knapp, L.B. and Canham, C.D., Invasion of an old-growth forest in New York by *Ailanthus altissima*: Sapling growth and recruitment in canopy gaps, *J. Torrey Bot. Soc.*, 127, 307, 2000.

Knopf, F.L., Avian assemblages on altered grasslands, *Stud. Avian Biol.*, 15, 247, 1994.

Lichstein, J.W., Grau, H.R., and Aragon, R., Recruitment limitation in secondary forests dominated by an exotic tree, *J. Veg. Sci.*, 15, 721, 2004.

Loehle, C., Competitive displacement of trees in response to environmental change or introduction of exotics, *Environ. Manage.*, 32, 106, 2003.

Martin, P.H., Norway maple (*Acer platanoides*) invasion of a natural forest stand: Understory consequence and regeneration pattern, *Biol. Invasions*, 1, 215, 1999.

Matlack, G.R., Diaspore size, shape, and fall behavior in wind-dispersed plant species, *Am. J. Bot.*, 74, 1150, 1987.

Meiners, S.J., Seed and seedling ecology of *Acer saccharum* and *Acer platanoides*: A contrast between native and exotic congeners, *Northeast. Nat.*, 12, 23, 2005.

Milstein, R.L., Mackinac Island State Park, Michigan, in *Geological Society of America Centennial Field Guide—North Central Section*, Briggs, D.L., Ed., Geological Society of America, Boulder, CO, 285, 1987.

Moody, M.E. and Mack, R.N., Controlling the spread of plant invasions: The importance of nascent foci, *J. Appl. Ecol.*, 25, 1009, 1988.

Neubert, M.G. and Caswell, H., Demography and dispersal: Calculation and sensitivity analysis of invasion speed for structured populations, *Ecology*, 81, 1613, 2000.

Nowak, D.J. and Rowntree, R.A., History and range of Norway maple, *J. Arboriculture*, 16, 291, 1990.

Orwig, D.A. and Foster, D.R., Forest response to the introduced hemlock woolly adelgid in southern New England, USA, *J. Torrey Bot. Soc.*, 125, 60, 1998.

Paynter, Q., Downey, P.O., and Sheppard, A.W., Age structure and growth of the woody legume weed *Cytisus scoparius* in native and exotic habitats: Implications for control, *J. Appl. Ecol.*, 40, 470, 2003.

Radosevich, S.R., Stubbs, M.M., and Ghersa, C.M., Plant invasions—Process and patterns, *Weed Sci.*, 51, 254, 2003.

Reichard, S. and Hamilton, C., Predicting invasions of woody plants introduced into North America, *Conserv. Biol.*, 11, 1993, 1997.

Reinhart, K.O., Greene, E., and Callaway, R.M., Effects of *Acer platanoides* invasion on understory plant communities and tree regeneration in the northern Rocky Mountains, *Ecography*, 28, 573, 2005.

Reinhart, K.O., Maestre, F.T., and Callaway, R.M., Facilitation and inhibition of seedlings of an invasive tree (*Acer platanoides*) by different tree species in a mountain ecosystem, *Biol. Invasions*, 8, 231, 2006.

Rice, S.K., Westerman, B., and Federici, R., Impacts of the exotic, nitrogen-fixing black locust (*Robinia pseudoacacia*) on nitrogen-cycling in a pine-oak ecosystem, *Plant Ecol.*, 174, 97, 2004.

Richardson, D.M., Williams, P.A., and Hobbs, R.J., Pine invasions in the Southern Hemisphere: Determinants of spread and invadability, *J. Biogeogr.*, 21, 511, 1994.

Rilov, G., Benayahu, Y., and Gasith, A., prolonged lag in population outbreak of an invasive mussel: A shifting habitat model, *Biol. Invasions*, 6, 347, 2004.

Rouget, M., Richardson, D.M., Nel, J.L., and Van Wilgen, B.W., Commercially important trees as invasive aliens—Towards spatially explicit risk assessment at a national scale, *Biol. Invasions*, 4, 397, 2002.

Rouget, M., Richardson, D.M., Milton, S.J., and Polakow, D., Predicting invasion dynamics of four alien *Pinus* species in a highly fragmented semi-arid shrubland in South Africa, *Plant Ecol.*, 152, 79, 2004.

Sanford, N.L., Harrington, R.A., and Fownes, J.H., Survival and growth of native and alien woody seedlings in open and understory environments, *Forest Ecol. Manage.*, 183, 377, 2003.

Shigesada, N., Kawasaki, K., and Takeda, Y., Modelling stratified diffusion in biological invasions, *Am. Nat.*, 146, 229, 1995.

Silander, J.A., Jr. and Klepeis, D.M., The invasion ecology of Japanese barberry (*Berberis thunbergii*) in the New England landscape, *Biol. Invasions*, 1, 189, 1999.

Spooner, P.G., Lunt, I.D., and Briggs, S.V., Spatial analysis of anthropogenic disturbance regimes and roadside shrubs in a fragmented agricultural landscape, *Appl. Veg. Sci.*, 7, 61, 2004.

Suarez, A.V., Holway, D.A., and Case, T.J., Patterns of spread in biological invasions dominated by long-distance jump dispersal: Insights from Argentine ants, *Proc. Natl. Acad. Sci.*, 98, 1095, 2001.

Wangen, S.R. and Webster, C.R., Potential for multiple lag phases during biotic invasions: Reconstructing an invasion of the exotic tree *Acer platanoides, J. Appl. Ecol.*, 43, 258, 2006.

Wangen, S.R., Webster, C.R., and Griggs, J.A., Spatial characteristics of the invasion of *Acer platanoides* on a temperate forested island, *Biol. Invasions*, 8, 1001, 2006.

Webb, S.L., Dwyer, M.E., Kaunzinger, C.K., and Wyckoff, P.H., The myth of the resilient forest: Case study of the invasive Norway maple *(Acer platanoides), Rhodora*, 102, 332, 2000.

Webb, S.L. and Kauzinger, C.K., Biological invasion of the Drew University (New Jersey) Forest Preserve by Norway maple (*Acer platanoides* L.), *Bull. Torrey Bot. Club*, 120, 343, 1993.

Webb, S.L., Pendergast, T.H., and Dwyer, M.E., Response of native and exotic maple seedling banks to removal of the exotic, invasive Norway maple *(Acer platanoides), J. Torrey Bot. Soc.*, 128, 141, 2001.

Webb, S.L. and Scanga, S.E., Windstorm disturbance without patch dynamics: Twelve years of change in a Minnesota forest, *Ecology*, 82, 893, 2001.

Webster, C.R., Jenkins, M.A., and Jose, S., Woody invaders and the challenges they pose to forest ecosystems in the eastern United States, *J. Forestry*, 104, 366, 2006.

Webster, C.R., Nelson, K., and Wangen, S.R., Stand dynamics of an insular population of an invasive tree, *Acer platanoides, Forest Ecol. Manage.*, 208, 85, 2005.

Whitney, G.D., *Soil Survey of Mackinac County, Michigan*, USDA Natural Resource Conservation Service, Washington, DC, 1997.

With, K.A., The landscape ecology of invasive spread, *Conserv. Biol.*, 16, 1192, 2002.

Wyckoff, P.H. and Webb, S.L., Understory influence of the invasive Norway maple *(Acer Platanoides), Bull. Torrey Bot. Club*, 123, 197, 1996.

6 Invasive Species and the Resiliency of a Riparian Environment

James M. Dyer and C. Mark Cowell

CONTENTS

6.1 INTRODUCTION

Research into exotic species invasions has focused on both the life-history traits of the invaders and the site qualities that may make ecosystems prone to invasion. Successful invasion of an ecosystem is typically a result of conducive combinations of species and site characteristics (Alpert et al. 2000). Crull's Island, PA, aggressively colonized by two invasive species following the construction of an upstream dam, illustrates this interaction well. The island is used as a case study to highlight the dynamics of a highly *invasible* system. The process of invasion is discussed within the context of resiliency theory. Originally posited for ecological systems (Holling 1973), resiliency theory has been extended to address linked ecological and social systems (e.g., Carpenter et al. 2005; Walker et al. 2002). Although a few studies have applied the framework of resiliency theory to investigations of exotic species (e.g., Forys and Allen 2002), little has been done regarding invasive plants. Our examination of Crull's Island highlights the many dilemmas associated with natural area management in the eastern United States, where extensive human modification of landscape processes has combined with introduction of nonnative species to fundamentally alter the functioning of the region's ecosystems.

6.2 WHAT CONTRIBUTES TO SUCCESSFUL INVASION?

Invasive species tend to share many characteristics: the ability to disperse long distances to new sites; the ability to establish readily in areas modified by human activities; the ability to outcompete native species and spread rapidly from their initial point of establishment; and the tendency to be difficult to eradicate once established. One example of a *species-focused* explanation of invasiveness is the empty niche hypothesis, which proposes that exotic species become invasive because they are able to access resources in the community that native species do not utilize (Hierro et al. 2005).

On the other hand, alteration of the natural disturbance regime, which is often critical for the establishment of invasive species, provides an example of a *site-focused* explanation of invasiveness. Disturbance alters both resource availability and competitive interactions, and changes to the disturbance regime may favor exotic species with adaptive traits not present within the native community (explaining why exotics are often seen to create persistent monotypic stands; Prieur-Richard and Lavorel 2000).

6.3 CRULL'S ISLAND

Dozens of islands are situated in the Allegheny River of northwestern Pennsylvania; seven located between Warren and Tionesta (Figure 6.1) were designated the "Allegheny Islands Wilderness" by the U.S. Forest Service in 1984. The largest (39 ha) of these wilderness islands is Crull's Island (41°49′ N, 79°16′ W), which possesses some of the best examples of riverine forest in Pennsylvania (Smith 1989). The Forest Service also designated Crull's Island as a research natural area in 1988, and in accordance with its wilderness designation, only recreation, education, and research activities are permitted on the island.

Crull's Island, formed from glacial outwash and alluvial deposits, is overlain by loam and sandy loam soils (Cerutti 1985). Topographically, the island is composed of a terrace <3 m above the river banks, which is surrounded by a floodplain 1–110 m wide. A steep slope separates the terrace and floodplain in most areas, though on the upstream side of the island there is a more gradual transition, with an intermediate upper floodplain surface (Figure 6.2). The region is characterized by a humid continental climate, with mild summers and year-round precipitation. The island is situated in the Northern hardwoods–hemlock forest region, near the boundary with the Mesophytic region (Dyer 2006).

Crull's Island has a long history of agricultural land use. The island was originally purchased by the Crull family, who were farming most of it by the mid-nineteenth century (Babbitt 1855). Cultivation and livestock grazing continued on the island into the early twentieth century (Crull 1996, personal communication).

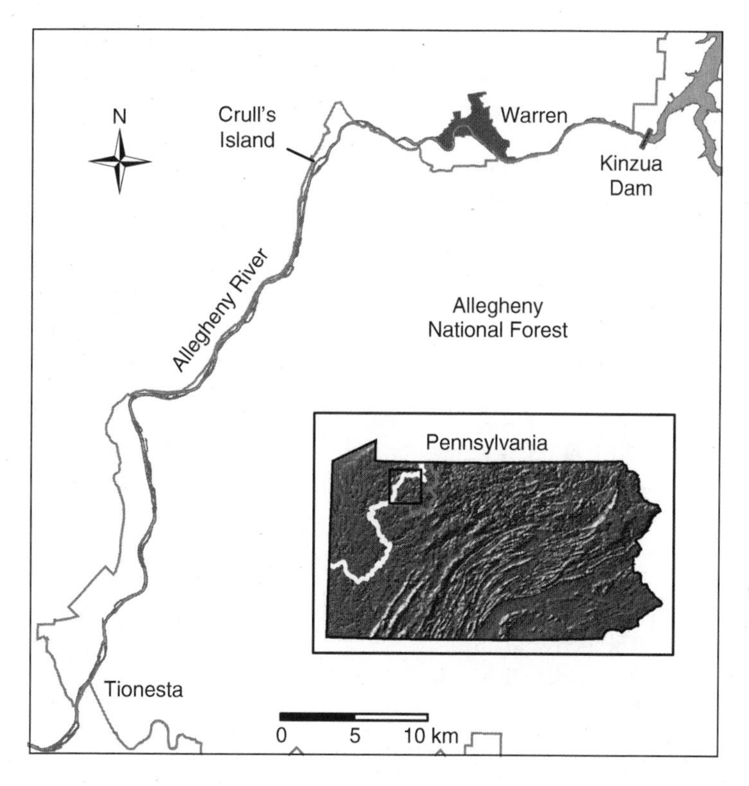

FIGURE 6.1 Location of Crull's Island. Inset map shows location of study area in Pennsylvania with respect to the Allegheny River and the Allegheny National Forest.

6.3.1 Invasive Species on Crull's Island

Crull's Island has been colonized by two problematic species with multiple *invasive* traits, the first being Japanese knotweed (*Polygonum cuspidatum*). Japanese

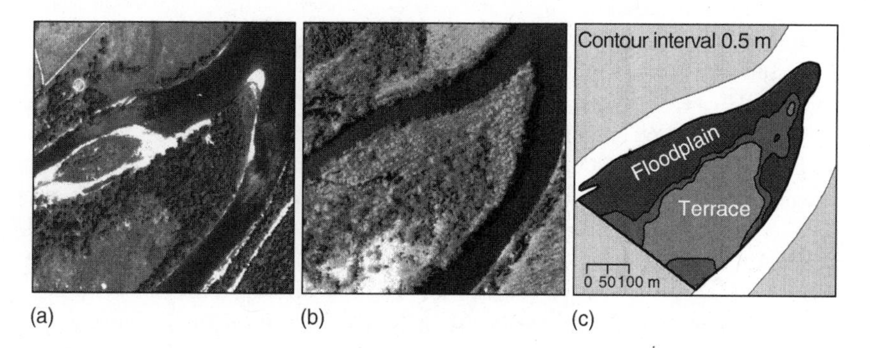

FIGURE 6.2 Vegetation and topography of Crull's Island: (a) 1939 USDA aerial photograph, (b) Present-day USGS digital orthophoto, (c) Contours and primary geomorphic surfaces. The upper floodplain, intermediate between the terrace and lower elevation floodplain, is evident at the head of the island.

knotweed is a shade-tolerant, rhizomatous perennial, 1–3 m in height. Although herbaceous, its erect stems can develop bamboo-like woody stems, explaining another of its common names, Mexican bamboo. In the United States, it is most frequently referred to as *P. cuspidatum*, whereas in the United Kingdom it is more commonly called *Fallopia japonica (Reynoutria japonica* is another synonym; Bram and McNair 2004). Introduced as an ornamental in the nineteenth century from its native East Asia, Japanese knotweed is considered invasive in both Europe and the United States, especially in New England and the Midwestern United States. It threatens native species, reducing both plant species diversity as well as habitat diversity (Weston et al. 2005). Although it does not seem capable of regeneration from seed in Europe (in contrast to the United States), it has demonstrated long-distance dispersal from points of introduction in both the United States and Europe (Beerling et al. 1994; Bram and McNair 2004; Hulme 2003). New colonization events can occur through the spread of rhizomes in soil transported by humans, and especially through rhizome dispersal by rivers (Pyšek and Prach 1993; Weston et al. 2005).

Japanese knotweed is most widespread on stream banks, in disturbed sites such as roadsides, as well as on the edges of, and within, open woodlands. It attains highest abundance in high light environments, and once established it is able to form dense patches. The dense clumps, dense canopy, and accumulation of stem litter exclude most other plants once Japanese knotweed establishes. It may also possess allelopathic properties. Japanese knotweed can spread via rhizomes, which can be >2 m in depth, and extend 15–20 m in length. Because of its extensive rhizomes, Japanese knotweed is very difficult if not impossible to control once established. Pulling, cutting, burning, and herbicide treatments have not proven entirely successful. Cutting may actually increase stem density and facilitate spread, and herbicide application is often problematic in riparian environments (i.e., adjacent to streams and rivers; Beerling et al. 1994; Bram and McNair 2004; Weston et al. 2005).

The second problematic species with multiple invasive traits on Crull's Island is reed canary grass (*Phalaris arundinacea*). Reed canary grass is a C3 grass that grows to a height of 1–2 m. Although native to both North America and Europe, European cultivars were introduced into the United States in the mid-nineteenth century. While the native variety was not considered aggressive, current populations may be hybrids and are considered invasive (Lindig-Cisneros and Zedler 2002). Despite the invasive potential of these cultivars, they continue to be planted for forage, stream bank erosion control, ditch stabilization, and phytoremediation (Gifford et al. 2002; Lavergne and Molofsky 2004). Reed canary grass seems to be able to occupy sites where native species perform poorly, and once established, it progressively displaces native species, especially in riparian environments (Apfelbaum and Sam 1987; Miller and Zedler 2003).

Reed canary grass is a prolific seeder, but can also spread from rhizomes or branch fragments. Once established, it spreads quickly through rhizomes to form dense stands (Gifford et al. 2002; Lavergne and Molofsky 2004; Lindig-Cisneros and Zedler 2002). Early-season growth is concentrated above ground, preempting the establishment of other species. Subsequent growth occurs below ground, favoring vegetative spread into adjacent plant canopies (Adams and Galatowitsch 2005).

In addition, reed canary grass exhibits seed dormancy, so that it becomes an important component of the seed bank (Lavergne and Molofsky 2004).

A facultative wetland species, reed canary grass is also found in riparian areas and some upland sites, and outcompetes native species under a wide range of moisture conditions from flooding to drought. It requires high light, and germinates best in moist to waterlogged soils following disturbance (Lavergne and Molofsky 2004; Lindig-Cisneros and Zedler 2002; Miller and Zedler 2003). Vegetative spread can be altered by changes in water depth, as well as frequency or duration of high water levels (Lavergne and Molofsky 2004). Herbicides have not been effective in treating reed canary grass, since the plant quickly reestablishes from seed and rhizomes (Perry and Galatowitsch 2004). As with other invasive species, control is difficult, and the best strategy would be to prevent the initial establishment. Disturbed areas seem most vulnerable to invasion, especially if sites have experienced multiple or interacting disturbances (Kercher and Zedler 2004). In riparian settings, downstream sites are at increased risk of invasion following initial establishment of both reed canary grass and Japanese knotweed.

Japanese knotweed and reed canary grass possess many typically invasive life-history traits. Both are capable of long-distance dispersal and readily colonize disturbed sites. Once established, they form dense stands, are able to spread vegetatively into new areas, and are very difficult to eradicate. Thus, positive feedbacks are established; even if a colonized site (such as Crull's Island) is returned to its preinvasion conditions, it is unlikely that these exotic species would relinquish their dominance.

6.3.2 Site Characteristics of Crull's Island

Successful invasion often results when exotic species traits interact with particular site characteristics (Alpert et al. 2000). Crull's Island has a number of properties making it prone to successful invasion, due to its riparian setting and the fact that it has been substantially influenced by human activity (Cowell and Dyer 2002).

Multiple traits of riparian areas in general confer vulnerability to invasion by exotic species, and wetland invasions are among the most aggressive of natural ecosystems in North America (Lavoie et al. 2003). The river itself serves as an agent of dispersal for rhizomes and seeds, and typically riparian areas contain many exposed banks, islands, and bars where new species might colonize. Additionally, riparian areas are subject to frequent disturbance through flooding, which removes biomass, frees up space, reduces competition, and results in high light levels. Many riparian settings experience favorable growing conditions for colonists: high moisture levels and frequent nutrient inputs. Once established, an invasive species can spread laterally from its point of introduction (Pyšek and Prach 1993; Tickner et al. 2001).

These *natural* characteristics which make riparian areas inherently invasible may be compounded by human activities, which can alter normal disturbance regimes and habitat conditions. For instance, riparian areas experience a high degree of land use change. Furthermore, alteration of a river's flow regime through human activities such as channelization or damming (which occurred upstream from Crull's Island) can result in changes to water table depths of riparian sites, soil moisture fluctuations, frequency of inundation, and changes in erosion and deposition of sediment. Such dramatic changes to the disturbance regime and habitat conditions at riparian sites make them especially vulnerable to exotic species invasion.

The hydrologic regime generated by dammed rivers is often associated with reduced species diversity (Nilsson and Berggren 2000). A fundamental effect of many dams is a reduction in the variability of stream flow, with fewer large floods leading to greater stability in riparian habitats (Naiman et al. 1993). The principal impact of dams on vegetation is often the maturation of riparian communities to late successional stages (Johnson 1994; Marston et al. 1995; Miller et al. 1995). Little analysis of the downstream impacts of dams on eastern U.S. deciduous forests exists, although Barnes (1997) and Knutson and Klaas (1998) demonstrate that forests along regulated rivers in Wisconsin and Minnesota have shifted in structure and composition from their presettlement condition. Upstream damming has significantly altered the hydrologic regime of Crull's Island.

In the nineteenth and twentieth centuries, the Allegheny River in the vicinity of Crull's Island experienced at least five major flooding events ($>1900\,m^3/s$). In 1965, Kinzua Dam was constructed 25 km upstream of Crull's Island to limit such flooding along the Allegheny River, particularly in urban areas such as Pittsburgh. The influence of dam construction on the geomorphology (and subsequently the vegetation) of Crull's Island has been dramatic. Prior to dam construction, highest flows affecting Crull's Island would occur in spring (mean maximum discharge $= 1269\,m^3/s$), associated with spring rains and snowmelt. Flows would gradually decrease until their summer minima (mean minimum discharge $= 12\,m^3/s$). Construction of the dam moderated annual extremes; now, mean maximum flow $= 688\,m^3/s$ and mean minimum discharge $= 29\,m^3/s$. As an example of this moderation, the highest recorded flow since dam construction (960 m^3/s in 1972) would have been common prior to the dam's existence, recurring on average every 1.5 years (Figure 6.3; Cowell and Stoudt 2002).

6.3.3 Changing Vegetation Patterns on Crull's Island

Riparian areas are naturally *patchy*, with spatial patterns in species assemblages reflecting the geomorphic features of the floodplain and terraces (Hupp and Osterkamp 1985, 1996). This pattern is driven by the decline in flood frequency and stream power away from the active channel, producing variation in the extent of erosive scouring and mechanical damage, the duration of anaerobic conditions, and the distribution of alluvial sediments (Hack and Goodlett 1960; Hupp 1986). Thus, disturbance and environmental conditions vary spatially and temporally in riparian settings, accommodating a variety of life-history strategies. Previously, we identified patches on Crull's Island in context of their historical development, especially with respect to construction of Kinzua Dam upstream (see Cowell and Dyer 2002 for details of the sampling methodology). Here, we focus specifically on the establishment and colonization of invasive species on the island.

Two distinctive plant associations—floodplain and terrace—segregate along the first axis of a Nonmetric Multidimensional Scaling (NMS) ordination. The segregation

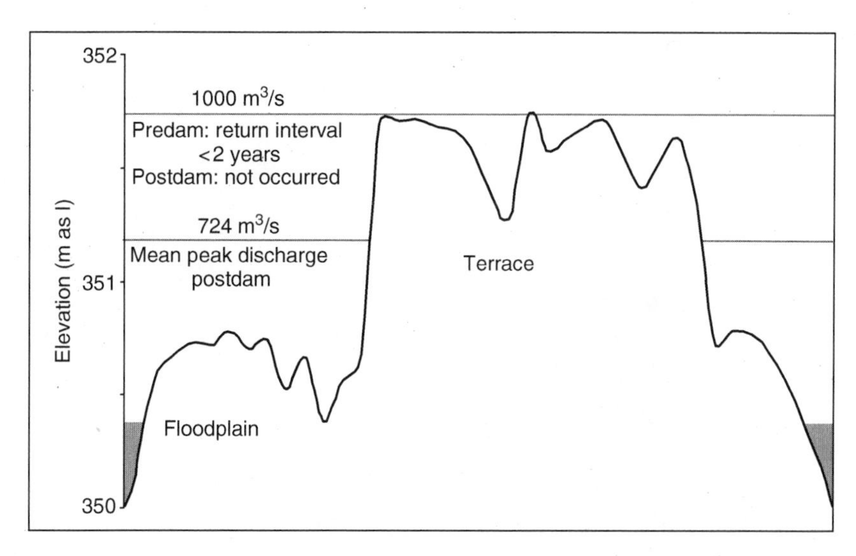

FIGURE 6.3 Elevation profile across a survey transect (vertical exaggeration is ~50×), showing pre and postdam discharge levels relative to floodplain and terrace of Crull's Island.

of these low-elevation floodplain sites and higher elevation terrace sites likely results from flooding-related impacts on the island's vegetation. The frequently inundated floodplain sites also had higher levels of base cations such as calcium and magnesium, and contained more silt than higher elevation terrace sites. Although vegetation patches are discernible on the island related to natural geomorphic settings, distinctive human imprints are observed within each patch.

6.3.3.1 Floodplain

Floodplain sites are dominated by sycamore (*Platanus occidentalis*) and silver maple (*Acer saccharinum*), and to a lesser extent American elm (*Ulmus americana*) and white ash (*Fraxinus americana*) in the smaller size classes (Figure 6.4). Although compositionally similar, two distinct patches are noted within the floodplain. Essentially, the lowest elevation sites included in the floodplain were largely bare before dam construction (Figure 6.2), and most individuals established in the decade following dam construction in 1965. In contrast, on the *upper floodplain*, many trees date back to the late nineteenth century, suggesting that predam scouring-associated mortality was not as frequent here as on the lower floodplain. The other especially notable difference between the upper and lower floodplain is the presence of invasive species: Japanese knotweed and reed canary grass are the dominant herbaceous species occurring on the lower floodplain.

Before the construction of Kinzua Dam, Crull's Island provided a classic example of vegetation mediated by the natural disturbance regime of its riparian setting. Flood events created early successional habitat by removing biomass, freeing up resources, and modifying substrate through scouring and deposition (Bendix and Hupp 2000; White and Jentsch 2001). Flow regulation via Kinzua

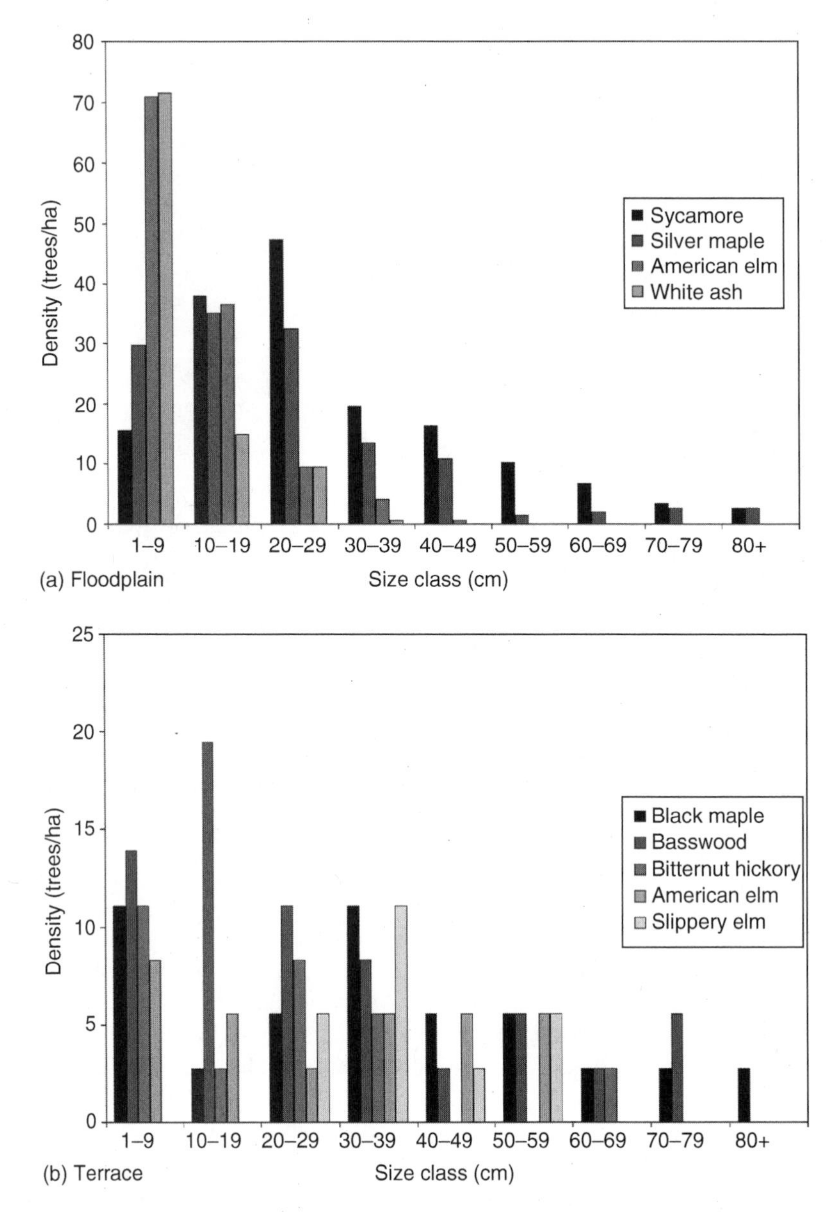

FIGURE 6.4 Size-class distribution for the two primary geomorphic surfaces on Crull's Island: the floodplain, and the older, mature forest of the terrace. Note the scale difference in the *y*-axes. Within the floodplain, tree density is higher on the younger, lower floodplain (349 trees/ha) compared with that on the older, higher floodplain surface (244 trees/ha).

Dam has caused an increase in the mean flow and the termination of the powerful flood events with the ability to destroy biomass and scour existing sites. Following dam construction, no early successional patches have been established on the island, and the lower floodplain has been colonized. Concomitant to this decrease in disturbance events has been the increased stress of anaerobic soil conditions associated with flood events of longer duration. Although the cohort of sycamore, silver maple, and other floodplain species that have established after flow regulation will likely maintain its dominance for decades, it is unlikely to be self-replacing in the absence of newly scoured sites. These early successional riparian species are intolerant of shade, and their establishment is tied to open sites created by flood disturbance. Reed canary grass and especially Japanese knotweed, however, are well suited to this modified environment.

6.3.3.2 Terrace

In contrast to the floodplain, forested areas of the higher elevation terrace are characterized by upland tree species. Like the floodplain, however, two distinctive patches are discernible within the terrace resulting from anthropogenic influences. The oldest trees of the terrace date to the period 1870–1890, corroborating historical accounts that agriculture was discontinued on this part of the island following a massive flood in 1865 (Crull 1996, personal communication). Forest in this part of the island is represented by trees in a range of size classes (suggesting continual establishment), especially black maple (*A. nigrum*) and basswood (*Tilia americana*), as well as American elm, slippery elm (*U. rubra*), and bitternut hickory (*Carya cordiformis*; Figure 6.4).

The highest ground on the terrace remained cultivated or pastured until around 1930, however. In terms of woody species, this area today contains mostly hawthorns (*Crataegus* sp.), which are able to establish in active pasture owing to their defensive thorns (Stover and Marks 1998). By and large, the hawthorns that are present today are visible on a 1939 air photo; succession has not proceeded within this pasture with the cessation of grazing, and a distinctive boundary still is evident between the maple–basswood terrace forest and the old field with its scattered hawthorns (Figure 6.2). The old field is still dominated by reed canary grass, which was likely planted for forage in the late nineteenth or early twentieth century. Not only is this invasive species precluding the establishment of other species, it also seems to be spreading into the adjacent forested area. As mature elm trees at the forest edge died, likely due to the arrival of Dutch elm disease in the 1970s (Stout 1999, personal communication), adjacent trees responded with increased growth rates; however, recruitment of juveniles into the newly opened site is not occurring. Instead, reed canary grass is expanding at the expense of the forest community.

On the higher island terrace, the older maple–basswood dominated forest does seem capable of gap-phase regeneration. All size classes seem to be represented in this stand, and as gaps are created upon the death of larger individuals, smaller trees are able to ascend into the canopy (Oliver and Larson 1996). On the terrace old field, however, succession to forest is not occurring, and reed canary grass maintains its dominance on the site. In addition, it encroaches into the adjacent forest stand.

6.3.4 Altered Abiotic Gradients and Competitive Relationships

Grime (2001) has characterized habitats on the basis of two primary gradients: productivity and disturbance intensity. These gradients form two sides of a triangle, and habitats occurring at the vertices of this triangle favor species with three distinctive life-history strategies. *Ruderals* are generally short-lived with high fecundity, and occur in productive sites that are subject to frequent disturbance. *Stress tolerators* occur in sites with infrequent disturbance but low productivity. *Competitors* are able to rapidly monopolize resources on productive sites, but are precluded by frequent disturbance. (These three strategies represent the extremes of life-history strategies; intermediate strategies [e.g., *competitive ruderal, stress-tolerant competitor*] are observed in habitats experiencing intermediate levels of disturbance and stress.) Both reed canary grass and Japanese knotweed would be considered competitors (Grime 2001), a strategy type apparently absent within the native community on Crull's Island with its frequent disturbance.

Human modification of the riparian environment on Crull's Island has significantly altered habitats of both terrace and floodplain environments. These changes create conditions favorable to novel life-history strategies according to Grime's (2001) categorization. A competitive ruderal life-history strategy, characterized by a long period of vegetative growth in this mesic, highly productive, nutrient-rich habitat, would have been favored on the floodplain prior to dam construction. Frequent flooding and associated scouring would have precluded dominance by competitors. The cessation of annual flooding after dam construction stabilized areas of the floodplain, however. Without frequent disturbance, the competitors reed canary grass and Japanese knotweed have been able to thrive at the expense of many resident species.

In contrast to the floodplain, the dominant human modification of the terrace involved land use change. A stress-tolerant competitor strategy, evidenced by later stages of forest succession compared with the floodplain with its recurring flooding, would be favored on this highly productive (but infrequently disturbed) environment. A permanent shift in vegetation composition has occurred with clearing of forest for agriculture and the introduction of the competitor reed canary grass.

Through alteration of the disturbance regime and abiotic gradients on Crull's Island, competitive relationships have changed, and nonnative species have been able to establish. The modified environment (e.g., cessation of scouring, but longer duration flood events) has created novel conditions (e.g., low light, anaerobic soil conditions on the floodplain) that the pool of native species does not seem able to exploit.

6.3.5 Management Concerns

The Ecological Society of America recently published policy and management recommendations dealing with invasive species (Lodge et al. 2006). These guidelines call for the strengthening of government authority and interagency coordination in the management of invasive species; they also elucidate the critical importance of intensive monitoring for known invasive species, and the importance of early control and eradication when populations are still localized. Of course, given the ability of invasive species to spread from their points of introduction, the policy recommendations also address the prevention of initial introductions, and the screening of nonnative organisms for their invasiveness risk.

A practical application of determining traits of species or sites that foster successful invasion is to identify and protect especially vulnerable ecological systems. For many sites where invasive species already dominate the native community, restoration to some earlier desired state is impractical in terms of available resources, costs, and potential for success. However, for many systems with perceived high ecological value, the control of exotic species invasion may be a high priority for management. For these systems, it is therefore useful to think about managing for natural integrity.

Crull's Island is the largest of several islands in Pennsylvania, federally designated as wilderness area. The wilderness designation attempts to address the trend of shrinking riparian forest habitat, a community type that has experienced dramatic declines in the eastern United States since Euro-American settlement. However, this chapter has highlighted the habitat changes on Crull's Island caused by dam construction as well as the introduction of exotic species with novel life-history traits. Viewed in the framework of resiliency theory, these changes seem to be steering Crull's Island to a new stable state, posing serious management concerns.

The outcome of succession following a disturbance event can be influenced by the nature of the disturbance event as well as predisturbance ecosystem conditions. However, resiliency theory acknowledges the fact that many ecological systems are observed to return to a *stable state* following disturbance (Mayer and Rietkerk 2004), with species composition, interactions, and ecosystem processes similar to predisturbance conditions. Conceptually, the landscape can be described as a series of basins representing stable ecosystem states. Under a given set of environmental conditions, a ball would come to rest at a point within a basin, and tend to return to this original state if perturbed. However, if the perturbation were severe enough, the ball might be pushed into another basin, coming to rest in a new stable state (Scheffer et al. 2001).

In ecological systems, resiliency refers to the level of disturbance a system can experience before shifting to a new stable state, characterized by a different set of dominant species. Species interactions create internal feedbacks that maintain the biotic and abiotic conditions of this new state; thus, an *alternative stable state* is attained (Holling 1973; Mayer and Rietkerk 2004). Because of these internal feedbacks, the longer the system exists in a particular state, the harder it is to *push* it back to a previous state (Mayer and Rietkerk 2004).

Examples frequently cited in treatments of resiliency and alternate stable states include the dramatic shift from clear to turbid lake conditions following human-induced eutrophication, or the conversion of grassland to woodland associated with a change in grazing intensity (e.g., Folke et al. 2004). These examples illustrate potential unexpected characteristics of a system shifting from one stable state to another. First is the possibility of nonlinear behavior of the change; environmental conditions may change gradually (e.g., an increase in nutrient loading to a lake) until a threshold is crossed, resulting in a rapid change of state. The existence of a critical threshold is often not known until it is crossed. (Note that nonlinear behavior—crossing a threshold in response to changing conditions—does not mean presence of alternate stable states.) Alternative stable states imply that for a given set of conditions, two or more persistent states must be possible (Mayer and Rietkerk 2004).

A second characteristic associated with these state changes is that, owing to feedbacks in the new state, merely changing conditions to *preshift* conditions may not be sufficient to change the system back to its original state (Mayer and Rietkerk 2004; Scheffer et al. 2001). In addition, human activity may affect system feedbacks (for example, by altering disturbance regimes or species relationships), which may reduce resiliency. The depth of the basin is made shallower, such that smaller perturbations can lead to an alternate stable state (Folke et al. 2004).

Elton (1958) suggested that high species richness conferred resistance to invasion, because fewer empty niches would be available for exotics to exploit in diverse communities. Although evidence for this hypothesis is equivocal (Richardson and Pyšek (2006) conclude that abiotic factors promoting high native species richness will also promote high exotic species richness), Allen et al. (2005) argue that system resilience is driven not by species identities within a system, but by the functions those species provide. If the same functional groups are represented in the invasive species that displace native species, system resilience may not necessarily be lost (Forys and Allen 2002; Prieur-Richard and Lavorel 2000). However, the introduction of an exotic species representing a new functional group (such as the nitrogen-fixing *Myrica faya* in Hawaii; Vitousek et al. 1987) can dramatically alter ecosystem interactions, and make the system more likely to shift to a new stable state (Folke et al. 2004).

On Crull's Island, it would appear that the invasive species represent a new functional group—the competitor strategists (Grime 2001). Reed canary grass and Japanese knotweed possess adaptive traits that enable them to be superior competitors when compared with native species. Once established, these species are able to aggressively maintain a site and spread into adjacent areas. Thus, internal feedbacks are established, which would make returning this system to its preshift condition extremely difficult without major inputs; merely restoring the predam flood regime may be insufficient to return the island to its previous state. Moreover, floods large enough to disturb the floodplain forest on Crull's Island are unlikely to be permitted due to their adverse economic impact.

Resilience can be defined as the ability of an ecological system to maintain its abiotic conditions and dominant species (and their interactions) when subjected to internal change and external disturbance (Cumming et al. 2005). On Crull's Island, human modification (land use change, altered flow regime, introduction of exotic species) has clearly acted as a driver, i.e., a force pushing the system from one basin of stability to another. The establishment of the invasive species reed canary grass and Japanese knotweed indicates that system resiliency is exceeded, leading to a shift to a new stable state on the island. Changes to plant species composition are ongoing, as riparian forest is replaced by monotypic stands of invasive species. Because of the superior competitive ability of these species once they become established, the state change will likely be irreversible without aggressive and continual management intervention. The changes in community composition call into question the notion that the forest of this wilderness island is an old-growth riparian community preserved from the effects of human influence (ANF 1996; Smith 1989; Wiegman and Lutz 1988). Although direct manipulation of the vegetation on Crull's Island is not permitted due to its wilderness designation (ANF 1996), the perception that the island's vegetation is a "natural community" with "no direct management needed to maintain the present ecological conditions" (Wiegman and Lutz 1988, p. 13) should be addressed.

Cronon (1996, 2003) has argued that "the time has come to rethink wilderness" since very few places are "untrammeled by man," but instead are experiencing a "rewilding" following past human activities. Although it is true that no place is free of human influence, especially in areas such as the eastern United States with very few *pristine* environments, we are nevertheless forced to address issues of ecosystem resilience in the face of aggressive invasive species. Vale (2005) has noted that creation of wilderness areas represents a protectionist impulse, and wilderness may be seen as a desired land use. To this end, it may be valid to think in terms of the need for more aggressive management in at least some wilderness areas. In the case of Crull's Island, the present system represents an obvious human imprint. It is not wilderness simply in the sense of free from human influence; it also fails in its goal to protect the riparian forest habitat. Biological diversity is of increasing concern today in the face of widespread extinction. Protected areas provide refugia of diversity as well as opportunities for visitor experience, enjoyment, and education. Biological diversity is a valued trait of many of our natural areas, yet it is threatened by invasive species (such as garlic mustard, purple loosestrife, and a host of others in addition to Japanese knotweed and reed canary grass) that have the potential to develop monotypic stands. In many areas, the inputs required to control invasive species exceed our ability to provide those inputs. Some areas, however, especially those with high intrinsic natural value, or representative of a threatened habitat type, may be considered a high priority for invasive species control regardless of wilderness designation. In the eastern United States, riparian areas like Crull's Island can be considered a high-priority conservation target. The extent of riparian forest has decreased by more than 80% due to floodplain development (Noss et al. 1995), and further declines in riparian biodiversity may be anticipated from increased modification of fluvial processes. Because of the historic loss of riparian habitat, restoration efforts may be required, shifting these systems back to some desired condition. Given the inherent variability in ecosystems, the benchmark used for the target of restoration efforts is debatable, though it is often sought to maintain the system within its natural range of variability (Sprugel 1991). Unfortunately, resiliency theory suggests that shifting from one stable state to another can be a nonlinear process, requiring a larger amount of time, effort, and resources (Mayer and Rietkerk 2004). Within this context, invasive species represent a major problem in natural areas where their establishment tips the system to a new steady state.

Control of natural disturbance to manage for a more consistent and predictable discharge has resulted in reduced resilience in this Allegheny River riparian ecosystem. Such unintended management consequences support the adoption of *natural disturbance-based management* approaches, which seek to emulate those processes that maintain habitat heterogeneity, both spatially and temporally, and thereby maintain ecosystem resilience (Drever et al. 2006). Given the difficulties with ecosystem restoration, when restoration involves moving the system out of one steady-state basin into another, it is obviously most advantageous to keep change from occurring in the first place. To prevent systems from undergoing undesirable state shifts, we have two routes: we can maintain system resilience (keep the sides of our stability basin steep), or we can control the perturbations that might trigger a state change (pushing the ball into a different stability basin). Variables that govern system resiliency, such as land use or soil properties, often change gradually and can be readily monitored; in contrast, events that trigger state shifts, such as the establishment of an invasive species, are often difficult to predict and control (Scheffer et al. 2001). For all practical purposes, successful colonization by invasive species may lead to a change of state that is irreversible. Not only must we consider exotic species invasions in terms of economic impacts, or as extinction threats to particular native species, but as a potential threat to entire ecosystems.

ACKNOWLEDGMENT

We thank Mary Dyer for many constructive comments on the manuscript.

REFERENCES

Adams, C.R. and Galatowitsch, S.M., *Phalaris arundinacea* (reed canary grass): Rapid growth and growth pattern in conditions approximating newly restored wetlands, *Ecoscience*, 12, 569, 2005.

Allegheny National Forest (ANF), *Allegheny National Wild and Scenic River: River Management Plan*, USDA Forest Service, Warren, PA, 1996.

Allen, C.R., Gunderson, L., and Johnson, A.R., The use of discontinuities and functional groups to assess relative resilience in complex systems, *Ecosystems*, 8, 958, 2005.

Alpert, P., Bone, E., and Holzapfel, C., Invasiveness, invisibility and the role of environmental stress in the spread of non-native plants, *Perspect. Plant Ecol. Evol. Syst.*, 3, 52, 2000.

Apfelbaum, S.I. and Sams, C.E., Ecology and control of reed canary grass (*Phalaris arundinaceae* L.), *Nat. Area. J.*, 7, 69, 1987.

Babbitt, E.L., *The Allegheny Pilot*, 1st ed., Babbitt, Freeport, PA, 1855.

Barnes, W.J., Vegetation dynamics on the floodplain of the lower Chippewa River in Wisconsin, *J. Torrey Bot. Soc.*, 124, 189, 1997.

Beerling, D.J., Bailey, J.P., and Conolly, A.P., *Fallopia japonica* (Houtt.) Ronse Decraene (*Reynoutria japonica* Houtt.; *Polygonum cuspidatum* Sieb. & Zucc.), *J. Ecol.*, 82, 959, 1994.

Bendix, J. and Hupp, C.R., Hydrological and geomorphological impacts on riparian plant communities, *Hydrol. Process.*, 14, 2977, 2000.

Bram, M.R. and McNair, J.N., Seed germinability and its seasonal onset of Japanese knotweed *(Polygonum cuspidatum), Weed Sci.*, 52, 759, 2004.

Carpenter, S.R., Westley, F., and Turner, M.G., Surrogates for resilience of social-ecological systems, *Ecosystems*, 8, 941, 2005.

Cerutti, J.R., *Soil Survey of Warren and Forest Counties, Pennsylvania*, USDA Soil Conservation Service, Washington, DC, 1985.

Cowell, C.M. and Dyer, J.M., Vegetation development in a modified riparian environment: Human imprints on an Allegheny River Wilderness. *Ann. Assoc. Am. Geogr.*, 92, 189, 2002.

Cowell, C.M. and Stoudt, R.T., Dam-induced modifications to upper Allegheny River streamflow patterns and their biodiversity implications, *J. Am. Water Resour. Assoc.*, 38, 187, 2002.

Cronon, W., The trouble with wilderness or, getting back to the wrong nature, *Environ. Hist.*, 1, 7, 1996.

Cronon, W., The riddle of the Apostle Islands: How do you manage a wilderness full of human stories? *Orion*, 22, 36, 2003.

Cumming, G.S., Barnes, G., Perz, S., Schmink, M., Sieving, K.E., Southworth, J., Binford, M., Holt, R.D., Stickler, C., and Van Holt, T., An exploratory framework for the empirical measurement of resilience, *Ecosystems*, 8, 975, 2005.

Drever, C.R., Peterson, G., Messier, C., Bergeron, Y., and Flannigan, M., Can forest management based on natural disturbance maintain ecological resilience? *Can. J. For. Res.*, 36, 2285, 2006.

Dyer, J.M., Revisiting the deciduous forests of eastern North America, *Bioscience*, 56, 341, 2006.

Elton, C.S., *The Ecology of Invasion by Animals and Plants*, Methuen, London, 1958, Chapter 6.

Folke, C., Carpenter, S., Walker, B., Scheffer, M., Elmqvist, T., Gunderson, L., and Holling, C.S., Regime shifts, resilience, and biodiversity in ecosystem management, *Annu. Rev. Ecol. Evol. Syst.*, 35, 557, 2004.

Forys, E.A. and Allen, C.R., Functional group change within and across scales following invasions and extinctions in the Everglades ecosystem, *Ecosystems*, 5, 339, 2002.

Gifford, A.L.S., Ferdy, J.B., and Molofsky, J., Genetic composition and morphological variation among populations of the invasive grass, *Phalaris arundinacea, Can. J. Bot.*, 80, 779, 2002.

Grime, J.P., *Plant Strategies, Vegetation Processes, and Ecosystem Properties*, 2nd ed., Wiley, Chichester, UK, 2001, Chapters 1 and 2.

Hack, J.T. and Goodlett, J.C., *Geomorphology and Forest Ecology of a Mountain Region in the Central Appalachians*, Geological Survey Professional Paper 347, U.S. Government Printing Office, Washington, DC, 1960.

Hierro, J.L., Maron, J.L., and Callaway, R.M., A biogeographical approach to plant invasions: The importance of studying exotics in their introduced and native range, *J. Ecol.*, 93, 5, 2005.

Holling, C.S., Resilience and stability in ecological systems, *Annu. Rev. Ecol. Syst.*, 4, 1, 1973.

Hulme, P.E., Biological invasions: Winning the science battles but losing the conservation war? *Oryx*, 37, 178, 2003.

Hupp, C.R., Upstream variation in bottomland vegetation patterns, northwestern Virginia, *B. Torrey Bot. Club*, 113, 421, 1986.

Hupp, C.R. and Osterkamp, W.R., Bottomland vegetation distribution along Passage Creek, Virginia, in relationship to fluvial landforms, *Ecology*, 66, 670, 1985.

Hupp, C.R. and Osterkamp, W.R., Riparian vegetation and fluvial geomorphic processes, *Geomorphology*, 14, 277, 1996.

Johnson, W.C., Woodland expansion in the Platte River, Nebraska: Patterns and causes, *Ecol. Monogr.*, 64, 45, 1994.

Kercher, S.M. and Zedler, J.B., Multiple disturbances accelerate invasion of reed canary grass (*Phalaris arundinacea* L.) in a mesocosm study, *Oecologia*, 138, 455, 2004.

Knutson, M.G. and Klaas, E.W., Floodplain forest loss and changes in forest community composition and structure in the upper Mississippi River: A wildlife habitat at risk, *Nat. Area. J.*, 18, 138, 1998.

Lavergne, S. and Molofsky, J., Reed canary grass (*Phalaris arundinacea*) as a biological model in the study of plant invasions, *Crit. Rev. Plant Sci.*, 23, 415, 2004.

Lavoie, C., Jean, M., Delisle, F., and Létourneau, G., Exotic plant species of the St. Lawrence River wetlands: A spatial and historical analysis, *J. Biogeogr.*, 30, 537, 2003.

Lindig-Cisneros, R. and Zedler, J.B., *Phalaris arundinacea* seedling establishment: Effects of canopy complexity in fen, mesocosm, and restoration experiments, *Can. J. Bot.*, 80, 617, 2002.

Lodge, D.M., Williams, S., MacIsaac, H.J., et al., Biological invasions: Recommendations for U.S. policy and management, *Ecol. Appl.*, 16, 2035, 2006.

Marston, R.A., Girel, J., Pautou, G., Piegay, H., Bravard, J.P., and Arneson, C., Channel metamorphosis, floodplain disturbance, and vegetation development: Ain River, France, *Geomorphology*, 13, 121, 1995.

Mayer, A.L. and Rietkerk, M., The dynamic regime concept for ecosystem management and restoration, *Bioscience*, 54, 1013, 2004.

Miller, J.R., Schulz, T.T., Hobbs, N.T., Wilson, K.R., Schrupp, D.L., and Baker, W.L., Changes in the landscape structure of a southeastern Wyoming riparian zone following shifts in stream dynamics, *Biol. Conserv.*, 72, 371, 1995.

Miller, R.C. and Zedler, J.B., Responses of native and invasive wetland plants to hydroperiod and water depth, *Plant Ecol.*, 167, 57, 2003.

Naiman, R.J., Décamps, H., and Pollock, M., The role of riparian corridors in maintaining regional biodiversity, *Ecol. Appl.*, 3, 209, 1993.

Nilsson, C. and Berggren, K., Alteration of riparian ecosystems caused by river regulation, *Bioscience*, 50, 783, 2000.

Noss, R.F., LaRoe, E.T., and Scott, J.M., Endangered ecosystems of the United States: A preliminary assessment of loss and degradation, Biological Report 28, U.S. Department of Interior, National Biological Service, Washington, DC, 1995.

Oliver, C.D. and Larson, B.C., *Forest Stand Dynamics*, McGraw Hill, New York, 1996.

Perry, L.G. and Galatowitsch, S.M., The influence of light availability on competition between *Phalaris arundinaceae* and a native wetland sedge, *Plant Ecol.*, 170, 73, 2004.

Prieur-Richard, A.H. and Lavorel, S., Invasions: The perspective of diverse plant communities, *Aust. Ecol.*, 25, 1, 2000.

Pyšek, P. and Prach, K., Plant invasions and the role of riparian habitats: A comparison of four species alien to central Europe, *J. Biogeogr.*, 20, 413, 1993.

Richardson, D.M. and Pyšek, P., Plant invasions: Merging the concepts of species invasiveness and community invisibility, *Prog. Phys. Geogr.*, 30, 409, 2006.

Scheffer, M.S., Carpenter, S.R., Foley, J.A., Folke, C., and Walker, B., Catastrophic shifts in ecosystems, *Nature*, 413, 591, 2001.

Smith, T.L., An overview of old-growth forest in Pennsylvania, *Nat. Area. J.*, 9, 40, 1989.

Sprugel, D.S., Disturbance, equilibrium, and environmental variability: What is "natural" vegetation in a changing environment? *Biol. Conserv.*, 58, 1, 1991.

Stover, M.E. and Marks, P.L., Successional vegetation on abandoned cultivated and pastured land in Tompkins County, New York, *J. Torrey Bot. Soc.*, 125, 150, 1998.

Tickner, D.P., Angold, P.G., Gurnell, A.M., and Mountford, J.O., Riparian plant invasions: Hydrogeomorphological control and ecological impacts, *Prog. Phys. Geog.*, 25, 22, 2001.

Vale, T.R., *The American Wilderness: Reflections on Nature Protection in the United States*, University of Virginia Press, Charlottesville, VA, 2005, Chapter 10.

Vitousek, P.M., Walker, L.R., Whiteaker, L.D., Mueller-Dombois, D., and Matson, P.A., Biological invasion by *Myrica faya* alters ecosystem development in Hawaii, *Science*, 238, 802, 1987.

Walker, B.S., Carpenter, S., Anderies, J., Abel, N., Cumming, G.S., Janssen, M., Lebel, L., Norberg, J., Peterson, G.D., and Pritchard, R., Resilience management in social-ecological systems: A working hypothesis for a participatory approach, *Conserv. Ecol.*, 6, 14, 2002 [online]. Available at http://www.ecologyandsociety.org/vol6/iss1/art14/print.pdf.

Weston, L.A., Barney, J.N., and DiTommaso, A., A review of the biology and ecology of three invasive perennials in New York State: Japanese knotweed (*Polygonum cuspidatum*), mugwort (*Artemisia vulgaris*) and pale swallow-wort (*Vincetoxicum rossicum*), *Plant Soil*, 277, 53, 2005.

White, P.S. and Jentsch, A., The search for generality in studies of disturbance and ecosystem dynamics, *Prog. Bot.*, 62, 399, 2001.

Wiegman, P.G. and Lutz, K.A., *Establishment Record for the Crull's Island Research Natural Area within the Allegheny National Forest, Warren County, Pennsylvania*, USDA Forest Service, Warren, PA, 1988.

7 Biogeography of Invasive Plant Species in Urban Park Forests

Robert E. Loeb

CONTENTS

7.1 INTRODUCTION

The National Research Council's Committee on the Scientific Basis for Predicting the Invasive Potential of Nonindigenous Plants and Plant Pests in the United States pointed out the importance of understanding the biogeography of invasive plants in the first conclusion of its 2002 report:

> After considering the history of invasions of plants and plant pests in the United States, reviewing the scientific knowledge about the factors associated with invasive species, and examining efforts to predict the potential species to invade, the committee reached . . .

> Conclusion 1. The record of a plant's invasiveness in other geographic areas is currently the most reliable predictor of its ability to establish and invade in the United States. The same is true for arthropods and pathogens if plants that they can use elsewhere occur in the United States.
>
> **National Academy Press 2002**

Research on the biogeography of invasive plant species has focused on the distribution of nonnative species across geographic regions. In the United States, descriptions of the distribution of nonnative plants have involved the analysis of reports of occurrence within a county (Robinson et al. 1994) or among states (McKinney 2006), using electronic databases from herbarium specimens. In Europe, the research has involved the analysis of vegetation samples of many different environments across a city (Kowarik 1995; Dickson 2000; Grapow et al. 2001), country (Haruska 1989; Kuhn et al. 2004), or region (Kowarik 1990; Pyšek 1998). In the aforementioned research, rather than focusing on invasive species and examining differences in groups of invasive species segmented by growth forms, nonnative plants have been used as a stand-in for invasive plants, and growth form was not part of the analysis. Previous research on nonnative plants has not focused on comparing similar environments that have specific characteristics favorable to hosting invasive plants. Locales of special importance are the places where invasive plants have been introduced and the centers for the spread of the invasive species.

7.1.1 Site Identification

McKinney (2006) provides insights into how to identify critical locations: "intentional and unintentional importation of species adapted to urban habitats, combined with many food resources imported for human use, often produces local species diversity and abundance that is often equal to or greater than the surrounding landscape." Clearly, all the invasive plants in an area did not arrive simultaneously. Therefore, consideration of the succession of plant introductions is vital to site selection. Sukopp (1990) noted "urban open spaces, even though they often represent completely new environments, are modifications of older ones. Careful historical analysis shows the links between present-day biotic communities and the former site conditions in the agrarian landscape."

Dehnen-Schmutz (1998) identified the place for plant invasion and spread during the Middle Ages, the medieval castle associated with a settlement: "The high portion of... usable plants (most of them medicinal and food plants)... emphasizes the reasons for their introduction in historic times. Later, when the usage changed, nonnative plants were mostly introduced because of aesthetic reasons. A fact which may explain the high amount of ornamentals...." Castles are refuges for nonnative species because they did not undergo changes, such as secondary succession or construction of roads and buildings, which remove the disturbed lands vegetation. In some cases, the opposite effect occurs today when new nonnative species are introduced in ornamental plantings for gardens or in landscape restoration projects which recreate past agricultural environments—perhaps including more species than were present in the original agricultural fields or gardens.

What locations in the twenty-first century new world have a history of invasive species introduction and landscape preservation similar to the medieval castle? A parallel environment can be discovered by examining the common environmental history of 10 large urban park forests in the mid-Atlantic region of the United States: Rock Creek Park, Washington, DC (Fleming and Kanal 1995); Oregon Ridge Park, Baltimore, MD (Redman 1999); Pennypack Park and Wissahickon Park, Philadelphia (Fairmont Park Commission 1999); Breezy Point, Jamaica Bay, and Wildlife Refuge, Brooklyn and Queens, NY (Tancredi 1983); Van Cortlandt Park, Bronx, NY (Profous and Loeb 1984); Pelham Bay Park, Bronx, NY (Loeb 1998); and Middlesex Fells, Middlesex, MA (Drayton and Primack 1996). The land comprising these 10 park forests was cultivated as farmland beginning as early as the mid-seventeenth century. Starting in the nineteenth century, estates and social institutions or commercial enterprises planted ornamental species. Concurrent with the creation of the large parks in the late nineteenth century, landscape plantings were completed with a variety of nonnative species. During the twentieth century, ornamental plantings continued, and disturbances, such as the building of recreational facilities, highways, and airports, caused these environments to be continuously disturbed. Like medieval castles, the 10 large park forests underwent a progression of human disturbances: farmlands, which introduced weeds associated with agriculture; horticultural showplaces, which introduced exotic ornamental species; and trampled environments, which host invasive species from surrounding home gardens. The urban location and large flow of people and vehicles through these parks make them centers for the spread of invasive species.

7.1.2 Research Questions

Urban park forests are the modern refuges for invasive plant species and potential origin points for the spread of invasive plant species. Several important questions arise regarding the study of these critical environments. What is the biogeography of invasive plant species in the large urban park forests of a vegetation region? What are the similarities and differences in the distribution of invasive species among urban park forests within a forest region? How many invasive species are in the park forests compared with the species listed as invasive in the region by government or scientific authorities? Does comparison of the invasive species in the park forests provide evidence of species homogenization or species saturation? Is there a geographic basis for the distribution of invasive plant species in the park forests?

7.2 METHODS

7.2.1 Research Sites

To examine the biogeography of invasive species in urban park forests, 10 large ($\geq$400 ha) urban parks with recorded floras were selected. The parks are located close to the coast in the mid-Atlantic region of the United States. These park forests are not close to the inner city but rather are located near the borders of Boston (Middlesex Fells), New York (Van Cortlandt, Pelham Bay, Breezy Point, Jamaica Bay, and Wildlife Refuge), Philadelphia (Pennypack and Wissahickon Creek), Baltimore (Oregon Ridge), and Washington, DC (Rock Creek). Table 7.1 shows

TABLE 7.1
Park Code, Park Name, Longitude, Latitude, County, State, Size, and Botanical Reference for the 10 Urban Parks

Code	Park	Longitude	Latitude	County	State	Size (ha)	Botanical Reference
MF	Middlesex Fells	71°06′21″ W	42°26′51″ N	Middlesex	MA	400	Drayton 1993
VC	Van Cortlandt	73°53′00″ W	40°53′52″ N	Bronx	NY	464	Natural Resources Group 1990
PB	Pelham Bay	73°48′28″ W	40°51′56″ N	Bronx	NY	1119	DeCandido 2001
BP	Breezy Point	73°54′37″ W	40°33′52″ N	Brooklyn	NY	420	Venezia and Cook 1991
JB	Jamaica Bay	73°50′07″ W	40°36′14″ N	Queens	NY	810	Venezia and Cook 1991
WR	Wildlife Refuge	73°50′30″ W	40°36′53″ N	Queens	NY	3604	Venezia and Cook 1991
PP	Pennypack	75°02′11″ W	40°03′23″ N	Philadelphia	PA	591	Horwitz et al. 2004
WC	Wissahickon	75°13′09″ W	40°07′00″ N	Montgomery	PA	748	Horwitz et al. 2004
OR	Oregon Ridge	76°41′38″ W	39°29′03″ N	Baltimore	MD	408	Redman 1999
RC	Rock Creek	77°02′47″ W	38°58′19″ N	District of Columbia	DC	713	Fleming and Kanal 1995

information regarding park spatial characteristics and botanical references. The floras were compiled by different researchers, and so a uniform procedure was not employed. The researcher for Middlesex Fells (Drayton 1993) did not identify graminoid species, as was done for the other parks.

Six of the 10 parks (Middlesex Fells, Van Cortlandt, Pennypack, Wissahickon Creek, Oregon Ridge, and Rock Creek) are similar in the absence of salt water environments in or near the parks. The parks were created by municipalities approximately a century ago, except for the Gateway National Recreational Area sites, Breezy Point, Jamaica Bay, and Wildlife Refuge, which are at least 50 years old. All 10 parks are located in Braun's (1950) oak-chestnut forest region. Three parks (Breezy Point, Jamaica Bay, and Wildlife Refuge) are located on glacial sands while the remaining seven parks are underlain by metamorphic or igneous rocks. The soils of the parks have been transformed to varying degrees by previous agriculture, construction, and landscaping activities.

7.2.2 Flora Analysis

A comparative flora of invasive vascular plant species reported in the 10 large urban park forests is provided in Table 7.2 with designation of growth form as tree, shrub, herbaceous, or graminoid. The species are listed alphabetically by genus and species names. Nomenclature follows Bailey et al. (1976) and Kartesz (1999). The designation of a species as an invasive plant is from the Mid-Atlantic Alien Plant Invaders List (United States National Parks Service Alien Plant Working Group 2007). Identification of the presence of a species in the four states and the District of Columbia where the 10 parks are located is from the Invasive and Noxious Weeds List (United States Department of Agriculture Natural Resources Conservation Service 2007).

To analyze the biogeography of the invasive species in the park forests, summary values for species similarity for each growth form were calculated with the Jaccard similarity measure (Pielou 1984). In other research (McKinney and Lockwood 2005), regression analysis of distances between parks and Jaccard values was used to search for patterns. However, this method does not reveal relationships among individual parks. To analyze the biogeography of the park forests, the Jaccard similarity measure was used as the clustering criteria for a hierarchical cluster analysis of the four growth forms in the 10 park forests. The biogeography pattern diagrams (see Figures 7.1 and 7.2) graphically display the species similarities as linkages among the urban park forests.

7.3 RESULTS

7.3.1 Species Distribution

The 278 species from the Mid-Atlantic Alien Plant Invaders List (United States National Parks Service Alien Plant Working Group 2007) were classified by growth form (tree, shrub, herbaceous, and graminoid); half were herbaceous, nearly a quarter were trees, and both graminoid and shrub species each comprised ~13% (Table 7.3). The number of species in each growth form group not present in any of the 10 parks ranged from 27% (tree species) to 31% (shrub species) of the total. For only the 195

TABLE 7.2
Presence of Invasive Plant Species from the Mid-Atlantic Alien Plant Invaders List in the Urban Park Forests

Abutilon theophrasti Medic. H; PB, OR; not listed
Acer palmatum Thunb. T; WC, OR, RC; DC
A. platanoides L. T; MF, PB, VC, BP, WR, PP, WC, OR, RC; MA, NY, PA, MD, DC
A. pseudoplatanus L. T; PB, VC, BP, WR, JB, PP, WC; PA
Aegopodium podagraria L. H; PB, PP, WC; PA
Aesculus hippocastanum L. T; PB, VC, WC; PA
Agrostis gigantea Roth. G; PB, VC, OR; NY
Ailanthus altissima (Mill.) Swingle. T; MF, PB, VC, BP, WR, JB, PP, WC, OR, RC; MA, NY, PA, MD, DC
Ajuga reptans L. H; OR, RC; MD
Akebia quinata (Houtt.) Decne. S; VC, WC; PA, MD, DC
Albizia julibrissin Durazz. T; PB, VC, JB, RC; MA, DC
Alliaria petiolata (Bieb.) Cavara & Grande. H; MF, PB, VC, PP, WC, OR, RC; MA, NY, PA, MD, DC
Allium vineale L. H; PB, VC, BP, WR, JB, PP, WC, OR, RC; PA, MD
Alnus glutinosa (L.) Gaertn. T; MF, PB, WR, RC; NY, PA
Alternanthera philoxeroides (Mart.) Griseb. H; not reported; not listed
Ampelopsis brevipedunculata (Maxim.) Trautv. G; PB, VC, WC, RC; MA, NY, PA, MD, DC
Anthoxanthum odoratum L. G; PB, VC, BP, JB, PP, RC; MD, DC
Aralia elata (Miq.) Seem. T; WC; PA
Arctium minus (J. Hill) Bernh. H; MF, PB, VC, PP, WC, OR, RC; PA, MD
Arrhenatherum elatius (L.) Beauvois ex J. Pres & K. Presl. G; PB; not listed
Artemisia stelleriana Bess. H; BP, WR, JB; not listed
A. vulgaris L. H; MF, PB, VC, BP, WR, JB, PP, WC, OR, RC; NY, PA, MD
Arthraxon hispidus (Thunb.) Makino. G; not reported; PA, MD
Arundo donax L. G; OR; MD
Bambusa vulgaris Schrad. Ex J.C. Wendl. G; not reported; DC
Berberis thunbergii DC. S; MF, PB, VC, BP, WR, JB, PP, WC, OR, RC; MA, NY, PA, MD, DC
B. vulgaris L. S; BP, JB; MA, NY, PA
Betula pendula Roth. T; not reported; MD
Brassica rapa L. H; MF, WR; MD
Bromus sterilis L. G; PB, JB, RC; MD
B. tectorum L. G; PB, BP, JB; PA
Broussonetia papyrifera (L.) Venten. T; PB, PP, WC; PA, MD, DC
Buddleja davidii Franch. S; OR; PA
Callitriche stagnalis Scop. H; not reported; not listed
Capsella bursa-pastoris (L.) Medikus. H; MF, PB, VC, BP, WR, JB, RC; MD
Cardamine impatiens L. H; PP, WC; not listed
Cardiospermum halicacabum L. S; not reported; not listed
Carduus acanthoides L. H; not reported; not listed
C. crispus L. H; not reported; not listed
C. nutans L. H; PB, OR, RC; PA, MD
Carex kobomugi Ohwi. G; not reported; MD
Carlina vulgaris L. H; not reported; not listed
Catalpa ovata G. Don. T; not reported; MA, MD
C. speciosa Warder. T; PB, VC; MD

TABLE 7.2 (continued)
Presence of Invasive Plant Species from the Mid-Atlantic Alien Plant Invaders List in the Urban Park Forests

Celastrus orbiculatus Thunb. S; PB, VC, PP, WC, OR, RC; MA, NY, PA, MD, DC
Centaurea biebersteinii DC. H; not reported; MA, NY, PA, MD
C. cyanus L. H; PB, WR, JB; MD
C. jacea L. H; PB, VC; not listed
C. solstitialis L. H; not reported; not listed
C. transalpina Schleich. ex DC. H; not reported; not listed
Cerastium biebersteinii DC. H; not reported; not listed
C. vulgatum L. H; MF, PB, WR, JB, OR, RC; MD
Chelidonium majus L. H; MF, PB, VC, PP, WC, OR; MD
Chenopodium ambrosioides L. H; PB, WR, JB, PP, RC; MD
Chrysanthemum leucanthemum L. H; PB, BP, WR, JB, OR, RC; MA, NY, PA, MD, DC
Cichorium intybus L. H; MF, PB, VC, BP, WR, JB, WC, OR, RC; PA, MD
Cirsium arvense (L.) Scop. H; PB, WR, JB, PP, OR, RC; PA, MD
C. vulgare (Savi.) Ten. H; MF, PB, VC, WR, JB, RC; MA, PA, MD
Clematis terniflora DC. H; PB, VC, WC, OR, RC; MD, DC
Commelina communis L. H; PB, VC, BP, WR, OR, RC; PA, MD
Conium maculatum L. H; PB; PA, MD
Convolvulus arvensis L. H; MF, PB, VC, WR, JB, OR; PA
Coronilla varia L. H; PB, WR, JB, OR; MD
Cruciata laevipes Opiz. H; not reported; not listed
Cynodon dactylon (L.) Pers. G; VC, OR; not listed
Cyperus amuricus Maxim. G; not reported; not listed
C. iria L. G; BP, PP; not listed
C. rotundus L. G; not reported; not listed
Cytisus scoparius (L.) Link. G; WR, JB; MA
Dactylis glomerata L. G; PB, VC, BP, WR, JB, PP, OR, RC; MD
Datura stramonium L. H; PB, VC, WR, JB; PA, MD
Daucus carota L. H; MF, PB, VC, BP, WR, JB, OR, RC; MD
Deutzia scabra Thunb. S; WC, RC; PA, MD, DC
Dianthus armeria L. H; MF, PB, WR, JB, OR, RC; MD
Dioscorea batatas Decne. H; RC; MD
Dipsacus fullonum L. H; not reported; not listed
D. laciniatus L. H; not reported; MD
Duchesnea indica (Andr.) Focke. H; PB, VC, PP, WC, OR, RC; PA, MD
Echium vulgare L. H; PB, WR, JB, OR; MD
Eichhornia crassipes (Mart.) Solms-Laub. H; not reported; not listed
Elaeagnus angustifolia L. S; PB, BP, WR, JB; NY, PA, MD
E. pungens Thunb. S; RC; not listed
E. umbellata Thunb. S; PB, WR, OR, RC; NY, PA, MD, DC
Elodea densa (Planch.) Casp. H; not reported; not listed
Elytrigia repens (L.) Desv. ex Nevski. G; PB, WR, JB, OR, RC; MD
Eragrostis capillaris (L.) Nees. G; JB, OR; not listed
E. curvula (Schrader) Nees. G; JB; MD
Euonymus alatus (Thunb.) Siebold. S; MF, PB, VC, PP, WC, OR; MA, PA, MD

(continued)

TABLE 7.2 (continued)
Presence of Invasive Plant Species from the Mid-Atlantic Alien Plant Invaders List in the Urban Park Forests

E. europaeus L. S; PB; MA
E. fortunei (Turcz.) Hand.-Mazz. S; PB, PP, WC, OR, RC; MD, DC
Euphorbia cyparissias L. H; PB, VC, WR; MA, NY
E. esula L. H; not reported; not listed
Festuca elatior L. G; VC, BP, JB, PP, OR, RC; not listed
Foeniculum vulgare Mill. H; not reported; not listed
Fragaria vesca L. H; MF, VC; not listed
Froelichia gracilis (Nutt.) Moq. H; not reported; not listed
Galega officinalis L. H; not reported; PA
Galinsoga parviflora Cav. H; PB; MD
G. quadriradiata Ruiz & Pavon. H; PB, VC, WR, OR, RC; MD
Galium mollugo L. H; PB, VC; not listed
G. verum L. H; not reported; not listed
Genista tinctoria L. S; PB; MA, DC
Geranium columbinum L. H; not reported; MD
Glechoma hederacea L. H; MF, PB, VC, WR, JB, PP, WC, OR, RC; PA, MD, DC
Hedera helix L. S; PB, VC, WC, OR, RC; PA, MD, DC
Helianthus petiolaris Nutt. H; WR; not listed
Hemerocallis fulva (L.) L. H; MF, PB, VC, PP, WC, OR, RC; PA, MD, DC
H. lilioasphodelus L. H; WR, JB; MD, DC
Heracleum mantegazzianum Sommier & Levier. H; not reported; NY, PA
Hesperis matronalis L. H; PB, PP, WC; PA, MD
Hibiscus syriacus L. S; PB, VC, OR, RC; PA
Hieracium pilosella L. H; not reported; not listed
Holcus lanatus L. G; PB, VC, WR; not listed
Humulus japonicus Siebold & Zucc. H; VC, PP, WC, RC; PA, MD, DC
Hydrilla verticillata (L.f.) Royle. H; not reported; MD, DC
Hypochaeris radicata L. H; not reported; not listed
Ilex aquifolium L. S; OR; MD
I. crenata Thunb. S; WC, OR, RC; DC
Imperata cylindrica (L.) Beauv. G; not reported; not listed
Ipomoea hederacea Jacq. H; OR, RC; not listed
I. lacunosa L. H; not reported; MD
I. purpurea (L.) Roth. H; PB, VC; not listed
Iris pseudacorus L. H; PB, VC, PP, OR; MD
Isatis tinctoria L. H; not reported; not listed
Kerria japonica (L.) DC. S; not reported; not listed
Kyllinga gracillima Miq. G; not reported; not listed
Lagerstroemia indica L. S; not reported; not listed
Lamium amplexicaule L. H; VC; MD
L. maculatum L. H; not reported; MD
L. purpureum L. H; PB, OR; MD
Lantana montevidensis (K. Spreng.) Briq. S; not reported; not listed
Lapsana communis L. H; PB; not listed
Lespedeza bicolor Turcz. S; OR; not listed

TABLE 7.2 (continued)
Presence of Invasive Plant Species from the Mid-Atlantic Alien Plant Invaders List in the Urban Park Forests

L. cuneata (Dum. Cours.) G. Don. S; JB, OR, RC; NY, MD, DC
L. thunbergii (DC.) Nakai. H; not reported; not listed
Ligustrum amurense Carr. S; not reported; not listed
L. obtusifolium Siebold & Zucc. S; PP, WC, OR, RC; PA, DC
L. ovalifolium Hassk. S; RC; PA
L. sinense Lour. S; not reported; MD
L. vulgare L. S; PB, VC, BP, WR, JB; PA, MD, DC
Linaria vulgaris Miller. H; MF, PB, VC, BP, WR, JB, PP, OR; not listed
Liriope spicata Lour. H; RC; MD
Lobelia chinensis Lour. H; not reported; not listed
Lolium pratense (Huds.) S.J. Darbyshire. G; PB, VC; MD, DC
Lonicera fragrantissima Lindl. & Paxt. S; PB, JB; not listed
L. japonica Thunb. S; MF, PB, VC, BP, WR, JB, PP, WC, OR, RC; MA, NY, PA, MD, DC
L. maackii (Rupr.) Maxim. S; VC, PP, WC, RC; PA, MD, DC
L. morrowii A. Gray. S; PB, VC, WC, OR, RC; NY, PA, MD, DC
L. standishii Jacques. S; not reported; PA
L. tatarica L. S; VC, WR, OR; MA, NY, PA, MD
Lonicera × bella Zabel. S; PB; NY, PA, MD
L. xylosteum L. S; JB; MA, NY
Lotus corniculatus L. H; PB, VC, BP, WR, JB; MA, NY
Lychnis flos-cuculi L. H; not reported; not listed
Lysimachia nummularia L. H; PB, PP, RC; PA, MD, DC
Lythrum salicaria L. H; MF, PB, VC, WR, JB, PP; MA, NY, PA, MD, DC
L. virgatum L. H; not reported; not listed
Maclura pomifera (Raf.) C.K. Schneider. T; PB, VC, RC; MD
Medicago lupulina L. H; MF, PB, BP, JB, OR, RC; not listed
Melia azedarach L. T; not reported; not listed
Melilotus officinalis (L.) Pallas. H; MF, PB, VC, BP, WR, JB, OR, RC; NY, MD
Microstegium vimineum (Trin.) A. Camus. G; PP, WC, OR, RC; MA, NY, PA, MD, DC
Miscanthus sinensis Andersson. G; WC, OR, RC; PA, MD, DC
Morus alba L. T; PB, VC, BP, WR, JB, PP, WC, OR, RC; NY, PA, MD, DC
Murdannia keisak (Hassk.) Hand.-Maz. H; not reported; not listed
Muscari botryoides (L). Miller. H; PB; MD
Myriophyllum aquaticum (Vell.) Verdc. H; not reported; MD
M. spicatum L. H; not reported; NY, PA
Nepeta cataria L. H; VC, OR; MD
Ornithogalum nutans L. H; RC; PA, MD, DC
O. umbellatum L. H; PB, VC, WR, JB, PP, WC, OR, RC; PA, MD, DC
Pachysandra terminalis Siebold & Zucc. H; PB, VC, WC, OR, RC; DC
Paspalum dilatatum Poir. G; not reported; MD
Pastinaca sativa L. H; PB, JB; PA
Paulownia tomentosa (Thunb.) Steudel. T; PB, VC, PP, WC, RC; PA, MD, DC
Perilla frutescens (L.) Britton. H; OR, RC; PA, MD, DC
Phalaris arundinacea L. G; PB, PP, WC, OR, RC; MA, NY, PA, MD

(continued)

TABLE 7.2 (continued)
Presence of Invasive Plant Species from the Mid-Atlantic Alien Plant Invaders List in the Urban Park Forests

Phellodendron amurense Rupr. T; not reported; MA, NY, PA
P. japonicum Maxim. T; not reported; not listed
Phleum pratense L. G; PB, VC, OR, RC; MD
Phragmites australis (Cav.) Trin. G; PB, VC, BP, WR, JB, PP, WC, OR; NY, PA, MD, DC
Phyllostachys aurea Carr. ex A.& C. Rivière. G; not reported; PA, MD
Picea abies (L.) Karsten. T; MF, PB, VC, WC, OR; MD
Pinus sylvestris L. T; PB, BP, WR, JB, WC, OR; MA, NY, PA
P. thunbergii Franco. T; BP, WR, JB, OR, WC, OR, RC; not listed
Pistia stratiotes L. H; not reported; not listed
Plantago lanceolata L. H; MF, PB, VC, BP, WR, JB, OR, RC; MD
P. major L. H; MF, PB, VC, BP, WR, JB, RC; MD
Poa compressa L. G; PB, VC, BP, JB, RC; PA
P. trivialis L. G; PB, RC; not listed
Polygonum caespitosum Blume. H; PB, VC, JB, PP, WC, OR; MD
P. cuspidatum Sieb. & Zucc. H; MF, PB, JB, PP, WC, OR, RC; MA, NY, PA, MD, DC
P. orientale L. H; not reported; not listed
P. perfoliatium L. H; PP, WC, OR, RC; MA, NY, PA, MD, DC
P. persicaria L. H; MF, PB, BP, OR, RC; MD
P. sachalinense F. Schmidt ex Maxim. H; not reported; MD
Populus alba L. T; PB, VC, BP, WR, JB, OR; MA, PA, MD, DC
Potamogeton crispus L. H; not reported; NY
Potentilla recta L. H; MF, PB, VC, BP, WR, JB, RC; MD
Prunus avium L. T; PB, VC, PP, WC, OR, RC; MA, NY, PA, MD
P. cerasus L. S; BP; NY, MD
P. mahaleb L. T; not reported; PA
P. padus L. T; not reported; PA
P. persica (L.) Batsch. T; PB, VC, JB; MD
Pseudosasa japonica (Sieb. & Zucc. Ex Steud.) Makino ex Nakai. G; not reported; PA, MD
Pueraria lobata (Willd.) Ohwi. S; PB, RC; PA, MD, DC
Pyrus calleryana Decne. T; PB, OR; PA, MD
P. malus L. T; MF, PB, VC, WR, JB, WC, OR; PA, MD
Quercus acutissima Carruth. T; RC; MD
Ranunculus acris L. H; MF, PB, VC; MD
R. bulbosus L. H; MF, PB, VC, PP, OR, RC; MD, DC
R. ficaria L. H; VC, PP, WC, OR, RC; PA, MD, DC
Raphanus raphanistrum L. H; PB, WR; not listed
Rhamnus cathartica L. S; MF, PB, VC; NY, PA, MD
R. frangula L. S; PB, VC, WR, JB; MA, NY, PA, MD
Rhodotypos scandens (Thunb.) Makino. S; PB, VC, PP, WC, OR, RC; NY, MA
Ribes rubrum L. S; not reported; not listed
Robinia hispida L. S; not reported; PA
R. pseudoacacia L. T; MF, PB, VC, JB, PP, WC, OR, RC; NY, PA, MD
Rosa bracteata J.C. Wendl. S; not reported; not listed
R. canina L. S; JB; PA, MD
R. carolina L. S; MF, WR, OR, RC; MA, NY, PA

TABLE 7.2 (continued)
Presence of Invasive Plant Species from the Mid-Atlantic Alien Plant Invaders List in the Urban Park Forests

R. gallica L. S; not reported; not listed
R. micrantha Borrer ex Sm. S; not reported; NY
R. multiflora Thunb. S; PB, VC, WR, JB, PP, WC, OR, RC; MA, NY, PA, MD, DC
R. rugosa Thunb. S; PB, VC, BP, WR, JB; MA, NY
R. wichuraiana Crep. S; not reported; not listed
Rubus armeniacus Focke. S; not reported; not listed
R. bifrons Vest ex Tratt. S; not reported; not listed
R. illecebrosus Focke. H; not reported; MD
R. laciniatus Willd. S; PB; PA
R. phoenicolasius Maxim. S; PB, VC, PP, WC, OR, RC; MA, NY, PA, MD, DC
Rumex acetosella L. H; MF, PB, VC, WR, JB, PP, OR, RC; NY
R. crispus L. H; MF, PB, VC, BP, WR, JB, PP, OR, RC; not listed
Salix alba L. T; PB, OR; NY
S. caprea L. T; not reported; PA
S. cinerea L. S; not reported; NY, MD
S. fragilis L. T; PB, VC, JB; MA, NY, PA
S. pentandra L. S; not reported; MA, PA, MD
S. purpurea L. S; not reported; MA, PA
Salix × sepulcralis Simonk. T; not reported; NY, DC
Salsola kali L. H; PB, JB; not listed
Sapium sebiferum (L.) Roxb. T; not reported; not listed
Saponaria officinalis L. H; MF, PB, VC, BP, WR, JB; MD
Setaria faberi R. Herrm. G; VC, OR, RC; PA
Solanum dulcamara L. H; MF, PB, VC, BP, WR, JB, WC, OR, RC; MA, PA, MD
S. nigrum L. H; MF, PB, VC, WR, OR, RC; not listed
Sorghum bicolor (L.) Moench. G; not reported; PA
S. halepense (L.) Pers. G; RC; PA, MD
Spergula morisonii Boreau. H; not reported; not listed
Spiraea japonica L.f. S; not reported; PA, MD
Stellaria aquatica (L.) Scop. H; PP, OR; PA, MD
S. media (L.) Villars. H; MF, PB, VC, BP, WR, JB, PP, WC, OR, RC; PA, MD
Tanacetum vulgare L. H; MF, PB, VC, OR; MD
Taraxacum laevigatum (Willd.) DC. H; not reported; not listed
T. officinale Weber. H; MF, PB, VC, BP, WR, JB, PP, WC, OR, RC; PA, MD
Taxus cuspidata Siebold & Zucc. S; PB, JB; not listed
Tragopogon pratensis L. H; MF, PB, VC, BP, WR, JB; MD
Trapa natans L. H; not reported; MA, NY, PA, MD
Trifolium aureum Pollich. H; WR, OR; not listed
T. campestre Schreber. H; PB, BP, WR, JB, OR, RC; MD
T. dubium Sibth. H; VC, WR, JB; MD
T. pratense L. H; MF, PB, VC, WR, JB, PP, OR, RC; MD
T. repens L. H; MF, PB, VC, WR, JB, PP, OR; MD
Tussilago farfara L. H; MF, PB, VC, OR, RC; not listed
Ulmus parvifolia Jacq. T; VC, WC, RC; DC

(continued)

TABLE 7.2 (continued)
Presence of Invasive Plant Species from the Mid-Atlantic Alien Plant Invaders List in the Urban Park Forests

U. pumila L. T; PB; MA, PA, MD
Verbascum blattaria L. H; MF, PB, VC, BP, WR, JB, OR, RC; not listed
V. thapsus L. H; MF, PB, BP, WR, JB, OR, RC; PA
Veronica beccabunga L. H; not reported; not listed
V. hederifolia L. H; PB, WC, OR, RC; MD
V. serpyllifolia L. H; PB, WR, PP, RC; MD
Viburnum dilatatum Thunb. S; PB, WC, RC; DC
V. opulus L. S; PB, WR; PA
V. plicatum Thunb. S; WC, RC; DC
V. sieboldii Miq. S; PB, VC, WC; NY, PA
Vinca major L. S; not reported; MD
V. minor L. S; MF, PB, VC, WR, WC, OR, RC; NY, PA, MD, DC
Wisteria floribunda (Willd.) DC. S; VC; MD, DC
W. sinensis (Sims) Sweet. S; PB, WC, RC; MA, NY, PA, MD, DC
Youngia japonica (L.) DC. H; RC; DC

Note: Designation of species as invasive is from United States National Parks Service Alien Plant Working Group, *Mid-Atlantic list of alien plant invaders*, http://www.nps.gov/plants/alien/list/midatlantic.htm, 2007.

The urban park forests include Middlesex Fells, Boston, MA (Drayton 1993)—MF; Pelham Bay, Bronx, NY (DeCandido 2001)—PB; Van Cortlandt, Bronx, NY (Natural Resources Group 1990)—VC; Breezy Point, Brooklyn, NY (Venezia and Cook 1991)—BP; Jamaica Bay, Queens, NY (Venezia and Cook 1991)—JB; Wildlife Refuge, Queens, NY (Venezia and Cook 1991)—WR; Pennypack, Philadelphia, PA (Horwitz et al. 2004)—PP; Wissahickon Creek, Philadelphia, PA (Horwitz et al. 2004)—WC; Oregon Ridge, Baltimore, MD (Redman 1999)—OR; Rock Creek, Washington, DC (Fleming and Kanal 1995)—RC; and not reported in any of the 10 parks—not reported.

Binomials and authorities follow Bailey et al. (1976) and Kartesz (1999).

Designation of growth form is tree, T; shrub, S; herbaceous, H; and graminoid, G.

Designation of species as invasive at the state level as listed in Invasive and Noxious Weeds List (United States Department of Agriculture Natural Resources Conservation Service 2007): Massachusetts—MA, New York—NY, Pennsylvania—PA, Maryland—MD, District of Columbia—DC, and not listed in any one of the four states or District of Columbia—not listed.

species (70%) present in the parks, the average number of parks reporting species in each growth form ranged from 4.63 for herbaceous species to 3.39 for shrub species. The averages for graminoid (3.48) and shrubs (3.39) were similar, as were the averages for herbaceous (4.63) and trees (4.56). The differences in the averages follow the pattern of the presence of relatively more herbaceous and tree species in 9 or 10 parks than graminoid and shrub species (Table 7.3).

A cross tabulation of species presence or absence in the parks with species designation as an invasive in one of the four states or the District of Columbia is presented in Table 7.4. Approximately 32% of the species were designated in only one state or the District of Columbia, and just over 29% of the species were not

TABLE 7.3
Cross Tabulation of Tree, Shrub, Herbaceous, and Graminoid Species

Number of Parks	Number of Species in Groups			
	Tree	Shrub	Herbaceous	Graminoid
0	10	21	42	10
1	3	12	11	4
2	3	8	18	5
3	7	7	10	6
4	2	7	9	3
5	2	5	8	3
6	3	2	17	2
7	3	2	9	0
8	1	1	7	2
9	2	0	5	0
10	1	2	3	0
Total	37	67	139	35

Note: Species present or absent in 10 urban park forests. Designation of species as invasive is from United States National Parks Service Alien Plant Working Group, *Mid-Atlantic list of alien plant invaders*, http://www.nps.gov/plants/alien/list/midatlantic.htm, 2007.

designated in any of the states nor the District of Columbia. However, nearly 13% of the species on the Mid-Atlantic Alien Plant Invaders List were not present in any of the parks but were designated as an invasive species in at least one of the states or the District of Columbia. Only 17 species (6%) had an exact matching of presence in parks with those in the four states or with the District of Columbia designation as an invasive species (i.e., no additional or fewer parks or states, including the District of Columbia). The four species reported in 9 or 10 parks and designated as invasive in the four states and the District of Columbia were Norway maple (*Acer platanoides* L.), tree of heaven (*Ailanthus altissima* (Mill.) Swingle.), Japanese barberry (*Berberis thunbergii* DC.), and Japanese honeysuckle (*Lonicera japonica* Thunb.).

7.3.2 Jaccard Similarity Indices

Jaccard similarity measures quantify the degree of commonality of species between two sites with values ranging from 0 (no similarity) to 1 (all common). The similarity values were organized by park from northern most (Middlesex Fells) to southern most (Rock Creek) in Tables 7.4 and 7.5. The major diagonal is blank in Tables 7.4 and 7.5 because the cells would contain the comparison of an urban park forest to itself. The Jaccard values are presented in half of a matrix: tree species above the major diagonal of the matrix in Table 7.5, shrub species below the major diagonal in

TABLE 7.4
Cross Tabulation of Species

	States and District of Columbia					
Parks	**0**	**1**	**2**	**3**	**4**	**5**
0	47	21	11	2	2	0
1	6	15	6	3	0	0
2	16	9	4	5	0	0
3	4	14	4	6	1	1
4	0	8	4	4	2	3
5	1	7	4	4	2	0
6	3	6	6	4	2	3
7	1	6	4	0	1	2
8	2	4	1	2	1	1
9	1	0	2	2	1	1
10	0	0	2	1	0	3
Total	81	90	48	33	12	14

Note: Species present or absent in 10 urban park forests and also in the four states and the District of Columbia that contain the 10 urban park forests. Designation of species as invasive is from United States National Parks Service Alien Plant Working Group, *Mid-Atlantic list of alien plant invaders*, http://www.nps.gov/plants/alien/list/midatlantic.htm, 2007.

Table 7.5, herbaceous species above the major diagonal in Table 7.6, and graminoid species below the major diagonal in Table 7.6.

For tree species, the lowest Jaccard value (0.16) occurred for the comparison between Rock Creek and Jamaica Bay, and the highest value (0.64) was between the Wildlife Refuge and Breezy Point. A cluster of three values above 0.54 occurred for the Wildlife Refuge, Jamaica Bay, and Breezy Point. The second highest Jaccard value (0.61) was found in the comparison of Van Cortlandt to Pelham Bay. The one value above 0.54 that did not apply to two parks in New York City was for Van Cortlandt and Wissahickon Creek (0.56).

With respect to the shrub species, the lowest Jaccard value (0.13) occurred for two comparisons: Breezy Point and Wissahickon Creek as well as Breezy Point and Rock Creek. The first and second highest values both occurred with comparisons of Rock Creek with Wissahickon Creek (0.68) and Oregon Ridge (0.56), respectively. The next two highest values were for the pairing of Wissahickon Creek with Van Cortlandt (0.55) and Pelham Bay (0.53). One other comparison (Van Cortlandt and Pelham Bay) had a value of 0.53.

For the herbaceous species, the lowest value (0.11) was associated with the Breezy Point and Pennypack pair, and the highest value (0.68) occurred for the pairing of Jamaica Bay and the Wildlife Refuge. Van Cortlandt was one part

TABLE 7.5
Jaccard Similarity Measures between Urban Park Forests for Tree Species (Above Main Diagonal) and Shrub Species (Below Main Diagonal)

Park Codes	MF (5)	PB (22)	VC (15)	BP (8)	WR (10)	JB (9)	PP (6)	WC (13)	OR (13)	RC (13)
MF (5)		0.23	0.25	0.18	0.36	0.17	0.22	0.29	0.29	0.20
PB (28)	0.18		0.61	0.30	0.45	0.41	0.27	0.40	0.46	0.35
VC (21)	0.24	0.53		0.28	0.32	0.41	0.40	0.56	0.33	0.40
BP (5)	0.43	0.18	0.24		0.64	0.55	0.40	0.31	0.31	0.17
WR (10)	0.25	0.23	0.29	0.50		0.58	0.33	0.35	0.44	0.28
JB (12)	0.21	0.25	0.22	0.42	0.47		0.25	0.29	0.29	0.16
PP (10)	0.25	0.31	0.35	0.15	0.18	0.16		0.46	0.27	0.36
WC (21)	0.18	0.53	0.55	0.13	0.15	0.14	0.48		0.44	0.37
OR (18)	0.21	0.39	0.50	0.15	0.22	0.20	0.40	0.50		0.30
RC (21)	0.14	0.48	0.45	0.13	0.15	0.18	0.35	0.68	0.56	

Note: The total number of tree species is given in parentheses following the park codes in the first row, and the total number of shrub species is given in parentheses following the park codes in the first column.

TABLE 7.6
Jaccard Similarity Measures between Urban Park Forests for Herbaceous Species (Above Main Diagonal) and Graminoid Species (Below Main Diagonal)

Park Codes	MF (39)	PB (77)	VC (54)	BP (28)	WR (54)	JB (47)	PP (31)	WC (24)	OR (55)	RC (52)
MF (0)		0.49	0.58	0.43	0.43	0.46	0.30	0.24	0.45	0.47
PB (17)	0.00		0.60	0.33	0.44	0.49	0.33	0.25	0.57	0.48
VC (13)	0.00	0.50		0.32	0.48	0.42	0.35	0.30	0.49	0.45
BP (8)	0.00	0.32	0.24		0.44	0.53	0.11	0.12	0.32	0.33
WR (5)	0.00	0.22	0.20	0.18		0.68	0.20	0.12	0.40	0.39
JB (11)	0.00	0.47	0.26	0.58	0.23		0.24	0.15	0.40	0.38
PP (7)	0.00	0.26	0.25	0.50	0.09	0.38		0.57	0.34	0.36
WC (3)	0.00	0.00	0.07	0.00	0.00	0.00	0.11		0.32	0.33
OR (12)	0.00	0.26	0.47	0.11	0.13	0.21	0.27	0.15		0.57
RC (15)	0.00	0.45	0.47	0.21	0.18	0.37	0.29	0.13	0.50	

Note: The total number of herbaceous species is given in parentheses following the park codes in the first row, and the total number of graminoid species is given in parentheses following the park codes in the first column.

of the second and third highest pairings, which were with Pelham Bay (0.60) and Middlesex Fells (0.58), respectively. Three pairings were at 0.57: Pelham Bay and Oregon Ridge, Pennypack and Wissahickon Creek, and Oregon Ridge and Rock Creek. One other pair, Breezy Point and Jamaica Bay, was at 0.53.

Since graminoid species were not identified for Middlesex Fells, only the Jaccard similarity values could be zero for the pairings with Middlesex Fells in Table 7.6. In addition, zero values occurred for the pairings of Wissahickon Creek with Pelham Bay, Breezy Point, Wildlife Refuge, and Jamaica Bay. Only the value for the pairing of Jamaica Bay and Breezy Point (0.58) was greater than 0.50. However, three pairings had a similarity value of exactly 0.50: Van Cortlandt and Pelham Bay, Breezy Point and Pennypack, and Oregon Ridge and Rock Creek.

7.3.3 Biogeography Patterns

Biogeography patterns are presented in a branching diagram by growth form group (Figures 7.1 and 7.2). The urban park forests are presented on the x-axis of the diagram, but not in the north to south gradient. Instead, the listing of urban park forests is organized following the hierarchical cluster analysis results. The y-axis of the diagram is marked with the rescaled distance of the hierarchical cluster analysis, which is the quantification of the degree of similarity. Note that as the rescaled distance increases the degree of similarity decreases. The final connection of urban park forests in the branching diagrams shows the least similar floras. The following analysis of the diagrams will refer to the levels of connection among the urban park forests, with the first level being the first connection among pairs of parks and the second level being either the connection of another park or pair of parks to a pair in the first level. The higher order levels connect progressively more urban park forests until the final level when all are connected.

The biogeography pattern analysis for the invasive tree species is presented in Figure 7.1. Three linkages occur at the first level: the Wildlife Refuge connects to Breezy Point, Van Cortlandt joins to Pelham Bay, and Rock Creek links to Pennypack. However, the rescaled distance of the third pairing indicates a much lower degree of similarity than the first and second pairings. The second-level linkages are the addition of Jamaica Bay to the Wildlife Refuge–Breezy Point pair and Wissahickon Creek to the Van Cortlandt–Pelham Bay pair. The third level continues to build connections across cities, Oregon Ridge in Baltimore joins the two Bronx, New York parks (Pelham Bay and Van Cortlandt), and the one park in Philadelphia (Wissahickon Creek). The connection made at the fifth level is the joining of the Rock Creek and Pennypack pair to the four parks linked at the fourth level. The sixth level joins the Breezy Point, Jamaica Bay, and Wildlife Refuge group with the linkages forming the fifth level. Finally, Middlesex Fells is joined to the other nine parks in the seventh level of the diagram.

The linkages for the invasive shrub species are shown in Figure 7.1. The first-level linkages are Wissahickon Creek to Rock Creek, Pelham Bay to Van Cortlandt, and the Wildlife Refuge to Breezy Point. At the second level, Oregon Ridge joins the Wissahickon Creek–Rock Creek pair, and Jamaica Bay links to the Wildlife Refuge–Breezy Point pair. The two connections at the third level occur at very

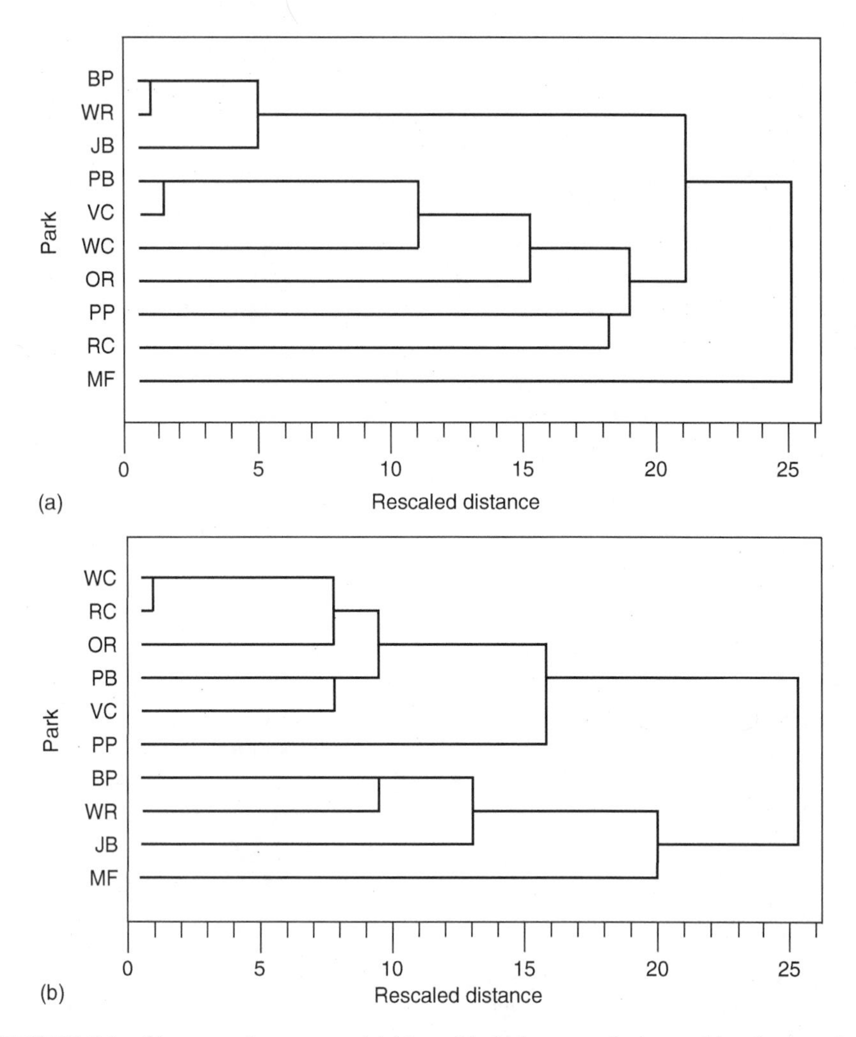

FIGURE 7.1 Biogeography pattern: (a) hierarchical cluster analysis combine for invasive tree species similarity and (b) hierarchical cluster analysis combine for invasive shrub species similarity.

different distances: the Pelham Bay–Van Cortlandt pair joins the Oregon Ridge–Rock Creek–Wissahickon Creek group before the rescaled distance value of 10, while Middlesex Fells joins the Brooklyn and Queens, New York group (Jamaica Bay–Wildlife Refuge–Breezy Point), after the rescaled distance value of 20. In the penultimate connection at level 5, Pennypack joins the group of five parks in the third level to form a combination of District of Columbia, Baltimore, Philadelphia, and Bronx, New York parks. It is not until the end, at level 6, that all five parks from New York are joined together.

The biogeography pattern analysis for the largest group of species, the invasive herbaceous species, is presented in Figure 7.2. There are four first-level linkages:

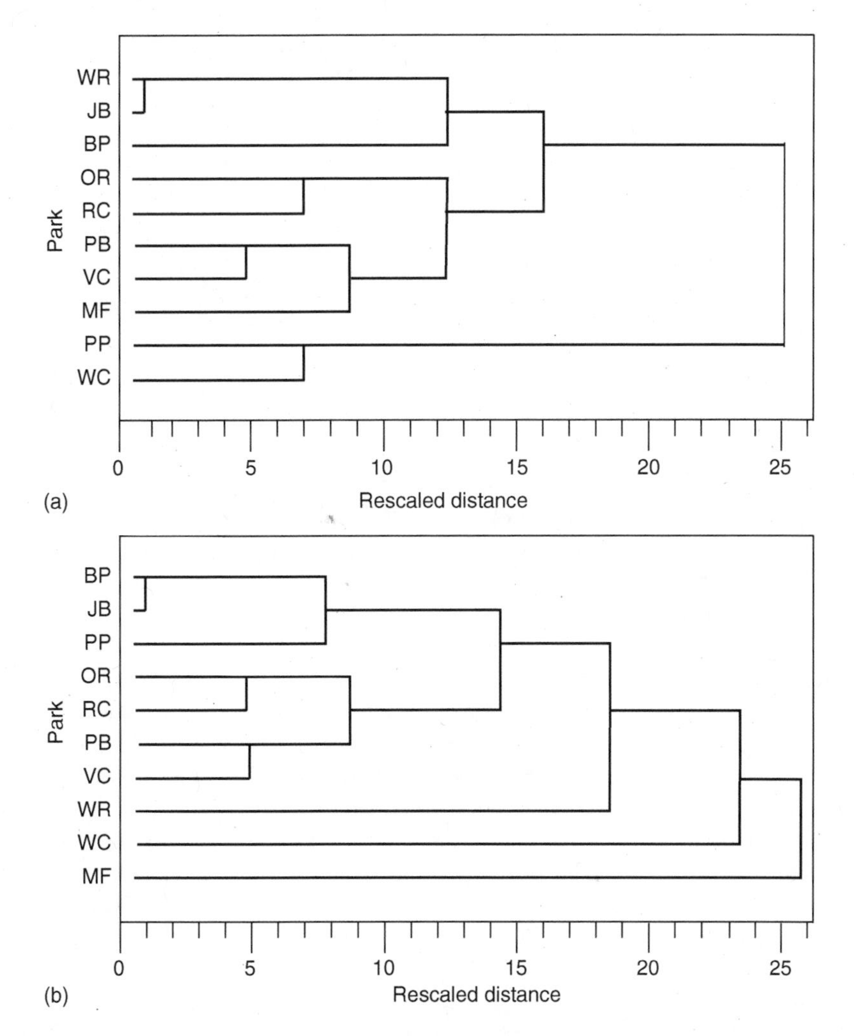

FIGURE 7.2 Biogeography pattern: (a) hierarchical cluster analysis combine for herbaceous species similarity and (b) hierarchical cluster analysis combine for invasive graminoid species similarity.

Jamaica Bay to Wildlife Refuge, Rock Creek to Oregon Ridge, Van Cortlandt to Pelham Bay, and Wissahickon Creek to Pennypack. At the second level, Middlesex Fells joins the Van Cortlandt–Pelham Bay pair and Breezy Point joins the Wildlife Refuge–Jamaica Bay pair. The third level joins the parks of Baltimore and Washington, DC, (Oregon Ridge–Rock Creek) to the Bronx and Boston parks (Pelham Bay–Van Cortlandt–Middlesex Fells). At the fourth level, the Queens and Brooklyn, New York parks (Breezy Point–Wildlife Refuge–Jamaica Bay), joins the group formed at the third level, which contains the remaining New York City parks. Finally, the Philadelphia parks (Wissahickon Creek and Pennypack) are joined to the other four cities at the fifth level.

Figure 7.2 has the diagram of linkages for the invasive graminoid species, which is the smallest group of species. At the first level, Jamaica Bay is connected to Breezy Point, Rock Creek is joined with Oregon Ridge, and Pelham Bay links with Van Cortlandt. The second-level connections include Pennypack joining the Breezy Point–Jamaica Bay pair and the link of the Bronx, New York pair (Pelham Bay–Van Cortlandt), and Baltimore and District of Columbia pair (Rock Creek–Oregon Ridge). Before any other parks are linked, the third-level connection is the joining of the two groupings at the second level. The linkages made at fourth, fifth, and sixth levels are the addition of Wildlife Refuge, Wissahickon Creek, and Middlesex Fells, respectively, to the groups at the previous levels.

7.4 DISCUSSION

7.4.1 BIOGEOGRAPHY

The species list for the 10 urban park forests (Table 7.2) clearly demonstrates that the invasive plant species identified for the mid-Atlantic region do not occur throughout the urban park forests of the region. Of great interest is the fact that with only 70% of the species on the Mid-Atlantic Alien Plant Invaders List (United States National Parks Service Alien Plant Working Group 2007) present in any of the 10 park forests, many opportunities exist for invasions. Furthermore, since only 6 species occurred in all 10 parks, the promise for more species becoming common to all 10 park forests is great. Although incomplete identification of species (such as unidentified graminoid species in Middlesex Fells) could explain the low number of common species, this argument is not supported by the distribution of unidentified species among the other growth forms. The four growth forms of invasive species not found in the park forests differed by no more than 4%. If there were a high percentage of graminoid species, for example, then the explanation of low recognition of graminoid species could be supported. Furthermore, since there were 10 invasive tree species not identified in any of the 10 urban park forests, the argument of systematic oversight is not supported.

A Jaccard value of at least 0.5 for all the growth forms only occurred for the pairing of Pelham Bay and Van Cortlandt, which are both in the Bronx. The two Bronx parks linked at the first level in the clustering for all four growth forms. The pairing of Breezy Point and Jamaica Bay (in Brooklyn, NY, and Queens, NY, respectively) and Oregon Ridge and Rock Creek (located in Baltimore, MD and Washington, DC, respectively) had only one of the four Jaccard values below 0.50. In contrast to the two Bronx park forests, there were only three first-level pairings (out of a possible eight) for the two pairs Breezy Point with Jamaica Bay and Oregon Ridge with Rock Creek. If the combination of both first and second-level pairings are considered, Breezy Point, Jamaica Bay, and Wildlife Refuge (in the adjacent counties Brooklyn, NY, and Queens, NY), join first for three of the four growth forms (not graminoid species). The two Philadelphia parks (Wissahickon Creek and Pennypack) are joined only at the first level for herbaceous species, a pairing that does not connect with the other eight parks until the final linkage (i.e., the herbaceous species in Philadelphia were differentiated from all the other cities). The biogeography pattern analysis does not indicate that geographic proximity of parks is the major predictor of species similarity among all the 10 urban park forests.

7.4.2 Historical Biogeography

An alternative explanation for the species distribution relationships among the park forests is that the history of species introductions in each urban park forest is the major predictor for presence and absence of invasive species. In the seventeenth and eighteenth centuries, European agriculture introduced invasive plant species to the United States, such as *Allium vineale* L., *Artemisia vulgaris* L., *Capsella bursa-pastoris* (L.) Medikus., *Carduus nutans* L., *Cerastium vulgatum* L., *Chelidonium majus* L., *Conium maculatum* L., *Daucus carota* L., *Elytrigia repens* (L.) Desv. ex Nevski., *Foeniculum vulgare* Mill., *N. cataria* L., *Plantago lanceolata* L., *P. major* L., *Pyrus malus* L., *Ranunculus ficaria* L., *Solanum dulcamara* L., *Stellaria media* (L.) Villars., *Tanacetum vulgare* L., *Taraxacum officinale* Weber., and *Verbascum thapsus* L. (Foy et al. 1983). The three herbaceous species (*A. vulgaris* L., *S. media* (L.) Villars., and *T. officinale* Weber.) in all 10 parks are among the aforementioned species, and the average number of parks containing the invasive herbaceous species in the earlier list is 6.65 as compared with 4.63 for all the invasive herbaceous species in the parks. Sources of nineteenth century horticultural materials were not exclusive to any particular park or city. Species planted in nineteenth century gardens (Leighton 1987) comprise 30% of the trees, 34% of the shrubs, and 15% of the herbaceous species (Leighton did not provide information concerning graminoid species) in the Mid-Atlantic Alien Plant Invaders List (United States National Parks Service Alien Plant Working Group 2007). If one counts only the species planted in nineteenth century gardens that are present in the 10 urban park forests, the average for the number of parks reporting species in each growth form was 6.18 for tree species (4.56 for all the tree species), 4.69 for herbaceous species (4.63 for all the herbaceous species), and 3.35 (3.39 for all the shrub species) for shrub species.

The longer a species is planted in a region, the more likely will it spread across a region and be found in multiple park forests. Unfortunately, the records needed to create the history of plant species introduction to a park are rare (see Loeb 1993 for Central Park, New York). However, mid-Atlantic region specific historical information on the species for sale from nurseries is available (Adams 2004). Pretwentieth century plantings comprise a third of the invasive tree, shrub, and herbaceous species in the park forests. All the tree and shrub species present in 10, 9, and 8 parks were available from nurseries before 1900. If one counts only the species available from nurseries before 1900 that are present in the 10 urban park forests, the average for the number of parks reporting species in each growth form was 5.38 for tree species (4.56 for all the tree species), 4.00 for herbaceous species (4.63 for all the herbaceous species), and 3.64 (3.39 for all the shrub species) for shrub species.

7.4.3 Monitoring Invasive Species Changes

The listing of species in Table 7.2 is the basis for monitoring changes in the establishment of invasive species in the urban park forests of the mid-Atlantic region. Two important perspectives concerning invasive species presence in the urban park forests of a region can be displayed in the form of a species homogenization chart (Figure 7.3) and a species saturation chart (Figure 7.4). The species homogenization chart (Figure 7.3) shows the opportunities for more invasive species to become established in the urban park forests as well as species expanding their ranges to include more urban park forests. Compared with the number of species in 1–6 urban park forests, the fewer species in 7–10 urban park forests indicates that the establishment of more invasive plant species across the mid-Atlantic region is an important concern to monitor. In this comparison of the 10 urban park forests for species saturation, the range is almost a threefold difference between the smallest and largest number of species. Clearly, these 10 urban park forests have not reached saturation for invasive plant species.

Attempting to predict what the species homogenization and species saturation charts will look like in the future requires an understanding of what changes will occur in several factors. First, new judgments about which species are designated as invasive in the region will cause the basis for comparisons in the future to be different. Second, current management plans (Van Cortlandt—Natural Resources Group 1990; Pelham Bay—Natural Resources Group 1988; Breezy Point, Jamaica Bay, and Wildlife Refuge—Cook and Tancredi 1990; Pennypack and Wissahickon—Fairmont Park Commission 1999; and Rock Creek—National Park Service, U.S. Department of the Interior 1996) target habitat rehabilitation for the reestablishment of native species, rather than invasive species extirpation. Therefore, if changes in management philosophy occur, there could be intentional losses of invasive species from the parks. Third, the nationalities of visitors to urban park forests are

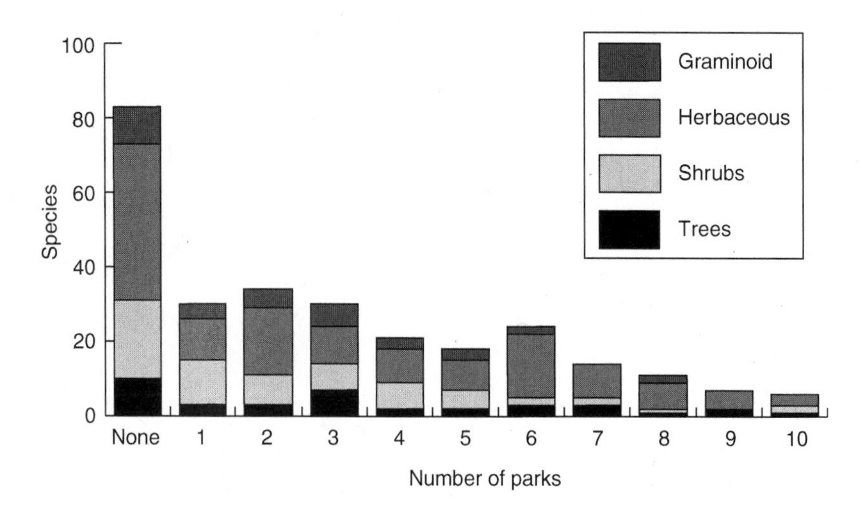

FIGURE 7.3 Species homogenization.

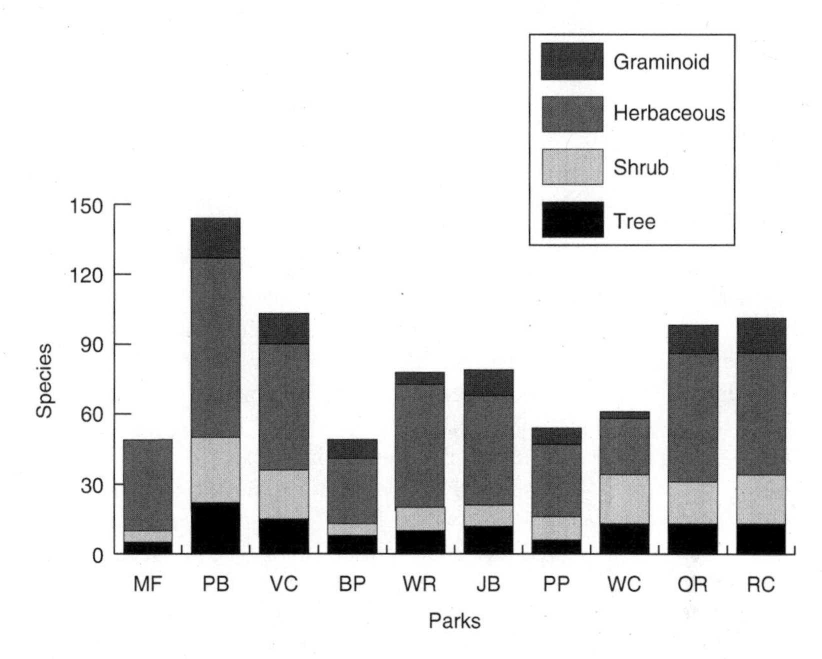

FIGURE 7.4 Species saturation.

becoming more globally diverse; this diversity may lead to the introduction of previously unidentified species from areas that are not traditionally looked to for horticultural specimens. Fourth, invasive plant species are more likely to be accepted for planting in urban park forests because these high foot-traffic environments need rapidly growing plant cover to prevent erosion in repeatedly disturbed areas. In the future, urban park forest managers will gain important information concerning species homogenization and saturation by monitoring the flora, especially for invasive species.

7.5 CONCLUSIONS

Urban park forests are disturbance communities well suited to host invasive plant species and are well located to serve as centers for the spread of invasive plant species. The presence and absence of invasive plant species in 10 urban park forests of 5 cities in the mid-Atlantic region of the United States indicate that homogenization in the urban park forests is not yet a concern: only 70% of the invasive species of the region appear in at least 1 of the 10 urban park forests and 2% of the invasive plants are common to all 10. The individual urban park forests are not close to attaining invasive species saturation as evidenced by the fact that only one park forest contains more than 50% of the invasive plant species in the region. The patterns of linkages from the hierarchical cluster analysis of Jaccard similarity values for all 10 park forests differed for trees, shrub, herbaceous, and graminoid species. Only two parks, Van Cortlandt and Pelham Bay, in Bronx County, New York City, formed a pair for all four growth forms. The biogeography pattern analysis of species similarity among the parks indicates there is no common pattern related to the distance separating the parks for the four growth forms.

The history of invasive plant species introductions in parks is suggested as an alternative explanation for the patterns in biogeography. The species common to all 10 park forests were available from nurseries in the region or were weeds of cultivation before 1800. Species available from horticultural nurseries in the region before 1900 were on average present in more of the park forests than all of the species. Information on garden plantings before 1900 revealed that some common invasive species may have been planted in the parks. The paucity of specific plantings information for the 10 urban park forests limited further historical biogeography pattern analysis. However, future research in other urban park forests should incorporate the available historical planting information to better understand changes in the biogeography of invasive species.

Since a large number of invasive plant species in the mid-Atlantic region are not represented in any of the 10 urban park forests examined in this research, monitoring changes in the flora for invasive plants is essential to the management of urban park forests. Additionally, the invasive plants present in the parks were significantly different among the 10 parks, which indicates further opportunities for the expansion of invasive plant species populations even if no new species became established in the parks. Urban park forest managers can employ the analyses of homogenization, saturation, and biogeography to monitor future changes in the distribution of invasive plant species.

ACKNOWLEDGMENTS

The listing of species for Pennypack Park and Wissahickon Creek Park was kindly provided by Richard Horwitz, Ann Rhoads, and Ernie Schuyler, with funding from the Natural Lands Restoration and Environmental Education Program of the Fairmont Park Commission under a grant from the William Penn Foundation. David Kunstler of the New York City Department of Parks and Recreation, Pelham Bay and Van Cortlandt Parks Administration provided the flora in Van Cortlandt Park. The efforts of Fei Guo and her advisor, Durland Shumay of the Statistical Consulting Center, Department of Statistics, The Pennsylvania State University, in performing the Jaccard similarity index calculations and hierarchical cluster analysis are appreciated. Thanks are extended to Janice Norris, Head Librarian, DuBois Campus, The Pennsylvania State University, for the information on latitude and longitude. I appreciate the reviews of the manuscript provided by my DuBois Campus (The Pennsylvania State University) colleagues G. Andrew Bartholomay and Susan Waitkus. The funding provided by the Robert and Joyce Umbaugh Faculty Development Fund to support this publication is deeply appreciated.

REFERENCES

Adams, D.W., *Restoring American Gardens*, Timber Press, Portland, OR, 419, 2004.

Bailey, L.H., Bailey, E.Z., and Staff of Liberty Hyde Bailey Hortorium, *Hortus Third*, Macmillan, New York, 1290, 1976.

Braun, E.L., *Deciduous Forests of Eastern North America*, Hafner, New York, 596, 1950.

Cook, R. and Tancredi, J., Management strategies for increasing habitat and species diversity in an urban national park, in *Environmental Restoration: Science and Strategies for Restoring the Earth*, Berger, E., Ed., Island Press, Washington, DC, 171, 1990.

DeCandido, R.V., Recent changes in plant species diversity in Pelham Bay Park, Bronx County, New York City, 1947–1998, Ph.D. Dissertation, City University of New York, New York, 2001.

Dehnen-Schmutz, K., Medieval castles as centers of spread of non-native plant species, in *Plant Invasions: Ecological Mechanisms and Human Responses*, Starfinger, U., Edwards, K., Kowarik, I., and Williamson, M., Eds., Backhuys, Leiden, 307, 1998.

Dickson, J.H., Interpretations of the patterns: City lovers, city haters and neutrals, in *The Changing Flora of Glasgow Urban and Rural Plants Through the Centuries*, Dickson, J.H., Macpherson, P., and Watson, K.J., Eds., Edinburgh University Press, Edinburgh, 290, 2000.

Drayton, B.E., Changes in the flora of the Middlesex Fells, 1894–1993, M.A. Thesis, Boston University, Boston, MA, 1993.

Drayton, B.E. and Primack, R.B., Plant species lost in an isolated conservation area in metropolitan Boston from 1894–1993, *Biol. Conserv.* 107, 30, 1996.

Fairmont Park Commission, Fairmount Park system: Natural lands restoration master plan, Vol. 1, Fairmont Park Commission, Philadelphia, PA, 1999 (unpublished report).

Fleming, P. and Kanal, R., Annotated checklist of vascular plants of Rock Creek Park, National Park Service, Washington, DC, *Castanea* 60, 283, 1995.

Foy, C.L., Forney, D.R., and Cooley, W.E., History of weed introductions, in *Exotic Plant Pests and North American Agriculture*, Academic Press, New York, Chap. 4, 1983.

Grapow, L., Di Marzio, C.P., and Blasi, C., The importance of alien and native species in the urban flora of Rome (Italy), in *Plant Invasions: Species Ecology and Ecosystem Management*, Brundu, G., Brock, J., Camarada, I., Child, L., and Wade, M., Eds., Backhuys, Leiden, 209, 2001.

Haruska, K., A comparative analysis of the urban flora of Italy, *Braun-Blanquetia*, 3, 45, 1989.

Horwitz, R.J., Rhoads, A., and Schuyler, A.E., Species list, Natural Lands Restoration and Environmental Education Program, Fairmont Park Commission, Philadelphia, PA, 2004 (unpublished report).

Kartesz, J.T., *A Synonymized Checklist of the Vascular Flora of the United States, Canada, and Greenland*, Timber Press, Portland, OR, 342, 1999.

Kowarik, I., Some responses of flora and vegetation to urbanization in central Europe, in *Urban Ecology*, Sukopp, H., Ed., SPB Academic, Amsterdam, 45, 1990.

Kowarik, I., On the role of alien species in urban flora and vegetation, in *Plant Invasions: General Aspects and Special Problems*, Pysek, P., Prach, K., Rejmanek, M., and Wade, P.M., Eds., SPB Academic, Amsterdam, 85, 1995.

Kuhn, I., Brandl, R., and Kotz, S., The flora of German cities is naturally species rich, *Evol. Ecol. Res.* 6, 749, 2004.

Leighton, A., *American Gardens of the Nineteenth Century*, University of Massachusetts Press, Amherst, MA, 395, 1987.

Loeb, R.E., Long-term arboreal change in a landscaped forest, Central Park, New York, *J. Arbor.* 19, 238, 1993.

Loeb, R.E., Evidence of prehistoric corn (*Zea mays*) and hickory (*Carya* spp.) planting in New York City: Vegetation history of Hunter Island, Bronx County, New York, *J. Torrey Bot. Soc.* 125, 74, 1998.

McKinney, M.L., Urbanization as a major cause of biotic homogenization, *Biol. Conserv.* 127, 247, 2006.

McKinney, M.L. and Lockwood, J.L., Community composition and homogenization: Evenness and abundance of native and exotic plant species, in *Species Invasions: Insights into Ecology, Evolution, and Biogeography*, Sax, S.V., Stachowicz, J.J., and Gaines, S.D., Eds., Sinauer, Sunderland, MA, 365, 2005.

National Academy Press, *Predicting Invasions of Nonindigenous Plants and Plant Pests*, National Academy Press, Washington, DC, 194, 2002.

National Park Service, U.S. Department of the Interior, Resource management plan, Rock Creek Park, Washington, DC, 466, 1996 (unpublished report).

Natural Resources Group, Natural areas management plan: Pelham Bay Park, Bronx, City of New York, Parks and Recreation, New York, 40, 1988 (unpublished report).

Natural Resources Group, Natural areas management plan: Van Cortlandt Park, Bronx, City of New York, Parks and Recreation, New York, 47, 1990 (unpublished report).

Pielou, E.C., *The Interpretation of Ecological Data: A Primer on Classification and Ordination*, Wiley, New York, 263, 1984.

Profous, G.V. and Loeb, R.E., Vegetation and plant communities of Van Cortlandt Park, Bronx County, New York, *Bull. Torrey Bot. Club*, 111, 80, 1984.

Pyšek, P., Alien and native species in central European urban floras: A quantitative comparison, *J. Biogeogr.*, 25, 155, 1998.

Redman, D.E., An annotated checklist of the vascular flora of Oregon Ridge Park, Baltimore County, Maryland, *Maryland Nat.*, 43, 1, 1999.

Robinson, G.R., Yurlina, M.E., and Handel, S.N., A century of change in the Staten Island flora: Ecological correlates of species losses and invasions, *Bull. Torr. Bot. Club*, 121, 119, 1994.

Sukopp, H., Urban ecology and its application in Europe, in *Urban Ecology*, Sukopp, H., Ed., SPB Academic, Amsterdam, 1, 1990.

Tancredi, J.T., Coastal zone management practices at an urban national park, *Environ. Manage.*, 7, 143, 1983.

United States Department of Agriculture Natural Resources Conservation Service, Invasive and noxious weeds list. Retrieved August 25, 2007 from http://plants.usda.gov/index.html.

United States National Parks Service Alien Plant Working Group, Mid-Atlantic list of alien plant invaders, 2007. Retrieved August 25, 2007 from http://www.nps.gov/plants/alien/list/ midatlantic.htm.

Venezia, K. and Cook, R.P., Flora of Gateway Natural Recreation Area, United States National Parks Service, Gateway National Recreation Area, Division of Natural Resources and Compliance, New York, 40, 1991 (unpublished report).

Section II

Ecological Impacts

8 Alien Invasive Plant Species and Their Effects on Hill Forest Ecosystems of Bangladesh

M.K. Hossain

CONTENTS

8.1 INTRODUCTION

Invasive plants and animals are a major threat to natural ecosystems and their species, second only to direct destruction of habitats by humans. The impacts of alien invasive species are immense, insidious, and usually irreversible. Alien invasive species are found in all taxonomic groups, including viruses, fungi, algae, mosses, ferns, higher plants, invertebrates, fishes, amphibians, reptiles, birds, and mammals. Extinctions of native species by aliens are enormous, and the ecological cost is the irretrievable loss of native species and ecosystems. The direct economic costs of alien invasive species are also remarkable. This problem is growing in severity and geographic extent as the volumes of international trade and travel increase. Communities all over the world are taking appropriate initiatives that contribute to better management practices and a reduced incidence of biological invasion.

Unfortunately, in Bangladesh the introduction of alien invasive species of flora and fauna was deliberate, primarily to increase productivity in order to support the needs of a huge population. Various types of economic plants were introduced at different times, of which there are no records (Ali 1991). The deliberate preferences of fast-growing, high-yielding cultivars eroded some of the native species and their genetic resources abruptly. Furthermore, we have very scarce information about the alien species in Bangladesh and their impacts on the ecosystem and the native species (Biswas et al. 2007; Barua et al. 2001; Hossain and Pasha 2004). However, this chapter describes some important invasive alien species that are so far known to occur in Bangladesh, and their impacts on the hill forest ecosystems. More than 300 exotic species are assumed to be either naturalized or cultivated as economic crops in Bangladesh (Hossain and Pasha 2004). Of them, the herbaceous plants and lianas are the most prevalent exotics, followed by trees and shrubs.

8.2 HILL FOREST ECOSYSTEMS OF BANGLADESH

8.2.1 Area, Coverage, and Use

The total area of forestland in Bangladesh is 2.46 million ha (Table 8.1). Of these, the tropical evergreen and semievergreen hill forests cover 0.67 million ha and are located in the districts of Sylhet, Cox's Bazar, Chittagong, and Chittagong Hill Tracts. Geographically, this region is located at the transition of Indo-Gangetic and Indo-Malayan subregions between the Himalayas and the Bay of Bengal, and is exceptionally rich in biodiversity. There is another 0.73 million ha of unclassified State Forestland virtually devoid of any potential tree cover. Both native and exotic

TABLE 8.1
Forest Types of Bangladesh with Location and Area

Forest Type	Location	Area (million ha)	Remarks
Tropical moist hill forest	Eastern part	0.67	Under the control of Forest Department
Unclassified state forest	Hill tracts division	0.73	Under the control of district administration
Tropical moist deciduous forest	Central and Northwest	0.12	Indigenous Sal and exotic short-rotation plantations
Mangrove forest	Southwest	0.57	World-known Sundarban mangrove forest
Coastal forest	All along the coast	0.10	Mangrove plantations
Village forest	All over the country	0.27	Homestead forests
Total forest		2.46	

grasses and weeds gradually invade these barren lands. The hilly areas of this forestland have also been in a gradual process of conversion into a production plantation forest through felling and artificial regeneration since 1871 (Hossain 1998). The decision to implement large-scale conversion of natural hill forests that have high biological diversity is based on the rationale that plantations would be well protected and would produce better and higher yields (ADB 1993). The objectives of the plantation programs for both native and exotic species are to increase forest resources so as to reduce the supply and demand gap, to improve the socioeconomic condition of the people, and overall, to improve the environment (Hossain 1998).

8.2.2 Natural Hill Forests and Species Diversity

The natural hill forests of Bangladesh are characterized by a large diversity of plant species. An estimated 5700 species of angiosperms alone are present in the forests of Bangladesh (including 68 woody legumes, 130 fiber-yielding plants, 500 medicinal plants, and 29 orchids). Of these, some 2260 species are reported from the Chittagong hilly regions, which fall between two major floristic regions of Asia (ADB 1993). It is also believed that Bangladesh has a rich biological heritage of flowering plants, mammals, birds, reptiles, amphibians, fishes, etc. But, there is no systematic study on the total biodiversity of these forest areas. Only some sporadic information that is available on the species diversity of these areas indicates that the hill forest areas are rich in native plant species (Khan and Afza 1968; Alam 1988; Khan et al. 1994; Nath et al. 1998; Hossain et al. 1997).

8.2.3 Extent of Problems in Hill Forests

Besides the denuded hills of unclassified State Forests, some reserve forests of hilly areas are denuded because of unauthorized shifting cultivation or jhuming practices, clearing of land for habitation, agriculture, illegal felling, and encroachments. The denuded hills are almost barren, practically devoid of any tree cover but full of aggressive weeds and climbers of both exotic and indigenous species. Some of the exotic species are growing so vigorously that the plantation programs are not successful due to the invasion of alien species into the desired plantations. Management cost is high for controlling the aggressive weeds and climbers, though some of the plantation tree species contribute significantly to the reforestation programs.

8.3 PLANTATION FORESTRY AND THE INTRODUCTION OF EXOTICS

In Bangladesh, plantations have some advantages over natural forests from the management and economic points of view, including concentrated production and the ability to choose species with desirable characteristics. Because of the acute shortage of timber and fuel wood in Bangladesh, a priority program of introducing fast-growing tree species was undertaken by the Forest Department. In order to increase productivity to fulfill the needs of 135 million people in an area of only 147,000 km^2, introduction of flora and fauna was somewhat deliberate in

Bangladesh. In plant introduction programs of recent years, some promising exotic species have been recommended for the large-scale afforestation and reforestation programs. However, the weediness hazards of some aliens, the management and policy issues presented by these invasives, and their impact on native biodiversity have become a vital concern.

Plant introduction is a very old practice and over a long period, plants of various types of economic importance have been introduced to Bangladesh (Islam 1991). Migration or introduction of plants from one place to another may be either natural or planned. Bangladesh, like many other countries, has a long history of plant introduction from different countries or geographic regions. Most of the plants were brought by settlers, invaders, seamen, and traders. In Bangladesh, there are no detailed records of exotic plants, except the common plants and a few cultivated ones. More than 300 exotic species are presumed to be either naturalized or cultivated as economic crops. Many of the exotic plants are considered to be beneficial economically, and some are vital as cash crops. However, a good number of exotic plants are weedy in nature. Most of them were first introduced as garden or ornamental plants and later on aggressively established elsewhere. Some of them are so well established that they are now dominant plants and have become noxious weeds of forests and wastelands, e.g., *Eupatorium odoratum, Lantana camara*, and *Mikania cordata*.

In the nineteenth century, the British contributed to the introduction of some economically important forest plants from almost all continents. In addition, to increase commercial production in agriculture and forestry, forest and agriculture managers introduced exotic species. Today, most of the tree species introduced are for commercial objectives. Previously introduced forest species include *Tectona grandis, Paraserianthes falcataria, Xylia kerrii*, and *Swietenia macrophylla*. In the twentieth century, this trend continued, and some Australian species (*Eucalyptus camaldulensis, Acacia mangium, A. auriculiformis*) are receiving preference in the plantation programs. *Leucaena leucocephala* (Tropical America) are also found all over the country, and pines (*Pinus oocarpa* and *P. caribaea*) are also being planted in the hilly areas. Of these, the *A. auriculiformis* is dominating in all the plantation programs and is growing well in all sorts of degraded lands.

8.4 BENEFITS OF THE INTRODUCTION OF EXOTICS

Successful plantation technologies have been developed in many countries (Evans 1992; Kanowski and Savill 1992). Light-demanding, colonizing exotic species have been the most successful in monocultures under plantation management (Hughes 1994). Tropical and subtropical plantation forestry has focused on a small number of fast-growing, colonizing species such as *Acacia, Eucalyptus, Gmelina, Pinus, Populus*, and *Tectona* (Evans 1992). These species have the ability to capture the site rapidly and tolerate harsh soil and climatic conditions and damage from animals, humans, fire, etc. These characteristics of exotics are prerequisites for success on what are often highly degraded sites. Higher yield advantages of exotics over indigenous species have been attributed to their greater tolerance of degraded sites and their escape from specialized pests and diseases. Thus, the diminishing natural forest resources are compensated for by rapid expansion of planted exotic trees worldwide (Davidson 1993). Successful exotics are planted in some other countries also, such as spruce in Europe, radiata pine in Australia and New Zealand, ipil-ipil in Philippines and Australia, and eucalypts in India, China, and Brazil (Evans 1992). *Lantana camara*, a native of South America, introduced as an ornamental plant in India, has become a serious invasive plant (Singh et al. 2003).

8.5 PERFORMANCE OF PLANTATIONS IN BANGLADESH

In Bangladesh, there has been a net loss of forest cover over the years, and the net yield per hectare of original planted area has dropped significantly (ADB 1992). The mean annual increment of present planting stock is extremely low by regional and international standards. There are also reports that 20%–30% of all plantations established during the last 30 years have been destroyed: of those surviving, the stock is much lower than the expected standards (ADB 1992). The plantations appear likely to produce well below the capacity of the site due to lack of proper maintenance. All this has contributed to very low growth of $2.5\,m^3\,ha^{-1}\,year^{-1}$ in areas where $7.5\,m^3\,ha^{-1}\,year^{-1}$ was the original norm (ADB 1992).

Teak, an economically important timber species, initially was found promising in the hill forest areas. However, subsequent criticisms arose stating that it depletes soil fertility, prevents undergrowth vegetation, and its productivity depends on specific suitable site conditions. Although the expected productivity for teak was more than $7–8\,m^3\,ha^{-1}\,year^{-1}$, in practice productivity is less than $2\,m^3\,ha^{-1}\,year^{-1}$ in a wide range of plantations (ADB 1992). However, *E. camaldulensis* experimental plantations showed excellent yield to $69\,m^3\,ha^{-1}\,year^{-1}$ in a closely spaced plantation (Davidson and Das 1985), but in hilly areas the production reduced to as low as $14\,m^3\,ha^{-1}\,year^{-1}$. The growth behaviors of some plantations in the hilly area were erratic and poor throughout the rotation cycle, which may be due to poor selection of seeds and site (Amin et al. 1995). Similarly, *A. mangium* suffers from heart rot disease, and because of this, the species is no longer used in the new plantation programs.

The remaining most commonly used species is *A. auriculiformis*, which is widely planted and has shown success in degraded sites (Ara et al. 1989; Hossain et al. 1997). Hossain et al. (1998) report the result of a provenance trial that leads to trees with single stems and good clear bole from a provenance trial in the Chittagong University area. However, improvement of the clean cylindrical bole and other management practices will need immediate attention for implementation of large-scale plantation programs in the selected provenance. Some *Acacia* plantations showed better survival and growth in different areas of the country, and the yield is $15–20\,m^3\,ha^{-1}\,year^{-1}$ at 10–12 years rotation, but some other successful species, viz., pine, *Gliricidia*, and ipil-ipil, are restricted to some localized plantations.

8.6 ALIEN SPECIES IN NATURAL ECOSYSTEMS

Exotic species growing and gradually becoming dominant in the natural hill ecosystems, crop fields, forests, fallow land, and marginal lands are categorized as tree, shrub, herb, and lianas (Hossain and Pasha 2004). Some of the species (Table 8.2)

TABLE 8.2
Alien Exotics in Bangladesh That Have a Detrimental Impact on Natural Hill Forest Ecosystems of Bangladesh

Species	Family	Origin	Status
Tree species			
A. auriculiformis	Mimosaceae	Australia	Extensively used in plantation and aggressively occupying the natural ecosystems
A. mangium	Mimosaceae	Australia/PNG	Ban for further plantings because of heart rot disease
E. camaldulensis	Myrtaceae	Australia	Introduced but now ban on further plantings because of its controversial impact on environment
Leucaena leucocephala	Mimosaceae	Central America	Cultivated, wild in coastal areas; suppressed the regeneration of other species
Weedy vegetation			
Acanthospermum hispidum	Asteraceae	South America	Common weed of cultivated fields
Cassia occidentalis	Caesalpiniaceae	Tropical America	Common weed of wasteland and roadside
Cestrum diurnum	Solanaceae	Tropical America	Weed of roadside and rail line
Lantana camara	Verbanaceae	Tropical America	Common weed of hill ecosystems; prevents regeneration of native species
Ageratum conyzoides	Asteraceae	South America	Common weed of waste and cultivated field; aeroallergic pollen species
Alternanthera flocoidea	Amaranthaceae	Brazil	Common weed of cultivated and wasteland
Atylosia scarabaeoides	Fabaceae	Australia	Common weed of wasteland
Chromolaena odoratum (Syn Eupatorium odoratum)	Asteraceae	Central America/ South America	Common weed of wasteland; suppressed the regeneration of other species in plantation programs
Commelina oblique	Commelinaceae	Java	Frequent weed in wasteland
Convolvulus arvensis	Convolvulaceae	Europe	Frequent weed of waste place
Croton bonplandianum	Euphorbiaceae	South America	Abundant weed of waste and cultivated land
Eichhornia crassipes	Pontederiaceae	Tropical America	Abundant aquatic weed; aggressive growth inhibits other aquatic flora
Evolvulus nummularius	Convolvulaceae	West Indies	Common weed in cultivated and open fields
Hyptis suaveolens	Lamiaceae	Tropical America	Common weed of hilly regions
Ipomoea carnea	Convolvulaceae	America	Common weed of all habitats
Ludwigia adscendens	Onagraceae	Central America	Common weeds in aquatic and marshy habitats
Mikania cordata	Asteraceae	Tropical America	Abundant weed of forest and wasteland; engulfs other economic crops by its aggressive growth
Mimosa pudica	Mimosaceae	South America	Common weed of cultivated and wasteland

have luxuriant growth and suppress the growth of other native species. This results in a loss of native floral diversity of the country. The threat posed to natural habitats by these alien invasive plants is becoming a major concern among the conservationists, ecologists, foresters, policy makers, and scientists, especially for the monoplantations of few exotic species. Although the undisturbed natural forests are resistant to such invasion, the degraded and secondary forest areas and wastelands are aggressively invaded by some of these exotic species. *Ageratum conyzoides* and *Lantana camara*, both alien invasive species of the Kangra Valley in western Indian Himalayas (Dogra et al. 2004), have reached the Eastern Himalayas, including Bangladesh.

8.7 RISKS OF THE INTRODUCTION OF ALIEN INVASIVE SPECIES

There is an increasing concern among foresters, ecologists, botanists, conservationists, and policy makers about the threat of uncontrolled introduction of aggressive tree species in the plantation programs. Invasion by exotics may cause major losses of biodiversity and species extinction due to either direct replacement by exotics or indirect effects on the ecosystem. Concern also exists about the degradation of the environment, for example, the controversial effect of eucalypts on environment. *Parthenium hysterophorus*, a serious alien invasive in India (Kohli et al. 2006), is a serious threat to Bangladesh. Similarly, *A. conyzoides*, a serious weed of agroecosystems in Indian hill tracts (Kohli et al. 2004), has already become a serious weed of Chittagong or even to the whole of Bangladesh. There are also risks of the decline of growth and yield in second and successive rotations, or the infestation of pests and diseases, e.g., Psyllids in *Leucaena leucocephala*, shoot borers in *Swietenia*, and leaf defoliators in *T. grandis*. We have to manage the alien invasive species for a number of reasons:

- Exotics disrupt the ecosystem, reduce native biodiversity, and jeopardize endangered plants and animals.
- They degrade habitats.
- Exotics are known to hybridize with native species, thus altering native genetic diversity and integrity.
- Their initial increased productivity may decrease in successive rotations, e.g., *Eucalyptus camaldulensis* in Bangladesh.
- Exotics may transmit diseases to native species, which may be disastrous to the ecosystem.

8.8 STRATEGIES AND ACTION PLANS FOR CONTROLLING FURTHER SPREAD OF EXOTICS

Invasive species are becoming a threat to the native flora and fauna of Bangladesh. Unfortunately, legislation and quarantine guidelines of the country are not strong enough to prevent introductions or to manage alien invasive species (Ameen 1999). People can easily bring seeds or other planting materials into Bangladesh from abroad. Similarly, some species are coming from neighboring countries along with commodities, containers, animals, or food grains. Again, people are not sufficiently aware about potential harmful effects of invasive species in ecosystems.

The issue is coming to the current National Biodiversity Strategy Action Plan. To prevent the adverse impacts of invasive plant species to the natural ecosystem, the possible recommendations include the following:

- Enhancing awareness among planters, growers, and the public about invasive species
- Developing a database on existing invasive species
- Developing environmentally sound eradication methods
- Requiring extreme care in the selection of species to be introduced in order to minimize impacts on native species
- Initiating attempts toward the restoration of indigenous flora and fauna to reduce native biodiversity loss
- Studying the autecology of the aliens (seed dispersal, reproductive ecology, and factors limiting its distribution and abundance in the natural habitat)
- Developing general screening tools to reduce future invasive plant introductions
- Strengthening the necessary quarantines, legislation, and regulations on introduction and spread of the invasive plants within the country (MoEF 2004)

8.9 FUTURE PROSPECTS

In Bangladesh, plantation programs are increasing day by day, and exotic species are receiving preferences over indigenous ones (ADB 1993). With deforestation, many degraded sites are available for plantation, though very little is known about how to manage such sites economically. Conversion of healthy prospective natural forests to plantations of alien species must be strongly discouraged. Productive plantations on previously deforested or nonforested sites may offer important options for conserving natural forests and arresting further deforestation activities. Another option of mixed species plantations of both indigenous and exotic species may also become an effective mechanism for sustainable resource management. Through the increase of productivity, pressures can be reduced on remaining natural forests. However, the emphasis must focus on existing plantations, new plantations on degraded or denuded land, agroforestry, and on community forestry programs rather than on natural forest.

The wastelands and denuded hillocks adjacent to human habitation, which are either covered with sungrass or overgrazed by leaving the sites with a thin vegetation cover and very little water retention capacity, may be reforested with noninvasive exotics like *Eucalyptus* species. However, as the techniques of successful teak, eucalypts, and *acacia* plantations became standard, environmentalists became conscious of the danger of growing these species in plantations. Considering the demands of the people, while artificial plantation programs may be beneficial, major questions still persist including the correct choice of species for the site, whether plantings are mixed or in pure stands, and the interactions of species and site, especially where exotic species are used.

REFERENCES

ADB (Asian Development Bank), *Forest Institutions, Forestry Master Plan*, ADB (TA No. 1355-Ban), UNDP/FAO BGD 88/025, Ministry of Environment and Forest, Government of Bangladesh, 191, 1992.

ADB (Asian Development Bank), *Forestry Master Plan: Main Plan-93/2012*, Vol. 1, Ministry of Environment and Forest, Government of Bangladesh, 162, 1993.

Alam, M.K., Annoted checklist of the woody flora of Sylhet forests, Bulletin 5, Plant Taxonomy Series, Bangladesh Forest Research Institute, Chittagong, 153, 1988.

Ali, S.B., Plant introduction—A continuous process, in *Two Centuries of Plant Studies in Bangladesh and Adjacent Regions*, Nurul Islam, A.K.M., Ed., Asiatic Society of Bangladesh, Dhaka, 195, 1991.

Ameen, M., Development of guiding principles for the prevention of impacts of alien species, Paper presented at a consultative workshop in advance of the fourth meeting of SBSTTA to the CBD, IUCN Bangladesh, Dhaka, May 25, 1999.

Amin, S.M.R., Ali, M.O., and Fattah, M.I.M., Eds., Eucalypts in Bangladesh, in *Proceedings of Seminar at Bangladesh Agricultural Research Council*, Dhaka, April 6, 1994, 73, 1995.

Ara, S., Gafur, M.A., and Islam, K.R., Growth and biomass production performance of *Acacia auriculiformis* and *Eucalyptus camaldulensis* reforested in the denuded hilly lands, *Bangladesh J. Bot.*, 18, 2, 187, 1989.

Barua, S.P., Khan, M.M.H., and Reza, A.H.M.A., The status of alien invasive species in Bangladesh and their impact on the ecosystem, in *Report of Workshop on Alien Invasive Species, GBF-SSEA, Colombo*, Balakrishna, P., Ed., IUCN Regional Biodiversity Programme, Asia, Colombo, Sri Lanka, 1, 2001.

Biswas, S.R., Choudhury, J.K., Nishat, A., and Rahman, M.M., Do invasive plants threaten the Sundarbans mangrove forest of Bangladesh? *For. Ecol. Manage.*, 245, 1, 2007.

Davidson, J., Ecological aspects of Eucalyptus plantations, in *Proceedings of the Regional Expert Consultation on Eucalyptus*, RAP publication, 1996/44, Vol. 1, 35–72, October 4–8, FAO Bangkok, Thailand, 1993.

Dogra, K.S., Kohli, R.K., Batish, D., and Singh, H.P., Status of exotic invasive plants in Kangra district of Himachal Pradesh (India), *Bull. Environ. Sci.*, I, II, 13, 2004.

Evans, J., *Plantation Forestry in Tropics*, 2nd ed., Clarendon, Oxford, 403, 1992.

Hossain, M.K., Role of plantation forestry in the rehabilitation of degraded and secondary hill forests of Bangladesh, in *Proceedings of the IUFRO Inter-Divisional Seoul Conference on Forest Ecosystem and Land use in Mountain Areas*, October 12–17, 1998, Seoul, 243, 1998.

Hossain, M.K. and Pasha, M.K., An account of the exotic flora of Bangladesh, *J. For. Environ.*, 2, 99, 2004.

Hossain, M.K., Islam, S.A., Zashimuddin, M., Tarafdar, M.A., and Islam, Q.N., Growth and biomass production of some *Acacia* and *Eucalyptus* species in degenerated sal forest areas of Bangladesh, *Indian For.*, 123, 3, 211, 1997.

Hossain, M.K., Hossain, M.S., and Aryal, U.K., Initial growth performance of eight provenances of *Acacia auriculiformis* at Chittagong University Campus, Bangladesh, *Indian For.*, 124, 4, 256, 1998.

Hughes, C.E., Risks of species introductions in tropical forestry, *Commonw. For. Rev.*, 73, 4, 243, 1994.

Islam, A.K.M.N., *Two Centuries of Plant Studies in Bangladesh and Adjacent Regions*, Asiatic Society of Bangladesh, Dhaka, 299, 1991.

Kanowski, P.J. and Savill, P.S., Forest plantations: Towards sustainable practice, in *Plantation Politics: Forest Plantations in Development*, Sargent, C. and Bass, S., Eds., Earthscan, London, Chapter 6, 121, 1992.

Khan, M.S. and Afza, S.K., A taxonomic report on the angiospermic flora of Teknaf and St. Martin's Island, *Dhaka Univ. Stud.*, 16, B, 25, 1968.

Khan, M.S., Rahman, M.M., Huq, A.M., and Mia, M.M.K., Assessment of biodiversity of Teknaf game reserve in Bangladesh focusing on economically and ecologically important plant species, *Bangladesh J. Plant Taxonomy*, 1, 1, 21, 1994.

Kohli, R.K., Batish, D.R., Singh, H.P., and Dogra, K.S., Status, invasiveness and environmental threats of three tropical American invasive weeds (*Parthenium hysterophorus* L. *Ageratum conyzoides* L., *Lantana camara* L.) in India, *Biol. Invasion*, 8, VII, 1501, 2006.

Kohli, R.K., Dogra, K.S., Batish, D.R., and Singh, H.P., Impact of invasive plants on the structure and composition of natural vegetation of northwestern Indian Himalayas, *Weed Technol.*, 18, 1296, 2004.

Ministry of Environment and Forest (MoEF), *National Biodiversity Strategy and Action Plan for Bangladesh*, Government of the People's Republic of Bangladesh, Bangladesh, 102, 2004.

Nath, T.K., Hossain, M.K., and Alam, M.K., Diversity and composition of trees in Sitapahar forest reserve of Chittagong Hill Tracts (South) forest division, Bangladesh, *Ann. For.*, 6, 1, 1, 1998.

Singh, H.P., Batish, D.R., and Kohli, R.K., Allelopathic interactions and allelochemicals: New possibilities for sustainable weed management, *Crit. Rev. Plant Sci.*, 22, 239, 2003.

9 Ecological Status of Some Invasive Plants of Shiwalik Himalayas in Northwestern India

Ravinder K. Kohli, Harminder P. Singh, Daizy R. Batish, and Kuldeep Singh Dogra

CONTENTS

9.1 INTRODUCTION

Invasion is not a novel phenomenon; however, it is one that has increased tremendously during the past few years because of rapidly expanding trade and transport among countries. It is one of the most important impacts humans have ever produced on the earth's ecosystems (Sharma et al. 2005b). In fact, it has led to globalization of world biota, further resulting in biotic homogenization (Drake et al. 1989). Invasive species belong to an array of taxonomic groups, and hence it is very difficult to classify them (Crawley 1997). Invasion may occur as a result of human introductions or because of accidental entries of species into alien habitats. Upon entry into the alien environment, these species bring about a number of changes in the native ecosystems, such as altering the structure and composition of plant communities; reducing agricultural productivity, wildlife, biodiversity, and fodder availability; changing soil structure; affecting health of human beings and livestock; and causing huge economic costs (Moore 2000; Sala et al. 2000; Pimentel et al. 2005; Herron et al. 2007; Khuroo et al. 2007). The ecological impact of invasive plants can be far reaching, and in addition to native plant communities, they may affect other components of the ecosystem. In fact, the impact of invasive species on global biodiversity is the second greatest after habitat fragmentation (Drake et al. 1989). It is one of the most significant ecological issues catching the attention of the people worldwide.

One of the major causes of invasion is the availability of fragmented or disturbed habitats that provide favorable sites for invasion. In such sites, the competitive ability of the native species is reduced, and this allows invasive species to spread (Moore 2000). Among the many hypotheses proposed for the successful invasion by the exotic species, the *Natural Enemies hypothesis* (absence of natural enemies such as predators, pests, and pathogens that cause considerable harm to the species) and *Novel Weapon hypothesis* or *Allelopathy* (through the release of chemicals in the environment that cause detrimental effects on the native flora) have been widely accepted for invasive weeds (Heirro and Callaway 2003).

Sharma et al. (2005b) highlighted various attributes of habitats prone to invasion. These include species scarcity, presence of poorly adapted species, absence of predators, availability of empty niches, and above all human disturbance. In addition, invasive species possess a set of biological traits that enable them to establish rapidly in the new environment. Some of these traits include faster growth rate, higher degree of adaptability, higher reproductive potential, efficient dispersal mechanisms, increased competitive ability, allelopathic property, and alternate methods of reproduction such as vegetative reproduction and phenotypic plasticity (Kohli et al. 2004, 2006; Sharma et al. 2005b). However, many times these characteristics do not correlate with the invasive potential of a species and often lead to ambiguity; this has been largely due to differences in sample size in various studies, stage of invasion, and other factors of invasion ecology (Herron et al. 2007).

Because of the alarming rate at which invasive species, particularly plants, are spreading worldwide and causing considerable harm to the structural and functional dynamism of invaded ecosystems, effective measures are required to manage and control these species. For the effective management of invasive species, it is pertinent to inventorise their potential habitats across the world, and to identify the species and understand their life histories and ecological features. In other words, there is a need to undertake studies on the status of invasive species in the vulnerable habitats at the regional level.

9.2 SHIWALIK REGION OF NORTHWESTERN HIMALAYAS AND ITS VULNERABILITY TO PLANT INVASION

The Himalayas constitute one of the important mountain ranges of the world and cover six countries of Asia, including India and China. Being an important biodiversity hotspot, these support rich floral and faunal diversity, including a number of endemic species. However, at present these are facing severe habitat loss due to excessive human interference. Among the human-induced changes, the introduction of invasive species, particularly in the foothills, has caused severe harm to its precious floral diversity (Kohli et al. 2004). However, detailed information regarding the impact of exotic invasive species on the native vegetation and communities of Himalayas is lacking. Previously, a study conducted in Himachal Pradesh (India) in Northwestern Himalayas revealed that exotic invasive species—*Ageratum conyzoides, Parthenium hysterophorus*, and *Lantana camara*—have caused havoc on the native flora, particularly in the lower and middle regions, including Shiwaliks (Kohli et al. 2004). However, a more elaborative and systematic study is required to have a complete information in the specified ecosystems. The present study focuses on the status and impact of some invasive plants in the Shiwalik ranges in Himachal Pradesh (northwestern India), located between 30°22′94″ N to 33°12′40″ N and 75°45′955″ E to 79°04′20″ E, and altitude varying between 244 and 6750 m from mean sea level (msl).

Himachal Pradesh in the northwestern Himalayas consists of three regions: the lower or outer Himalayas (altitude range of 244–1500 m from msl), the middle or inner Himalayas (1500–4500 m from msl), and the upper or greater Himalayas (>4500 m from msl). Of these, the outer or lower Himalayan range is also known as Shiwalik range, which literally means *tresses of Lord Shiva*. The Shiwalik range is geologically a younger hill range and basically consists of sub-Himalayan tracts or foothills. These form a continuous chain from Jammu to Sikkim through Punjab, Haryana, Himachal Pradesh, Uttarakhand, and Nepal. In Himachal Pradesh, the Shiwaliks cover several districts such as Kangra, Hamirpur, Una, Bilaspur, and lower parts of Mandi, Solan, and Sirmaur (Balokhra 1999). The soil of the region is very fragile, permeable and consists of loosely packed sandstone easily washed away with water (Balokhra 1999). Further, the region has a number of rivulets or choes that wash away the fertile soil while passing through the agricultural fields. More recently, the increasing anthropogenic activity, tourism, pollution, rapid industrialization and urbanization, and introduction of exotics have also put a severe stress on these ecosystems. Furthermore, the Gujjar and Gaddi communities—the main inhabitants of this region—have a livestock-dependent lifestyle that has also resulted in excessive deforestation or denudation of the area. The Shiwalik ranges have been chosen for this study because of their greater vulnerability to invasion.

9.3 STATUS OF EXOTIC AND INVASIVE PLANTS AND MECHANISM OF SPREAD IN THE SHIWALIK REGION OF NORTHWESTERN HIMALAYAS

In whole of the state of Himachal Pradesh, more than 40% of the flora is exotic: the majority of the exotic species are native to the American continent, followed by Eurasia, Europe, Asia, Africa, and Australia (Dogra 2007). The largest number of exotic plants belongs to the family Asteraceae followed by Fabaceae, Poaceae,

Lamiaceae, and Rosaceae. A total of 190 plants of exotic origin belonging to 51 families were encountered in Shiwalik range of Himachal Pradesh in the field surveys conducted during the years 2002–2004 (Dogra 2007). These plants belong to all life forms including herbs, shrubs, trees, grasses, vines, sedges, and even climbers. Among these, texa of family Asteraceae were most prevalent, closely followed by the families Fabaceae and Poaceae (Dogra 2007). During the survey, noxious invasive weeds such as *P. hysterophorus* L., *A. conyzoides* L., *Eupatorium odoratum* L., *L. camara* L., and *Tagetes minuta* L., which have created havoc in the region, were also encountered. The schematic mechanism of spread of the invasives is highlighted in Figure 9.1. The propagules (seeds, cuttings, underground propagules, nursery stocks, etc.) of invasive plants enter available niches such as overgrazed grasslands, deforested areas, open-canopy forests (heavily logged and grazed), eroded areas in the Shiwalik region through various natural agencies. Their entry may be accidental and facilitated by air, water, and transport systems or even through biotic agents such as animals and human beings. Alternatively, they may have been introduced purposely, e.g., *T. minuta* for essential oils. These propagules remain dormant for sometime until the favorable conditions for their growth exist, at which time they start geminating or sprouting and expand. Being competitively strong and coupled with allelopathic qualities, they perpetuate and spread rapidly and soon dominate the area at the expense of native vegetation. In the absence of natural enemies, their growth and spread is unchecked and they form monospecific stands (Kohli et al. 2004).

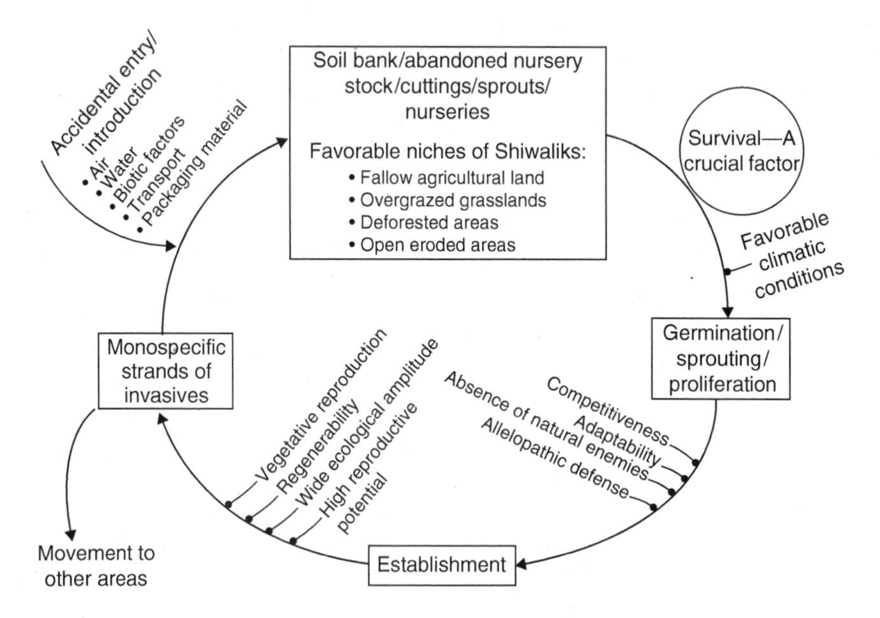

FIGURE 9.1 Mode of spread of invasive species in the Shiwalik region of western Himalayas.

9.4 SOME COMMONLY FOUND INVASIVE PLANTS IN SHIWALIKS OF HIMACHAL PRADESH

9.4.1 *AGERATUM CONYZOIDES* L. (FAMILY ASTERACEAE)

A native of Central and South America, commonly known as billy goat weed or billgoat weed, goatweed, chickweed, and whiteweed, this species now occurs worldwide, particularly in the tropics and subtropics (GISD, 2007; Wagner et al. 1999). It is spreading at an alarming rate in various parts of southeast Asia, including countries such as India and China. Although not included in the list of invasive alien species of Global Invasive Species Database (GISD), it is one of the major invasive plant species in India, China, Thailand, Indonesia, and even Australia. In India, the weed was introduced in the year 1860, probably as an ornamental (National Focal Point for APFISN, India 2005). Later, it acquired a weedy habit and became established almost throughout India.

In India, the weed is particularly prevalent in the hilly tracts of the Himalayas (up to 1800 m) including Shiwaliks (Figure 9.2a). In Shiwalik ranges of Himachal Pradesh, it prefers to grow in croplands, grasslands, forest–grassland ecotones, or any abandoned or vacant areas. It is particularly common in the fields of ratoon, maize, wheat, and rice fields (Angiras et al. 1988; Kohli et al. 2006). The agricultural fields left for fallowing are more prone to infestation by this weed. The native people, particularly farmers, in their local language call this weed destructive (*ujaro* in local dialect), since it causes huge loss of crop production, and heavy infestation results in severe fodder scarcity.

A. conyzoides is an annual aromatic herb reaching a height of about 1 m. It has a pubescent stem and ovate to ovate-rhomboid leaves. It bears characteristic purplish blue colored inflorescence and grows abundantly (forming monocultural stands) in invaded areas. At maturity, the weed produces a large number of small, lightweight seeds (*cypsela*; up to 40,000) that are dispersed by wind and water and can grow under wide range of conditions (Holm et al. 1977) or as contaminants of crops and fodder. In addition to seeds, the weed also multiplies vegetatively through stolons.

This weed is also known to be allelopathic and thus suppresses the growth of other plants (Kong et al. 1999; Okunade 2002; Singh et al. 2003). Almost all parts of the weed, viz. stem, roots, leaves, and all aboveground parts, including inflorescence and decaying and brown leaves lying on the ground as debris, are known to cause allelopathic effects. Several volatile (precocene I and precocene II, caryophyllene, farnesene, carene, cubebene, copaene, myrcene, pinene, limonene, etc.) and non-volatile allelochemicals (phenolic acids such as gallic acid, coumalic acid, protocatechuic acid, *p*-hydroxybenzoic acid, *p*-coumaric acid, sinapic acid, and benzoic acid) have been identified in this species (Kong et al. 1999, 2002; Xuan et al. 2004). Singh et al. (2003) reported that wheat growth was significantly reduced in the rhizosphere of this weed, and soil extracts and leaf residue extracts were inhibitory to radicle and shoot growth of wheat. A significant amount of phenolics was found in the rhizosphere soil and in the plant tissue, including leaves and debris. The allelochemicals of *Ageratum* affect not only the growth of crops but also the

FIGURE 9.2 Pictures showing infestation of various habitats of Shiwaliks invaded by invasive weeds—(a) *A. conyzoides*, (b) *E. odoratum*,

(*continued*)

nodulation (number and weight of nodules and content of leghemoglobin content) in the chickpea (*Cicer arietinum*) (Batish et al. 2006). The aforementioned studies show that this weed is strongly allelopathic and adversely affects crop productivity through its allelochemicals that accumulate in soil or air (volatile).

(c)

(d)

FIGURE 9.2 (continued) (c) *L. camara*, (d) *P. hysterophorus* and

(continued)

9.4.2 *Chromolaena odorata* (L.) King & Robinson (= *Eupatorium odoratum* L.; Family Asteraceae)

C. odorata (L.) King & Robinson is another weed recognized among the top 100 invasive species by the GISD (2007). The weed is commonly known as siam weed, chromolina, bitter bush, and triffid weed. It was introduced to India in 1840 as an

(e)

FIGURE 9.2 (continued) (e) *T. minuta.*

ornamental plant (National Focal Point for APFISN, India 2005). Now it is one of the major weeds of most southeast Asian countries. In India, it is a noxious weed of the northeastern area and of southern regions, particularly Western Ghats (Singh 1998). It is now a serious weed of several useful plantation crops such as coconut, rubber, coffee, and teak. Of late, the weed has been spreading alarmingly into the northwestern Himalayas, particularly in Shiwaliks, and it is of great concern (Figure 9.2b).

9.4.3 *Lantana camara* L. (Family Verbenaceae)

Lantana is another of the worst invasive weeds that has rapidly spread in Himachal Pradesh, particularly in the lower Shiwalik region (Figure 9.2c). According to the GISD (2007), the weed is one among the 100 worst invaders of the world. It is known by several names including largeleaf lantana, prickly lantana, and wild sage. It has over 650 varieties and has been reported from over 60 countries of the world (Day et al. 2003). The formation of so many varieties has resulted from intensive horticultural improvement and hybridization of the weed for ornamental purposes. Though a native of tropical America, the species is now found worldwide and is a serious weed in tropical and subtropical regions. It is a very troublesome weed in India, Australia, Africa, Philippines, Fiji, and Hawaii, where it occurs both in natural and agricultural ecosystems (Holm et al. 1977; Kohli et al. 2006).

In India, *L. camara* was introduced in the beginning of nineteenth century as an ornamental plant, but now it occurs as a weed throughout the country except at higher altitudes (>1700 m) (Kohli et al. 2004). It is a serious invader of forests, particularly those with open canopy, grasslands, agricultural land, and any vacant areas. In Shiwaliks, *L. camara* occurs in the overgrazed open canopy forests, roadsides, railway tracts, along canals and is still spreading to other uninvaded areas. Being perennial, woody, and bushy in habit, it forms dense impenetrable thickets that displace the native species, especially the economically important medicinal herbs (Kohli et al. 2006). The weed possesses several biological and ecological features that make it a successful invader. Sharma et al. (2005a) have listed various biological and ecological attributes such as fitness homeostasis, phenotypic plasticity, allelopathy, fire tolerance, and interaction with other animals that contribute to the invasiveness of *L. camara*. Kohli et al. (2006) have reported that the weed produces a very high number of seeds (10,000–12,000 per plant), owing to a long flowering period and each fruit containing 1–2 seeds. Seeds of the weed are widely dispersed through a variety of vertebrates (birds, domestic animals, and even human beings). These germinate quickly and form dense thickets that are fire prone. In addition, the weed also multiplies through suckering and layering. The burning and slashing of weed thickets stimulates seed germination and even suckering ability, thereby increasing its invasiveness (Ambika et al. 2003). *L. camara* is strongly allelopathic and interferes with the growth and development of a wide range of plants including vines, ferns, and agricultural crops (Achhireddy and Singh 1984; Mersie and Singh 1987; Wadhwani and Bhardwaja 1981; Singh and Achhireddy 1987). Various volatile and nonvolatile allelochemicals have been identified from this weed, and these adversely affect the growth of other plants, including its own populations (Arora and Kohli 1993; Ambika et al. 2003).

9.4.4 *PARTHENIUM HYSTEROPHORUS* L. (FAMILY ASTERACEAE)

P. hysterophorus L., commonly known as congress grass, feverfew, ragweed parthenium, parthenium weed, or white top is a noxious weed native to tropical America. It has now naturalized in several tropical and subtropical parts of the world, particularly in countries such as India, Australia, Pakistan, Taiwan, South Africa, and Ethiopia. (Kohli and Rani 1994). It is one of the most troublesome weeds and figures among the list of invasive species in the GISD (2007). It is believed to have entered several countries accidentally through contaminated food or pasture grains. In India, it was first reported in 1956 growing as a stray plant in the outskirts of Poona (Rao 1956); however, reports indicate that the weed occurred in India much earlier and was present in the early nineteenth century (Bennet et al. 1978). In the Shiwalik ranges, though its exact date of entry is not known, it has become one of the major and prevalent weeds (Figure 9.2d) during the last two decades and is still continuing to spread in uninvaded areas (Kohli et al. 2004).

P. hysterophorus is an annual, erect herb, reaching a height of 1.0–1.5 m or more during favorable conditions. The stem has a tendency to become woody with age and also becomes profusely branched. During the initial growth stages, the leaves are larger and form a rosette. After bolting, a number of relatively smaller, highly dissected leaves are formed. The stem is whitish-green, angular, densely branched, and hairy. Inflorescence is capitulum (both terminal and axillary) and is formed soon after the growth. The leaves, stem, and inflorescence are covered with hairy structures—trichomes, which are rich in parthenin, a major chemical constituent of *P. hysterophorus* (Kohli and Rani 1994).

It is a strong aggressive colonizer of disturbed lands, particularly overgrazed pastures. It forms dense colonies wherever it invades. Several factors such as fast growth rate, high reproductive potential, wide ecological amplitude, strong adaptability, and allelopathic properties contribute to its success as an invader. The weed is strongly allelopathic and allelochemicals such as phenolic acids (caffeic acid, *p*-coumaric acid, *p*-hydroxybenzoic acid, anisic acid, and ferulic acid) and sesquiterpene lactones (parthenin and coronopilin) have been identified in this species (Kohli and Rani 1994). The weed causes a number of agricultural, ecological, and health hazards (Kohli and Rani 1994). It is known to reduce floristic diversity of the given area because of its allelopathic and aggressive nature (Kohli et al. 2004, 2006).

9.4.5 *Tagetes minuta* L. (Family Asteraceae)

T. minuta L., commonly known as wild or Mexican marigold, stinkweed, or tall khaki weed, is a native of South America, introduced to various parts of the world including India for its essential oil and medicinal use. Holm et al. (1997) have listed *T. minuta* as a noxious invasive weed found in over 35 countries worldwide. It has reportedly encroached wastelands, orchards, and croplands and is a troublesome weed of maize, barley, beans, peas, coffee, potatoes, cotton crops, etc. (Holm et al. 1997). Additionally, the weed is known to possess toxic properties (Soule 1993). In India, it was introduced for its essential oil. However, it seems to have escaped from cultivated areas and acquired a weedy habit. It is often seen growing luxuriantly in disturbed sites of the Shiwalik ranges, particularly in abandoned sites, dry areas, construction sites, and along roadsides (Figure 9.2e). Its fast spread is a cause of concern since it is spreading at the cost of native species.

T. minuta is an erect, annual, aromatic herb attaining a height of 1–2 m or more. Its leaves are slightly glossy, green, and pinnatisect with serrated margins and on the under surface have a number of small, punctate, multicellular orange colored glands, which upon rupturing produce a licorice-like aroma (Soule 1993). The heads or capitula are light yellow colored, small and are borne in corymbose clusters. The fruits are dark brown or black colored (Soule 1993). The plant is aromatic due to the presence of essential oil in its aboveground parts, particularly the leaves and inflorescence, and is rich in ocimenone, ocimene, dihydrotagetone, tagetone, limonene, linalool, α-pinene, myrcene, 1,8-cineole, and isoeugenol (Singh et al. 1992; Bansal et al. 1999). The tendency of this weed to spread rapidly may be attributed to allelopathic properties of the species. The volatile essential oil of the weed inhibits the root growth of maize; this has been attributed to the presence of ocimenone, α-pinene, and limonene in the oil (Scrivanti et al. 2003). Even the aqueous extracts of the weed have suppressed the seed germination of *Lotus corniculata* and *Lactuca sativa*, and also inhibited callus induction of some crop plants (Kil et al. 2002; Lee et al. 2002).

9.5 CONCLUSIONS AND THE WAY FORWARD

From the earlier discussion, it is clear that the Shiwalik range of Indian Himalayas has become a favorable haven for invasive plants such as *A. conyzoides, L. camara*, and *P. hysterophorus*, while others such as *C. odorata* and *T. minuta* are fast spreading in the region. These have not only displaced the native plant species but also deteriorated the quality of native ecosystems. Mack et al. (2000) have indicated that, if current rate of biotic invasion continues, most of our productive and vital ecosystems will be at the risk of homogenization and be unable to sustain themselves. Long-term strategies are thus required for checking the ever-increasing risk of biotic invasion. Unfortunately, few studies are available, and there is a lack of information on the status of invasive species in different parts of world. The Himalayas, one of the global biodiversity hotspots, is also facing this human-driven global change. Although the vital information on the status of invasive species in this region is lacking, a few useful but scanty studies are available. Kohli et al. (2004) have reported that invasive species such as *A. conyzoides, L. camara*, and *P. hysterophorus* are spreading at a very fast rate in the northwestern Himalayas, particularly in the lower Shiwalik region, and are causing much harm to the ecology of the region. Therefore, timely action is required to save these vital ecosystems. Various biological, mechanical, chemical, and cultural methods used for their control have not been successful, and thus an integrated approach including effective community participation could be very useful (Batish et al. 2004; Kohli et al. 2006). Khurro et al. (2007) have pointed out that in the Kashmir Himalayas, 571 (29%) of the species are aliens, and of these, 17% are invasive aliens. This systematic study provides vital information on invasive plant status and can be useful in devising management strategies in order to check their fast spread. To save Himalayas and Shiwaliks, in particular, from invasive plants the following strategies could be useful if implemented:

- Compiling comprehensive information on the invasive plant species including their status, mode of entry, and spread in the area.
- Determining possible mode(s) of entry of invasive plants in order to control their further spread in the area.
- Understanding the biological and ecological attributes of the invasive plants, and determining their life-history patterns so as to devise effective and corrective control measures.
- Determining the socioeconomic and ecological impact of invasive plants in the area and disseminating this information to the public so as to encourage effective community participation.
- Strengthening integrated and eco-friendly approaches to control invasive plants.
- Devising preventive measures for areas free of invasive weeds in order to protect them.

REFERENCES

Achhireddy, N.R. and Singh, M., Allelopathic effects of lantana (*Lantana camara*) on milkweed vine (*Morrenia odorata*), *Weed Sci.*, 32, 757, 1984.

Ambika, S.R., Poornima, S., Palaniraj, R., Sati, S.C., and Narwal, S.S., Allelopathic plants. 10. *Lantana camara* L., *Allelopathy J.*, 12, 147, 2003.

Angiras, N.N., Tripathi, B., and Singh, C.M., Studies on chemical control of *Ageratum* sp., in *Proceedings of Seminar on Control of Lantana and Ageratum*, Singh, C.M. and Angiras, N.N., Eds., Himachal Pradesh Krishi Vishvwidyalaya, Palampur, 66, 1988.

Arora, R.K. and Kohli, R.K., Autotoxic impact of essential oils extracted from *Lantana camara* L., *Biol. Plant.*, 35, 293, 1993.

Balokhra, J.M., *The Wonderland: Himachal Pradesh*, H.G. Publications, New Delhi, 1999.

Bansal, R.P., Bahl, J.R., Garg, S.N., Naqvi, A., Sharma, S., Muni, R., and Kumar, S., Variation in quality of essential oil distilled from vegetative and reproductive stages of *Tagetes minuta* crop grown in north Indian plains, *J. Essent. Oil Res.*, 11, 747, 1999.

Batish, D.R., Singh, H.P., Kaur, S., and Kohli, R.K., Phytotoxicity of *Ageratum conyzoides* towards growth and nodulation of *Cicer arietinum, Agric. Ecosyst. Environ.*, 113, 399, 2006.

Batish, D.R., Singh, H.P., Kohli, R.K., Johar, V., and Yadav, S., Management of invasive exotic weeds requires community participation, *Weed Technol.*, 18, 1445, 2004.

Bennet, S.S.R., Naithani, H.P., and Raizada, M.B., *Parthenium* L. in India: Review and history, *Indian J. For.*, 1, 128, 1978.

Crawley, M.J., Biodiversity, in *Plant Ecology*, Crawley, M.J., Ed., Blackwell Scientific, Oxford, UK, 595, 1997.

Day, M.D., Wiley, C.J., Playford, J., and Zalucki, M.P., *Lantana: Current Management, Status and Future Prospects*, Australian Centre for International Agricultural Research, Canberra, 2003.

Dogra, K.S., Impact of some invasive species on the structure and composition of natural vegetation of Himachal Pradesh, Ph.D. Thesis, Panjab University, Chandigarh, 2007.

Drake, J.A., Mooney, H.A., di-Castri, F., Groves, R.H., Kruger, F.J., Rejmanek, M., and Williamson, M., *Biological Invasions: A Global Perspective*, Wiley, Chichester, UK, 1989.

GISD, *Global Invasive Species Database*. Available online at http://www.issg.org/database, 2007.

Heirro, J.L. and Callaway, R.M., Allelopathy and exotic plant invasion, *Plant Soil*, 256, 29, 2003.

Herron, P.M., Martine, C.T., Latimer, A.M., and Leicht-Young, S.A., Invasive plants and their ecological strategies: Prediction and explanation of woody plant invasion in New England, *Divers. Distrib.*, 13, 633, 2007.

Holm, L., Doll, J., Holm, E., Pancho, J., and Herberger, J., *World Weeds. Natural Histories and Distribution*, Wiley, New York, 1997.

Holm, L.G., Plucknett, D.L., Pancho, J.V., and Herberger, J.P., *The World's Worst Weeds: Distribution and Biology*, The University of Hawaii, Honolulu, HI, 1977.

Khuroo, A.A., Rashid, I., Reshi, Z., Dar, G.H., and Wafai, B.A., The alien flora of Kashmir Himalaya, *Biol. Invasions*, 9, 269, 2007.

Kil, J.-H., Shim, K.-C., and Lee, K.-J., Allelopathy of *Tagetes minuta* L. aqueous extracts on seed germination and root hair growth, *Korean J. Ecol. Sci.*, 1, 171, 2002.

Kohli, R.K., Batish, D.R., Singh, H.P., and Dogra, K.S., Status, invasiveness and environmental threats of three tropical American invasive weeds (*Parthenium hysterophorus* L., *Ageratum conyzoides* L., *Lantana camara* L.) in India, *Biol. Invasions*, 8, 1501, 2006.

Kohli, R.K., Dogra, K.S., Batish, D.R., and Singh, H.P., Impact of invasive plants on the structure and composition of natural vegetation of northwestern Indian Himalayas, *Weed Technol.*, 18, 1296, 2004.

Kohli, R.K. and Rani, D., *Parthenium hysterophorus*—A review, *Res. Bull. (Sci.) Panjab Univ.*, 44, 105, 1994.

Kong, C., Hu, F., Xu, T., and Lu, Y., Allelopathic potential and chemical constituents of volatile oil from *Ageratum conyzoides, J. Chem. Ecol.*, 25, 2347, 1999.

Kong, C., Hu, F., and Xu, X., Allelopathic potential and chemical constituents of volatile oil from *Ageratum conyzoides* under stress, *J. Chem. Ecol.*, 28, 1173, 2002.

Lee, S.Y., Shim, K.C., and Kil, J.H., Phytotoxic effects of aqueous extracts and essential oils from southern marigold (*Tagetes minuta*), *N.Z. J. Crop Hort. Sci.*, 30, 161, 2002.

Mack, R.N., Simberloff, D., Lonsdale, W.M., Evans, H., Clout, M., and Bazzaz, F.A., Biotic invasions: Causes, epidemiology, global consequences and control, *Ecol. Appl.*, 10, 689, 2000.

Mersie, W. and Singh, M., Allelopathic effects of *Lantana* on some agronomic crops and weeds, *Plant Soil*, 98, 25, 1987.

Moore, P.D., Alien Invaders, *Nature*, 403, 492, 2000.

National Focal Point for APFISN, India, *Stocktaking of National Forest Invasive Species Activities, India (India Country Report 101005)*, Ministry of Environment and Forests, New Delhi, 2005.

Okunade, A.L., *Ageratum conyzoides* L. (Asteraceae), *Fitoterapia*, 73, 1, 2002.

Pimentel, D., Zuniga, R., and Morrison, D., Update on the environmental and economic costs associated with alien invasive species in the United States, *Ecol. Econ.*, 52, 273, 2005.

Rao, R.S., Parthenium a new record for India, *J. Bombay Nat. Hist. Soc.*, 54, 218, 1956.

Sala, O.E., Chapin, F.S., III, Armesto, J.J., et al., Global biodiversity scenarios for the year 2100, *Science*, 287, 1770, 2000.

Scrivanti, L.R., Zunino, M.P., and Zygadlo, J.A., *Tagetes minuta* and *Schinus areira* essential oils as allelopathic agents, *Biochem. Syst. Ecol.*, 31, 563, 2003.

Sharma, G.P., Raghuvanshi, A.S., and Singh, J.S., Lantana invasion: An overview, *Weed Biol. Manage.*, 5, 157, 2005a.

Sharma, G.P., Singh, J.S., and Raghuvanshi, A.S., Plant invasions: Emerging trends and future implications, *Curr. Sci.*, 88, 726, 2005b.

Singh, B., Sood, R.P., and Singh, V., Chemical composition of *Tagetes minuta* L. oil from Himachal Pradesh (India), *J. Essent. Oil Res.*, 4, 525, 1992.

Singh, M. and Achhireddy, N.R., Influence of *Lantana* on growth of various citrus rootstocks, *HortScience*, 22, 385, 1987.

Singh, H.P., Batish, D.R., Pandher, J.K., and Kohli, R.K., Assessment of allelopathic properties of *Parthenium hysterophorus* residues, *Agric. Ecosyst. Environ.*, 95, 537, 2003.

Singh, S.P., A review of biological suppression of *Chromolaena odorata* K & R in India, in *Proceedings of the Fourth International Workshop on Biological Control and Management of Chromolaena odorata*, Ferrar, P., Muniaapan, R. and Jayanth, K.P., Eds., University of Guam, Guam, 86, 1998.

Soule, J.A., *Tagetes minuta*: A potential new herb from South America, in *New Crops*, Janick, J. and Simon, J.E., Eds., Wiley, New York, 649, 1993.

Wadhwani, C. and Bhardwaja, T.N., Effect of *Lantana camara* L. extracts on fern spore germination, *Experientia*, 37, 245, 1981.

Wagner, W.L., Herbst, D.R., and Sohmer, S.H., *Manual of the Flowering Plants of Hawaii*, Revised ed., Two volumes, Bernice P. Bishop Museum Special Publication, University of Hawaii Press/Bishop Museum Press, Honolulu, HI, 1999.

Xuan, T.D., Shinkichi, T., Hong, N.H., Khanh, T.D., and Min, C.I., Assessment of phytotoxic action of *Ageratum conyzoides* L. (billy goat weed) on weeds, *Crop Prot.*, 23, 915, 2004.

10 Invasive Species and Their Impacts on Endemic Ecosystems in China*

Jichao Fang and Fanghao Wan

CONTENTS

* This chapter is based on the presentation "Biological and environmental impacts and management of exotic invasive species in China" in the International Seminar on Biological Invasions: Environmental Impacts and the Development of a Database for the Asian-Pacific Region, Tsukuba, Japan, 2003 (some data have been supplemented and updated).

10.1 INTRODUCTION

Invasive species are nonnative species that are capable of establishing and spreading in disturbed habitats or natural communities. Weeds are one category of invasive species, as are invasive microorganisms that act as pathogens as well as invasive insect pests. In contrast, many beneficial exotic species in China that are introduced and maintained by humans are not considered invasive species. Herein, the term *invasive pest* refers to a species that is introduced, either naturally or by humans intentionally or unintentionally, into a new habitat where it becomes established and causes damage to the natural ecosystem, native species, and even human health. Invasion by exotic pests is becoming a problem of greater concern in China as the global economy becomes increasingly integrated and international trade continues to grow.

Many invasive pests such as the pinewood nematode (*Bursaphelenchus xylophilus*) in China have become expensive biopollutants. Once an invasive pest population is established, it is much more difficult to clean up than is chemical pollution. Of more than 200 invasive pests reported in China, several well-known species such as the serpentine vegetable leaf miner (*Liriomyza sativae*), the pinewood nematode, the rice water weevil (*Lissorhoptrus orysophilus*), and common cordgrass (*Spartina anglica*) cost the country more than US$7 billion per year. China is an extensive country with a span of about 5,200 km from east to west and 50 latitudinal degrees across tropical, subtropical, and temperate zones from south to north. A wide range of habitats and environmental conditions has made this country especially vulnerable to establishment of invasive species originating in other countries. Invasive species from most regions of the world may find suitable habitats somewhere in China. As invasive species replace native flora and fauna, the result is not only a huge economic cost but also a great loss of biodiversity and function of endemic ecosystems.

10.2 SCOPE AND DISTRIBUTION OF MAJOR INVASIVE PESTS IN CHINA

10.2.1 INVASIVE WEEDS

More than 800 species of exotic plants occur in China, including 380 species of invasive plants and 108 species of important weeds, belonging to 23 families and 76 genera (Table 10.1). Among the important weeds, 15 species such as water hyacinth (*Eichhornia crassipes*), common ragweed (*Ambrosia artemisiifolia*), eupatorium (*Ageratina adenophora*), siam weed (*Chromolaena odorata* L.), darnels (*Lolium* spp.), and alligator weed (*Alternanthera philoxeroides*) are well-known malignant weeds spreading nationwide. It is noteworthy that 62 species (58%) of the important invasive weeds were intentionally introduced as pasture, forage, agricultural, ornamental, medicinal, or landscape plants, which then escaped into the wild (Qiang and Cao 2000).

According to data from the Ministry of Agriculture in China, invasive weeds have caused various economic costs and huge losses in agriculture, forestry, livestock breeding, and aquaculture. Yield losses of rice, wheat, corn, sweet potato, and lettuce

TABLE 10.1
Important Invasive Plants Documented in Mainland China

Common Name	Scientific Name
Huisache	*Acacia farmesiana* (L.) Willd
Bristly starburr	*Acanthospermun australe* (Loef.) Kuntze
Sticky snakeroot	*Ageratina adenophora* (Spreng.) King & Robins.
Tropic ageratum	*Ageratum conyzoides* L.
Mexican ageratum, floss flower	*A. houstonianum* Mill.
Alligator weed	*Alternanthera philoxeroides* (Mart.) Griseb
Spinyflower alternanthera	*A. pungens* H.B. Kunth
Amaranth	*Amaranthus viridis* L.
Spiny amaranth	*A. spinosus* L.
Joseph's coat, fountain plant	*A. tricolor* L.
Slim amaranth	*A. chlorostachys* Willd.
Tumble pigweed	*A. albus* L.
Redroot amaranth, rough pigweed	*A. retroflexus* L.
Mat amaranth, prostrate pigweed	*A. blitoides* S.Watson
Common ragweed	*Ambrosia artemisiifolia* L.
Giant ragweed	*A. trifida* L.
Heartleaf madeira vine	*Anredera cordifolia* (Tenore) van Steenis
Thin-leaved celery	*Apium leptophyllum* (Pers.) F.J. Muell.
Annual saltmarsh	*Aster subulatus* Michx.
Wild oat	*Avena fatua* L.
Carpet-grass	*Axonopus compressus* (Swartz) Beauv.
Devil's beggartick	*Bidens frondosa* L.
Hairy beggarticks; Spanish needles	*B. pilosa* L.
Washington grass	*Cabomba caroliniana* Gray
Hemp	*Cannabis indica* L.
Sensitiveplant-like senna	*Cassia mimosoides* L.
Coffee senna	*C. occidentalis* L.
Sickle senna	*C. tora* (L.) Roxb.
Bear grass	*Cenchrus echinatus* L.
Cornflower	*Centaurea cyanus* L.
Night-blooming cereus	*Cereus grandiflorus* (L.) Mill.
Mexican tea, wormseed	*Chenopodium ambrosioides* L.
Siam weed, paraffin weed	*Chromolaena odorata* (L.) King & Robins.
Common chicory	*Cichorium intybus* L.
Ivy gourd	*Coccinia grandis* (L.) Voigt
Flax-leaf fleabane	*Conyza bonariensis*
Horseweed	*C. Canadensis* (L.) Cronq.
Broadleaf fleabane, tall fleabane	*C. sumatrensis* (Retz) Walker
Golden tickseed	*Coreopsis tinctoria* Nutt.
Lance coreopsis	*C. lanceolata* L.
Hawksbeard velvet-plant	*Crassocephalum crepidioides* (Benth.) Moore
Dodder	*Cuscuta* spp.
Purple nut sedge	*Cyperus rotundus* L.
Thorn apple	*Datura stramonium* L.

(*continued*)

TABLE 10.1 (continued)
Important Invasive Plants Documented in Mainland China

Common Name	Scientific Name
Wild carrot	*Daucus carota* L.
Golden dewdrop	*Duranta repens*
Water hyacinth	*Eichhornia crassipes* (Mart.) Solms
Wire grass, goose grass	*Eleusine indica* (L.) Gaertn.
Ramose scouring rush	*Equisetum ramosissimum* Desf.
Daisy fleabane	*Erigeron annuus* (L.) Pers.
Canada fleabane	*E. canadensis* L.
Philadelphia fleabane	*E. philadelphicus* L.
Long-stem pipewort	*Eriocaulon melanocephalum* H.B.Kunth
Foecid eryngo	*Eryngium foetidum* L.
Crofton weed	*Eupatorium adenophorum* Spreng
Odor eupatorium	*E. odoratum* L.
Sun spurge	*Euphorbia helioscopia* L.
Cats hair, asthma weed	*E. hirta* L.
Spotted spurge	*E. maculata* L.
Tall fescue	*Festuca arundinacea* Schreb.
Small flower galinsoga, quickweed	*Galinsoga parviflora* Cav.
Carolina	*Geranium caroliniamum* L.
Gomphrena weed	*Gomphrena celosioides* Mart.
Sneezeweed	*Helenium autumnale* L.
Common heliotrope	*Heliotropium europaeum* L.
Bulbous barley	*Hordeum bulbosum* L.
Rombic bushmint	*Hyptis rhomboidea* Mart. & Gal.
Wild spikenard, pignut	*H. suaveolens* (L.) Poit
Ivy-leaved morning glory	*Ipomoea cairica* (L.) Sweet.
White-edge morning glory	*I. nil* (L.) Roth.
Common morning glory	*I. purpurea* (L.) Roth
Tubeleaf kalanchoe	*Kalanchoe tubifolia* L.
Common lantana	*Lantana camara* L.
Weeping lantana	*L. montevidensis* (Spreng.) Briq.
Three-nerved duckweed	*Lemna trinervis* Small
Field pepperweed	*Lepidium campestre* L.
Clasping pepperweed	*L. perfoliatum* L.
Virginia pepperweed	*L. virginicum* L.
Horse tamarind	*Leucaena leucocephala* (Lam.) de Wit.
Italian ryegrass	*Lolium multiflorum* Lam.
Persian darnel	*L. persicum* Boiss & Hohen
Annual ryegrass	*L. rigidum* Gaud.
Flax ryegrass	*L. remotum* Schrank
Darnel ryegrass	*L. temulentum* L.
Common cat's claw vine	*Macfadyena unguiscati* (L.) A. Gentry
Coromandel coast false mallow	*Malvastrum coromandelianum* (L.) Garcke
Alfalfa	*Medicago sativa* L.
White sweet clover	*Melilotus albus* Medik
Yellow sweet clover	*M. officinalis* (L.) Lam.

TABLE 10.1 (continued)
Important Invasive Plants Documented in Mainland China

Common Name	Scientific Name
Mile-a-minute weed	*Mikania micrantha* H.B. Kunth
Sensitive plant	*Mimosa pudica* L.
Giant sensitive plant	*M. invisa* Mart.
Spinyless Brazilian sensitive plant	*M. invisa* var. *inermis* Adelb.
Four-o'clock	*Mirabilis jalapa* L.
Water cress	*Nasturtium officinale* L.
Slender false garlic	*Nothoscordum gracile* (Ait.) Stearn
Red flower primrose	*Oenothera orseus*
Sweet prickly pear	*Opuntia ficus* (L.) Mill.
Prickly pear	*O. monacantha* (Willd.) Haw.
Rest pear	*O. stricta* (Haw.) Haw.
Branched broomrape	*Orobanche brassicae*
Corymb wood sorrel	*Oxalis corymbosa* DC.
Guinea grass	*Panicum maximum* Jacq.
Torpedo grass	*P. repens* L.
Iceland poppy, Arctic poppy	*Papaver nudicaule* L.
Guayule	*Parthenium hysterophorus* L.
Virginia creeper	*Parthenocissus quinquefolia* (L.) Planch
Caterpillar grass, dallisgrass	*Paspalum dilatatum* Poir.
Blue passion flower	*Passiflora coerulea* L.
Weed passion flower	*P. foetida* L.
West African pennisetum, Kikuyu grass	*Pennisetum clandestinum* Chiov.
Shiny peperomia	*Peperomia pellucida* (L.) H.B. Kunth
Common timothy	*Phleum pretense* L.
Husk tomato	*Physalis pubescens* L.
Common pokeweed	*Phytolacca americana* L.
Artillery clearweed	*Pilea microphylla* (L.) Liebm.
Water lettuce	*Pistia stratiotes* L.
Hoary plantain	*Plantago virginica* L.
Hairy leafcup	*Polymnia uvedalia* L.
Vasey's pondweed	*Potamogeton vaseyi* J.W. Robbins
Castor-oil plant	*Ricinus communis* L.
Beaked bulrush	*Ryhnchospora submarginata* L.
Vacourinha, sweetbroom	*Scoparia dulcis* L.
Groundsel	*Senecio dubilobilis* L.
Common groundsel	*S. vulgaris* L.
Knotroot bristlegrass	*Setaria geniculata* (Lam.) Beauv.
Palm grass	*S. palmifolia* (Koenig) Stapf.
Himalaya nightshade	*Solanum aculeatissimum* Jacq.
Soda-apple nightshade	*S. capsicoides* All.
Mullein nightshade	*S. erianthun* D. Don
Wild tomato	*S. torvum* Swartz
Tall goldenrod	*Solidago altissima* L.
Canada goldenrod	*S. canadensis* L.

(*continued*)

TABLE 10.1 (continued)
Important Invasive Plants Documented in Mainland China

Common Name	Scientific Name
Camomileleaf soliva	*Soliva anthemifolia* (Juss.) R. Br.
Johnson grass	*Sorghum halepense* (L.) Pers.
Sudan grass	*S. sudanense* L.
Common cordgrass	*Spartina anglica* C.E. Hubbard
Smooth cordgrass	*S. alterniflora* Loisel.
Corn spurry	*Spergula arvensis* L.
Buttonweed	*Spermacoce latifolia* Aubl.
Jamaica Falsevalerian	*Stachytarpheta jamaicensis* (L.) Vahl.
Lesser chickweed	*Stellaria pallida* Dumort.
Cinderella weed, nodeweed	*Synedrella nodiflora* (L.) Gaertn.
Marigold	*Tagetes erecta* T. Patula
Jewels of opar	*Talinum paniculatum* (Jacq.) Gaertn.
Blue sky vine	*Thunbergia grandiflora* Roxb. ex Rottl.
White clover	*Trifolium repens* L.
Clasping venus' looking-glass	*Triodanis perfoliata* (L.) Nieuwl.
Cow cockle	*Vaccaria segetalis* Garcke ex Asch.
Persian speedwell	*Veronica persica* Poir.
Gray field speedwell	*V. polita* Fries
Vetiver	*Vetiveria zizanioides* (L.) Nash
Sleepy morning	*Waltheria indica* L.
Wedelia	*Wedelia chinensis* (Osb.) Merr.
Trilobe wedelia, creeping daisy	*W. trilobata* (L.) Hitch.
Spiny cocklebur	*Xanthium spinosum* L.

damaged by widespread alligator weed average 45%, 36%, 19%, 63%, and 47%, respectively. An asthma disease of horses and sheep, derived from a toxin of eupatorium weeds, caused a loss of 60,000 sheep or US$2.5 million in the autonomous canton of Liang'shan, Sichuan province in 1996 (Wan and Guan 1993). The annual cost of removing water hyacinth mats in the Shanghai metropolitan area alone exceeded US $7.5 million in 2001 and US$23 million in 2002. The cost of national and provincial research programs on invasive weeds in China was estimated to be up to US$50 million in 2001 and US$160 million in 2004.

10.2.2 Invasive Insect Pests and Plant Pathogens

Invasive insect pests recorded in mainland China include 32 species, including the serpentine vegetable leaf miner, the rice water weevil, and the fall webworm. The main invasive plant pathogens include 23 species, including *Xanthomonas oryzicola, Ceratocystis fimbriata,* and *Fusarium vasifectum* (Table 10.2). It is estimated that control of some newly invasive insects costs China US$200 million annually, and the new invaders have caused economic and environmental losses of billions of dollars per year.

TABLE 10.2
Invasive Pests Documented in China

Category	Common Name (Scientific Name)
Insect	Serpentine vegetable leaf miner (*Liriomyza sativae* Blanchard)
	Rice water weevil (*Lissorhoptrus oryzophilus* Kuschel)
	Banana moth (*Opogona sacchari*)
	Chinese citrus fly (*Tetradacus citri* Chen)
	Oriental fruit fly (*Bactrocera dorsalis* Hendel)
	Hessian fly (*Mayetiola destructor* Say)
	Codling moth (*Laspeyresia pomonella* Linne)
	Wooly apple aphid (*Eriosoma lanigerun* Hausmann)
	Grape phylloxera (*Viteus vitifolii* Fitch)
	Khapra beetle (*Trogoderma granarium* Everts)
	Kapok borer (*Heterobostrychus aequalis* Waterhouse)
	Graham bean beetles (*Callosobruchus analis* Fabricius)
	Cowpea seed beetle (*C. maculatus* Fabricius)
	Mexican bean weevil (*Zabrotes subfasciatus* Boheman)
	False indigo weevil (*Acanthoscelides pallidipennis* Motschulsky)
	Bean weevil (*A. obtectus* Say)
	Broad bean weevil (*Bruchus rufimanus* Boheman)
	Soybean pod gall midge (*Asphondylia* sp.)
	Potato tuber moth/tobacco splitworm (*Phthorimaea operculella* Zeller)
	Cottony cushion scale (*Icerya purchasi* Maskell)
	Sweetpotato weevil (*Cylas formicarius* Fabricius)
	Colorado potato beetle (*Leptinotarsa decemlineata* Say)
	Granary weevil (*Sitophilus granarius* Linnaeus)
	Pine bast scale (*Matsucoccus matsumurae* Kuwana)
	Pine greedy scale (*Hemiberlesia pitysophila* Takagi)
	Pine longicorn beetle (*Monochamus altenatus* Hope)
	Loblolly pine mealybug (*Oracella acuta* (Lobdell) Ferris)
	Fall webworm (*Hyphantria cunea* Drury)
	Red turpentine bark beetle (*Dendroctonus valens* Leconte)
	Mango seed weevil (*Sternochetus olivieri* Faust)
	Mango pulp weevil (*S. frigidus* F.)
	Coconut leaf beetle (*Brontispa longissima* Gestro)
	Xiaoi pine stem weevil (*Hylobitelus xiaoi* Zhang)
	Cotton pink bollworm (*Pectinophora gossypeilla* Saunders)
	Japanese beetle (*Popillia japonica*)
	American cockroach (*Periplaneta americana*)
	German cockroach (*Blattella germanica*)
	Rubber tree termite (*Copotermes Curvignathus* Holmgren)
	Taiwan subterranean termite (*C. formosanus* Shiraki)
	Western drywood termite (*Incisitermes minor* Hagen)
	Erythrina gall wasp (*Quadrastichus erythrinae* Kim)
Nematode	Pine wilt disease (*Bursaphelenchus xylophilus* Steiner & Buhrer)
	Banana root burrowing nematode (*Radopholus similis* (Cobb) Thorne)
	Stem and bulb nematodes (*Ditylenchus* spp.)

(*continued*)

TABLE 10.2 (continued)
Invasive Pests Documented in China

Category	Common Name (Scientific Name)
Plant pathogen	Cotton fusarium wilt (*F. oxysporum* f. sp. *vasinfectum* (Atk.) Snyder & Hansen)
	Cotton verticillium wilt (*Verticillium dahliae* Kleb.)
	Bacterial leaf streak of rice (*Xanthomonas campestris pv. oryzicola* Dowson)
	Black rot of sweet potato (*Ceratocystis fimbriata* Ell. et halst)
	Soybean blight (*Phytophthora megasperma* f. sp. *glycinea* Kuan & Erwin)
	Dwarf bunt of wheat (*Tilletia controversa* Kuhn)
	Fusarium wilt or Panama disease (*F. oxysporum* f. sp. *cubense*)
	Verticillium wilt of alfalfa (*V. alboatrum* Reinke & Berthold)
	Downy mildew of maize or sorghum (*Peronosclerospora* spp. Shaw)
	Potato wart disease (*Synchytrium endobioticum* (Schilberszky) Percival)
	Citrus canker (*X.* campestris *pv. citri* (Hasse) Dye)
	Bacterial blight of cassava (*X. campestris pv. manihotis*)
	Tobacco ring spot virus
	Prunus necrotic ring-spot virus
	Bacterial canker of tomato (*Clavibacter michiganensese* subsp. *mishiganen* Davis et al.)
Mammal	Brown hare (*Lepus capensis* Linnaeus)
	Nutria (*Myocastor coypus*)
	Musk rat (*Ondatra zibethicus*)
	Brown rat (*Rattus norvegicus*)
Bird	Çanada goose (*Anser candensis*)
	Sulfur-crested cockatoo (*Cacatua sulpurea*)
	Rainbow lorikeet (*Trichoglossus haematodus*)

10.2.3 Distribution of Major Invasive Pests in Mainland China

With the increase of imported plants and plant products into China, and transportation of the products within China, the rate of invasion and spread of exotics has increased in recent years (Wang 2002). Only one invasive species was introduced in all of the 1970s, but 2 were introduced in the 1980s, 10 in the 1990s (including *Phytophthora megasperma* f. sp. *Glycinea* introduced into Helongjiang, 1991, *Leptinotarsa decemlineata* into Yi'li, Xinjiang, 1993, *Liriomyza sativae* into Hainan and Guangdong, 1994, *Opogona sacchari* into Guangdong and Beijing, 1995, *Sternochetus olivieri* Faust into Guangxi, 1997, *Acanthoscelides obtectus* Say into Zhejiang, 1998, and prunus necrotic ring-spot virus into Shuangxi, 1999), and 3 in 2001. The rice water weevil was first found in Tang'shan city (Tang'hai county) in 1988, and had spread to 7 provinces by 1996 and to 12 provinces by 2005. The serpentine vegetable leaf miner has spread over most of this country, from 280 ha in 1993 when it was first found in San'ya city, Hainan province, China (Table 10.3).

10.3 MEANS OF INTRODUCTION OF INVASIVE PESTS

Invasive species have arrived in China by several avenues. Most biological invasions have been related to human activities, including intentional and unintentional

TABLE 10.3
Distribution of Major Invasive Pests in Mainland China in 2005

Category	Common Name (Scientific Name)	Distribution
Insect and nematode	Serpentine vegetable leaf miner (*Liriomyza sativae*)	Most of China (except HLJ and Tibet)
	Rice water weevil (*Lissorhoptrus orysophilus*)	HEB/TJ/LN/JL/FJ/HEN/ZJ/AH/BJ/JS/SD/SH (508 $kham^2$)
	Colorado potato beetle (*Leptinotarsa decemlineata*)	XJ (Yi'li & Ta'cheng)
	Pine bast scale (*Matsucoccus matsumurae*)	LN/JL/SD (104 khm^2)
	Pine greedy scale (*Hemiberlesia pitysophila* Takagi)	GD/FJ (72 counties, 1375 khm^2)
	Loblolly pine mealybug (*Oracella acuta*)	GD/GX (60 counties, 148 khm^2)
	Banana moth (*Opogona sacchari*)	GD/BJ/others (over 20 provinces)
	Fall webworm (*Hyphantria cunea*)	SD/TJ/HEB/BJ/LN/SUX/SH (130 khm^2)
	Red turpentine bark beetle (*Dendroctonus valens*)	SX/HEN/HEB/SD/SUX/GS (500 khm^2, 3.52 million pines died)
	Xiaoi pine stem weevil (*Hylobitelus xiaoi*)	JX/HN/GX (66 counties, 57.4 khm^2)
	Coconut leaf beetle (*Brontispa longissima*)	GD
	Pinewood nematode (*Bursaphelenchus xylophilus*)	JS/AH/ZJ/SD/GD/HB/SH/GX/CQ (80.4 khm^2, 50 million pines damaged)
	Banana root burrowing nematode (*Radopholus similes* Thorne)	FJ/GD/HAN/GX/YN/SC. Eradicated by 1991
Weed	Mile-a-minute weed (*Mikania micrantha*)	Shen'zhen, Zhu'hai and Guang'zhou cities in GD
	Common cordgrass (*Spartina anglica*)	FJ and other coastal regions
	Common ragweed (*Ambrosia artemisiifolia*), giant ragweed (*A. trifida*)	LN/HLJ/JL/SD/HN/HB/JS/ZJ/AH/JX/SH
	Alligator weed (*Alternanthera philoxeroides*)	Nearly all over China
	Water hyacinth (*Eichhornia crassipes*)	South of China
	Bearded darnel (*Lolium temulentum*)	Over 20 provinces
	Johnson grass, Arabian millet (*Sorghum halepense*)	Around large cities of south, east and southwest of China
	Eupatorium (*Ageratina adenophora*)	Southwest of China (7780 khm^2)
	Siam weed (*Chromolaena odorata*)	South and southwest of China
Plant pathogen	Cotton fusarium wilt (*F. vasifectum*)	Most of China
	Cotton verticillium wilt (*V. dahliae*)	Most of China
	Bacterial leaf streak of rice (*X. oryzicola*)	GD/GX/HN/SC/YN/GZ/JS/ZJ/FJ/JX/AH/HB/HAN
	Black rot of sweet potato (*Ceratocystis fimbriata*)	Most of China

(*continued*)

TABLE 10.3 (continued)
Distribution of Major Invasive Pests in Mainland China in 2005

Category	Common Name (Scientific Name)	Distribution
	Soybean blight (*P. megasperma* f. sp. *glycinea*)	HLJ/FJ/NM
	Fusarium wilt or Panama disease (*F. oxysporum* f. sp. *cubense*)	GD/FJ/GX/HAN
	Verticillium wilt of alfalfa (*V. alboatrum*)	5 cities in north of XJ
	Downy mildew of maize or sorghum (*Peronospòra* spp.)	GX/YN
	Potato wart disease (*Synchytrium endobioticum*)	YN/SC/GZ
	Citrus yellow shoot disease (*Citrus huanglungbin*)	GD/GX/FJ/YN/ZJ/JX/HN/SC/HAN
	Citrus canker (*X. citri*)	JX/HN/GX/SC/YN/JS/SH/GD/ZJ/GZ/FJ/SUX
	Bacterial blight of cassava (*X. campestris pv.manihotis*)	HN/GD
	Tobacco ring-spot virus	FJ, SC
	Bacterial canker of tomato (*lavibacter michiganensese* subsp. *mishiganen*)	HEB/IMG/JL/LN/BJ/HAN/HLJ

Note: GD, Guangdong; GX, Guangxi, HN, Hunan; SC, Sichuan; CQ, Chongqin; YN, Yunnan; GZ, Guizhou; JS, Jiangsu; ZJ, Zhejiang; FJ, Fujian; JX, Jiangxi; AH, Anhui; HB, Hubei; HAN, Hainan; SH, Shanghai; SUX, Shuangxi; BJ, Beijing; LN, Liaoning; XJ, Xinjiang; GS, Gansu; HLJ, Heilongjiang; HEN, Henan; SD, Shandong; SX, Shanxi; HEB, Hebei; TJ, Tianjin; IMG, Inner Mongolia; JL, Jielin.

introduction by transportation facilities, trade goods, and travelers. A few species such as eupatorium (*A. adenophora*) naturally dispersed into China after breeding and establishing themselves in adjacent countries.

10.3.1 Intentional Introduction

China has a long history of introducing nonnative species, especially those that are beneficial, ornamental, or productive. However, many exotics we now consider pests were originally introduced with the best of intentions, yet later escaped from cultivation. This is particularly true of plants and animals originally imported for agriculture, horticulture, forestry, aquaculture, or animal husbandry. It is estimated that 837 species of exotic plants, in 263 families, were intentionally introduced into China by 1970, and more have been introduced since. Water hyacinth, common cordgrass, Canadian goldenrod, and cacti (*Cactaceae* spp.) are examples of environmental, economic, and ornamental plants that then escaped and became troublesome weeds. It has been documented that more than 50% of the invasive weeds in China were intentionally introduced. Typical examples of intentional introduction of vertebrates and invertebrates include rainbow lorikeet (*T. haematotus*) and live-bearers (*Poeciliidae*) for recreation, and nutria (*Myocastor coypus*), bullfrog (*Rana catesbeiana*), and Amazonian snail (*Ampullaria gigas*) for animal husbandry.

10.3.2 Unintentional Introduction

Trade goods, travelers, and natural factors can serve as carriers for unintentional introductions of invasive pests. Incidental invasion is thought to be a major route for wood-boring insects in incompletely treated wood and marine organisms in ship ballast. Container freight brought the serpentine vegetable leaf miner and banana moth (*O. sacchari*) to China just before 1995. Common ragweed and giant ragweed that is distributed in the area along railways and widely spread across 15 provinces of China originally came from North Korea by train. Many invasive halobios (mussels and tidal organisms) were initially carried in ship ballast water, and now threaten the aquaculture and associated water systems. Wheat breeding stock from the United States harbored the darnel ryegrass, which closely resembles wheat seeds. Pinewood nematodes arrived in wooden pallets, and became established first in Nanjing, Jiangsu province and then became widespread.

10.3.3 Natural Dispersal

It is estimated that about 1.3% of invasive species arrived in China successfully through natural dispersal. Eupatorium (*A. adenophora*), which annually produces up to 100,000 lightweight seeds, was most likely introduced to the Yunnan province in southwest China from Burma by wind dispersal (Liu and Xie 1985). Mile-a-minute weed (*Mikania micrantha*) may have been carried into Shenzhen Special District, Guangdong province by air currents from Hongkong or Southeast Asia. The rice water weevil (*Lissorhoptrus orysophilus*) probably was transported to Dan'dong, Liaoning province from North Korea by way of air currents. It is important to realize that successful introduction of an invasive pest probably occurs due to several pathways.

10.4 IMPACTS OF INVASIVE PESTS ON NATIVE ECOSYSTEMS IN CHINA

The tide of exotic invasions threatens both ecosystem services and the native character of landscapes, and incurs huge economic costs. Invasive pests have to a great extent depressed populations of native species, and sometimes have replaced them, because they are often aggressive competitors. Of eight surveyed cities, including the northeastern city of Sheng'yang, an index of common ragweed dominance within communities ranged from 0.85 to 1.0 while an index of the community biodiversity was only 0–0.62 on isolated hills, riverbanks, and roadsides of railways and highways (Wan and Guan 1993). Some invasive plants and fish threaten the native wildlife of China by competitively occupying niches or eating the food of native species, impairing the native species with secreted toxins (such as toxins from common ragweed and mile-a-minute weed) (Li 1994; Wang 1995; Huan and Cao 2000), or even directly preying on the natives. Many roadsides in tropical southwest

FIGURE 10.1 Individuals of the plant *M. micrantha* overwhelming an ecosystem in Shen'zhen city, GD.

China have become jungles of invasive siam weed and eupatorium, which have replaced natives such as a knotweed (*Polygonum perfoliatum* L.). As a result, native phytophagous butterflies such as *Timandra griseata* (Petersen) and *Cletus schmidti* (Kiritsch.) are severely threatened. Mile-a-minute weed has grown rapidly and overshadowed native plants in the Inner Lingding (Lonely) Island of Shen'zhen city, the National Nature Reserve in Guangdong province (Figure 10.1), to exclude the native wild animals that feed there. Canada goldenrod (*Solidago canadensis* L.) has overwhelmed other grasses in partial suburbs of Shanghai (Che and Guo 1999) and Jiangsu (Wu et al. 2005). Water hyacinth mats have outcompeted native hydrophytes in the Dian'chi Lake (Figure 10.2), so that the diversity of higher plants there fell from 16 species in the 1970s to 3 species in 1992 (Wu 1993). After 13 invasive species of fish such as *Ctenogobius cliffordpopei* were introduced into the

FIGURE 10.2 *Eichhornia crassipes* mat covering the Dian'chi lake in Yunnan province from shore to shore. Removal of a portion of *E. crassipes* mat (*left*). (Photo courtesy of Jinglun Wu.)

Er'hai Lake in southern Yunnan province, 5 of 17 native species, including a carp endemic to the lake and a schizothoracine fish (*Schizothorax kozlovi*), were on the brink of extinction, and a number of the other native populations were decreasing significantly (Xie and Li 2006).

Invasive pests have posed a major threat to the natural functioning of our local ecosystems. As the inventory of invasive species expands in China, we run the risk that many of our ecosystems will no longer be able to deliver important services such as erosion prevention, water and air purification, and climate amelioration. Eucalyptus trees were introduced from Australia and planted more than 130,000 ha in many forested areas of Hainan Island and Le'zhou peninsula, Guangdong province in southern China. By taking up water through their roots, these trees made the soil arid and infertile and subsequently prone to erosion, and unsuitable for nearly all other plants. Malignant invasive weeds such as common ragweed and alligator weed impede forest regeneration from fire suppression in northeast China. In some parts of northwest China, these weeds impede restoration of overgrazed grasslands that are no longer suitable for most grazing animals.

Biological invasions have also seriously impacted the genetic diversity of local ecosystems. Many invasive pests genetically erode native species by crossbreeding with adjacent native stock. Additionally, as invasive populations fragment, surround, and penetrate native populations, the native plants gradually begin to inbreed within the smaller fragments, leading to genetic drift within the native populations. Different genera of invasive Canada goldenrod (*Solidago canadensis*) and native China aster (*Aster ptarmicoides*) crossbred in Shanghai, as did stocks of Chinese larch (*Larix*) and adjacent Japanese larch (*L. kaempferi*) in northeast and north China. Hybridization of Chinese *Sonneratia* and petal-free *S. apetala* from Bengal, both planted together in Hainan province, carries the risk of genetically swamping the native stock. It is notable that an established invasive species generally poses various impacts when it becomes integrated over a sustained period in a native ecosystem.

Invasive species have concurrently posed serious economic losses in China. The annual direct loss imposed by several notorious invasive species was estimated to be US$7 billion. The annual cost of total invasive species in China has not yet been calculated, but it probably runs in dozens of billions of U.S. dollars.

10.5 EXAMPLES OF INVASIVE PESTS

10.5.1 MILE-A-MINUTE WEED

The mile-a-minute weed is a native liana grass of the composite family in Central and South America. It first occurred in Hong Kong early in 1919. By the early 1980s, its climbing and twining stems overtook almost all other plants on the Inner Lingding Island National Nature Reserve in Shen'zhen city, Guangdong province. The Nature Reserve originally contained species such as banana, litchi, longan, and wild orange as well as other shrubbery and arbor trees such as mangroves fed on or inhabited by more than 600 rhesus macaques, pangolins, boa constrictors, and other protected animals. Because of its high reproductive ability, the weed has spread to over 80% of the 500 ha island, so that now the ecosystem is at a huge risk and the macaques and other animals there face severe shortages of their food source of natural fruits. All over the Pearl River delta, the creeper has rapidly spread over more than 2,700 ha of woodland.

10.5.2 Water Hyacinth

This floating plant native to South America is the worst aquatic weed in the world, and a major problem in southern China, covering up to several million hectares. Boat traffic on dozens of rivers including the Pearl River in Guangdong province, the Huangpu River in Shanghai, and the Min River in Fujian province has been halted; additionally, hundreds of lakes and ponds, such as the well-known Dian'chi lake in Kun'ming city, Yunnan province, have been covered from shore to shore (Figure 10.3). Although it is thought that water hyacinth cannot survive winters in northern China, it does survive freezing conditions in the extensive middle provinces of China where it has become established. This invasive species was intentionally introduced in 1901 as an ornamental plant with its unusual appearance and attractive flowers, and in the 1950s–1960s was encouraged to proliferate nationwide as forage for pigs. Whatever benefits this invasive plant provided were greatly overshadowed by the environmental invasiveness and economic cost of this noxious species, especially in the warmer south of China. It has tremendous growth and reproductive ability,

FIGURE 10.3 Forests infested by the pine nematode and its vectors *Monochamus* spp. (*upper right*) in Nanjing (*upper left* and *middle*), Jiangsu province, and Guangzhou (*lower right*), Guangdong province.

reproducing sexually through seeds and vegetatively through daughter plants, which causes substantial problems. In the noted Dian'chi lake, a small population of water hyacinth was introduced intentionally to remove excess nutrients in the water and to provide animal forage. The plant subsequently spread over all 1,000 ha of the lake, with dense cover at the water surface that interfered with navigation, recreation, and irrigation. This mat also competitively excluded native submersed and floating-leaved plants and further excluded native fish, thereby preventing sunlight and oxygen from getting into the water. Just in the last couple of years, more than US$20 million was spent in Shanghai city and US$220,000 by the Shuikou Power Plant, vital to the Fujian provincial economy, in order to clear water hyacinth mats floating in rivers, entangling tugboats and impeding water flow. However, the effectiveness of control measures, mainly through harvesting and use of aquatic herbicides, are limited because of the plant's prodigious reproductive ability.

10.5.3 Serpentine Vegetable Leaf Miner

The serpentine vegetable leaf miner (*Liriomyza sativae*), an insect that feeds on vegetable mesophyll tissue between the upper and lower epidermis of leaves, was first found in the southernmost city of San'ya, Hainan province in 1993 and spread to 280 ha in Hainan and Guangdong provinces in 1994. To date, it has rapidly spread nationwide in China (except in the areas of Tibet and Heilongjiang provinces), and causes as much as US$600 million in losses annually. The adult fly has 9–10 generations of offspring, and its larval maggots feed on 68 species of vegetables, including cowpea, kidney bean, haricot bean, spinach, lettuce, towel gourd, cucumber, melon, tomato, and eggplant in the Yangtse River delta. The damage that resulted from the leaf miner activity to the yield of host plants was about 10%–15% on average in Jiangsu and neighboring provinces, with some extreme cases of 100% loss. Feces of the maggots in serpentine tunnels have contaminated leafy tissue intended for consumption. In vegetable crops not marketed for their foliage, such as legumina and cucumber, stunting occurs due to reduction of the photosynthetic leaf surface area. For control of this pest, it is important to detect leaf miner damage as early as possible. Implementation of control measures must be done before the leaf miner enters the leaf where it will then be protected by the plant itself. Deep plowing in early spring to destroy infested weeds such as plantain and nightshades and plant material from the previous season reduces the severity of the leaf miner outbreaks. Three parasitic wasps, *Opims* sp., *Halticoptera circulus*, and *dacmusa* sp., and several predators such as *Propylaea japonica*, *Harmonia axyridis*, *Lycosa pseudoannulagu*, and *Eubrellia pallihe* provide some natural control of the leaf miner and other five species of leaf miner flies in Jiangsu province. Because of the protected habit of the maggots within the plant, insecticidal control is difficult. Chemical insecticides such as Abamectin and Imidacloprid are applied for control of the maggots, with efficacy of about 85%.

10.5.4 Rice Water Weevil

The rice water weevil has been an important biological constraint on rice yields in Liaoning province of northeast China, mainly by way of root pruning by its larvae.

It was first found in Japan in 1976, Korea in 1988, and North Korea in 1989. In China, the weevil was first found in Tang'hai county of Tianjin metropolis in 1988, Dong'gang city of Liaoning province in 1989 (the source of this introduction was North Korea), and Ji'an city of Jilin province in 1993. To date, it has spread to 12 provinces (Tianjin, Beijing, Hebei, Shandong, Jiangsu, Zhejing, Fujian, Anhui, Henan, Shanghai, Jilin, and Liaoning) in mainland China. It was also found in Tao'yuan county, Taiwan in 1990. The extent of the pest on the mainland exceeds 7,000,000 ha, mainly concentrated in northeast China where this species has become the most important insect pest. Typical losses of rice yield there due to weevil damage exceed 10% and can approach 30% or more. This pest is controlled primarily by pyrethroid, organophosphorus insecticides such as cyhalothrin and isofenphos-methyl, fipronil, imidacloprid (Yang et al. 1997), and by cultural measures in the northeast. The goal of ongoing research is the development of a cost-effective and environmentally friendly program for integrated management of the weevil based on a thorough understanding of the biology of this important pest and its interaction with varieties of rice plants that are resistant to and tolerant of the weevil damage (Fang 2001).

10.5.5 Pinewood Nematode

The pinewood nematode is vectored by cerambycid beetles in the genus *Monochamus*, which are opportunistic and largely saprophytic insects. It has caused serious damage to the pine forests of China, Japan, and Korea, many of which are susceptible to the nematode. However, it generally does no damage to pines native to North America because the pines are not susceptible (He 2000). The nematode was first found in Sun Yat-sen's cemetery scenery zone, Nanjiang, Jiangsu province in 1982, and since then its range has steadily spread in China (Figure 10.3). As the population has grown, its rate of spread has even increased. Since 1995, the nematode-infested area has doubled from 40,000 to 80,400 ha (Forestry Society of China 2002). In 2002, the annual economic loss from nematode damage alone to pine forests approached US$2.2 billion. In five provinces, over 20 million pine trees have died, and 50 million infested pine trees have been cleared from the forests. To make matters worse, the nematode is closing in on some well-known Chinese national parks such as the Huang'shan mountains and the Zhang'jia'jie national park, which is attracting great concern nationwide. By now, quite effective integrated management measures for the beetles and the nematode have been developed and used successfully to control their further spread and damage in Jiangsu and Guangdong provinces. These measures include clearing and destroying the infested pines as thoroughly as possible in all epidemic forest areas, imposing quarantines on infested pines, and replanting resistant pines such as *Pinus taeda* and *P. echinata*. An effective trap baited with volatile lures was also developed and used to manage the vector beetles. Use of certain parasites such as *Scleroderma guani, Dastarcus longulus*, and *Ibalia leucospoides* to control the beetles has been explored (Wang et al. 2001; He 2000). In addition, *S. guani* has been manually propagated in mass and released at 5,000–10,000 wasps/ha to the forests in Ju'rong, Ni'shui, Wu'xi

FIGURE 10.4 A mass of *Scleroderma guani* adults reproduced artificially (*upper* photo), released on pine trees (*lower left*) in the infested area by the sawyer beetles *Monochamus*, and then hunting for (*lower middle*) and laying eggs on the beetle larvae (*lower right*) in undermined pine tunnels in Jiangsu, China. (Photo courtesy of Fuyuan Xu.)

cities, and etc., of to Jiangsu province with 48.7%–84.2% of beetle larvae ultimately parasitized (Figure 10.4). More than US$600 million has been spent to control the spread of this pest. The nematode remains the number one forestry pest in China and poses a devastating threat to pine forests in southern China.

10.5.6 Red Turpentine Beetle

The red turpentine beetle, *Dendroctonus valens* LeConte, is a common pest in North America. Yet despite the abundance and wide distribution of this beetle, outbreaks have not been extensive or severe in that region. In China, however, since its first outbreak in pine forests of Shanxi province in 1999, the invasive pest has spread rapidly to five other adjacent provinces, Hebei, Henan, Shuanxi, Shandong, and Gansu, and infested over half a million hectares of pine stands. In addition, it has caused severe mortality of the Chinese pine, *P. tabulae*. More than 3 million of the Chinese pines, as well as some other pines such as *P. bungeana*, had been killed by 2001 (Li et al. 2001; Miao et al. 2001). The historical record has shown that the pest was introduced into China in early 1980s when unprocessed logs were imported from the west coast of the United States. Several consecutive years of drought conditions have severely stressed its primary host, *P. tabulae* and contributed significantly to the sudden outbreak. With the use of pines as a major reforestation species in China, and *P. tabulae* widely planted across a large portion of the country, the potential range and damage by this invasive beetle is overwhelming (Li et al. 2001). The beetle has been found colonizing pine roots extensively and has overwintered inside the roots in China. Overlapping generations have made it hard to treat the beetle effectively with chemicals, because there are larvae present inside the roots year around (Miao et al. 2001). Therefore, an alternative and more environmentally friendly method of control needs to be developed.

10.6 MANAGEMENT OF INVASIVE PESTS IN CHINA

Responding to the urgency of the status of invasive pests occurring in China, priorities have focused on promoting the national capacity of administration, research, and management for these pests. This is being accomplished by comprehensively intercepting invasive pests from neighboring countries and major trade partners; enforcing quarantines on imported goods that may contain pests; identifying all mechanisms of invasion by invasive pests; understanding biological bases of invasive pests; and using environmentally friendly methods for control of pests and recovery of affected habitats.

A key national research program and other linked national integrated research and management programs on major and potential exotics in China were launched from early to late 2001, based on specific items within several national research programs focusing on single invasive pests since 1993. Five-hundred and ten national monitoring spots for invasive pests have been set up and put into effect. Six national integrated management projects in forestry have made great progress in areas of concern such as the spread of the pinewood nematode and restoration of disturbed and damaged habitats so that they are more resistant to further invasion by and establishment of many other invasive pests. Complete elimination of invasive pests, although a laudable goal, does not seem biologically or economically feasible in most cases (Bomford and O'Brien 1995). Some invasive pests such as serpentine vegetable leaf miner and alligator weed are too widespread and too firmly established to be eliminated. Nevertheless, it is important to realize that every alien species carries risks in various parts of China.

We must provide detailed biological information to guide management decision making. Monitoring techniques based on the biology of invasive pests need to be developed to improve detection at port facilities. Inspectors simply cannot keep up with the flow of trade using existing techniques. New advances in the field of molecular biology look promising, but need further development to make these techniques practical for large-scale implementation. A broad spectrum of insect traps for port facilities is needed because it is impossible to identify and build individual traps for all possible threats. And of course, trained taxonomists are much needed to identify the insects captured in such traps.

ACKNOWLEDGMENTS

We thank Professors Jinglun Wu, Fuyuan Xu, and Jiaxin Zhang for providing some photos of invasive pests and their damage in China, and Dr. Kazuo Hirai for his valuable comments and revision suggestions for this manuscript.

REFERENCES

Bomford, M. and O'Brien, P., Eradication or control for vertebrate pests? *Wildl. Soc. Bull.*, 23, 249, 1995.

Che, J.D. and Guo, X.H., Canada goldenrod, *Solidago Canadensis, Chin. Weed Sci.*, 1, 17, 1999.

Fang, J.C., Advances in strategy and techniques of integrated pest management on rice, in *Developmental Strategy of Plant Protection Facing the 21st Century*, Chinese Science and Technology Press, Beijing, 326, 2001.

Forestry Society of China, Status of forest insect pests and diseases occurring in 2001 and tendency of the pests in 2002, *Newsl. Entomol. Soc. Jiangsu Prov.*, 1, 6, 2002.

He, R.Q., Pest status of pine wood nematode and its management, *For. Sci. Technol.*, 4, 5, 2000.

Huan, Z.L. and Cao, H.L., Occurring and damage status of mile-a-minute weed in different habitats and forest, *Chin. J. Trop. Subtrop. Plant*, 8, 2, 131, 2000.

Li, D.Q., Effect of ragweed on biodiversity, in *Abstract of the First National Biodiversity Protection and Sustainable Utilization*, Entomological Society of China, Beijing, 107, 1994.

Li, J.S., Chang, G.B., Song, Y.S., Wang, Y.W., and Chang, B.S., Management of red turpentine beetle, *Chin. For. Pest Dis.*, 20, 4, 41, 2001.

Liu, L.H. and Xie, S.C., Distribution, damage and control approach of *Ageratina adenophora* in China, *Acta Ecol. Sin.*, 5, 1, 1, 1985.

Miao, Z.W., Zhou, W.M., Huo, F.Y., et al., Biology of red turpentine beetle, *Shanxi For. Sci. Technol.*, 1, 34, 2001.

Qiang, S. and Cao, X.Z., Investigation and analysis on alien invasive weeds in China, *Chin. J. Plant Resour. Environ.*, 9, 4, 34, 2000.

Wan, F.H. and Guan, G.Q., *Ragweed and Integrated Management for the Ragweed*, Chinese Science and Technology Press, Beijing, 1993.

Wang, C.L., Current status and management strategy of exotic invasive insects and diseases, *Newsl. Entomol. Soc. Jiangsu Prov.*, 1, 4, 2002.

Wang, D.L., Review on plant chemical sensitiveness in ragweed, *Chin. J. Ecol.*, 14, 4, 48, 1995.

Wang, M.X., Cheng, L.C., and Song, Y.S., Pest risk analysis of pine wood nematode (*Bursaphelenchus xylophilus*) on Hunan forestry and ecological environment, *For. Pest Dis.*, 20, 2, 42, 2001.

Wu, J.L., Li, Y.F., Li, G., et al., Occurring and management of Canadian goldenrod (*Solidago canadensis* L.) in Jiangsu, *Jiangsu J. Agric. Sci.*, 21, 1, 11, 2005.

Wu, K.Q., Ecological equilibrium in the Dian'chi Lake area, *Newsl. Domest. Lake Collab. Netw.*, 1, 47, 1993.

Xie, Y. and Li, Z., Characteristics of alien invasive species in China. Available at http://www.baohu.org (in Chinese), 2006.

Yang, H.J., Xia, L.R., and Fang, J.C., Effect and application technique of imidacloprid against rice water weevil, *Jiangsu J. Agric. Sci.*, 5, 47, 1997.

11 Invasive Plants in Australian Forests, with an Emphasis on Subtropics and Tropics

J. Doland Nichols and Mila Bristow

CONTENTS

11.1 FOREST COVER IN AUSTRALIA

Australia is both a continent and a country, with a total land area of 768 million ha. Centered over the southern hemisphere subtropical high-pressure system, the continent is mostly quite dry. The distribution of effective rainfall to most of Australia imposes limitations on where forests, and hence forest weeds, can grow. Using a very liberal definition of forest as a vegetation type that has a height of woody plants of at least 2 m and overstory crown cover of greater than 20%, one may say that ~21% or 164 million ha of Australia are forests (Bureau of Rural Science 2003).

Of this area, ~127 million ha, or 78% of all forest, is eucalypt forests of various types. Of this, only about 93,000 ha is in closed-canopy eucalypt forest (>70% cover), and the remainder is classified as either open (canopy) forest (30%–70% cover) or woodland. About 10% of the total forest area is in acacia stands, 4% in *Melaleuca*, and 3% in rainforest. There is a total of 1.6 million ha of plantations, including nearly 1 million ha in exotic pines (*Pinus radiata, P. elliottii*, and *P. elliotii* × *P. caribaea* hybrids). The remaining 600,000 ha are planted with hardwoods (National Forest Inventory 2005), mainly eucalypts, with hardwood plantations expanding at around 80–100,000 ha per year.

11.2 DEFINITION OF WEED

An array of confusing terms exists in relation to exotic, invasive, or weedy plants. In Australia, the National Weed Strategy (NWS) recognizes that virtually any plant can be a weed (Department of Agriculture Fisheries and Forestry 1999). A common definition used is "any plant in the wrong place" and as such is a subjective term. Some more refined definitions of weeds are listed in Table 11.1. Groves et al. (2003) discuss some varying criteria for the definitions and use of terms. One definition describes a weed as a suite of characteristics including fecundity, vigor and persistence, and adaptability to different environmental conditions. Not coincidently, these attributes also apply to a suitable forest tree or hardy garden plant, and such attributes are generally sought after when selecting species for commercial introduction by these industries (Bristow and Skelton 2004).

TABLE 11.1
Common Weed Nomenclature Used in Australia

Weed Type	Definition	Source
Noxious weed	A plant that has the ability to grow quickly, spread widely, and compete strongly	Department of Agriculture Fisheries and Forestry (1999)
Environmental weed	A plant that has less direct impact on humans, but is detrimental to the conservation of our natural resources	Williams and West (2000)
Naturalized species (or taxa)	An introduced species that has established self-supporting, viable populations in the wild	Department of Agriculture Fisheries and Forestry (1999) and Glanznig (2005)
Invasive species	A naturalized species that is spreading	Australian Bureau of Statistics (2002) and Glanznig (2005)
Declared weed	A pest plant species that has, or could have, serious economic, environmental or social impacts, and hence is declared under State legislation and targeted for control	Department of Natural Resources and Water (2006)
Sleeper weeds	Weeds that are established or newly arrived but are not as yet a widespread problem	Australian Bureau of Statistics (2002)
Ruderal species	A plant species that colonizes disturbed areas (such as road verges)	

Since most of the weeds in Australia originated from other countries, some agencies and NGOs consider all exotic (alien, nonnative, or introduced) plants to be weeds, or to have weed potential or risk (Grice and Setter 2003; Werren 2003; Wilson et al. 2004). Approximately 2700 introduced plants that are known to be naturalized (10% of total plant introductions, estimated by Virtue et al. 2004) have been categorized into major threats (30%) or minor threats (53%) to natural ecosystems and agriculture in Australia (Groves et al. 2003). The term weed is not limited to introduced plants; native plants can be considered weeds when they establish outside their natural habitat or increase in abundance as a result of human disturbance (see *Acacia dealbata* in forestry plantations, Hunt et al. 2006; and umbrella tree (*Schefflera actinophylla*) and cadaghi (*Eucalyptus torelliana*) rainforest trees in urban areas, Anonymous 2002).

Websites such as "A Global Compendium of Weeds" (Randall 2002) contain lists of potential weed species, and in some cases, the source material is not readily traced (Bristow and Skelton 2004). Regardless of veracity in these listings, in the absence of long-term studies to develop understanding of the ecology of weed species, commercial growers of some introduced plants are advocating responsible use policies; for example, the use of African mahogany (*Khaya* spp.) in plantation forestry (Bristow and Skelton 2004) and *Leucaena* spp. for cattle fodder (RUMP 2004; The Leucaena Network 2004; Emms et al. 2005).

Distinguishing exactly which plants are entirely native and which are naturalized and/or weedy is not always clear-cut. For example, in the single genus *Erythrina* in the Faboideae, there are some 150 species present pantropically. There are 5 species in this genus in Australia: 3 endemic and 2 naturalized. In the state of New South Wales is *E. vespertilio* Beth, a native, *E. X sykesi* Barneby & Krukoff, which is a cultivated rarely naturalized hybrid of origin probably in New Zealand, and *E. crista-galli* L., a native to South America, cultivated and widely naturalized.

It is challenging even to define which species among the many naturalized ones are invasive. In an attempt to prioritize environmental weeds, a project in southeast Queensland sought to rank invasive naturalized plants via a panel of weed scientists and botanists (Batianoff and Butler 2002). They began with a list of 1060 naturalized species. On the basis of field observations and understanding of biological performance of species, they reduced this to a group of 200 invasive naturalized species for listing as environmental weeds. One-third of these were categorized as highly invasive. Although such a weed list can be useful, the authors explain that the process of weed invasion is ongoing, with an average 87 new naturalizations recorded per decade in the region. Thus, new information and reassessment of priorities will constantly be required.

Woody plants in the highly invasive category had been in residence for an average of 90 years, including top-ranked weed *Lantana camara* (present for 120 years). Nonarboreal plants, herbs and vines, had an average time of 42 years since first specimen record, suggesting that it takes longer for woody species to become established as weeds. Many of the current environmental weeds in southeast Queensland had been established in association with Arbor Days in the 1890s. In the late 1800s, acclimatization societies emerged around Australia with the aim of introducing exotic plants and animals, whether useful or ornamental, primarily for dispersion to suitable parts of the colony. In short, these regional societies were fixated with creating a New England landscape by altering the natural landscape of Australia (Australian Bureau of Statistics 2002).

Our focus here is mainly on environmental weeds, defined as "plants that invade natural ecosystems and are considered to be a serious threat to nature conservation" (Williams and West 2000). Weeds cost Australia at least AU$3 billion per year (Vranjic et al. 2000). But this estimate is only for the cost of control and of losses due to agricultural and pastoral weeds. If somehow the costs of environmental weeds could be established, then the costs to the country would be much greater, from damage to water quality, loss of biodiversity, tourism, recreation, and general aesthetic values (Goosem 2003).

11.3 WEEDS OF TROPICAL RAINFORESTS

The rainforests of Australia are discontinuously distributed across the coastal and adjacent ranges of eastern and northern Australia. Occurring in a variety of climatic and edaphic ecotypes, rainforests are structurally and floristically diverse. Some are so diverse, like those in north Queensland's wet tropics bioregion, that they have been accorded international significance with World Heritage listing (Werren 2003). These wet tropical rainforests contain over 4600 plant species, 50% of Australia's bird species, a large proportion of Australia's invertebrate species, and many endemic vertebrate animal taxa (Goosem et al. 1999). As with rainforests worldwide, Australian rainforests have been extensively cleared and/or disturbed over the past century; with about 30% of pre-European extent now cleared. There is concern that remaining areas of rainforest may be degraded, losing biodiversity and structure, because of fragmentation and habitat loss, and that the high perimeter-to-area ratios will result in remnants being more prone to weed invasion and altered fire regimes (Erskine 2002; Fox et al. 1997; Goosem 2003; Tucker 2000).

Rainforests are important for their biological diversity, their provision of ecosystem services, and their direct and indirect economic importance (Gould 2000; Grice and Setter 2003; Kanowski et al. 2005; Lamb et al. 2005), and the threat posed by weeds to these beneficial aspects is not well known. However, significant infestations of weeds are caused by a variety of life forms (i.e., vines, grasses, forbs, shrubs, trees, and aquatic plants) (Werren 2003), and are spread by a range of dispersal mechanisms (Westcott and Dennis 2003) into, and around the edges of, rainforests and associated ecosystems (Goosem 2003). With 510 exotic plant species naturalized in Queensland's wet tropics rainforests, there are increasingly more examples of weeds invading and threatening these forests (Werren 2003; Wet Tropics Management Authority 2006).

11.3.1 Case Study: Pond Apple

There are many rainforest ecosystems classified as *endangered* or *of concern*. This is related in part to the history of extensive clearing and conversion of these forest types, and also to the risk from future disturbance and exotic plant invasion. Tropical lowland freshwater paperbark (*Melaleuca* spp.) swamps are one such threatened ecosystem, and are rapidly invaded by pond apple, *Annona glabra* (Werren 2003). With buoyant fruit and seeds, pond apple is water dispersed. Its fruit, however, is also attractive to vertebrate frugivores such as the rare and threatened southern cassowary, *Casuarius casuarius*, and the introduced, feral pig, *Sus scrofa* (Westcott and Dennis 2003). Pond apple is a woody, semiaquatic tree that establishes along streams, disrupts the hydraulic conductivity, traps sediments, and blocks waterways. Because of complex dispersal mechanisms, rapid growth, and their resultant impairment of ecosystem function, it is a particularly problematic weed of riparian edges of rainforest.

11.4 WEEDS OF SUBTROPICAL RAINFORESTS

Zones of high rainfall in the subtropics generally have soils that are too infertile to support rainforest. But there are a few areas, centered near the NSW–Queensland border and on soils derived from relatively recent volcanic activity, that support highly diverse subtropical rainforests, similar in appearance and biodiversity levels to true tropical rainforests farther north. Since these rainforests are close to heavily settled areas, the Gold Coast, Brisbane, and northern NSW, they are susceptible to human interference and invasions by weeds.

For example, one 75,000 ha area of subtropical rainforest, the Big Scrub, was almost entirely cleared for agricultural and pastoral uses in the late nineteenth century. Weeds present two different kinds of problems in efforts to maintain or restore this particular rainforest:

1. Weeds, especially vines, can easily invade and overcome remnants of this forest type.
2. Weeds are a hindrance, and their elimination poses a large expense in efforts to restore rainforest.

Field days on which volunteers clear weeds have been important in preservation of remnants. Since rainforest seedlings are costly and most areas to be planted are in pastures, restoration plantings are expensive to establish. In this local region, the efforts of nongovernmental organizations have also been important in developing a good understanding of the key problem presented by weeds and of the techniques for controlling them (Big Scrub Rainforest Landcare Group 2000). Among the most dominant of exotic weeds in this area is the tree from China, *Cinnamomum camphora* (L.) Nees and Eberm.

11.4.1 CASE STUDY: CAMPHOR LAUREL

Many environmental weeds in Australia started off originally as garden trees with desirable aesthetic traits, for example, *Jacaranda mimosifolia* (jacaranda), *Celtis sinensis* (Chinese elm), and *Schefflera actinophylla* (umbrella tree). All of these vigorously growing species are able to reproduce and spread in the apparent absence of insects and pathogens that may control their spread in their home range (Adkins and Walker 2000). Another weed of this type, *Cinnamomum camphora*, known as camphor laurel, has been in Australia since at least 1854 (Firth 1980). In the subtropics, camphor laurel was reputedly not a major problem until the 1970s (ANPWS 1991). At 40 years of age and beyond, trees produce large number of fruits, which are dispersed by native birds. This has led to the plant reaching an expansion phase of spread. This development has also coincided with the abandonment of large areas of pasture, which subsequently were easily colonized by camphor laurel.

A large spreading tree, camphor laurel now dominates much of the landscape on better *ferrosol* soils, originally occupied by subtropical rainforest, in northeast NSW and southeast Queensland. Popularly promoted in urban areas in Arbor Days from the 1890s, this tree is common in older parts of towns and cities. Although bringing prices as high as AU$500/m^3 of wood for uses including bench tops, tables, boxes, and breadboards, most trees have several stems and do not contain easily exploitable wood. Because this species sprouts readily, many camphor-laurel-occupied environments contain thousands of stems per hectare. Unless seedlings are successfully uprooted or larger trees completely ringbarked, plants tend to revive, making elimination difficult.

Many species of rainforest birds have been recorded in camphor laurel, including at least 10 species with high potential to distribute rainforest seeds (Neilan et al. 2006). As part of an environment to encourage rainforest regeneration, camphor laurel has been found to be more hospitable than the vegetation that usually precedes it, namely open paddocks. Neilan et al. (2006) found that many rainforest tree seeds are able to germinate and produce seedlings beneath a camphor laurel canopy. They argued that the combination of seed-feeding birds and camphor-laurel-altered environments could assist subtropical rainforest succession.

Camphor laurel has at least two chemotypes (Stubbs and Brushett 2001) and contains a dozen or more essential oils (Schenk unpublished PhD thesis). In the popular press, the species is frequently alleged to be both toxic (to fish and frogs, for example) and allelopathic to other plants, though neither of these properties has been substantiated (Low 1999; Schenk unpublished PhD thesis). Camphor laurel, like eucalypts, produces a commercial oil, albeit one with a low price, AU$3–5 per kilogram (Stubbs et al. 1999). Proposals have been made for using cleared camphor laurel trees for energy production with oil as a by-product.

Ultimately, the case of camphor laurel illustrates the pros, cons, and the ambiguity that can be associated with an invasive plant. It is true that camphor laurel aggressively colonizes areas that could support rainforest or mesic eucalypt forests, forming multistemmed thickets of trees with little commercial value. On the other hand, camphor laurel colonizes bare land and degraded pastures, creating conditions amenable for rainforest regeneration. In addition, it is a large tree providing shade in many towns where summer temperatures are frequently over 40°C. Further, slabs of camphor laurel are a high-value speciality timber, and it may well be used for other purposes.

11.4.2 Case Study: Lantana

L. camara L. (Verbenaceae) is a profusely branching, scrambling scrub, usually 2–4 m tall and wide. Lantana is a species complex that originated in tropical

America, and includes some 650 named varieties. According to the NSW National Parks and Wildlife Service (NSW NPWS 2006), most taxa in Australia are tetraploids and even though ploidy level and ecological correlates have not been identified, "the genetic and chromosomal levels of diversity, within the species, its propensity for somatic mutation, and the interfertility of many of the taxa, give it a high level of potential adaptability as a pest species."

Lantana generally grows well when there is moderate to good summer rainfall and sites are well drained. It is found along the east coast of Australia from southern NSW to Cape York, mainly in NSW and Queensland, having invaded 4 million ha. This plant is an invader of disturbed sites, colonizing forest edges and canopy gaps, and often is dominant in regenerating pastures. It forms dense thickets that suppress native vegetation and carry intense fires.

Lantana has been in Australia for more than 160 years. By 1879, Lantana was reported as a *most troublesome weed*. It has gained status as a top-twenty weed of national significance (Thorp and Lynch 2000; WONS 2004). To meet this classification, it needed to meet the standard statement of a *top* weed: "... one of the worst weeds in Australia because of its invasiveness, potential for spread, and economic and environmental impacts."

The environmental impacts of lantana are quite varied. Lantana is particularly notorious because it grows in and threatens moist eucalypt forests and rainforests, habitats with high biodiversity. The NSW NPWS estimates that at least 80 species of plants are threatened by the extensive coverage of the landscape by lantana (NSW NPWS 2006). However, lantana is arguably a keystone species for many species of animals. It provides shelter for marsupials like wallabies and bandicoots, forage (in the form of leaves and fruits) for many bird species, and nectar for rare native bird species. Some vulnerable species like the black-breasted button quail are dependent on a native forest type—in this case, dry rainforest, which is mostly gone. This species has therefore become dependent on lantana for shelter.

11.5 BELL-MINER-ASSOCIATED DIEBACK: WOODY WEED, BIRD, INSECT, AND LOSS OF FOREST CANOPY

Bell-Miner-Associated Dieback (BMAD) is a forest health problem that results from the complex interaction of a variety of factors. It most commonly occurs in moist eucalypt forests, often ones dominated by *Eucalyptus saligna*, but usually including a mix of eucalypts. The actual cause of foliage loss, and ultimately tree crown dieback, is feeding by psyllids of various species but predominantly of the genus *Glycaspis*. BMAD is spreading widely in moist eucalypt forests throughout the state of NSW (Billyard 2004). A recent estimate of the affected area of mixed rainforest *E. saligna* in northeast NSW is 20,000 ha (C. Stone 2005, personal communication), with a further 2.5 million ha at risk.

A series of factors are involved in changing the conditions in the canopies of these forests so that psyllid outbreaks occur (Stone 1996, 1999, 2005; Florence 2005; Wardell-Johnson et al. 2006). Among these are disturbances, including fire and clearing for grazing or agriculture, varying levels of nutrients in leaves, and changes in forest structure. The last factor, the development of an understory that provides

FIGURE 11.1 Where lantana understory is providing bell miner nesting habitat, dieback from feeding by psyllids overoccurs in crowns of *Eucalyptus saligna*. (Photo courtesy of J.D. Nichols.)

cover for the bird bell miner (*Manorina melanophyrys*), may be the most critical in the development of dieback. This bird species is highly territorial and forms colonies in dense understories, particularly of lantana and rainforest shrubs and vines (Figure 11.1). When bell miners have dominance over a site, apparently other bird species that predate on psyllids are driven off; therefore, psyllid populations build up to high levels, causing loss of leaves and eventually of tree crowns. Without an overstory canopy and thus with higher light levels at ground level, lantana thrives and expands its range. Bell miner colonies follow and the dieback syndrome spreads.

Biological control in Australia has both a negative reputation (for an introduction like the cane toad) and a positive one, particularly for the successful introduction of an insect, *Cactoblastis cactorum*, to control prickly pear, *Opuntia stricta*. Many potential biocontrol agents have been introduced with lantana as a target, including the recent introduction of a Mexican sucking insect *Aconophora compressa* (Day et al. 2003a,b). To date, none have had much impact. Most people would agree that, even though the exact role of lantana in providing shelter to birds that then presumably allow psyllid populations to break out is not clearly understood, nevertheless an organism effective at controlling lantana would probably help lesser the spread of BMAD.

11.6 CASE STUDY: ARE PINES WEEDS?

Pine plantations in Australia, of which there are ~1 million ha, are an important source of construction timber. *Pinus radiata* dominates in the southern temperate areas of NSW, Victoria, and South Australia, while subtropical plantations like those near Brisbane are mainly *P. elliotti, P. caribaea*, or hybrids of those two. Groves et al. (2003) rank the tropical and subtropical *P. elliottii* and *P. caribaea* as "known to be a major problem at 3 or fewer locations within a State or Territory," with *P. elliottii* listed as a controlled plant in Queensland. The southern pines, *P. radiata* and *P. pinaster*, rank higher, "known to be a major problem at four or more locations within a State or Territory" (Groves et al. 2003). In a recent review of *P. radiata* in Australia, Williams and Wardle (2007) note that all *Pinus* species can be considered "invasive" and "successfully naturalised."

Pines tend to be invasive on disturbed sites rather than on occupied sites and are frequently seen on roadcuts near pine plantations. They are less invasive in established forests, where light understorey burns can control them. Richardson et al. (1994) found that the extent of invasion by a variety of pine species in the Southern Hemisphere was positively correlated with residence period. Further, they ranked ground cover categories according to their vulnerability to invasion as follows: forest < shrubland < grassland < dunes < bare ground. The general conclusion is that pines take some time to establish and can be relatively easily controlled, and so classifications such as that by Blood (2001), which state that pines are "highly invasive," need to be considered critically. Williams and Wardle (2007) conclude that wilding populations need to be continuously monitored. They also recommend monitoring trends in plantation establishment, with, for example, a tendency to plant more small-scale woodlots with higher boundary-to-size ratios, possibly increasing weed invasions.

11.7 AUSTRALIAN INSTITUTIONS AND INVASIVE PLANTS

In Australia, a bewildering array of institutions, public and private, commonwealth (federal), state, and local, address the many problems posed by invasive plants. These institutions range from centralized agencies with large budgets to small, local, mostly volunteer groups. Many of the State and Commonwealth government agencies involved with weed identification and management in Australia have contributed to a National Strategy on Weeds. This plan has three goals:

- To prevent the development of new weed problems
- To reduce the impact of existing weed problems of national significance
- To provide the framework and capacity for ongoing management of weed problems of national significance (Department of Agriculture Fisheries and Forestry 1999)

The NWS is advisory and has no statutory basis; however, it is administered through State Government agencies. Under the NWS, 20 introduced plants were identified as Weeds of National Significance (WONS) (Table 11.2) based on their

TABLE 11.2
Weeds of National Significance

Common Name	Scientific Name	Type of Weed	% of Australia That Could Potentially Be Occupied
Alligator weed	*Alternanthera philoxeroides*	Aquatic	0.4
Athel pine	*Tamarix aphylla*	Arid lands, shrub or tree	1.0
Boneseed, bitou bush	*Chrysanthemoides monilifera*	Shrub widely adapted, special problem in coastal sand dunes	3.0
Blackberry	*Rubus fruticosus* species aggregate	Temperate climates with annual precipitation >700 mm	9.0
Bridal creeper	*Myrsiphyllum asparagoides*	Southern Australia, especially woodlands	5.0
Cabomba/fanwort	*Cabomba caroliniana*	Aquatic	0.5
Chilean needle grass	*Nassella neesiana*	Temperate climate with >500 mm precipitation, mainly pastures	0.2
Gorse	*Ulex europaeus*	Mainly temperate, >650 mm precipitation, Tasmania and Victoria	3.0
Hymenachne	*Hymenachne amplexicaulis*	Grass in tropical wetlands and waterways	1.0
Lantana	*L. camara*	High rainfall areas in tropical, subtropical, and temperate areas	5.1
Mesquite, mimosa, parkinsonia	*Prosopis* spp.	Dense impenetrable thickets in pastoral country	5.3
	Mimosa pigra	Prickly shrub in tropical wetlands	1.0
	Parkinsonia aculeata	Aggressive shrub in semiarid ranges and wetlands	12.4
Parthenium weed	*Parthenium hysterophorus*	Annual herb in pastures	5.6
Pond apple	*Annona glabra*	Small tree in wet tropics wetlands	0.4
Prickly acacia	*Acacia nilotica*	Woody thorny weed in 6.6 million ha of arid and semiarid Queensland	2.3
Rubber vine	*Cryptostegia grandiflora*	Shrub or vine; loss of grazing	7.7
Salvinia, serrated	*Salvinia molesta*	Aquatic weed	5.0
Tussock	*N. trichotoma*	Perennial grass with no grazing value	2.2
Willows	*Salix* spp.	Damage to streambeds	0.8

Source: From Sinden, J., Jones, R., Hester, S., Odom, D., Kalisch, C., James, R. and Cacho, O., in *The Economic Impact of Weeds in Australia*, CRC for Australian Weed Management, Glen Osmond, 2004; From Weeds of National Significance (WONS), Weeds in Australia, http://www.deh.gov.au/biodiversity/ invasive/weeds/wons.html, 2006.

invasiveness, impacts, potential for spread, and their socioeconomic and environmental values (Department of Natural Resources and Water 2006). National management strategies have been published for all of these species. Despite this, an article in recent issue suggests that most Australian weed legislation is generally ineffective at stopping importation of new weed plants and at preventing spread of existing weeds (Glanznig 2005).

In the research sector, the Centre for Cooperative Research (CRC) for Weed Management coordinates activities among many member institutions, including universities and land management agencies. CRCs are typically 7 year programs that support studies, and dissemination of results, on a select group of topics.

11.7.1 Case Study: Rubber Vine

One of the most serious tropical WONS is rubber vine (*C. grandiflora* R.Br.). Rubber vine is a perennial woody climbing vine native to Madagascar, which was introduced to Queensland as an ornamental plant in the 1940s, and is now found across 34 million ha of northeastern Australia, with at least 700,000 ha heavily infested (ARMCANZ and ANZECC 2001) (Figure 11.2).

Rubber vine has impacts on the environment, primary production, and tourism. Its main impact is on pastoralism through the loss of grazing country, which in 1995 was estimated to cost the Queensland beef industry AU$18 million via increased costs of mustering and fencing.

FIGURE 11.2 Rubber vine is a climbing weed that *aggressively colonizes* native vegetation. (Photo courtesy of M. Bristow.)

Research into the control of rubber vine has been conducted for a long time. The leaf rust, *Maravalia cryptostegiae*, and rubber vine moth, *Euclasta gigantalis*, both released in 1995 have had significant impact on plants. Fire is the most effective management tool (Bebawi et al. 2000), and is recommended to be used in conjunction with chemical, mechanical, and biological control (ARMCANZ and ANZECC 2001).

11.8 MANAGEMENT PRINCIPLES AND CHALLENGES

In less than 100 years, Australia has changed as a country with an emphasis on establishing primary production industries across the countryside and recreating European landscapes to one dominated by an urban population that places an increasing value on native ecosystems and plants. In this context, the weed legacy left by the colonial period, and the continued dominance and invasion by a variety of weeds, have left both the agricultural and environmental sectors with major challenges.

The main aims of integrated weed management (IWM) for environmental weeds are as follows:

1. To contain effectively the spread of existing weeds
2. To manage the environment to prevent the spread of new weeds
3. To rehabilitate the disturbed ecosystem as well as possible (Vranjic et al. 2000)

To achieve these aims, it is of course important to identify the life cycles of weeds and understand how they spread. In Australia, this has been generally well done for existing weeds by commonwealth, state, local, and nongovernmental groups (ANPWS 1991; ARMCANZ 1999, 2001; Blood 2001; ANBG, 2006; Department of Natural Resources and Water 2006; WONS 2006).

Identifying potentially damaging invasive plants early and directing intensive elimination exercises at them can be much more effective than allowing individual species to build up uncontrollable populations. This means being aware of *sleeper* weeds (Table 11.1) and taking rapid effective action as soon as possible after introduction. Quarantine is a part of this process, with the Australian quarantine agency, AQIS being aware of weed problems and aggressively attempting to keep new weeds from entering the country. Among the many control techniques suggested, the use of herbicides often predominates. Chemical control needs to be applied with caution, lest damage be done to introduced biocontrol agents or to targeted native plants. A key Southern Hemisphere family, Proteaceae, is notoriously sensitive to herbicides, including glyphosate, and an endangered species, *Grevillea iaspicula*, was damaged by herbicides applied to the widespread temperate zone weed, blackberry (*R. fructicosus*).

One important principle that has emerged is that there is no point in eliminating a weed if there is no effective plan for how to replace it with native vegetation. For example, where the target restored vegetation is moist eucalypt forest, weed control

FIGURE 11.3 *L. camara* (foreground) and *Cinnamomum camphora*, camphor laurel, Northern New South Wales, Australia. (Photo courtesy of J.D. Nichols.)

should be timed to take advantage of large eucalypt seeding events. Even if the target weed itself does not regenerate through vegetative means or from a stored seed bank, there will often be many other weed species in the area ready to colonize a site.

In the subtropical rainforest area, lantana and camphor laurel often grow in close proximity (Figures 11.3 and 11.1). Even when cover and shade have been established, shade-tolerant weeds like camphor laurel and large-leaved and small-leaved privets (*Ligustrum lucidum* and *L. sinense*) can take hold and persist, becoming large understory or even canopy trees. Thus, any weed treatment needs adequate follow-up to be effective.

11.9 CONCLUSIONS

Weeds in Australia not only create large costs for agricultural and pastoral activities, but environmental weeds, including invasive plants in forests, continually degrade ecosystem services. As the climate changes, and new weeds are constantly introduced, and as society's focus shifts from primary production and recreating a European landscape to restoring native ecosystems, Australia faces an ongoing challenge in managing its massive weed problem.

The obsession with converting native *scrubland* to domesticated countryside has given way, in some areas, to an obsession with native species that at its most extreme would even slow down the natural processes of dispersal of plant propagules carried out by bats and birds. As Zavaleta et al. (2001) point out, many exotic species are here to stay, with no realistic possibility that they will be eliminated entirely. Further, invasive plants in Australian forests are now parts of complex new ecosystems and in many cases in fact provide benefits, including the preservation of native flora and fauna

REFERENCES

Adkins, S.W. and Walker, S.R., Challenges and future approaches to weed management, in *Australian Weed Management Systems*, Sindel, B.M., Ed., Cooperative Research Centres for Weed Management Systems, R.B. and F.J. Richardson, Melbourne, VIC, 481, 2000.

Agriculture and Resource Management Council of Australia and New Zealand (ARMCANZ), Australian and New Zealand Environment and Conservation Council and Forestry Ministers, The National Weeds Strategy: A strategic approach to weed problems of national significance, National Weeds Strategy Executive Committee, Canberra, ACT, 1999.

Anonymous, Southeast Queensland environmental weeds strategy 2000–2005, Department of Natural Resources, Brisbane, QLD, 2002.

ARMCANZ and ANZECC, Weeds of national significance: Rubber vine (*Cryptostegia grandiflora*) strategic plan, National Weeds Strategy Executive Committee, Launceston, 2001.

Australian Bureau of Statistics, Measuring Australia's progress, ABS Catalogue No. 1370.0, Commonwealth of Australia, Canberra, ACT, 2002.

Australian National Botanic Gardens (ANBG), Environmental weeds in Australia, http://www.anbg.gov.au/weeds/weeds.html, 2006.

Australian National Parks and Wildlife Service (ANPWS), Plant invasions: The incidence of environmental weeds in Australia, Part 1: A status review and management directions, Commonwealth of Australia, Canberra, ACT, 1991.

Batianoff, G.N. and Butler, D.W., Assessment of invasive naturalised plants in south-east Queensland, *Plant Prot. Q.*, 17, 1, 27, 2002.

Bebawi, F.F., Campbell, S.D., Lindsay, A.M., and Grice, A.C., Impact of fire on rubber vine (*Cryptostegia grandiflora* R.Br.) and associated pasture and germinable seed bank in a sub-riparian habitat of north Queensland, *Plant Prot. Q.*, 15, 2, 62, 2000.

Big Scrub Rainforest Landcare Group, Common weeds of northern NSW rainforest: A practical manual on their identification and control, Big Scrub Rainforest Landcare Group, Bangalow, NSW, 2000.

Billyard, R., BMAD Working Group perspective on forest decline and potential management solutions, in *Fundamental Causes of Eucalypt Forest Decline and Possible Management Solutions*, White, T.C.R. and Jurskis, V., Eds., Proceedings of Colloquium, November 18–19, 2003, Batemans Bay, NSW, State Forests of NSW, Sydney, 31, 2004.

Blood, K., *Environmental Weeds, A Field Guide for SE Australia*, CH Jerram, Mt Waverly, VIC, 2001.

Bristow, M. and Skelton, D., Preliminary weed risk assessment for *Khaya senegalensis* in plantations in northern Australia, in *Prospects for High-Value Hardwood Timber Plantations in the 'Dry' Tropics of Northern Australia*, Bevege, I., Bristow, M., Nikles, D.G. and Skelton, D., Eds., published as a CD ROM by Private Forestry North Queensland Association, Kairi, QLD, 2004.

Bureau of Rural Science, National Forest Inventory, *Australia's State of the Forests Report, 2003*, BRS, Canberra, ACT, 408, 2003.

Day, M.D., Wiley, C.J., Playford, J., and Zalucki, M.P., *Lantana: Current Management Status and Future Prospects*, Australian Centre for International Agricultural Research, Canberra, ACT, 2003a.

Day, M.D., Broughton, S., and Hannan-Jones, M.A., Current distribution and status of *Lantana camara* and its biological control agents in Australia, with recommendations for further biocontrol introductions into other countries, *Biocontrol News Inform.*, 24, 63N, 2003b.

Department of Agriculture Fisheries and Forestry, Revised National Weeds Strategy: A strategic approach to weed problems of national significance, Commonwealth of Australia, Canberra, ACT, 1999.

Department of Natural Resources and Water, Weeds and pest animal management, Department of Natural Resources and Water, Brisbane, QLD, http://www.nrm.qld.gov.au/pests/weeds/wons/index.html, 2006.

Emms, J., Virtue, J.G., Preston, C., and Bellotti, W.D., Legumes in temperate Australia: A survey of naturalisation and impact in natural ecosystems, *Biol. Conserv.*, 125, 3, 323, 2005.

Erskine, P.D., Land clearing and forest rehabilitation in the Wet Tropics of north Queensland, Australia, *Ecol. Manage. Restor.*, 3, 2, 135, 2002.

Firth, D.J., History of introduction of *Cinnamomum camphora* (camphor laurel tree) to Australia, *J. Aust. Inst. Agric. Sci.*, 46, 244, 1980.

Florence, R., Bell-miner-associated dieback: An ecological perspective, *Aust. For.*, 68, 4, 263, 2005.

Fox, B.J., Taylor, J.E., Fox, M.D., and Williams, C., Vegetation changes across edges of rainforest remnants, *Biol. Conserv.*, 82, 1, 1997.

Glanznig, A., Making State weed laws work, WWF-Australia Issues Paper, WWF-Australia, Sydney, 2005.

Goosem, S., Landscape processes relevant to weed invasion in Australian rainforests and associated ecosystems, in *Weeds of Rainforests and Associated Ecosystems*, Grice, A.C. and Setter, M.J., Eds., CRC Tropical Rainforest Ecology and Management and CRC Australian Weed Management, Cairns, QLD, 24, 2003.

Goosem, S., Morgan, G., and Kemp, J.E., Wet tropics, in *The Conservation Status of Queensland's Bioregional Ecosystems*, Sattler, P. and Williams, R., Eds., Environmental Protection Agency, Queensland Government, Brisbane, QLD, 1999.

Gould, K., A historical perspective on forestry management, in *Securing the Wet Tropics*? McDonald, G. and Lane, M., Eds., The Federation Press, Sydney, 85, 2000.

Grice, A.C. and Setter, M.J., Eds., *Weeds of Rainforests and Associated Ecosystems*, Cooperative Research Centre for Tropical Rainforest Ecology and Management, CRC Rainforest, Cairns, 116, 2003.

Groves, R.H., Hosking, J.R., Batianoff, G.N., et al., *Weed Categories for Natural and Agricultural Ecosystem Management*, Bureau of Rural Sciences, Canberra, ACT, 2003.

Hunt, M.A., Battaglia, M., Davidson, N.J., and Unwin, G.L., Competition between plantation *Eucalyptus nitens* and *Acacia dealbata* weeds in north-east Tasmania, *For. Ecol. Manage.*, 233, 260, 2006.

Kanowski, J., Catterall, C.P., and Wardell-Johnson, G.W., Consequences of broadscale timber plantations for biodiversity in cleared rainforest landscapes of tropical and subtropical Australia, *For. Ecol. Manage.*, 208, 359, 2005.

Lamb, D., Erskine, P., and Parrotta, J.A., Restoration of degraded tropical forest landscapes, *Science*, 310, 1628, 2005.

Low, T., *Feral Future*, Viking, Ringwood, VIC, 1999.

National Forest Inventory, Bureau of Rural Science, Australia's forests at a glance, 2005, Department of Agriculture, Fisheries and Forestry, Canberra, ACT, 2005.

Neilan, W., Catterall, C.P., Kanowski, J., and McKenna, S., Do frugivorous birds assist rainforest succession in weed dominated oldfield regrowth of subtropical Australia? *Biol. Conserv.*, 129, 3, 393, 2006.

New South Wales National Parks and Wildlife Service (NSW NPWS), *Lantana camara*—Key threatening process declaration, http://www.nationalparks.nsw.gov.au/npws.nsf/Content/lantana_ktp, 2006, accessed September 19, 2006.

Randall, R.P., A global compendium of weeds, http://www.hear.org/gcw/index.html, 2002.

Responsible Use and Management of Plants (RUMP), Policy to reduce the weed threat of Leucaena, Department of Primary Industries and Fisheries, Environmental Protection Agency, Department of Natural Resources and Mines, Brisbane, QLD, 2004.

Richardson, D.M., Williams, R.A., and Hobbs, R.J., Pine invasions in the Southern Hemisphere: Determinants of spread and invadability, *J. Biogr.*, 21, 511, 1994.

Schenk, J., Allelopathy in the ecology of camphor laurel, Southern Cross University, Lismore, NSW, SA, unpublished PhD thesis.

Sinden, J., Jones, R., Hester, S., Odom, D., Kalisch, C., James, R., and Cacho, O., *The Economic Impact of Weeds in Australia*, CRC for Australian Weed Management, Glen Osmond, SA, 2004.

Stone, C., Assessment and monitoring of decline and dieback of forest eucalypts in relation to ecologically sustainable forest management: A review with a case study, *Aust. For.*, 62, 1, 51, 1999.

Stone, C., Bell-miner-associated dieback at the tree crown scale: A multi-trophic process, *Aust. For.*, 68, 4, 237, 2005.

Stone, C., The role of psyllids (Hemiptera:Psyllidae) and bell miners (*Manorina melanophrys*) in canopy dieback of Sydney blue gum (*Eucalyptus saligna* Sm.), *Aust. J. Ecol.*, 21, 450, 1996.

Stubbs, B.J. and Brushett, D., Leaf oil of *Cinnamomum camphor* (L.) Nees and Eberm., *East. Aust. J. Essent. Oil Res.*, 13, 51, 2001.

Stubbs, B.J., Cameron, D.M., O'Neill, M., and O'Neill, R., From pest to profit: Prospects for the commercial utilisation of camphor laurel (*Cinnamomum camphora*) in the Northern Rivers Region of New South Wales, Centre for Coastal Management, Southern Cross University, Northern NSW Forestry Services, Lismore, NSW, 1999.

The Leucaena Network, A code of practice for the sustainable use of Leucaena-based pasture in Queensland, The Leucaena Network, Yeppoon, QLD, 2004.

Thorp, J.R. and Lynch, R., The determination of weeds of national significance, Commonwealth of Australia and National Weeds Strategy Executive Committee, Launceston, TAS, 2000.

Tucker, N.I.J., Linkage restoration: Interpreting fragmentation theory for the design of a rainforest linkage in the humid Wet Tropics of north-eastern Queensland, *Ecol. Manage. Restor.*, 1, 1, 35, 2000.

Virtue, J.G., Bennett, S.J., and Randall, R.P., Plant introductions in Australia: How can we resolve 'weedy' conflict of interest? in *Weed Management: Balancing People, Planet and Profit*, Sindel, B.M. and Johnson, S.B., Eds., Proceedings of the 14th Australian Weeds Conference, Weed Society of New South Wales, Sydney, NSW, 42, 2004.

Vranjic, J.A., Groves, R.H., and Willis, A.J., Environmental weed management systems, in *Australian Weed Management Systems, Cooperative Research Centres for Weed Management Systems*, Sindel, B.M., Ed., R.B. and F.J. Richardson, Meredith, VIC, 329, 2000.

Wardell-Johnson, G., Stone, C., Recher, H., and Lynch, J.J., Bell miner associated dieback (BMAD), Independent scientific literature review: A review of eucalypt dieback associated with bell miner habitat in north-eastern New South Wales, Australia, Occasional Paper DEC 2006/116, Department of Environment and Conservation, Coffs Harbour, NSW, 2006.

Weeds of National Significance (WONS), Weeds in Australia, http://www.deh.gov.au/biodiversity/invasive/weeds/wons.html, 2006.

Weeds of National Significance (WONS), Lantana control manual—Current management and control options for lantana (*Lantana camara*) in Australia, Natural Heritage Trust, Department of Natural Resources, Mines and Energy, Queensland, New South Wales Department of Agriculture, Brisbane, QLD, 2004.

Werren, G.L., A bioregional perspective of weed invasion of rainforests and associated ecosystems: Focus on the Wet Tropics of North-East Queensland, in *Weeds of Rainforests and Associated Ecosystems*, Grice, A.C. and Setter, M.J., Eds., Cooperative

Research Centre for Tropical Rainforest Ecology and Management, CRC Rainforest, Cairns, QLD, 9, 2003.

Westcott, D.A. and Dennis, A.J., The ecology of seed dispersal in rainforests: Implications for weed spread and a framework for weed management, in *Weeds of Rainforests and Associated Ecosystems*, Grice, A.C. and Setter, M.J., Eds., Cooperative Research Centre for Tropical Rainforest Ecology and Management, CRC Rainforest, Cairns, QLD, 19, 2003.

Wet Tropics Management Authority, Pressures on the world heritage area: Invasive weeds, http://www.wettropics.gov.au/mwha/mwha_weeds.html, 2006.

Williams, J.A. and West, C.J., Environmental weeds in Australia and New Zealand: Issues and approaches to management, *Aust. Ecol.*, 25, 425, 2000.

Williams, M.C. and Wardle, G.M., *Pinus radiata* invasion in Australia: Identifying key knowledge gaps and research directions, *Aust. Ecol.*, 32, 721, 2007.

Wilson, G.W., Waterhouse, B.M., and Werren, G.L., The use of exotic species in dry tropics forestry: Assessments, potential conflicts of interests and the application of the precautionary principle, in *Prospect for High-Value Hardwood Timber Plantations in the 'Dry' Tropics of Northern Australia*, Bevege, D.I., Bristow, M., Nikles, D.G. and Skelton, D., Eds., Published as a CD ROM by Private Forestry North Queensland Association, Kairi, QLD, 2004.

Zavaleta, E.S., Hobbs, R.J., et al., Viewing invasive species removal in a whole-ecosystem context, *Trends Ecol. Evol.*, 16, 8, 454, 2001.

12 Risk Analysis for Alien Plants in European Forests, Illustrated by the Example of *Prunus serotina*

Gritta Schrader and Uwe Starfinger

CONTENTS

12.1 INTRODUCTION

The introduction* of invasive alien plants into new ranges may have significant consequences for habitats and ecosystems. Although many invasive species have had detrimental effects on native species and ecosystems, not all invasive plants cause serious ecological harm. Many plants are unable to establish permanently, and others become naturalized into the environment, without causing any observable problems. Consequences of invasion can also be beneficial. The "tens rule" (Williamson 1996), stating that 10% of imported species spread, 10% of these establish, and 10% of the established species cause problems (= 0.1%), is generally applicable to alien plants. The challenge is to identify the 0.1% of species that can be harmful among the plant species introduced into a country or a region. If negative impacts occur, in the majority of cases these threats are caused by indirect damage that affects plants primarily through processes such as competition for space and resources or alteration of ecosystem properties, for example, the alteration of soil chemistry or water regimes. The effects and consequences of these processes, however, are often difficult to evaluate and to quantify in detail.

If introductions of alien plants are intentional, a pest risk analysis (PRA) should be done before the species is accepted for introduction and planting (Schrader and Unger 2003). For unintentional introductions, it is necessary to find out how an alien plant can enter a country. Therefore, pathways have to be analyzed and a pathway-initiated PRA should be conducted for the relevant situations. Pathways for alien plants are, for example, soil and seeds or grain, as these can be contaminated with weed seeds. For species-based or pathway-based risk-assessment procedures, international standards in the framework of plant health are available from the International Plant Protection Convention (IPPC 1997). The first International Standard on Phytosanitary Measures (ISPM) with regard to PRA, ISPM No. 2, *Guidelines for Pest Risk Analysis*, was developed in 1995 (FAO 1996) and recently has been revised and adopted by the governing body of the IPPC, the Committee on Phytosanitary Measures (FAO 2007). The other standard relevant in this context is ISPM No. 11 *Pest Risk Analysis for Quarantine Pests, Including Analysis of Environmental*

* This term and others related to plant health are defined in a glossary at the end of this chapter.

Risks and Living Modified Organisms (FAO 2004). Both are available from the IPPC website (www.ippc.int). These standards have been transformed into a user-friendly scheme by the European and Mediterranean Plant Protection Organisation (EPPO; Schrader 2004), which is updated annually. The scheme is available on the EPPO website (http://www.eppo.org/QUARANTINE/quarantine.htm).

This chapter will focus on intentional introductions and the procedures of PRA for alien plants based on these international standards—special attention will be paid to the situation in forests. The objectives are threefold: (1) to describe an approach to evaluate the probability of establishment and spread of invasive alien plants and the magnitude of the associated potential economic and environmental consequences (risk assessment); (2) to discuss, on the basis of the pest risk assessment, how to deal with the identified risk, what to do if it is deemed unacceptable and whether options for risk management would need to be suggested; and (3) to describe the proper techniques of identifying and choosing appropriate management options.

Provided that sufficient information is available, alien plants introduced into European forests can be evaluated by using the IPPC standards and the EPPO scheme. For clear cases, where negative impacts have already been observed elsewhere in comparable climates and under similar conditions, risks and impacts can be predicted with some certainty (Williamson 1999). However, general predictions of invasion risk and impact are hampered by the lack of broad scientific criteria. Therefore, case-by-case studies are necessary. Several publications deal with difficulties related to prediction of invasiveness (Reichard and Hamilton 1997; Kolar and Lodge 2001; Williamson 2001; Heger and Trepl 2003; Krivánek and Pyšek 2006).

In Europe as well as in other parts of the world, there is a long tradition of using alien tree species in forestry for timber production and other purposes. Soon after the discovery of the Americas, many American plant species were brought to Europe, at first primarily for ornamental purposes and for botanical curiosity (Kowarik 2003a). In the late 1800s in particular, a large number of alien tree species were tested for their capacity to produce valuable timber more effectively than the native European trees. Most of the species that were imported were never submitted to risk analysis. Today, a significant proportion of commercial forests in Europe consist of alien species (Haysom and Murphy 2003). These include the American *Pseudotsuga menziesii* used for its quick growth and valuable timber production, the North American *Robinia pseudoacacia* that is capable of growing in dry climate and on sandy soils, and a number of alien pines (*Pinus* spp.) or *Eucalyptus* spp. in the Mediterranean part of Europe.

12.2 EXAMPLE OF *PRUNUS SEROTINA*

P. serotina Ehrh. (Figure 12.1) is a good example of the introduction of an alien tree into Europe. It was planted into European forests in the nineteenth century for a variety of purposes, and it has spread considerably (Starfinger 1991, 1997). The ability of *Prunus serotina* to invade and alter ecosystem properties has caused it to receive much attention from foresters, ecologists, economists, and nature conservationists. In particular, the history of the species and its impacts in forests exemplifies the necessity to assess the consequences of using an alien species, as these may differ greatly from the reasons foresters had in mind (such as timber production or erosion control) when introducing and planting them (Starfinger et al. 2003).

FIGURE 12.1 Dense shrub layer of *P. serotina* in a pine (*Pinus sylvestris*) plantation. Forest near Eberswalde, Germany in September 2006. (Photo courtesy of U. Starfinger.)

Because comprehensive information is available on *P. serotina*, it has been chosen in this work as a model plant to describe the procedure of PRA. The application of an analysis such as the one described here might have helped to avoid the large-scale negative consequences in Europe caused by the introduction of this species.

12.3 PRA IN THE FRAMEWORK OF PLANT HEALTH

The purpose of plant health is to secure common and effective action to prevent the spread and introduction of pests of plants and products, and to promote appropriate measures for the control of these pests (IPPC 1997). PRA is a key element of plant health, because it helps to identify pests of plants and plant products and options for adequate pest management. The introduction of several invasive alien plant pests into Europe in the nineteenth century—such as the fungus-like *Phytophthora infestans*, which caused the disastrous famine in Ireland in the 1840s by destroying the Irish potato harvest, and the grape phylloxera *Daktulosphaira vitifoliae*, which ruined the vineyards in Europe on a large scale in the second part of the nineteenth century—led to the development of plant quarantine measures and plant health services. Much research has since been done to identify pathways and to develop appropriate phytosanitary measures to prevent the introduction and spread of alien plant pests. With the goal of coordinating these measures, and to protect plants, plant products, and other regulated articles against plant pests, the IPPC was developed in the 1920s.

12.4 STAGES OF RISK ANALYSIS FOR ALIEN PLANTS

The International Standard on Phytosanitary Measures (ISPM) No. 2 *Guidelines for Pest Risk Analysis* in its revised version describes the basic concept of risk analysis within the framework of the IPPC. It introduces the three stages of PRA—stage 1: initiation, stage 2: risk assessment, and stage 3: risk management—and the components for collecting, recording, and communicating information, which are carried out during the entire PRA process (Figure 12.2). The initiation stage is explained in detail in this standard. The other stages are outlined in detail in ISPM No. 11. The following section explains the PRA procedure, and illustrates its individual steps using the model species *Prunus serotina*.

12.4.1 Stage 1: Initiation

"The aim of the initiation stage is to identify the pests and pathways which are of quarantine concern and should be considered for risk analysis in relation to the identified PRA area" (FAO 2004, ISPM No. 11). The meaning of the term "quarantine concern" is described in this chapter. One important point is that quarantine concern, in the context of plant health, is tied to the fact that an organism is injurious to plants (or plant products). These include cultivated plants as well as wild flora, provided that there is an interest to protect them. Plant injury can be direct or indirect (FAO 1998). According to ISPM No. 11 (FAO 2004), pathogens, parasites, and herbivores that directly damage, infect, or feed on plants, plant products, or other regulated articles are direct pests. Indirect pests include invasive

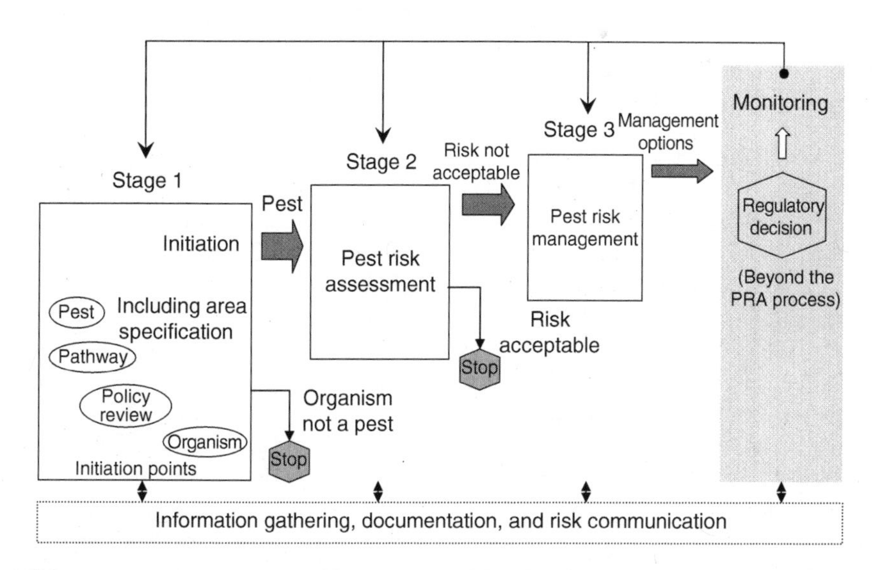

FIGURE 12.2 PRA flowchart, illustrating the PRA process. (From Food and Agriculture Organisation, Appendix 1 to ISPM No. 2 (rev.) *Framework for Pest Risk Analysis*, FAO, Rome, 2007.)

plants competing for light, water, or minerals that injure other plants indirectly (e.g., weeds in agricultural crops). In addition, quarantine concern includes other organisms that affect plants even more indirectly via other biotic factors, for example, when pollinators are depleted by a pest and as a result, there is reduced pollination of the plant species. When intentionally introduced and established in designated habitats (i.e., gardens and parks), introduced plants may pose a risk of unintentionally spreading to other habitats (i.e., arable land, natural, or seminatural habitats) and causing injury to cultivated or wild native or nonnative plants that are protected for certain reasons.

When a proposal is made to import a new plant species or a variety for cropping, ornamental, or environmental purposes, this plant may be considered for PRA, and the process starts by determining whether the plant may be a pest. The process of plant prescreening includes determining the following: taxonomic identity, predictive indicators for establishment and spread, previous introductions, ecological adaptability, competitive ability, dispersal ability, seed set, propagule pressure, and allelopathy. From the information gathered in the prescreening, a prediction is made if the organism is likely to become a pest. Usually, this is relatively straightforward with regard to direct pests as viruses, nematodes, bacteria, etc. For alien plants, this is more difficult, because there are no consistent rules for predicting invasiveness (Goodwin et al. 1999; Rejmánek et al. 2005). In addition, time lags from initial planting to initial spread of introduced plant species may be substantial (Kowarik 1995). Negative impacts may consequently become evident only after naturalization and spread have begun. Harmful effects may thus remain hidden for a prolonged period of time. Time lags may be inherent in the nature of population growth and range expansion, may result from genetic factors connected with fitness, and, in particular, may be caused by environmental factors related to improving ecological conditions for the organism (Crooks and Soulé 1999).

12.4.1.1 *P. serotina*—Initiation Stage

The taxonomic identity of *P. serotina* is quite clear: early confusions with *P. virginiana* are no longer important; its specific rank is without controversy (Fowells 1965). If this plant were to be imported to a country where it is currently absent, a prescreening would reveal that it indeed meets many of the warning criteria of the prescreening process. *P. serotina* has been shown to build up dense shrub layers in forests, which shade out plant species of the ground layer and also affect the regeneration of forest trees. The ability to colonize different climatic regions and different soils has proven *P. serotina*'s ecological adaptability. An allelopathic potential was reported, though its exact effects remain unclear. In addition, life-history traits such as copious seed production and efficient dispersal have frequently been reported from its native and introduced range (van den Tweel and Eijsackers 1987; Marquis 1990; Uchytil 1991; Starfinger 1997).

The potential of *P. serotina* to have various negative impacts became evident as early as the 1960s and is well documented (Starfinger et al. 2003). The prescreening stage can thus identify the plant as a pest, and the PRA procedure continues to the next stage.

12.4.2 Stage 2: Pest Risk Assessment

The process for pest risk assessment involves several steps. In pest categorization, it is determined whether the pest has the characteristics of a quarantine pest. In the next step, the probability of introduction (which includes entry and establishment) and spread is assessed. Subsequently, economic and environmental impacts are examined. With regard to intentional introduction of alien plants, it is especially important to assess the plants' potential to become a threat to ecosystems, habitats, or species in the PRA area. The main questions are as follows: Does the plant have a high potential for establishment and spread and does it damage or threaten biodiversity? What effects does the plant have on other plants, habitats, and ecosystems? Comparisons with similar situations or experience from previous PRAs may help to answer these questions. An invasion is often triggered by planting large numbers of a plant species, and by repeated and secondary introductions (see e.g., Kowarik 2003b); therefore, it is also necessary to consider intended use and volume of the introduction.

The endangered area is identified and the degree of uncertainty is determined. In conclusion, the overall pest risk is summarized on the basis of assessment results regarding introduction, spread, and potential economic impacts.

12.4.2.1 Pest Categorization

Risk assessment starts with pest categorization, which consists of the following steps: checking presence or absence of the species in the PRA area, reviewing the regulatory status, evaluating the potential for establishment and spread, and evaluating the economic and environmental consequences in the PRA area.

12.4.2.2 *P. serotina*—Pest Categorization

P. serotina occurs as an invasive alien species in several European countries, including Germany, Poland, Belgium, the Netherlands, and Denmark, where negative ecological and economic impacts have been observed (Starfinger 2006; Verheyen et al. 2007). Efforts for controlling *P. serotina* have been made for a long time, but have proven to be extremely expensive and have achieved little success. In several of these countries, forest authorities have policies that forbid any further planting of the species. In the German federal state of Lower Saxony, for example, an ordinance was issued that prohibits further planting in forestry (MELF 1990). With regard to pest categorization, *P. serotina* is present and regulated by forest authorities in several European countries, still rare in some countries (e.g., Sweden and Estonia), and has not been reported in other countries, including Finland and Lithuania. Consequently, the potential exists for the species to spread to new countries and expand its distribution in countries where it now occurs.

12.4.2.3 Probability of Introduction and Spread

The next step in the PRA process is to evaluate the probability of introduction. This includes an assessment of the probability of entry and establishment (see definition of "Introduction" in the Glossary), the movement along the pathway, the assembly of a list of potential pathways, and the probability of spread.

12.4.2.3.1 Entry and Movement along PRA Pathway

For alien species intended for planting purposes, it is not necessary to assess whether and how they can enter a country because they are intentionally imported, traded, and planted. Instead, it is important to examine the volume of imports and the frequency of movement along the pathway, as well as the distribution of the plant throughout the PRA area.

12.4.2.3.2 Volume of Imports and Planting of P. serotina

In the past, the techniques and objectives of planting *P. serotina* have varied greatly. Although the use of single individuals for ornamental purposes or the planting of small experimental plots in forestry tests was the rule when the species was first introduced, large numbers were used in more large-scale plantings during the twentieth century. Little information is available regarding the volume and frequency of transport of plants and seeds of *P. serotina*. However, the large numbers of individuals planted per area during reforestation efforts in Germany and the Netherlands suggest that the transport volume was considerable. Up to 2,700 *P. serotina* plants per hectare were used in newly planted pine and larch plantations in northwestern Germany (Starfinger et al. 2003). In the Netherlands an estimated area of 100,000 ha was planted with *P. serotina* (Olsthoorn and van Hees 2002).

Regarding the PRA area, it is believed that most of *P. serotina* stands in Europe resulted from planting. The occurrence of the species over vast areas from France to Poland and from Denmark to Austria reflects the scale of plantings of the species in the past.

12.4.2.3.3 Establishment and Suitable Habitats

In the case of introduced plants intended for permanent outdoor planting, the habitats chosen are deemed to be suitable for survival. Establishment (i.e., in the context of the IPPC standards, the perpetuation for the foreseeable future) in the intended habitat is very probable and even desired. For risk assessment, in accordance with ISPM No. 11, it is necessary to take the pathway from the intended to the unintended habitat and the probability of establishment in the unintended habitat into account. In other words, can the plant escape from where it has been sown or planted and can it survive after its escape? This assessment requires the consideration of the following: (1) climatic and other abiotic factors, (2) reproductive strategy of the plant species, (3) means of dispersal, (4) possible prevention of establishment by natural enemies or competition from other species present in the PRA area, and (5) the likelihood of successfully mitigating impacts after introduction through eradication, containment, or control.

To predict a plant's potential for establishment in a given area, a comparison of the relevant ecological factors within its native and introduced range is needed. The most general approach to this is climate matching, which is well developed and can be assisted by existing computer tools (e.g., CLIMEX, Creative Research Systems 2004). However, climate matching is a relatively weak predictor, and some plants have become successful invasive aliens far outside their original climatic ranges, for example, the tropical *Eichhornia crassipes* in nontropical California (Calflora 2007). On the other hand, even a detailed positive climate match will not automatically lead to a new invasion. Where this information is available, the relation of a plant to other factors such as soil texture and chemistry, nutrient availability, and biotic habitat features should be included in the analysis in as much detail as possible.

12.4.2.3.4 *Probability of Establishment for* P. serotina

In the case of *P. serotina*, a solid baseline of information exists (e.g., Fowells 1965; Marquis 1990; Starfinger 1997). The plant is native to eastern North America. Similarity of North American and European climates is sufficient to support the occurrence of the species in its alien distribution range. *P. serotina* can occur in a wide range of precipitation and temperature levels as is illustrated by its natural occurrence from southern Canada through Georgia and from the Atlantic Coast to the American Midwest.

P. serotina is able to germinate and establish across a wide range of conditions, including dry to wet and coarse to moderately fine soils (Fowells 1965; Uchytil 1991). Although it is an early successional species that is intolerant of shade, seedlings of *P. serotina* can survive for several years under dense shade. However, overstory disturbance is needed for seedlings to develop and reach the canopy. In the native as well as in the introduced range, *P. serotina* is known to form dominant stands that shade out other plant species. Seed production is abundant, and seeds are dispersed by generalist frugivorous animals. Insects and viruses are known as natural enemies but do not severely impact the reproduction, growth, or survival of the species. *P. serotina* is often replaced in later successional stands by shade-tolerant species. Eradication, containment, and control are difficult once the species has established. On the basis of this information, establishment in forest ecosystems, in particular in light canopied forest on sandy soils, is probable.

12.4.2.3.5 *Assessment of* P. serotina's *Potential to Become a Threat to Ecosystems, Habitats, or Plant Species in PRA Area*

One of the possible difficulties in assessing the risks of alien plant species is the identification of the plant's potential to become a threat to ecosystems, habitats, or plant species. A key issue within PRA is determining whether the focal species has intrinsic attributes that may lead to invasion and subsequent harm to plants or plant communities. A prescreening for these attributes of plants has already been completed in stage 1 (initiation). In stage 2, this prescreening is evaluated in detail and supported with relevant information and references (see, for example, Rejmánek and Richardson 1996; Rejmánek 2000; Heger and Trepl 2003).

12.4.2.3.6 P. serotina's *Intrinsic Attributes That Make It a Likely Pest*

Previous research of *P. serotina* has led to a wealth of valuable and readily applicable information. On the basis of this current knowledge base, the prediction of invasive behavior seems straightforward.

Across both its native and invasive ranges, *P. serotina* is known as an early successional tree that can attain dominance in certain stages of succession (Uchytil

1991; Starfinger 1997). Its life-history strategy is adapted to this successional niche through a number of characteristics:

- Its flowers are pollinated by unspecialized insect pollinators.
- It produces seeds early in life and abundantly.
- Seeds are dispersed by a large number of unspecialized birds and mammals.
- Germination requirements are broad.
- Growth can be plastic: the plant can survive in deep shade without much height growth, but growth rates can increase when light becomes available.
- Establishment and growth are not strongly dependent on specific environmental factors.

These characteristics make *P. serotina* a successful invader: they enable it to spread from plantations into neighboring forests as well as into open habitats such as bogs, heathland, dry grasslands, etc. The effects on the invaded ecosystems result from its tendency to form dense layers of shrubs or trees, which keep light from reaching the lower levels of vegetation. These effects are most pronounced in formerly open areas where light-demanding plants are unable to survive with reduced light availability. It also applies to forests where ground layer vegetation, including seedlings of other tree species, is affected. These effects may persist, because *P. serotina* is able to regenerate in its own shade. Seedlings germinate in the shade and can survive some time, and sprouts also form a reservoir (Closset-Kopp et al. 2007). Trees and younger plants of *P. serotina* will copiously sprout from the base of the trunk when the aboveground part of the plant is damaged. This makes mechanical control of the species particularly difficult. Further spread into areas not yet colonized may be predicted on the basis of these characteristics.

Invasibility of an area will also depend on the land-use history (Heger and Trepl 2003). Often, highly disturbed sites are especially prone to invasion. Depending upon the alien species requirements, nutrient-rich or nutrient-poor soils (or waters) may be preferred.

P. serotina is known to respond positively to human-made disturbance. In pine plantations, its speed of invasion increased after thinning, because thinning enhances the availability of light and exposed soil, and thus facilitated the establishment of seedlings (Rode et.al. 2002).

12.4.2.4 Primary Consequences Resulting from Establishment and Spread

When assessing the potential of a plant to become a threat to ecosystems, habitats, or species, the next step is to identify the likely consequences resulting from establishment and spread. In this context, primary and secondary consequences have to be considered. Regarding environmental risks, important primary consequences would include the reduced abundance of keystone plant species, which define an ecosystem, or are the main drivers for the development of an ecosystem. In addition, species that are major components of ecosystems may require focused management efforts because their loss or large-scale reduction in numbers may alter ecosystem structure or function. Negative impacts on endangered native plant species also have to be prevented in order to protect biodiversity. Furthermore, protection of other plant species from significant reduction, displacement, or elimination is taken into account, provided there is an interest in their protection.

12.4.2.5 Secondary Consequences Resulting from Establishment and Spread

Examples for secondary consequences according to ISPM No. 11 (FAO 2004) relate to

> significant effects on plant communities, significant effects on designated environmentally sensitive or protected areas, significant change in ecological processes and the structure, stability, or processes of an ecosystem (including further effects on plant species, erosion, water table changes, increased fire hazard, nutrient cycling, etc.), effects on human use (e.g., water quality, recreational uses, tourism, animal grazing, hunting, fishing), and costs of environmental restoration.

If, for example, the symbiotic nitrogen-fixing black locust (*Robinia pseudoacacia*) invades species-rich grasslands, it may not only have a significant effect on the composition of the vegetation, but also may have severe impacts on animal communities as was demonstrated for spiders and ground beetles by Platen and Kowarik (1995).

Secondary effects may also result from interactions with higher trophic levels, i.e., when feeding preferences of animals lead to different feeding pressures on plant species. Low palatability of alien plants to herbivores compared with European forest species can result in a competitive advantage for the alien plants. But the attraction of animals may also have deleterious effects on other plant species. For example, the flowers of the alien *Impatiens glandulifera* are so attractive to hymenopterans that neighboring individuals of the native *Stachys palustris* may suffer reduced seed set due to limited pollination of their own flowers (Chittka and Schürkens 2001). Hybridization is another negative impact of invasive alien species. For example, hybridization of *Fallopia japonica* and *F. sachalinensis* has resulted in the highly invasive hybrid *F.* × *bohemica* (Pyšek et al. 2003). In addition, hybridization and introgression with alien plants can severely threaten the genetic identity of native species, especially if they are rare and threatened by other factors. Introgression from widely planted *Populus* × *canadensis*, for example, may threaten the endangered European native *P. nigra*. Similarly, the rare *Malus sylvestris* and *Pyrus pyraster* are threatened by introgression from domesticated forms.

12.4.2.6 Consequences Resulting from Establishment and Spread of *Prunus serotina*

In the case of *P. serotina*, primary and secondary consequences from establishment and spread vary with the particular habitat type. The main change to invaded ecosystems is an altered light climate due to dense shrub or tree layers, which leads to a decrease in species richness in the ground vegetation. In forests, the invasion of this species typically does not affect species of special concern because the pine–oak communities that are most frequently invaded are relatively species-poor and do not typically contain rare or threatened species. However, changes in these light-canopied forests may alter the course of succession for extended periods of time, as *P. serotina* layers can impede the natural regeneration of shade-intolerant forest trees (Starfinger 1997). Invasion of grasslands or bogs can, in contrast, lead to complete degradation of the habitat as the community begins to succeed to a forest dominated by *P. serotina*. Low palatability of *P. serotina* to deer compared with European forest species enhances its competitive advantage. Effects on other trophic levels of the affected ecosystem are little studied but can be expected. The low C/N ratio in *Prunus* litter may significantly alter soil chemistry in pine–oak forests (Rode et al. 2002).

12.4.2.7 Assessment of Economic Consequences

In this part of the PRA, it is assessed if the pest will have unacceptable economic consequences in the PRA area. Unacceptable economic impact is described in ISPM No. 5 Glossary of phytosanitary terms, Supplement No. 2: guidelines on the understanding of potential economic importance and related terms (FAO 1996). It may be possible to do the assessment very simply, if sufficient evidence is already available or the risk presented by the pest is widely agreed.

For a valuation of the environment, ISPM No. 11 provides different methodologies, including the consideration of "use" and "non-use" values. Use values can be separated into consumptive (e.g., fishing in a lake) and nonconsumptive (e.g., using forests for leisure activities). Non-use values can be divided into option value (value for use at a later date), existence value (knowledge that an element of the environment exists), and bequest value (knowledge that an element of the environment is available for future generations). To assess these values, methods are available referring to market-based approaches, surrogate markets, simulated markets, and benefit transfer. In addition, the assessment can be based on nonmonetary valuations, such as number of species affected or water quality. In any case, these methodologies are best applied in consultation with experts in economics. The procedures should be documented, consistent and transparent, and environmental values should be clearly categorized.

12.4.2.8 Economic Effects of *P. serotina*

A variety of negative economic effects of *P. serotina* has been demonstrated in the extensive literature on the species (e.g., van den Tweel and Eijsackers 1987; Späth et al. 1994; Reinhardt et al. 2003; Starfinger 2006). They fall into four categories:

1. Direct yield loss of forest trees due to competition for water or light (Dik and Jager 1970).
2. Increased labor cost of forest workers due to impenetrable shrub layers. This added expense was estimated at 8%–15% of revenue from affected tree stands (Reinhardt et al. 2003).

3. Costs of control that is necessary for new timber tree plantations (Olsthoorn and van Hees 2002).
4. Costs of general control where a policy of reducing *P. serotina* exists.

These direct costs add up to several millions of Euros for individual countries, as was calculated for the Netherlands (Olsthoorn and van Hees 2002) and Germany (Reinhardt et al. 2003). In addition, there are indirect costs through the loss of biodiversity, but how these costs should be calculated is a matter of ongoing discussion (see Pimentel et al. 2001).

Costs have to be viewed against the possible benefits of the species. *P. serotina* is a highly valuable timber species in part of its native range where it is used in the manufacture of fine furniture. However, for climatic and edaphic reasons the tree does not grow to any merchantable size in most European forests but remains small with forked or crooked trunks that make it unsuitable for furniture making. With few exceptions, timber production will not influence the cost–benefit calculation for this species to a considerable degree.

Long-term costs may depend on the potential to control or eradicate an alien species. *P. serotina* has long been subjected to a variety of control measures. It is particularly resistant to mechanical control since cutting results in vigorous resprouting. Several herbicides have been used against the species in the past with varying success. More recently, foresters have tried to control *P. serotina* with silvicultural methods such as subplanting with beech. However, the lack of success in controlling *P. serotina* suggests that it cannot be easily eradicated once it is established (Verheyen et al. 2007) and local control can only be achieved at high costs (control costs were estimated in the 1980s as being between EUR 150 and 1500 per hectare) (Späth et al. 1994).

12.4.2.9 Identification of Endangered Area

The endangered area is defined as the part of the PRA area where the presence of the pest will result in economically or environmentally important losses. In a pest risk assessment, it is necessary to identify the part of the PRA area where presence of host plants or suitable habitats and ecological factors favor the establishment and spread of the pest to define the endangered area. The endangered area may be the whole of the PRA area, or part or parts of the area. It can be defined ecoclimatically, geographically, by crop or by production system (e.g., protected cultivation such as glasshouses) or by types of ecosystems.

12.4.2.10 Area Endangered by *P. serotina*

Using *P. serotina* as an example, and focusing only on forest-related costs whose calculation is more feasible, the relevant area can roughly be tied to the extent of sandy soils in the European countries, since sandy soils usually support forests rather than agricultural fields due to their lack of fertility. In addition, forests on better soils tend to be made up of shade-tolerant trees and consequently suffer less severely from invasion of *P. serotina*. In a cost estimation for *P. serotina* in Germany, the endangered area was calculated as 10,000 km^2 by using this approach (Reinhardt et al. 2003).

12.4.2.11 Uncertainty

Assessing the risk of invasive plant species involves many uncertainties. Often, the level of uncertainty is greater in the assessment of risks to the environment than of risks to cultivated plants. Uncertainty arises because of the lack of information, additional complexity associated with ecosystems, and variability associated with pests, hosts, or habitats. For the identification of management options, it is important to consider the degree of uncertainty. Generally, measures implemented to stop the introduction and spread of harmful organisms are designed to account for uncertainty, but should also avoid being "... more trade restrictive than necessary. Measures should be applied to the minimum area necessary for the effective protection of the endangered area" (FAO 2004).

12.4.2.12 Uncertainty Regarding *P. serotina*

In the case of *P. serotina*, the prediction that it will spread and affect native species, habitats, and ecosystems can safely be made based on the broad evidence. The actual speed of the spread and consequently the magnitude of its impact on individual species or habitats depend on the specific situation and are thus more difficult to assess.

12.4.2.13 Conclusions of Pest Risk Assessment

As a result of the pest risk assessment, the evaluated organism may be considered appropriate for pest risk management (stage 3). The endangered area to which this pest risk management is applied should have been identified in the risk assessment. The estimate of the probability of introduction and the estimate of economic and environmental consequences are given, and the uncertainties are outlined. These results represent the basis for pest risk management.

12.4.2.14 Conclusions of Pest Risk Assessment for *P. serotina*

The conclusions for *P. serotina* are that it can enter, establish, and spread in the PRA area. It is evident that it can become a major threat to biodiversity and to economic goals. Therefore, management options should be analyzed to reduce the risks posed by this plant.

12.4.3 Stage 3: Pest Risk Management

If the assessment of an organism for which the PRA is being done reveals an unacceptable risk to plants in the PRA area, management options must be identified to reduce or exclude these risks. With intentionally introduced plants, management options are quite different in comparison with unintentional introductions. The management section of ISPM No. 11 does not give detailed guidance on how to proceed with the import of invasive or potentially invasive plants. In the framework of EPPO, a standard for the import of alien plants has, therefore, been drafted. This standard was adopted by the EPPO Council in autumn 2006 (EPPO 2006). Important points to consider are: the surveillance after planting, the preparation of control or emergency plans if a plant is found outside its intended habitat and spreads to an unacceptable degree, the restriction on import, sale, holding, and planting (including authorization of intended habitats, prohibition of planting in unintended habitats, required growing conditions for plants), the notification before import, restrictions on movement (e.g., prevention of movement to specified areas), and the obligation to report findings. In any case, the intended use of the plant influences the choice of management measures.

For plants new to a given area, it is difficult to predict their ability to invade. If an invasive behavior has never been observed before, but some characteristics or attributes of the plants and their potential habitats raise suspicion for potential threat to ecosystems, habitats, or species, an option could be not to take measures at import, but to apply surveillance or other procedures after entry, and to monitor plants after import and planting.

Such a close monitoring of invasive behavior seems justified for alien forest plants in general. Forest plants are usually long lived, with long periods between planting and harvest. Many alien trees planted in forestry have become invasive with significant impacts after a relatively long term of unobtrusive behavior (Kowarik 1995). Specific risk management measures such as the selection of habitats that are not prone to invasion, implantation of a spread barrier (strong plastic foil, etc.), and only using either male or female plants of a dioecious species may be options to avoid negative biodiversity impacts without completely refraining from using or testing exotic tree species.

12.4.3.1 Management of *P. serotina*

Most of the management options mentioned earlier are not applicable to *P. serotina*. Since the species has severe negative impacts through its forming of tall and dense stands as well as its rapid spread, any large-scale use of the species will lead to severe negative impacts. As control is also hardly feasible and the beneficial effects of *P. serotina* are comparatively small, the result of the risk analysis should be a cancellation of any intended plantings, at least in and near forests.

Differentiation between the intended use of a species, e.g., for gardening (within urban areas) or for landscaping (planted in large amounts and many different locations, in the countryside), can also influence the selection of possible risk measures. This leaves a small area where *P. serotina* can still be used relatively safely. The species has long been favored as an ornamental in gardens and parks due to its shiny leaves, cream-white flowers, and varied fall coloring. Intensive management and care in gardens are usually sufficient to suppress regeneration and spread within the garden. A continued use as a garden ornamental may thus be feasible. Birds may spread the seeds to nearby natural parks and forests. However, since most seeds are deposited in the vicinity of the parent trees (Pairon et al. 2006), dispersal into forests decreases strongly with distance (cf. Deckers et al. 2005). In consequence, plantings should be restricted to areas where no woodlands or other invasible habitats are in close vicinity.

12.4.3.2 Finalization of Pest Risk Management Stage

At the end of the pest risk management stage, all potential management options will be summarized and their effectiveness indicated, based on expert judgment.

Uncertainties must be identified. The conclusion will be whether or not appropriate phytosanitary measures adequate to reduce the pest risk to an acceptable level are available, cost-effective, and feasible. Different measures could be combined, for example, with an emergency plan to be used if the plant is found outside its intended habitats. As already mentioned, the phenomenon of *time lag* has to be considered. Because the public may not understand the need for control measures, information should be provided in an accessible format that explains the need for invasive alien species control. Measures may be more readily accepted for existing species that present an obvious problem than for species where only a potential risk has been identified.

12.4.3.3 Finalization of Pest Risk Management for *P. serotina*

Further planting in the vicinity of forests and other vulnerable habitats should be avoided. Planting may be acceptable in certain places (e.g., in urban areas without forests). Additional safety may be gained by information campaigns aimed at a broad general public and garden owners in these settings.

12.5 CONCLUSIONS

In light of the vast consequences that biological invasions can have, precaution is clearly reasonable when new species are introduced. The risk is particularly high when these species are to be used and released in large quantities, as several examples have shown. Accordingly, risk analysis should precede any change in the use of plant species, be it for import of new species or the extension of the plantation of trees already present. The standards and schemes for PRA by IPPC and EPPO introduced in this chapter provide the necessary elements for a substantial risk analysis and form the basis for risk assessment and management procedures that should be applied to planned import or use of new tree species. The assessment of risks posed by alien plants to a PRA area is a difficult task because of high levels of complexity in ecosystems, uncertainty about threats to biodiversity, and pressure arising from globalization including trade and tourism.

Prediction of impacts has, so far, not produced satisfactory certainty for many cases. However, a case-by-case evaluation and a systematic approach to risk analysis may better help in decision making.

12.5.1 Conclusions on *P. serotina*

The example of *P. serotina* has shown how careless planting of alien trees can lead to a variety of negative impacts on biodiversity and economic values. Forestry in Europe has long benefited from the use of nonnative tree species. Many forest authorities still believe in the economic necessity of continuing to use alien species. Consequently, reference to the ecological and economic damage posed by alien trees is not sufficient to influence policies of forest authorities and forest owners. Therefore, risk analysis is all the more important in the case of alien trees in forests and a number of different options may result from risk analysis:

- Alternative species with similar benefits but lower damage potential may be chosen for planting.
- Risk management may be found necessary and feasible so that species with a damage potential may still be used for planting.
- In this case, management measures have to be selected in proportion to the risk.
- Risk analysis may demonstrate that obvious and short-term economic benefits may be exceeded by costs inferred through long-term damages.

In summary, risk analysis may help to control economically and environmentally harmful plant species.

GLOSSARY*

Area: An officially defined country, part of a country, or all or parts of several countries (based on the World Trade Organization Agreement on the Application of Sanitary and Phytosanitary Measures).

Endangered Area: An area where ecological factors favor the establishment of a pest whose presence in the area will result in economically important loss.

Establishment: Perpetuation, for the foreseeable future, of a pest within an area after entry.

International Standard for Phytosanitary Measures: An international standard adopted by the Conference of FAO, the Interim Commission on phytosanitary measures, or the Commission on phytosanitary measures, established under the IPPC.

International Standards: International standards established in accordance with Article X paragraphs 1 and 2 of the IPPC.

Introduction: The entry of a pest resulting in its establishment.

Official Control: The active enforcement of mandatory phytosanitary regulations and the application of mandatory phytosanitary procedures with the objective of eradication or containment of quarantine pests or for the management of regulated nonquarantine pests (see Glossary Supplement No. 1).

Pathway: Any means that allows the entry or spread of a pest.

Pest: Any species, strain or biotype of plant, animal or pathogenic agent injurious to plants or plant products.

Pest Categorization: The process for determining whether a pest has or has not the characteristics of a quarantine pest or those of a regulated nonquarantine pest.

---

* *Source:* From Food and Agriculture Organisation, *Glossary of Phytosanitary Terms*, FAO, Rome, 2007.

Pest Risk (for Quarantine Pests): The probability of introduction and spread of a pest and the magnitude of the associated potential economic consequences (see Glossary Supplement No. 2).

Pest Risk Analysis (Agreed Interpretation): The process of evaluating biological or other scientific and economic evidence to determine whether an organism is a pest, whether it should be regulated, and the strength of any phytosanitary measures to be taken against it.

Pest Risk Assessment (for Quarantine Pests): Evaluation of the probability of the introduction and spread of a pest and the magnitude of the associated potential economic consequences.

Pest Risk Management (for Quarantine Pests): Evaluation and selection of options to reduce the risk of introduction and spread of a pest.

Pest Status (in an Area): Presence or absence, at the present time, of a pest in an area, including where appropriate is its distribution, as officially determined using expert judgment on the basis of current and historical pest records and other information.

Phytosanitary Measure (Agreed Interpretation): Any legislation, regulation, or official procedure with the purpose to prevent the introduction and/or spread of quarantine pests, or to limit the economic impact of regulated nonquarantine pests.

Plant Quarantine: All activities designed to prevent the introduction and/or spread of quarantine pests or to ensure their official control.

PRA Area: Area in relation to which a PRA is conducted.

Quarantine Pest: A pest of potential economic importance to the area endangered thereby and not yet present there, or present but not widely distributed and officially controlled.

Regulated Article: Any plant, plant product, storage place, packaging, conveyance, container, soil, and any other organism, object, or material capable of harboring or spreading pests, deemed to require phytosanitary measures, particularly where international transportation is involved.

Spread: Expansion of the geographical distribution of a pest within an area.

Standard: Document established by consensus and approved by a recognized body that provides, for common and repeated use, rules, guidelines, or characteristics for activities or their results, aimed at the achievement of the optimum degree of order in a given context.

ACKNOWLEDGMENTS

Thanks are due to Shibu Jose for inviting us to contribute to this volume and to M.A. Jenkins and R. Collins for useful comments on an earlier version.

REFERENCES

Calflora, Information on California plants for education, research and conservation (web application), The Calflora Database (a nonprofit organization), Berkeley, CA, available at http://www.calflora.org (accessed: September 21, 2007), 2007.

Chittka, L. and Schürkens, S., Successful invasion of a floral market. An exotic Asian plant has moved in on Europe's river-banks by bribing pollinators, *Nature* 411, 653, 2001.

Closset-Kopp, D., Chabrerie, O., Valentin, B., Delachapelle, H., and Decocq, G., When Oskar meets Alice: Does a lack of trade-off in r/K-strategies make *Prunus serotina* a successful invader of European forests? *For. Ecol. Manage.*, 247, 120, 2007.

Creative Research Systems, CLIMEX—Software to predict the effects of climate on species, available at http://www.climatemodel.com, 2004.

Crooks, J. and Soulé, M.E., Lag times in population explosions of invasive species: Causes and implications, in *Invasive Species and Biodiversity Management*, Sandlund, O.T., Schei, S.J., and Vikens, A., Eds., Kluwer Academic, The Netherlands, 103, 1999.

Deckers, B., Verheyen, K., Hermy, M., and Muys, B., Effects of landscape structure on the invasive spread of black cherry *Prunus serotina* in an agricultural landscape in Flanders, Belgium, *Ecography*, 28, 99, 2005.

Dik, E.J. and Jager, K., De involved van bestrijding van loofhout op de groei van Japanse lariks, *Nederlands Bosbouw Tijdschrift*, 42, 95, 1970.

European and Mediterranean Plant Protection Organisation (EPPO), Guidelines for the management of invasive alien plants or potentially invasive alien plants which are intended for import or have been intentionally imported, *EPPO Bull.*, 36, 3, 417, 2006.

Food and Agriculture Organisation (FAO), ISPM No. 2, *Guidelines for Pest Risk Analysis*, International Standard for Phytosanitary Measures Publication No. 2, Food and Agriculture Organisation of the United Nations, Rome, 1996.

Food and Agriculture Organisation (FAO), Interim Commission on Phytosanitary Measures: Excerpts from the Report of the Conference of FAO (C97/REP), 29 Session, Rome, November 7–18, 1997, Appendix I: Interpretations as agreed by the fourteenth session of the Committee on Agriculture, ICPM-98/INF/1, Food and Agriculture Organisation of the United Nations, Rome, 1998.

Food and Agriculture Organisation (FAO), ISPM No. 11, *Pest Risk Analysis for Quarantine Pests Including Analysis of Environmental Risks*, International Standards for Phytosanitary Measures Publication No. 11 (rev. 1), Food and Agriculture Organisation of the United Nations, Rome, 2004.

Food and Agriculture Organisation (FAO), ISPM No. 2 (rev.), *Framework for Pest Risk Analysis*, International Standard for Phytosanitary Measures Publication No. 2, Food and Agriculture Organisation of the United Nations, Rome, 2007.

Fowells, H.A., *Silvics of Forest Trees of the United States*, USDA Forest Service, 1965.

Goodwin, B.J., McAllister, A.J., and Fahrig, L., Predicting invasiveness of plant species based on biological information, *Conserv. Biol.*, 13, 2, 422, 1999.

Haysom, K.A. and Murphy, S.T., The status of invasiveness of forest tree species outside their natural habitat: A global review and discussion paper, Forest Health and Biosecurity Working Paper FBS/3E, Forestry Department, FAO, Rome (unpublished, available at http://www.fao.org/docrep/006/j1583e/J1583E04.htm), 2003.

Heger, T. and Trepl, L., Possibilities to predict biological invasions, *Biol. Invasions*, 5, 4, 313, 2003.

IPPC, International Plant Protection Convention, new revised text, FAO, Rome, 1997.

Kolar, C.S. and Lodge, D.M., Progress in invasion biology: Predicting invaders, *Trends Ecol. Evol.*, 16, 4, 199, 2001.

Kowarik, I., Time lags in biological invasions with regard to the success and failure of alien species, in *Plant Invasions—General Aspects and Special Problems*, Pyšek, P., Prach, K., Rejmánek, M., and Wade, M., Eds., SBP Academic, Amsterdam, 15, 1995.

Kowarik, I., *Biologische Invasionen: Neophyten und Neozoen in Mitteleuropa*, Ulmer, Stuttgart, 2003a.

Kowarik, I., Human agency in biological invasions: Secondary releases foster naturalisation and population expansion of alien plant species, *Biol. Invasions*, 5, 293, 2003b.

Krivánek, M. and Pyšek, P., Predicting invasions by woody species in a temperate zone: A test of three risk assessment schemes in the Czech Republic (Central Europe), *Divers. Distrib.*, 12, 319, 2006.

Marquis, D.A., *Prunus serotina* Ehrh. Black Cherry, in *Silvics of North America, Vol. 2: Hardwoods*, Burns, R.M. and Honkala, B.H., Eds., Agriculture Handbook 654, U.S. Department of Agriculture, Forest Service, Washington, DC, 594, 1990.

Ministerium für Ernährung Landwirtschaft und Forsten in Niedersachsen (MELF), *Wald und Forstwirtschaft in Niedersachsen*, MELF, Hannover, 113, 1990.

Olsthoorn, A. and van Hees, A., 40 years of Black Cherry (*Prunus serotina*) control in the Netherlands: Lessons for management of invasive tree species, in *Biologische Invasionen*, Starfinger, U. and Kowarik, I., Eds., Herausforderung zum Handeln? *Neobiota* 1, 339, 2002.

Pairon, M., Jonard, M., and Jacquemart, A.-L., Modeling seed dispersal of black cherry, an invasive forest tree: How microsatellites may help? *Can. J. For. Res.*, 36, 6, 1385, 2006.

Pimentel, D., McNair, S., Janecka, J., Wightman, J., Simmonds, C., O'Connell, C., Wong, E., Russel, L., Zern, J., Aquino, T., and Tsomondo, T., Economic and environmental threats of alien plant, animal, and microbe invasions, *Agric. Ecosyst. Environ.*, 84, 1, 2001.

Platen, R. and Kowarik, I., Pflanzen-, Spinnen- und Laufkäfergemeinschaften innerstädtischen Bahnbrachen, *Verh. Ges. f. Ökologie*, 24, 431, 1995.

Pyšek, P., Brock, J.H., Bimova, K., Mandak, B., Jarosik, V., Koukolikova, I., Pergl, J., and Stepanek, J., Vegetative regeneration in invasive *Reynoutria* (Polygonaceae) taxa: The determinant of invasibility at the genotype level, *Am. J. Bot.*, 90, 10, 1487, 2003.

Reichard, S.H. and Hamilton, C.W., Predicting invasions of woody plants introduced into North America, *Conserv. Biol.*, 11, 193, 1997.

Reinhardt, F., Herle, M., Bastiansen, F., and Streit, B., Ökonomische Folgen der Ausbreitung von gebietsfremden Organismen in Deutschland, *UBA-Forschungsbericht*, 201, 86, 211, 2003.

Rejmánek, M., Invasive plants: Approaches and predictions, *Aust. Ecol.*, 25, 5, 497, 2000.

Rejmánek, M. and Richardson, D.M., What attributes make some plant species more invasive? *Ecology*, 77, 6, 1655, 1996.

Rejmánek, M., Richardson, D.M., Higgins, S.I., Pitcairn, M., and Grotkopp, E., *Ecology of Invasive Plants: State of the Art*, in Mooney, H.A., McNeely, J.A., Neville, L., Schei, P.J., and Waage, J., Eds., Invasive alien species: A new synthesis, Island Press, Washington, DC, 104, 2005.

Rode, M., Kowarik, I., Müller, T., and Wendebourg, T., Ökosystemare Auswirkungen von *Prunus serotina* auf norddeutsche Kiefernforsten, *Neobiota*, 1, 135, 2002.

Schrader, G., A new working program on invasive alien species started by a multinational European organisation dedicated to protecting plants, *Weed Technol.*, 18, 1342, 2004.

Schrader, G. and Unger, J.G., Plant quarantine as a measure against invasive alien species: The framework of the International Plant Protection Convention and the plant health regulations in the European Union, *Biol. Invasions*, 5, 4, 357, 2003.

Späth, I., Balder, H., and Kilz, E., Das Problem mit der Spätblühenden Traubenkirsche in den Berliner Forsten, *Allgemeine Forst- und Jagdzeitung*, 11, 234, 1994.

Starfinger, U., Population biology of an invading tree species—*Prunus serotina*, in *Species Conservation: A Population Biology Approach*, Seitz, A. and Loeschke, V., Eds., A. Birkhäuser Verlag, Basel, 171, 1991.

Starfinger, U., Introduction and naturalization of *Prunus serotina* in central Europe, in *Plant Invasions: Studies from North America and Europe*, Brock, J.H., Wade, M., Pyšek, P., and Green, D., Eds., Backhuys, Leiden, 161, 1997.

Starfinger, U., NOBANIS—Invasive alien species fact sheet—*Prunus serotina*, Online Database of the North European and Baltic Network on Invasive Alien Species, NOBANIS, available at http://www.nobanis.org (accessed September 25, 2007), 2006.

Starfinger, U., Kowarik, I., Rode, M., and Schepker, H., From desirable ornamental plant to pest to accepted addition to the flora? The perception of an alien plant species through the centuries, *Biol. Invasions*, 5, 323, 2003.

Uchytil, R.J., *Prunus serotina*, in Fire effects information system, U.S. Department of Agriculture, Forest Service, Rocky Mountain Research Station, Fire Sciences Laboratory, available at http://www.fs.fed.us/database/feis/ (accessed October 1, 2007), 1991.

van den Tweel, P.A. and Eijsackers, H., Black cherry, a pioneer species or "forest pest," *Proc. K. Ned. Akad. Wet. Ser.* C, 90, 59, 1987.

Verheyen, K., Vanhellemont, M., Stock, T., and Hermy, M., Predicting patterns of invasion by black cherry (*Prunus serotina* Ehrh.) in Flanders (Belgium) and its impact on the forest understorey community, *Divers. Distrib.*, 13, 487, 2007.

Williamson, M., *Biological Invasions*, Chapman & Hall, London, 1996.

Williamson, M., Invasions, *Ecography*, 22, 5, 1999.

Williamson, M., Can the impacts of invasive plants be predicted? in *Plant Invasions. Species Ecology and Ecosystem Management*, Brundu, G., Brock, J.H., Camarda, I., Child, L.E., and Wade, P.M., Eds., Backhuys, Leiden, 11, 2001.

13 Monitoring and Assessment of Regional Impacts from Nonnative Invasive Plants in Forests of the Pacific Coast, United States

Andrew Gray

CONTENTS

13.1 INTRODUCTION

13.1.1 Problem Statement

Invasions of nonnative plants into new regions have a tremendous impact on many natural and managed ecosystems, affecting their composition and function. Nonnative invasive species have a large economic impact through lost or degraded land use and eradication costs, and are a primary cause of extinction of native species (Vitousek et al. 1996; Mooney and Hobbs 2000; Pimentel et al. 2005). Nonnative invasive plants can affect ecosystems and land use by competitively excluding desired species and altering disturbance regimes (D'Antonio and Vitousek 1992). As a result, characterizing the prevalence of invasive species is a key element of several efforts to assess ecosystem health and sustainable management (Anonymous 1995; National Research Council Committee to Evaluate Indicators for Monitoring Aquatic and Terrestrial Environments 2000; Heinz Center 2002).

Despite their importance, few data on the abundance, distribution, and impact of nonnative invasive plants are available (Blossey 1999). Information is often incomplete (e.g., quantifying distribution, but not abundance) and available for only a few species in a few areas for selected time periods. As a result, it is currently not possible to provide a comprehensive assessment of the abundance and impacts of nonnative invasive plants in the United States (National Research Council Committee to Evaluate Indicators for Monitoring Aquatic and Terrestrial Environments 2000; Heinz Center 2002). The objective of this chapter is to describe recent efforts to monitor nonnative invasive plants for strategic forest inventories in the Pacific coastal states of California, Oregon, and Washington, and evaluate the utility of the information developed. A general introduction to various monitoring approaches and common challenges is presented, along with brief discussion of the pros and cons of each. Analyses of two types of vegetation data from the extensive forest inventory of the Forest Inventory and Analysis (FIA) Program are presented as case studies to illustrate some of the difficulties, strengths, and tradeoffs faced in the design and implementation of an effective monitoring program.

13.1.2 Monitoring Objectives

Natural resource monitoring undertaken without clearly defined questions and objectives can lead to unusable data or ambiguous results (Morrison and Marcot 1995). There are many different objectives for collecting data on invasive plant populations, including:

1. Detecting species early enough in their invasion to facilitate control and eradication
2. Assessing the effectiveness of control efforts on detected populations
3. Evaluating species' impacts on selected habitats
4. Quantifying species' distribution and abundance
5. Quantifying changes in species' distribution and abundance over time
6. Predicting species' future spread

Addressing any single monitoring objective adequately and efficiently places unique demands on the type, quality, and quantity of data that must be collected. As a corollary, no single monitoring strategy can adequately address all objectives.

13.2 MONITORING APPROACHES

A wide range of approaches and types of data have been used to inventory and monitor invasive plants. Some of the common approaches are presented later and their suitability for addressing different objectives discussed.

Published floras are critical for documenting the presence of an invasive species in a region and facilitating the reliable identification of suspected plants. Detecting recently-arrived invasive species can be hampered if they are not included in the accepted floristic references for the area, so keeping these references up to date would help the monitoring of invasive species. A global information service that compiled information on invasive species' detections, traits, and habitat preferences would be very useful, particularly for early detection, perhaps even before species escape from their original vector into an area (Ricciardi et al. 2000). The descriptions of species' ranges and habitat preferences in flora are generally too brief and ambiguous to characterize species abundance and distribution, however.

Records and specimens maintained by herbaria are very useful for documenting changes in a region's flora. Because plant specimens are usually well preserved, identifications can be updated to match current taxonomic references, and confidence in species' identifications is high. However, herbarium records result from many different types of sampling, which vary in geographic, temporal, and taxonomic intensity. To account for some of the variations, it is possible to adjust the number of records of the species of interest by the total number of herbarium records or by records of species found in similar habitats (Delisle et al. 2003). Herbarium data have been primarily useful in the retrospective analysis of changes in invasive plant distributions over large geographic areas (Pyšek and Hulme 2005).

Intensive research projects are another important source of information that usually focuses on single species and specific attributes of a species' distribution, reproduction, dispersal, community interactions, or other life-history traits. This information provides valuable insights that inform the design of management and monitoring efforts. But results can rarely be extrapolated to assess a species' overall distribution, abundance, and impact on plant communities because intensive efforts are usually limited in space and time. However, some landscape-scale studies (e.g., Parendes and Jones 2000), if documented sufficiently, could be repeated after several years and used to assess change.

Field surveys focused on invasive plants are conducted by many local, state, and federal government agencies in the United States, although scope and strategy vary widely among efforts. For instance, surveys may be focused on specific species or specific areas (or both) of the landscape. Some can be quite focused on examining suitable habitat for specific invaders to detect new infestations as soon as possible and facilitate control. To promote greater standardization, many agencies in the United States have adopted a set of rules developed by the North American Weed Management Association in 2002 for collecting and storing data on invasive plants. These standards specify the data fields to be used to describe the location, size of area, and cover within the assessed area, for the surveyed species. This information should be most useful for land managers interested in early detection and monitoring of eradication efforts. However, with little information provided concerning the sampling strategy (i.e., area searched, species searched for) or on the spatial and temporal sampling intensity, it is not possible to determine how representative the data are for a species, a habitat type, or a management unit. The standards also lack quality control and quality assurance provisions for plant identification, cover estimation, and area mapping. It might be possible to assess change over time by reassessing previously mapped patches, but it would be difficult to develop reliable change estimates for larger areas of interest (e.g., management units). Land managers do have options for collecting more reliable data on management units, for example, with the Forest Service stand exam sampling approach.

Other types of surveys proposed for national action would focus monitoring on the primary vectors of invasive plant introductions. Most of these vectors are associated with the horticultural trade (e.g., arboreta, nurseries, seed companies); intensive monitoring of these locations could prevent potentially invasive species from being released into the environment in the first place (Reichard and Hamilton 1997; Reichard and White 2001). The proposed global information service mentioned above would help monitoring programs to identify potential species of concern (Ricciardi et al. 2000).

Strategic inventories collect resource information across large regions using relatively stable and well-defined procedures and probability-based sampling. The USDA Forest Service's FIA program is the most comprehensive and consistent inventory and monitoring program in the United States (National Research Council Committee to Evaluate Indicators for Monitoring Aquatic and Terrestrial Environments 2000). Field sample points are installed on forestlands across all ownerships and measurements taken periodically. Data are usually appropriate for the state or multicounty level; the spatial density of sample points is generally insufficient for providing accurate information for any but the largest landowners. Recent efforts across the country have included a variety of approaches to monitoring invasive plants as part of the FIA inventory (Rudis et al. 2005). Results and the strengths and weaknesses of this kind of monitoring are the primary focus of this chapter.

13.3 CHALLENGES TO MONITORING

There are several challenges to monitoring nonnative invasive plant species in a way that provides high-quality data in a repeatable fashion. The challenges can be grouped into difficulties with species selection and logistical difficulties with collecting data.

Determining which species are "nonnative" to an ecosystem or region is not always straightforward. Plant species' distributions change continuously in response to changes in climate, land management, urbanization, and natural disturbance frequency and intensity. Assuming a species' origins can be determined, large geographic areas (continent or subcontinent) are usually used as criteria for determining nonnativity; for example, Eurasian plant species found in North America, or the barred owl (*Strix varia*) invasion of western North America from the east. In some cases, new genotypes of a species are introduced, which are not morphologically distinguishable from local populations, and which may hybridize with them.

For example, *Achillea millefolium* is one of the most abundant species in disturbed western forestlands, but it is unknown if this is a trait common to the native genotypes or a result of the spread of the introduced genotypes (Hitchcock et al. 1955; Hurteau and Briggs 2003).

Determining which nonnative species to consider "invasive" is not always clear. In its broadest application, the term invasive applies to "alien species whose introduction does or is likely to cause economic or environmental harm or harm to human health" (U.S. Executive Office, 1999). Quantifying economic and environmental harm is not always easy, and different interest groups tend to give greater weight to different kinds of impacts. For agriculture and production forestry, substantial research has been dedicated to quantifying the impact of noncrop plants on crop yields and the economic costs of control. Most states in the United States maintain noxious weed lists of the species of greatest concern. Because these lists often include legal requirements for control or eradication by landowners, however, adding species to these lists can be a contentious political process. Most state lists tend to be dominated by species that impact agriculture and grazing (and often include some undesirable native species in addition to nonnatives).

There have been some efforts to formalize the criteria for ranking the invasiveness of nonnative species that are focused on nonagricultural plant communities (e.g., Morse et al. 2004). These expert-opinion approaches give high rankings to species that tend to dominate the cover of invaded ecosystems, particularly if those ecosystems are rare or otherwise threatened. Rankings are also higher for plants that have lasting presence or impact (e.g., allelopathy) or are difficult to control and eradicate. This approach to ranking is sensible from a biodiversity conservation standpoint, but has some interesting implications when applied to forest ecosystems. For instance, even though early seral stages for most forest types are transitory, they can be dominated by nonnative species. However, early seral stages are generally not a concern of most biodiversity conservation efforts, so nonnatives that primarily occur in them are generally not ranked highly in terms of invasiveness. Similarly, plants that may be ubiquitous in common forest types may be ranked lower than plants covering much less total land area that impact a rare forest type. Because specific impacts of nonnative species on native ecosystems (e.g., competition for resources, competition/suppression of pollinators, redirection of grazing impacts, and allelopathic effects) are usually unknown, species' rate of spread and cover levels are used by default as the primary criteria in ranking invasiveness.

The sheer number of plant species presents many challenges to monitoring. There are 4139 nonnative vascular plant species thought to be introduced to the United States (USDA NRCS 2000). Although not all of these species would be found in any particular region or land type, an effort to identify invasive species on forestlands in California, Oregon, and Washington (which is discussed further in the next section) initially resulted in a list of 245 species. Reliable identification of even a portion of that number of species in the field requires considerable expertise, especially considering the need to distinguish them from the numerous and varied native species (e.g., 3400 vascular plant species are listed for the state of Oregon alone).

The taxonomic skills needed to reliably identify invasive species can be difficult to acquire. Some species can be very distinctive, but many invasive species belong to families that require special skills to reliably key out and identify (e.g., *Compositae* or *Gramineae*). Focused inventories of a single or a few selected species, however, can effectively use nonspecialist personnel by providing some training and thorough guides for plant identification. These guides can usually incorporate more tips on identification and distinction from similar species than those available in the formal keys of published flora.

Monitoring approaches may need to be tailored to the expected habitats and abundance of the species of interest. Early in the invasion process, a species may occur on a small part of the landscape, requiring extensive travel and searching to locate populations. Later in the invasion process, plants may be more abundant, making sampling and quantification easier (but making control more difficult). Where species occur and how rapidly they disperse are functions of climatic tolerances, seed production, seed dispersal, competitive ability, and microsite requirements. Therefore, any species' spatial and temporal pattern of invasion is bound to be individualistic. However, many nonnative invasives are ruderal in many environments, and tend to occur in areas that are disturbed or have high levels of resources (e.g., roads, clear-cuts, and riparian areas). Therefore, monitoring that targets specific habitats or is stratified by different types could be more efficient than systematic sampling. Stratified sampling does have risks, however. A species may be shade intolerant and limited to cut areas or roads in wetter forest types, but may also invade high-light understories in dry forest types, or dry areas of otherwise moist regions (e.g., ridgetops). Some ruderal species with high dispersal abilities are also able to establish in small disturbed patches of otherwise intact forests (e.g., tree-fall gaps) and establish a seed bank for future spread. There are also many examples of shade-tolerant nonnative species that are invading intact forests. For these reasons, any stratified sampling strategy would ideally include some measurements in areas that are not preferred habitats for selected species.

Species are not equally identifiable at all times of the year. Phenology can have a big effect on the reliability of monitoring. Some species may be unambiguously identified for only a few weeks during the year when the diagnostic plant parts are present (e.g., flowers or fruits). The timing of those periods may not be very predictable in years when weather patterns differ substantially (e.g., a cool wet spring vs. a warm dry one). For monitoring programs in mountainous regions (e.g., western North America), it can be logistically impossible to sample locations in each elevation zone during the peak phenological time. For multiple-species monitoring, the peak phenological period may not coincide on a given site; for example, herbs may flower and senesce before grasses are fully developed. Specific sites can also be affected by other things, like intense grazing, which make plant sampling difficult. Phenology also affects the amount of cover a species has on a site at the time it is sampled, although plant cover in some forest types can be quite stable for extended periods. Although it is possible to sample most locations when most plant species are identifiable, phenology can be an important source of measurement error.

13.4 FOREST INVENTORY OF INVASIVE PLANTS

13.4.1 Forest Inventory Design

The FIA employs three "phases" of data collection to inventory and monitor the forestlands of the United States. The first phase uses remotely sensed images (historically aerial photo points, currently satellite images) to poststratify field plot data and improve the accuracy of inventory estimates (Bechtold and Patterson 2005). The second phase consists of a systematic grid of field locations on a 4.9 km spacing (1 point per 2,400 ha), hereafter referred to as the *standard* plots. The third phase consists of 1 out of every 16 standard locations (or 1 point per 38,800 ha), hereafter referred to as the *forest health* plots. The plot grids extend across all lands and all ownerships. The number of forested plots per Pacific state ranges from 4,068 to 8,170 for standard plots and from 256 to 428 for forest health plots (Table 13.1). With the implementation of the *annualized* approach in the Pacific states starting in 2001, 1 out of every 10 plots is sampled each year and evenly distributed across each state (Gillespie 1999). Earlier inventory designs did not sample all ownerships and/or used different plot designs in different areas. Although not discussed here, the data have been useful for assessing some invasive species (Gray 2005).

Vegetation data are collected at all standard plot locations on lands defined as "forestland" (i.e., land areas $\geq$0.4 ha in size that support, or recently supported, 10% stocking or 5% canopy cover of tree species and are not primarily managed for a nonforestland use) that are accessible (i.e., permission granted by owners and not hazardous to sample). (On land managed by the National Forest System in the Pacific states, vegetation is also sampled on accessible nonforest areas.) Each plot consists of a cluster of four 0.017 ha (7.32 m radius) subplots distributed over a 1 ha area. Because the plot design is fixed around the systematic plot location, plots can sample multiple land-use conditions, vegetation types, and stand age classes, termed "condition classes," which are distinguished in the field using an elaborate set of criteria. All collected data are identified to the condition class on which they were sampled. Travel is a large portion of the cost of inventory plot measurement, so crews strive to complete measurements of each plot in a single day.

TABLE 13.1
Land Area and Numbers of Standard and Forest Health Inventory Plots in the Pacific States, by State and Land Class

	California	Oregon	Washington
Total land area (ha)	40,393,282	24,863,054	17,234,832
Standard plot locations	16,860	10,356	7,276
Forested standard plots	7,170	5,514	4,068
Forest health plot locations	1,049	634	453
Forest health plots	428	334	256

Source: From U.S. Census Bureau, Census 2000 gazetteer of counties of the United States, http://www.census.gov/tiger/tms/gazetteer/county2k.txt, 2000.

Tree size and status information are collected at standard plots across the nation on trees $\leq$12.5 cm diameter at breast height (DBH, at 1.37 m height) with 2.07 m radius *microplots* at each subplot, and larger-diameter trees on the 7.32 m radius subplot. Sampling of other vegetation on standard plots is optional for each FIA unit, but most units across the country collect some information on invasive species (Rudis et al. 2005). In the Pacific coastal states of California, Oregon, and Washington, understory vegetation is sampled on subplots by recording the cover of individual species with $\geq$3% cover or species that are among the three most abundant of their growth form (i.e., tree, shrub, forb, or graminoid) on the subplot. Standard plots are measured during a long sample window (April to October) by crews with general forest resource measurement skills. Crew members are expected to be able to identify all trees and most shrubs encountered, as well as the most common forbs and ferns. Therefore, many forbs, ferns, and most graminoids may be identified to genus or growth form instead. To allow sufficient time for other inventory measurements, crews are limited to 15 min per subplot to record understory vegetation measurements.

On the subsample of forest health plots, all vascular plant species found in forestland conditions are recorded. A nested sampling design is used (Mueller-Dombois and Ellenberg 1974; Stohlgren et al. 1995), with measurements taken at three 1 m^2 quadrats within each subplot, and on the subplot as a whole (Stapanian et al. 1997; Gray and Azuma 2005). The measurements taken at both scales have changed slightly over the last decade; the currently stable protocols are described here and by Schulz (2003). Each species with canopy cover within 0–1.8 m above each quadrat is recorded. Plant cover estimates made on standard and forest health FIA plots use the standard cover definitions of Daubenmire (1959).

Cover of each species found on the subplot is also recorded. Vegetation on forest health plots is only measured during the summer season of peak phenology by experienced botanists. Plants that botanists cannot confidently identify to species and are deemed to be potentially identifiable (e.g., are sufficiently developed, or have flowers or fruits) are collected for later identification by herbarium experts. To allow completion of the protocol in a day, the subplot search for species and cover estimation is limited to 45 min per subplot. The forest health vegetation protocol has not been fully implemented nationally; data have only been collected for a few years in selected states.

13.4.2 Invasive Plant Lists

The PNW-FIA program began a collaborative project with the University of Washington in autumn 2005 to develop a prioritized list of invasive species to monitor on forestlands in California, Oregon, and Washington. The project will also develop the guides and training aids needed to ensure reliable sampling. Although the work is still in progress, the challenges faced to date are instructive.

The first step was to compile existing lists of invasive species from state and federal agencies and nongovernmental organizations in each state. This resulted in 421 species that were present on at least one list. Of those, 11 species were either native to the region or were thought to have both native and nonnative populations;

these were excluded from further consideration. Forty-three species were added to the list that were nonnative and had already been recorded on an FIA forest health plot or more than five standard FIA plots. The next step was to determine whether the species had been found in forestland habitats. This was difficult to ascertain with much certainty because habitat descriptions in most flora and herbarium sources were very general. Only species whose habitat descriptions included "forest," "woodland," or "riparian" were selected, resulting in 245 species.

Since 245 species was too many for a targeted monitoring list, it was necessary to rank the species' invasiveness. Some organizations in the Pacific states have adopted the expert-opinion approach described in Section 13.3 (Morse et al. 2004); these rankings were used where available. To avoid undue emphasis on highly localized or potential threats, species were also ranked according to available information about how widespread they were.

The ranking process resulted in a list of 95 species with a combined invasiveness and distribution ranking of medium to high. This list was sent to botanists and invasive plant experts in state and federal agencies in the three states for feedback on which species were important in their areas and how they would prioritize the lists. The steps yet to be completed are to develop regionalized lists (as opposed to a single list for all three states) of 30–40 species to monitor, assemble identification guides for reliable field sampling, and elicit support from cooperators for final implementation (or not).

13.4.3 Invasive Plant Inventory Results

The data used for the analyses in this chapter were collected on forestlands in California, Oregon, and Washington, USA, between 32.8° and 49.0° N latitude and 116.5° and 124.8° W longitude (Figure 13.1). Strong gradients in climate and forest vegetation occur across this region, ranging from moist Sitka spruce (*Picea sitchensis*) and redwood (*Sequoia sempervirens*) forests along the Pacific Ocean, oak (*Quercus*) woodlands in the interior valleys, and mountain hemlock (*Tsuga mertensiana*) and red fir (*Abies magnifica*) forests in the Cascade and Sierra Nevada ranges, to the Ponderosa pine (*Pinus ponderosa*) and pinyon–juniper (*P. monophylla–Juniperus occidentalis*) forests of the dry interior basins (Franklin and Dyrness 1973; Munz and Keck 1959). Some of these variations can be captured using geographic zones defined as *ecoregions* (Omernik 1987), shown in Figure 13.1.

13.4.3.1 Forest Health Monitoring Results

The forest health vegetation measurements were taken on 110 plots in Oregon in 2000 and 2001 and 91 plots in Washington in 2004 and 2005. Repeatability of sampling and some invasive species results from the Oregon plots have been presented elsewhere (Gray and Azuma 2005). Species were identified as nonnative to the United States based on the PLANTS database list (USDA NRCS 2000); species for which some populations were thought to be native (e.g., *Achillea millefolium*) were not included. Frequency points were summed for each species by assigning a score of 3 for each quadrat record, and a score of 1 for each additional subplot-search record, for a maximum possible 40 points per species per plot.

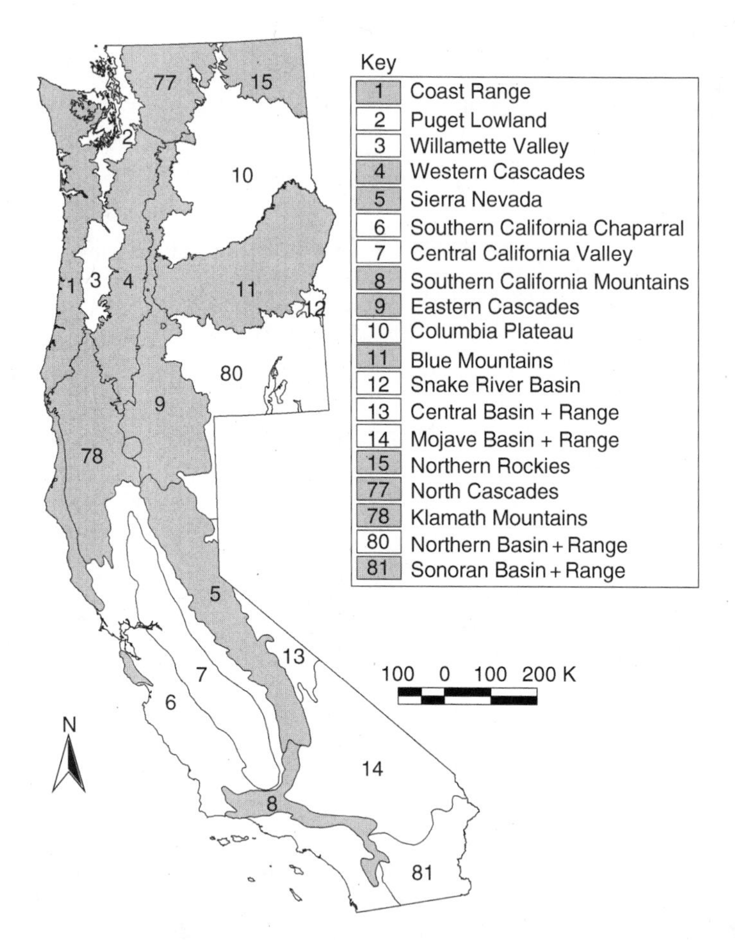

FIGURE 13.1 Map of California, Oregon, and Washington showing the ecoregions; ecoregions that are predominantly forestland sampled by the FIA program are shaded. (From Omernik, J.M., *Ann. Assoc. Am. Geogr.*, 77, 118, 1987.)

Plot-level measures of nonnative plant importance used proportions of nonnative to total species, and were calculated for species richness, summed frequency points, and summed cover. Data were grouped by ecoregion, forest type, and stand size class for analysis. Species summaries were compiled with plot counts, sums of frequency points, and means of plot-level cover from plots where species were recorded. One-way ANOVAs were performed to test for differences in nonnative proportion of species richness, frequency points, and total cover within different plot-level categories. The proportions were transformed with the arcsine square-root transformation to meet the normality requirements for ANOVA.

One or more nonnative species were recorded on 63% of all sampled plots in Oregon and Washington (Table 13.2). This percentage varied among ecoregions, from 100% for the Northern Basin and Range to 33% for the North Cascades; the latter was the only ecoregion where more than half the plots had no nonnative

TABLE 13.2
Summary of Nonnative Plant Abundance by Ecoregion

Ecoregion	Number of Plots	Plots with Nonnatives		Number of Species	Nonnative Species		Nonnative Percentages	
		Number	%		Number	%	Frequency	Cover
Coast Range	35	18	51.4	31.5	3.2	7.5	4.9	4.2
Puget Lowland	5	3	60.0	23.2	1.8	6.4	5.4	6.5
Willamette Valley	5	4	80.0	37.8	11.4	25.3	35.7	25.4
Western Cascades	41	25	61.0	36.5	2.4	6.1	3.3	3.8
Eastern Cascades	24	15	62.5	23.0	1.8	7.2	7.8	6.6
Blue Mountains	34	29	85.3	36.1	4.0	10.7	12.8	7.3
Northern Rockies	15	11	73.3	51.7	4.3	7.6	5.7	6.8
North Cascades	27	9	33.3	31.1	1.0	2.7	2.3	2.8
Klamath Mountains	9	5	55.6	28.8	1.3	5.2	1.0	0.7
Northern Basin and Range	6	6	100.0	22.2	1.3	6.7	5.8	3.5
Total	201	127	63.2					

Note: Table shows proportions of forest health plots with one or more nonnative species, and means of proportions of species, plot-level frequency points, and plot cover by nonnatives.

species recorded. The mean percentage of species on a plot that were nonnative differed by ecoregion ($p = .0006$), but was less than 10% except for the Willamette and Blue Mountains ecoregions. The nonnative proportions by ecoregion of summed frequency points ($p = .0001$) and plot-level cover ($p = .0061$) generally followed the same patterns found with species richness, but suggest some intriguing differences in the plot-level scale of dominance of nonnatives among ecoregions.

As expected, proportional richness, frequency, and cover of nonnative species differed by forest stand size class, a surrogate for time since stand-replacing disturbance ($p = .0001$, .0026, and .0019, respectively) (Figure 13.2). The highest nonnative proportions for species richness, frequency points, and cover were found in the smallest size classes, and the lowest proportions were found in the largest size classes. Both size class and ecoregion were significant factors when included in the same ANOVA model for proportional nonnative species richness ($p = .0001$), but the interaction was not significant, suggesting that the stand-size trend was similar across ecoregions.

The most important nonnative species on forestland in Oregon and Washington was *Bromus tectorum* (Table 13.3), a species well known in the region for its marked impacts on rangelands, and the most common nonnative species in the eastern Cascades, Blue Mountains, and Northern Basin and Range ecoregions. The next two most commonly encountered species (*Mycelis muralis* and *Tragopogon dubius*), however, have not received much attention on the region's invasive lists. *M. muralis* was the most common nonnative in the western and north Cascades ecoregions, while *T. dubius* was important in the Blue Mountains and northern Rockies

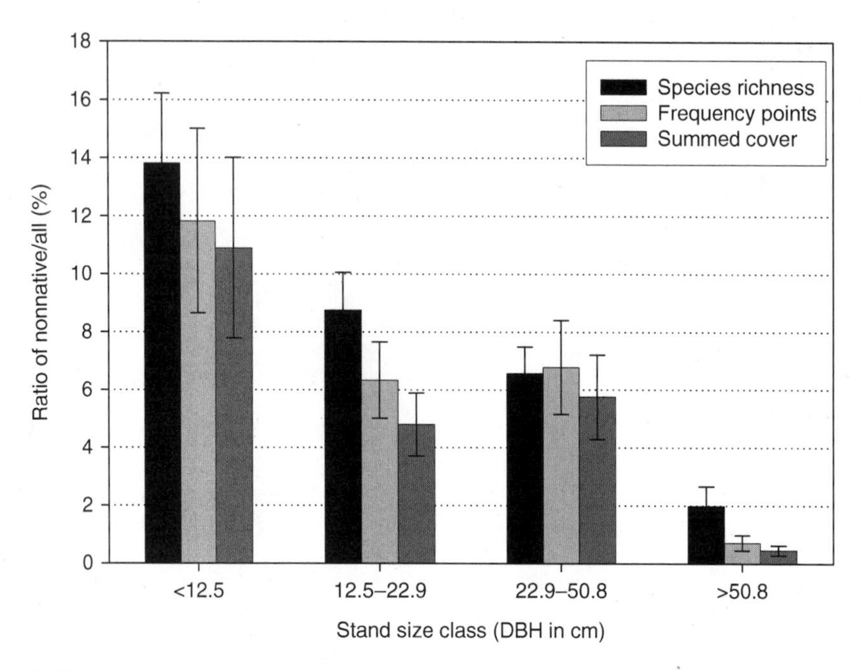

FIGURE 13.2 Importance of nonnative species by forest stand size class, as measured by the proportion of plot-level species richness, frequency points, and cover by nonnatives.

TABLE 13.3
Nonnative Species Found on Five or More Forest Health Plots

Scientific Name	Common Name	Number of Plots	Frequency Points	Cover	Number of Invasive Lists
B. tectorum	Cheatgrass	40	688	7.11	4
Mycelis muralis	Wall-lettuce	27	220	1.17	0
Tragopogon dubius	Yellow salsify	24	100	0.43	1
Hypericum perforatum	Common St. John's wort	21	156	1.73	6
Digitalis purpurea	Purple foxglove	20	124	1.89	3
Cirsium vulgare	Bull thistle	19	120	2.31	6
Dactylis glomerata	Orchard grass	18	102	1.55	2
Rumex acetosella	Common sheep sorrel	18	95	0.43	1
Hypochaeris radicata	Hairy catsear	17	139	3.18	3
Rubus laciniatus	Cutleaf blackberry	17	135	2.90	0
Senecio jacobaea	Stinking willie	16	86	1.09	7
Holcus lanatus	Common velvet grass	15	199	17.02	2

TABLE 13.3 (continued)
Nonnative Species Found on Five or More Forest Health Plots

Scientific Name	Common Name	Number of Plots	Frequency Points	Cover	Number of Invasive Lists
R. discolor	Himalayan blackberry	15	165	7.21	6
Leucanthemum vulgare	Oxeye daisy	14	96	0.88	4
Lactuca serriola	Prickly lettuce	14	88	0.25	2
Verbascum thapsus	Common mullein	12	52	0.43	4
Cynosurus echinatus	Bristly dogstail grass	10	56	0.98	2
Cirsium arvense	Canada thistle	10	53	4.18	7
Poa bulbosa	Bulbous bluegrass	9	85	3.27	0
Phleum pretense	Timothy	9	46	2.13	0
Cerastium fontanum	Common mouse-ear chickweed	8	18	0.31	0
Ranunculus repens	Creeping buttercup	8	13	0.83	2
Trifolium repens	White clover	7	28	0.60	0
Ilex aquifolium	English holly	7	13	1.43	4
B. japonicus	Japanese brome	6	69	0.38	0
Agrostis capillaries	Colonial bent-grass	6	33	6.67	0
Plantago lanceolata	Narrowleaf plantain	6	26	0.53	2
B. secalinus	Rye brome	5	56	1.01	0
Cytisus scoparius	Scotch broom	5	29	6.42	7
Lolium arundinaceum	Tall fescue	5	27	2.89	0
Torilis arvensis	Spreading hedge parsley	5	22	0.50	1
Agropyron cristatum	Crested wheatgrass	5	17	0.50	0

Note: Table shows the number of plots the species was recorded on (out of 201), sum of frequency points, mean characteristic cover at the plot level, and the number of invasive lists the species was found on (out of 8).

ecoregions. Some of the common nonnative species were found on several invasive species lists (e.g., *Hypericum perforatum* and *Cirsium vulgare*), but other prevalent species were on few or no lists (e.g., *Rubus laciniatus* and *Holcus lanatus*).

13.4.3.2 Standard Inventory Results

The Pacific states FIA program is investigating regional invasive species lists for inclusion in their standard plot inventory. However, the existing procedures of recording abundant, readily identifiable species provide valuable information for selected species and an indication of the kind of information that could be developed in the future for species selected to be part of an invasive list. Vegetation measurements were taken on 7558 forestland plots in California, Oregon, and Washington in 2001–2005. The same criteria were used to identify nonnative species in the database as with the forest health plot data. Plot-level frequencies were calculated as the proportion of measured subplots where each species occurred, and mean cover per species was calculated at the plot level. Nonnative occurrence was summarized by ecoregion, and mean frequency, mean cover, and plot counts were summarized for each species.

Logistic regression was used to assess relationships between the occurrence of selected nonnative species and climatic, topographic, and stand variables. The dependent variable was the odds ratio of species frequency, or the number of subplots occupied by a species over the total number of subplots sampled in the stand (GENMOD procedure, SAS Institute Inc. 1999). The analysis employed the same techniques that were used on older data from a portion of western Oregon (Gray 2005). The climate variables for this analysis were selected by intersecting plot locations with grids of the same climatic variables used by Ohmann and Gregory (2002), except that the base climate data were derived from the DAYMET model (Thornton et al. 1997). Regression models were built manually by including the strongest variable and assessing the strength of additional (uncorrelated, $r < .5$) variables. Additional variables were only included if the sign and parameter of existing variables did not change markedly and if visual examination of residuals indicated that the relationship was not determined by a few outliers.

The percentage of standard plots with nonnative species recorded was 26%, much lower than the 63% detected with the forest health plots (Table 13.2). This is not surprising given that only the most abundant species on a plot were recorded, ability of crews to identify composites and graminoids was limited, and some plots were sampled during suboptimal times of the year. Percentages of standard plots with nonnatives by ecoregion were more similar to those found on forest health plots in those ecoregions where the nonnatives were readily identifiable and tended to be dominant where found (e.g., 74% in the Willamette Valley and 58% in the Puget Lowland, where *R. discolor* is important). The most frequently recorded nonnative species—found on more than 100 standard plots—in descending order were *B. tectorum, R. discolor, C. vulgare, Hypericum perforatum, Digitalis purpurea*, and *Cynosurus echinatus*. Comparison of this list with the top of Table 13.3 indicates that some of the nonnative species were not sampled well with the standard plot protocols. Logistic regression models were developed for the most frequently recorded species.

The importance of climate and stand variables for describing nonnative frequency differed among species. *B. tectorum* was most frequently recorded in the Eastern Cascades and Blue Mountains ecoregions, but was well distributed in the eastern areas of the three Pacific states (Figure 13.3). In this region, the frequency of *B. tectorum* was primarily associated with low annual precipitation and low tree basal area (Table 13.4). In contrast, *Cirsium vulgare* was well distributed throughout the region, and its frequency was primarily associated with low basal area and secondarily associated with climate variables. *Cynosurus echinatus* was primarily found in the Klamath and southern California Chaparral ecoregions; its frequency was primarily associated with high annual temperatures and annual variation in precipitation. Most of the *D. purpurea* was recorded in the Coast Ranges; high annual vapor pressure and low tree basal area were the most important variables associated with its frequency. Although *H. perforatum* was well distributed in the region, its frequency was primarily associated with annual variation in precipitation and low elevations. *R. discolor* was well distributed in the

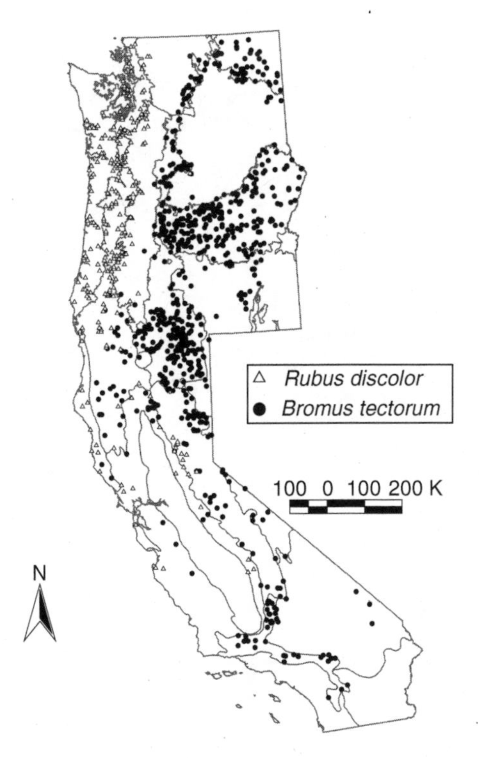

FIGURE 13.3 Distribution of the two most abundant nonnative species found on standard FIA plots, showing ecoregion outlines.

TABLE 13.4
Variables Selected for Regression Models of Species' Frequency, Their Parameter Estimates, and Significance of Parameter

Variables	Estimate	$F_{(1,7505)}$
B. tectorum ($N = 676$)		
Annual precipitation	−1.6969	960.7
Live tree basal area	−0.0143	707.1
Mean minimum temperature, December	−0.0533	55.5
C. vulgare ($N = 214$)		
Live tree basal area	−0.0141	764.7
CV of summer and winter precipitation	0.0808	181.9
Mean temperature, May to September	−0.1504	148.0
Cynosurus echinatus ($N = 111$)		
Annual temperature	0.3689	926.1
CV of summer and winter precipitation	0.0513	204.5
Aspect, cosine-transformed (SW = 0)	−0.3248	47.0

(*continued*)

TABLE 13.4 (continued)
Variables Selected for Regression Models of Species' Frequency, Their Parameter Estimates, and Significance of Parameter

Variables	Estimate	$F_{(1,7505)}$
D. purpurea (*N* = 146)		
Annual vapor pressure	0.0102	2803.2
Live tree basal area	−0.0149	1667.0
Mean maximum temperature, August	−0.2490	798.7
H. perforatum (*N* = 159)		
CV of summer and winter precipitation	0.1570	801.0
Elevation above sea level	−0.0016	653.4
Live tree basal area	−0.0086	385.4
R. discolor (*N* = 257)		
Elevation above sea level	−0.0043	3355.3
Live tree basal area	−0.0081	531.4
Mean temperature, May to September	−0.0446	19.8

Note: *P* values for all variables were less than 0.0001.
N is the number of plots with each species (out of 7558).

western parts of the region (Figure 13.3), but was primarily associated with low elevations. Although some of these relationships were similar to those found in a portion of western Oregon (Gray 2005), this analysis described a broader range of conditions, and climate tended to be more important. Despite the variety of climate variables available for the models in both analyses, elevation was an important factor for several species, which may reflect proximity to large plant populations on nonforest areas of human habitation and agriculture.

13.5 CONCLUSIONS

The results from the strategic forest inventory data on nonnative invasive species in the Pacific coast states illustrate the power of having a comprehensive assessment with consistent protocols and sampling effort. The high percentages of plots with nonnative species could be quite surprising to policy makers and the general public, many of whom regard most of the regions' forestlands as rather pristine and consider invasive species to still be an emerging threat. The lack of bias in sample location and sampling effort provides high confidence in the representativeness of the data. Indeed, having a consistent plot size and spatial plot design is critical to plant sampling because the probability of species' detections and species richness is very sensitive to plot size and shape and is not easily comparable across different sampling schemes. Inventory results are applicable to the entire sample population of forestland in a region, and regional or subregional estimates (e.g., by county or owner group) and associated statistical errors can be readily calculated.

By systematically sampling all vascular plants, the forest health monitoring design not only provides a comprehensive evaluation of nonnative invasive species, but also makes it possible to assess the importance and potential impact of invasives on the rest of the species in a stand. Results also document the considerable importance of many nonnative species that for whatever reason have not made it onto many agencies' invasive lists. Although the protocol is not part of the base FIA program, the data could be quite valuable if fully implemented. Analyses based on the full forest health plot grid would have more information than the partial set presented here and could take greater advantage of the other stand data collected on the same plots (e.g., disturbance history, stand structure, and topography).

The standard FIA plot grid provides a much higher density of points with which to assess the distribution and abundance of nonnative invasive species. This sample can be used to identify and map large invasions and regional hotspots and explore plausible relationships among associated attributes, including disturbance and management history (Gray 2005). It is not logistically feasible, without substantial increases in funding, to sample a large number of (or all) plant species on the standard plot grid, however. Several FIA regions have developed lists of invasive species that crews are trained to detect on standard plots (Rudis et al. 2005). If adopted in the Pacific states, an invasive list would need to be limited to fewer than 40 species to be feasible without a large increase in resources. Unfortunately, selecting these lists given the current absence of reliable data on invasive plant abundance and impacts introduces the risk of missing important species or searching for species that are not a threat in forestland. Although the list could be adapted to new information, this temptation should be weighed against the value of collecting long-term trend data.

The FIA inventory sample will rarely be useful for monitoring programs whose primary goal is early detection of new invasions. With a standard grid density of one plot per 2430 ha and a sample area of 0.067 ha, inventory plots sample one-36,000th of the landscape. The number of plots falling in rare forest types or those that occupy small portions of the landscape (e.g., riparian stands) can be quite small, particularly for small regions (e.g., a county). Therefore, a species would have to be fairly widespread before many detections would occur on inventory plots. The inventory data do supplement existing knowledge of species distributions (indeed, new county records in Oregon have been established for some vouchered species from forest health plots). The primary value from strategic monitoring of nonnative invasive species is probably the unique ability to quantify the abundance and impact of multiple species of concern on our forestlands. This information could aid managers' efforts to allocate scarce resources, inform policy makers about the magnitude of the issue, and provide a baseline of forest condition with which to judge the success of future prevention and control efforts.

ACKNOWLEDGMENTS

Thanks to B. Shulz for protocol design and quality assurance; K. Barndt for invasive ranking; J. Morefield, J. Pijoan, J. Belsher-Howe, and C. Grenz for data collection; and S. Jovan for manuscript edits.

REFERENCES

Anonymous, Statement on criteria and indicators for the conservation and sustainable management of temperate and boreal forests, *J. For.*, 93, 18, 1995.

Bechtold, W.A. and Patterson, P.L., Eds. The enhanced Forest Inventory and Analysis program-national sampling design and estimation procedures. USDA For. Serv. Gen. Tech. Rep. SRS-80, USDA Forest Service, Asheville, NC, 2004.

Blossey, B., Before, during and after: The need for long-term monitoring in invasive plant species management, *Biol. Invasions*, 1, 301, 1999.

D'Antonio, C.M. and Vitousek, P.M., Biological invasions by exotic grasses, the grass/fire cycle, and global change, *Ann. Rev. Ecol. Syst.*, 23, 63, 1992.

Daubenmire, R., A canopy-coverage method of vegetational analysis, *Northwest Sci.*, 33, 43, 1959.

Delisle, F., Lavoie, C., Jean, M., and Lachance, D., Reconstructing the spread of invasive plants: Taking into account biases associated with herbarium specimens, *J. Biogeogr.*, 30, 1033, 2003.

Franklin, J.F. and Dyrness, C.T., Natural Vegetarian of Oregon and Washington, USDA For. Serv. Gen. Tech. Rep. PNW-8, USDA Forest Service, Portland, OR, 1973.

Gillespie, A.J.R., Rationale for a national annual forest inventory program, *J. For.*, 97, 16, 1999.

Gray, A., Eight nonnative plants in western Oregon forests: Associations with environment and management, *Environ. Monit. Assess.*, 100, 109, 2005.

Gray, A.N. and Azuma, D.L., Repeatability and implementation of a forest vegetation indicator, *Ecol. Indicat.*, 5, 57, 2005.

Heinz Center (H. John Heinz III Center for Science, Economics, and the Environment), *The State of the Nation's Ecosystems: Measuring the Lands, Waters, and Living Resources of the United States*, Cambridge University Press, Cambridge, UK, 2002.

Hitchcock, C.L., Cronquist, A., Owenby, M., and Thompson, J.W., *Vascular Plants of the Pacific Northwest*, University of Washington Press, Seattle, Part 5, 1955.

Hurteau, M.D. and Briggs, R., Common yarrow *Achillea millefolium* L. plant fact sheet, United States Department of Agriculture, http://plants.usda.gov, 2003.

Mooney, H.A. and Hobbs, R.J.H., *Invasive Species in a Changing World*, Island Press, Washington, DC, 2000.

Morrison, M.L. and Marcot, B.G., An evaluation of resource inventory and monitoring program used in national forest planning, *Environ. Manage.*, 19, 147, 1995.

Morse, L.E., Randall, J.M., Benton, N., and Hiebert, R., An invasive species assessment protocol: Evaluating non-native plants for their impact on biodiversity, Version 1, NatureServe, Arlington, VA, http://www.natureserve.org/library/invasiveSpeciesAssessmentProtocol.pdf, 2004.

Mueller-Dombois, D. and Ellenberg, H., Aims and methods of vegetation ecology, Wiley, New York, 1974.

Munz, P.A. and Keck, D.D., *A California Flora*, University of California Press, Berkeley, CA, 1959.

National Research Council Committee to Evaluate Indicators for Monitoring Aquatic and Terrestrial Environments, *Ecological Indicators for the Nation*, National Academy Press, Washington, DC, 2000.

Ohmann, J.L. and Gregory, M.J., Predictive mapping of forest composition and structure with direct gradient analysis and nearest neighbor imputation in coastal Oregon, U.S.A., *Can. J. For. Res.*, 32, 725, 2002.

Omernik, J.M., Ecoregions of the conterminous United States, Map (scale 1:7,500,000), *Ann. Assoc. Am. Geogr.*, 77, 118, 1987.

Parendes, L.A. and Jones, J.A., Role of light availability and dispersal in exotic plant invasion along roads and streams in the H.J. Andrews Experimental Forest, Oregon, *Conserv. Biol.*, 14, 64, 2000.

Pimentel, D., Zuniga, R., and Morrison, D., Update on the environmental and economic costs associated with alien-invasive species in the United States, *Ecol. Econ.*, 52, 273, 2005.

Pyšek, P. and Hulme, P.E., Spatio-temporal dynamics of plant invasions: Linking pattern to process, *Ecoscience*, 12, 302, 2005.

Reichard, S. and Hamilton, C.W., Predicting invasions of woody plants introduced into North America, *Conserv. Biol.*, 11, 193, 1997.

Reichard, S. and White, P., Horticulture as a pathway of invasive plant introductions in the United States, *Bioscience*, 51, 103, 2001.

Ricciardi, A., Mack, R.N., Steiner, W.M., and Simberloff, D., Toward a global information system for invasive species, *Bioscience*, 50, 239, 2000.

Rudis, V.A., Gray, A., McWilliams, W., O'Brien, R., Olson, C., Oswalt, S., and Schulz, B., Regional monitoring of non-native plant invasions with the Forest Inventory and Analysis program, in *Proceedings of the Sixth Annual Forest Inventory and Analysis Symposium*, McRoberts, R.E., et al., Eds., USDA For. Serv. Gen. Tech. Rep. WO-70, USDA Forest Service, Washington, DC, 49, 2005, SAS Institute Inc., SAS/STAT User's Guide, Version 8, Carny, NC, 1999.

Schulz, B., Forest inventory and analysis: Vegetation indicator, FIA fact sheet series, USDA Forest Service, Washington, DC, http://fia.fs.fed.us/library/fact-sheets/p3-fac sheets/vegetation.pdf, 2003.

Stapanian, M.A., Cline, S.P., and Cassell, D.A., Evaluation of a measurement method for forest vegetation in a large-scale ecological survey, *Env. Mon. Assess.*, 45, 237, 1997.

Stohlgren, T.J., Falkner, M.B., and Schell, L.D., A modified-Whittaker nested vegetation sampling method, Vegetation, 117, 113, 1995.

Thornton, P.E., Running, S.W., and White, M.A., Generating surfaces of daily meteorology variables over large regions of complex terrain, *J. Hydrol.*, 190, 214, 1997.

U.S. Census Bureau, Census 2000 gazetteer of counties of the United States, http://www.census.gov/tiger/tms/gazetteer/county2k.txt, 2000.

U.S. Executive Office, Executive Order 13112 regarding invasive species, Federal Register 64, 6183, Washington, DC.

USDA NRCS, *The PLANTS Database*, National Plant Data Center, Baton Rouge, LA, http://plants.usda.gov, 2000.

Vitousek, P.M., D'Antonio, C.M., Loope, L.L., and Westbrooks, R., Biological invasions as global environmental change, *Am. Sci.*, 84, 468, 1996.

14 *Imperata cylindrica*, an Exotic Invasive Grass, Changes Soil Chemical Properties of Forest Ecosystems in the Southeastern United States

Alexandra R. Collins and Shibu Jose

CONTENTS

14.1 INTRODUCTION

How and why certain nonindigenous species become successful invaders has been a central question in invasion ecology. Many hypotheses have emerged; yet there is still little consensus as to what the primary driving mechanisms are for invasion establishment, colonization, and proliferation. The ability of invasive species to alter soil chemical properties, both through nutrient acquisition and allelopathy, remains a relatively unanswered question and could offer important insights into potential mechanisms for invasion success.

Invasions by exotics are changing large areas of North American ecosystems, but their biogeochemical impacts are not well characterized (Hook et al. 2004). For many successful plant invasions, at least in part, it appears that soil biochemistry can play an important role in the ability of a plant to establish and colonize a new area. Past research has shown that introduction of a new plant species, such as an exotic invasive, has the potential to change many components of the carbon (C), nitrogen (N), water, and other cycles of an ecosystem (Duda et al. 2003; Ehrenfeld 2003).

The fluctuating resource hypothesis predicts that invasive species will invade areas of high resource availability and often invade areas that are subject to frequent disturbance (Davis et al. 2000). Disturbance such as fire or clear cutting may increase soil resource levels and may allow for increased spread. Environmental disturbance that is able to alter soils could facilitate future invasions and decrease native biodiversity. It is predicted that invasive species have high resource requirements and will be less likely to invade resource-poor areas.

Exotic plants alter soil nutrient dynamics by differing from native species in biomass and productivity, tissue chemistry, symbiotic associates, plant morphology, and phenology. Available data suggest that invasive plants frequently increase biomass and net primary productivity, alter nutrient availability, and produce litter with higher decomposition rates than do the co-occurring native species (Ehrenfeld 2003). Differences in litter mass or the litter decomposition are often, but not always, accompanied by changes in organic matter. For example, Windham (1999) found that despite having large differences in standing crop biomass and litter dynamics, there was no difference in soil organic matter content in *Phragmites*-invaded marshes compared with noninvaded *Spartina patens* marshes.

Plant invasions do not result in consistent changes in soil properties, even for the same invasive species. For example, a recent study by Hook et al. (2004) indicated that *Centaurea maculosa*, a perennial Eurasian forb, might increase or decrease soil C and N pools in native grasslands in Montana, United States. Available data suggest a number of trends with respect to soil nutrients and plant invasions. Invasions have been associated with increases (Rutherford et al. 1986; Stock et al. 1995; Vitousek and Walker 1989; Witkowski 1991), decreases (Feller 1983; Versfeld 1986), or no change in soil N (Belnap and Phillips 2001). Howard et al. (2004) surveyed 44 sites in southeastern New York to examine the relationships between plant community characteristics, soil characteristics, and invasions by a number of exotic invasive plant species. Their study indicated that soil nitrogen mineralization and nitrification rates were strongly related to the degree of site invasion. Across broad environmental gradients and community types, invasive species were more commonly found in communities associated with higher resource levels.

Exotic plant invasions have also been shown to have an effect on a variety of other elements, including P, K, Ca, and Mg. Decreases in soil extractable pools may be associated with high uptake rates of these elements, which is driven by a large biomass or high tissue nutrient concentration of the exotic species. For example, studies conducted by Suding et al. (2004) showed that reduction of soil P weakened the ability of *C. diffusa*, an exotic invasive species, to tolerate neighbor competition proportionately more than other native species in grazed mixed-grass prairie. Consequently, under low P conditions, *C. diffusa* lost its competitive advantage and tolerated neighbor competition similar to other native species.

Changes in soil pH have also been reported to result from a variety of exotic plant invasions. Conflicting results of increases (Ehrenfeld et al. 2001; Kourtev et al. 1998, 1999), decreases (Boswell and Espie 1998; Scott et al. 2001), or no change in pH in response to invasion have been reported. Studies of two newly invading exotic plant species in hardwood forests of New Jersey, *Berberis thunbergii* and *Microstegium vimineum*, revealed that the soil pH in invaded areas were significantly higher than in the noninvaded areas, and the litter and organic horizons were thinner (Kourtev et al. 1998). In contrast, studies conducted by Scott et al. (2001) found that *Hieracium* species, invading New Zealand's tussock grasslands, increased total soil C and N and lowered the soil pH and mineral N relative to the adjacent herb-field vegetation.

In the present study, soil chemical properties were examined in *Imperata cylindrica* (hereafter cogongrass) invaded and noninvaded areas. As an opportunistic invasive C_4 perennial grass, cogongrass possesses characteristics that make it more competitive compared with native species. Cogongrass can colonize disturbed or undisturbed areas from a large number of seeds produced and also from an extensive belowground rhizome network. It forms dense monospecific stands that compete with and displace native species.

The successful invasion of cogongrass in many areas can also be attributed to its ability to tolerate a wide range of soil conditions. The habitats where cogongrass invades are diverse, ranging from the coarse sands of shorelines, to the fine sands or sandy loams of swamps and river margins, and to the clay soils of reclaimed phosphate settling ponds (MacDonald 2004). Cogongrass has extremely efficient nutrient uptake (Saxena and Ramakrishnan 1983) and associations with mycorrhizae, which may help explain its competitiveness on unfertile soils (Brook 1989). Brewer and Cralle (2003) suggested that cogongrass is a better competitor for phosphorus than are native pine-savanna plants, especially legumes, and that short-lived, high-level pulses of phosphorus addition reduce this competitive advantage without negatively affecting native plant diversity. Species richness of native plants was negatively correlated with final aboveground biomass of cogongrass in control and P-fertilized plots. In addition, cogongrass also has allelopathic properties, which may add to its ability to outcompete other species (Eussen and Soerjani 1975).

The objective of our study was twofold: (1) to determine how invaded and noninvaded cogongrass patches differ for soil chemical properties and (2) to examine how soils differ in nutrient availability between two sites: a disturbed and undisturbed site. In Florida, cogongrass infestation occurs primarily in disturbed areas (Willard et al. 1990). Two sites were chosen that differed in their disturbance history: a heavily disturbed clear-cut site as well as an undisturbed longleaf pine forest site. Under the fluctuating resource hypothesis, we would predict that resource levels should be greater in the logged rather than unlogged sites.

We predict that (1) soil organic matter will increase in invaded areas because of the higher biomass of cogongrass in comparison to the native flora, (2) soil nutrients

(NO_3–N, P, K, Ca, and Mg) will decline in cogongrass-invaded patches because of its extensive rhizome and root network and rapid accumulation of aboveground biomass, and (3) if allelopathy exists, root exudates into the rhizosphere will alter soil pH in invaded areas.

14.2 MATERIALS AND METHODS

Field studies were conducted at two sites in Santa Rosa County, Northwest Florida, USA: a logged site and an unlogged site. The logged site, owned by a paper company (30°50′ N, 87°10′ W), was a cutover site that was under 17 year old loblolly pine prior to clear cutting. The unlogged site, in Blackwater River State Forest, is one of the largest contiguous longleaf pine forest tracts in the southeast (30°50′ N, 86°50′ W). Four patches of cogongrass were randomly chosen from each site for a total of eight patches. Two soil surveys were conducted, one in the spring of 2004 and the other in the fall of 2004.

Soil samples were collected from rectangular plots ($4\,m \times 1\,m$) established at random around the perimeter of each patch at both sites. Plots were established so that $\sim 1\,m^2$ of the $4\,m \times 1\,m$ plot was initially cogongrass dominated and $3\,m^2$ was native vegetation. Two composite samples of 10 soil cores (15 cm deep) each were extracted pairwise from a total of 75 plots in the native and cogongrass monoculture areas using a soil auger. Soil samples were taken along the edge of the cogongrass patch to assess how cogongrass invasion affects soil chemical properties in newly invaded areas.

All soil composites were analyzed for organic matter, NO_3–N, P, K, Ca, Mg, and soil pH. Soil organic matter was determined by loss on ignition (500°C) (Storer 1984). Soil samples were sieved through a 2 mm sieve prior to extractions for elemental analysis. Soil NO_3–N was determined by extracting 20 g subsamples with 100 mL 1 *M* KCl. The extractant was gravity filtered and then frozen in 20 mL scintillation vials until analysis by Analytical Research Laboratory of the University of Florida (ARL-UF) using an Alpkem Flow Solution IV semiautomated spectrophotometer. P, K, Ca, and Mg were also analyzed (EPA method 200.7) by ARL-UF using an inductively coupled plasma atomic emission spectrometer following extraction of 5 g subsamples with 20 mL Mehlich-1 extractant solution.

Soil acidity was measured in a 1:2 soil–water solution (Kalra 1995). Soil texture was determined based on a soil composite sample from each site (Waters Agricultural Laboratories, Camilla, GA).

14.3 STATISTICAL ANALYSIS

Each patch was considered to be a replicate (total of eight patches) and plots as subplot descriptors. Soil organic matter, NO_3–N, P, K, Ca, Mg, and soil pH between invaded and native patches were analyzed using a mixed linear model, which took into account the within-plot variation (pairwise sampling) and also the between-patch variation using analysis of variance (SAS PROC MIXED) (SAS Institute 2002). Treatment effects were considered significant at $\alpha = 0.10$.

14.4 RESULTS

14.4.1 Soil Organic Matter

Soil organic matter content was the same between invaded and noninvaded areas at both sites (3.05% invaded and 2.88% native; $P = .46$). There were also no differences in soil organic matter (0–15 cm) between logged and unlogged sites ($P = .68$).

14.4.2 Nutrient Pool

NO_3–N levels of the invaded patches were 2.7% lower than those of native areas for the first soil survey ($P = .010$) (Figure 14.1). NO_3–N levels were significantly lower at logged site compared with the unlogged site ($P = .019$). The second soil survey (fall) showed a significant difference in NO_3–N between sites ($P = .10$), but did not show a significant difference between native and cogongrass-infested areas ($P = .60$). Average soil NO_3–N at the logged site was 0.1676 mg/L whereas it was 0.1751 mg/L at the unlogged site.

The amount of available P did not differ between the native and cogongrass patches in both spring ($P = .30$) and fall ($P = .53$) samplings (Figure 14.2). There was also no significant difference between the two sample sites for both soil surveys ($P = .12$ for spring and $P = .12$ for fall). Similarly, there was no significant site difference in available K for both sampling dates ($P = .75$, $P = .95$). However, in the spring measurement, soil available K was 6% lower ($P = .0001$) in cogongrass patches than in native patches (Figure 14.2). The fall sampling yielded similar results with 4.9% less K in the cogongrass patches ($P = .038$) (Figure 14.2) than in the native areas.

There was no significant difference between the invaded and native patches for available Mg during the spring sampling ($P = .71$), or the fall sampling. There was no significant difference in Mg concentration between the two sites ($P = .31$) for any of the sampling dates.

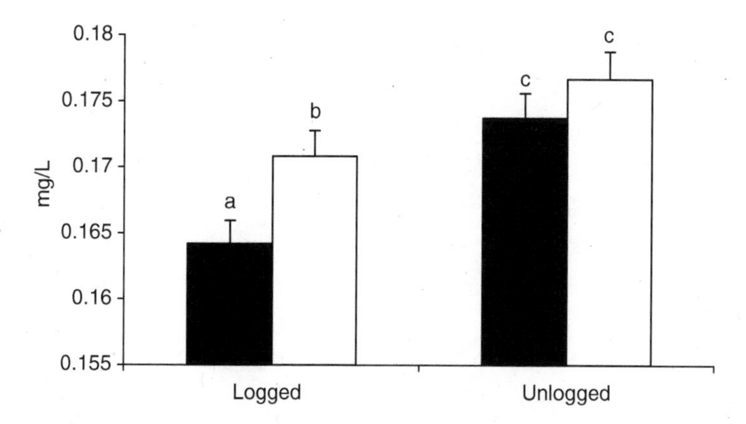

FIGURE 14.1 Mean concentrations (mean + 1 SE) of available NO_3–N for the logged ($n = 4$ patches) and unlogged ($n = 4$ patches) sites for the spring 2004 soil survey. Different letters indicate significant differences between cogongrass patches (■) and native (control) patches (□) ($P < .05$).

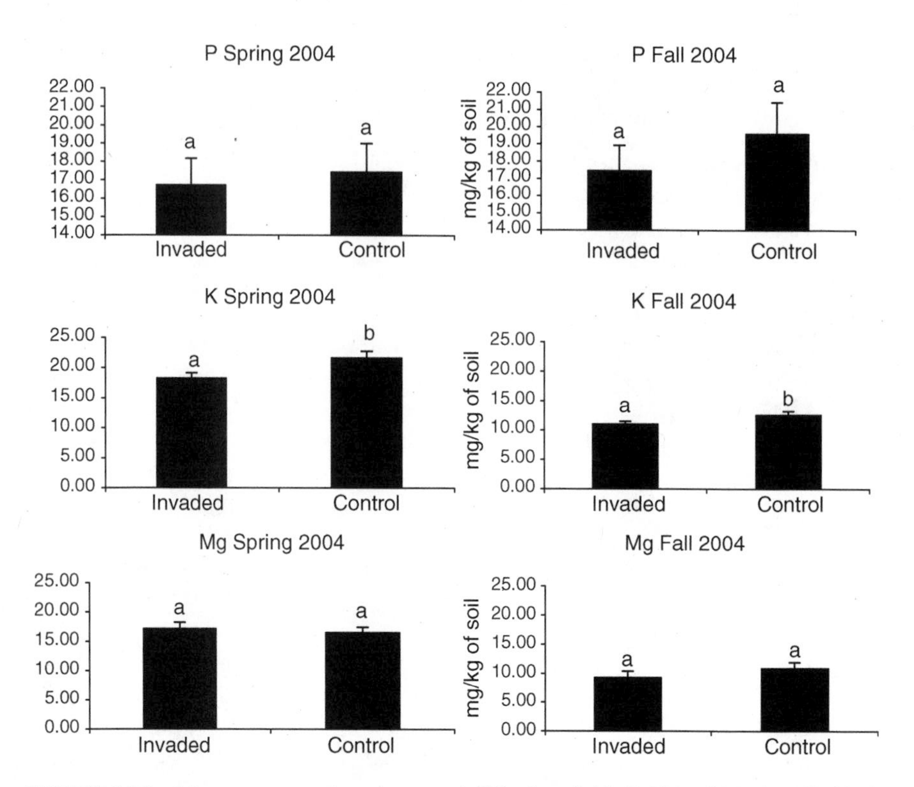

FIGURE 14.2 Mean concentrations (mean + 1 SE) of available P, K, and Mg ($n = 8$) for the combined logged and unlogged sites for the spring and fall 2004. Different letters.

Ca concentration did not differ between the invaded and native patches for both samplings. Both sampling dates showed significantly less available Ca at the unlogged sites compared with the logged sites ($P = .033$ for spring, $P = .031$ for fall).

14.4.3 Soil pH

Soil pH was lower in the cogongrass patches compared with the native patches for the summer (5.02 for cogongrass and 5.11 for native) and fall (4.40 for cogongrass and 4.62 for native) sampling (Table 14.1). There was no difference between the two sites for both sampling dates.

TABLE 14.1
Mean Soil pH in Invaded and Native Patches for Soil Surveys in Spring 2004 and Fall 2004 for the Logged and Unlogged Sites Combined

Season	Invaded	Control	SD	F(dfn, dfd)	Probability Value
Spring 2004 ($n = 8$)	5.02	5.11	0.13	18.71 (1, 140)	<.001
Fall 2004 ($n = 8$)	4.40	4.62	0.49	8.38 (1, 138)	<.01

14.5 DISCUSSION

Despite the extensive rhizome network and increased aboveground biomass in cogongrass-invaded areas (Ramsey et al. 2003), no difference in soil organic matter was observed between the invaded and native patches sampled. Differences in litterfall mass interact with differences in the litter decomposition rate to affect the net flux of C into the soil. The slow decomposition rate of cogongrass litter may ultimately be the reason for no observed difference in soil organic matter between treatments. This is supported by research of Hartemink and O'Sullivan (2001) who tested the decomposition and nutrient release patterns of cogongrass in the humid lowlands of Papua New Guinea. These authors determined that cogongrass leaf litter decomposed much slower, and half-life values exceeded the period of observation. The differences among decomposition patterns were best explained by the lignin + polyphenol:N ratio, which was highest for cogongrass (24.7).

There are several reasons to believe that cogongrass played a role in lowering soil NO_3–N in the invaded patches. All patches were similar with regard to soil texture, color, and disturbance history at each site. The lower NO_3–N availability found in invaded patches may have resulted from cogongrass' aggressive growth pattern, extensive rhizome network, and longer growing season. A companion study by Daneshgar et al. (2005) at the logged site showed that belowground biomass of cogongrass was 10 times greater than that of native vegetation. Lower NO_3–N levels in cogongrass patches may be the result of efficient nutrient uptake by the dense root and rhizome systems. Cogongrass is also known to have mycorrhizal associations, which may also explain the lower nitrate availability in invaded patches (Brook 1989). Mycorrhizae improve nutrient availability to host plants and alter their morphology, physiology, and competitive ability (Bray et al. 2003). As a result of lowering nutrient levels, specifically N, cogongrass may also be able to impede colonization and survival of native species, and facilitate its own persistence. Lower nitrate levels may also indicate lower ammonium and reduced nitrification; however, these variables were not tested. This represents a limitation of our study, and future work to measure NH_4^+ and total available nitrogen pools would be beneficial.

Differences in soil NO_3–N among soil surveys are attributed to differences in active growth and nutrient uptake during the growing season (spring 2004) compared with the slow growth in the fall. Following active vegetative growth early on, cogongrass growth and biomass accumulation reach a plateau in late summer (Shilling et al. 1997). The difference in soil NO_3–N between sampling dates, perhaps, reflects the difference in nutrient uptake patterns at different times of the year. Similar conclusions have been drawn by Wolf et al. (2004), whose study of invasion of a nitrogen-fixing nonnative species, *Melilotus officinalis*, yielded a similar trend for soil nitrate. In this case, NO_3–N was significantly lower in invaded patches sampled in May and then progressively decreased in both invaded and control patches so that no significant difference between the areas was evident when sampled in August.

Other nutrients (P, K, Ca, and Mg) exhibited varying trends with respect to cogongrass invasion. There was no significant difference in soil extractable pools for P for both surveys. This was contrary to previous research that has shown that levels of P and N are often related (Evans and Belnap 1999). Because our sampling scheme allowed only the newly invaded edges to be sampled, perhaps a significant relationship was not observed because of temporal constraints. Changes in soil nutrient pools may require longer periods of time to show differences in comparison with native patches. Future research could examine P differences between cogongrass patches that have been established for longer periods of time and native vegetation.

Lower levels of K were observed in invaded patches than in native patches for both soil surveys. Many grass species with fine, fibrous root systems are able to exploit K held in clay interlayers and near the edges of mica and feldspar crystals of clay and silt size particles (Brady and Weil 2002). The extensive belowground rhizome network (Daneshgar et al. 2005) as well as association with mycorrhizae (Brook 1989) may account for the ability of cogongrass to exploit soil K more efficiently than native species. Elephant grass (*Pennisetum purpureum* Schum), has been shown to obtain K from sand-sized particles, a form usually considered unavailable to plants (Brady and Weil 2002). Potassium is known to affect cell division, formation of carbohydrates, translocation of sugars, some enzyme actions, resistance of some plants to certain diseases, cell permeability, and several other functions (Plaster 1992). Thus, decreases in soil K in cogongrass areas could have serious implications for recruitment and growth of native plant species.

Lower pH was found in invaded patches in relation to native patches for both soil surveys. The mechanisms for decreases in pH in response to exotic species invasion have been attributed to increased nitrification, high rates of uptake of NH_4^+, and/or changes in litter quality (more acidic, base-poor litter) (Ehrenfeld 2003). Although NH_4^+ was not tested directly, decreases in NO_3–N were observed in cogongrass areas, and lower nitrate levels may also indicate lower ammonium levels. The preferential uptake of NH_4^+ ions from cogongrass-infested areas releases H^+ ions, resulting in a lower pH in the rhizosphere soil immediately surrounding the plant root. Differences in NH_4^+ uptake between cogongrass and native plants could account for the differences in soil pH in such close proximity. It is unclear at this point whether our measured pH differences are biologically relevant and what the implications are for future spread.

Although we do not have direct evidence of any mechanism responsible for lowering soil pH in cogongrass-invaded patches, past research findings may also point to allelopathy as a potential mechanism. In addition to the possibility of acidic root exudates, allelochemicals produced by cogongrass may also make the soil more acidic. Phenolic compounds present in foliage, roots, and rhizomes of cogongrass may be responsible for the allelopathic inhibition of germination and seedling development of other species (Inderjit and Dakshini 1991). Koger and Bryson (2003) suggest that allelopathic substance(s) provide cogongrass its extreme invasive and competitive abilities. However, the specific phenolic compounds in cogongrass tissues have not been identified and tested for allelopathic properties, and any research on potential allelopathy by cogongrass is still preliminary and inconclusive.

Decreased pH in cogongrass patches may also have implications for other soil extractable pools in the long term. At low pH, the cation exchange capacity is in general lower, with only the permanent charges of the 2:1 type clays and a small portion of the pH-dependent charges on organic colloids, allophane, and some 1:1 type clays holding exchangeable ions. Strongly acidic soils hold H^+ and hydroxy aluminum ions tightly to the soil surface. This tight association prevents K and other elements from being closely associated with the colloidal surfaces, which reduces their susceptibility to fixation (Brady and Weil 2002). Continuous acidic conditions may eventually reduce many soil nutrient pools, greatly reducing the success of surrounding native vegetation as well as transforming ecosystem biogeochemical properties.

Differences in resource availability between sites do not appear to play a large role for invasion success for the variables we measured. Ca was the only nutrient examined that showed lower levels in the unlogged site. Howard et al. (2004) found a significant positive relationship between soil Ca levels and site invasibilty for forests in southeastern New York. Cogongrass most commonly invades disturbed areas in Florida, and differences in Ca may increase invasibility. However, disturbance at the logged site may also provide open areas with reduced competition that increase colonization rates.

14.6 CONCLUSIONS

This research offers valuable insight for researchers and managers as to whether soil effects may be due to or responsible for cogongrass invasion and may generate important research questions for future experimentation. Future research could continue to examine how cogongrass changes soil properties in the area it is invading. Further variables such as nitrogen mineralization, total available nitrogen pools, and soil moisture would be valuable to consider. Cogongrass invasion in the southeastern United States continues to be a large economic and ecological problem. Extensive study of soil processes would be a valuable tool before expensive resource-demanding control programs are undertaken (Zavaleta et al. 2001).

ACKNOWLEDGMENTS

The authors express their sincere thanks to Drs. Francis E. Putz, Barry Brecke, Gregory MacDonald, and Deborah L. Miller for their critical review of an earlier version of the manuscript. Funding provided by International Paper, Plum Creek Timber Company and the Institute of Food and Agricultural Sciences is greatly appreciated.

REFERENCES

Belnap, J. and Phillips, S., Soil biota in an ungrazed grassland: Response to annual grass (*Bromus tectorum*) invasion, *Ecol. Appl.*, 11, 1261, 2001.

Boswell, C.C. and Espie, P.R., Uptake of moisture and nutrients by *Hieraceium pilosella* and effects on soil in a dry sub-humid grassland, *N Z J. Agric. Res.*, 41, 251, 1998.

Brady, N.C. and Weil, R.R., *The Nature and Properties of Soils*, 13th ed., Pearson Education, Upper Saddle River, NJ, 629, 2002.

Bray, S.R., Kitajima, K., and Sylvia, D.M., Mycorrhizae differentially alter growth, physiology, and competitive ability of an invasive shrub, *Ecol. Appl.*, 3, 565, 2003.

Brewer, J.S. and Cralle, S.P., Phosphorus addition reduces invasion of longleaf pine savanna (Southeastern USA) by a non-indigenous grass (*Imperata cylindrica*), *Plant Ecol.*, 167, 237, 2003.

Brook, R.M., Review of literature on *Imperata cylindrica* (L.) Raeuschel with particular reference to South East Asia, *Trop. Pest Manage.*, 35, 12, 1989.

Daneshgar, P., Jose, S., Ramsey, C.L., and Collins, A., Loblolly pine seedling response to competition from exotic vs. native plants, in *Proceedings of the Thirteenth Biennial Southern Silvicultural Research Conference*, Memphis, TN. USDA Forest Service Gen. Tech. Rep. SRS-92, Asheville, NC, pp. 50–52, 2005.

Davis, M.A., Grime, P., Thompson, K., Fluctuating resources in plant communities: A general theory of invasibility. *Journal of Ecology*, 88, 528, 2000.

Duda, J.J., Freeman, D.C., Emlen, J.M., Belnap, J., Kitchen, S.G., Zak, J.C., Sobek, E., Tracy, M., and Montante, J., Differences in native soil ecology associated with invasion of the exotic annual chenopod, *Halogeton glomeratus*, *Biol. Fertil. Soils*, 38, 72, 2003.

Ehrenfeld, J.G., Effects of exotic plant invasions on soil nutrient cycling processes, *Ecosystems*, 6, 503, 2003.

Ehrenfeld, J.G., Kourtev, P., and Huang, W., Changes in soil functions following invasions of exotic understory plants in deciduous forests, *Ecol. Appl.*, 11, 1287, 2001.

Eussen, J.H.H. and Soerjani, M., Problems and control of "alang-alang" [*Imperata cylindrica* (L.) Beauv.] in Indonesia, in *Proc. 5th Ann. Conf. Asian-Pacific Weed Sci. Soc.*, 5, 58, 1975.

Evans, R.D. and Belnap, J., Long-term consequences of disturbance on nitrogen dynamics in an arid ecosystem, *Ecology*, 80, 150, 1999.

Feller, M.C., Effects of an exotic conifer (*Pinus radiata*) plantation on forest nutrient cycling in southeastern Australia, *For. Ecol. Manage.*, 7, 77, 1983.

Hartemink, A.E. and O'Sullivan, J.N., Leaf litter decomposition of *Piper aduncum, Gliricidia sepium* and *Imperata cylindrica* in the humid lowlands of Papua New Guinea, *Plant Soil*, 230, 115, 2001.

Hook, P.B., Olson, B.E., and Wraith, J.M., Effects of the invasive forb *Centaurea maculosa* on grassland carbon and nitrogen pools in Montana, USA, *Ecosystems*, 7, 686, 2004.

Howard, T.G., Gurevitch, J., Hyatt, L., Carreiro, M., and Lerdau, M., Forest invasibility in communities in southeastern New York, *Biol. Invasions*, 6, 393, 2004.

Inderjit and Dakshini, K.M.M., Investigations on some aspects of chemical ecology of cogongrass, *Imperata cylindrica* (L.) Beauv., *J. Chem. Ecol.*, 17, 343, 1991.

Kalra, Y., Determination of pH of soils by different methods: Collaborative study, *J. AOAC Int.*, 78, 310, 1995.

Koger, C.H. and Bryson, C.T., Effects of cogongrass (*Imperata cylindrica*) residues on bermudagrass (*Cynodon dactylon*) and Italian ryegrass (*Lolium multiflorum*), *Proc. South Weed Sci. Soc.*, 56, 341, 2003.

Kourtev, P., Ehrenfeld, J.G., and Huang, W., Effects of exotic plant species on soil properties in hardwood forests of New Jersey, *Water Air Soil Pollut.*, 105, 493, 1998.

Kourtev, P., Ehrenfeld, J.G., and Huang, W., Differences in earthworm densities and nitrogen dynamics under exotic and native plant species, *Biol. Invasions*, 1, 237, 1999.

MacDonald, G.E., Cogongrass (*Imperata cylindrica*)—Biology, ecology and management, *Crit. Rev. Plant Sci.*, 23, 367, 2004.

Plaster, E.J., *Soil Science and Management*, 2nd ed., Delmar, Albany, NY, 1992.

Ramsey, C.L., Jose, S., Miller, D.L., Cox, J., Portier, K.M., Shilling, D.G., and Merritt, S., Cogongrass [*Imperata cylindrica* (L.) Beauv.] response to herbicide and disking on a cutover site in a mid-rotation pine plantation in Southern USA, *For. Ecol. Manage.*, 179, 195, 2003.

Rutherford, M.C., Pressinger, F.M., and Musil, C.F., Standing crops, growth rates and resource use efficiency in alien plant invaded ecosystems, in *The Ecology and Management of Biological Invasions in Southern Africa*, Macdonald, I.A.W., Kruger, F.J. and Ferrar, A. A., Eds., Oxford University Press, Cape Town, 189, 1986.

SAS Institute, *SAS User's Guide*, 9th ed., SAS Institute, Cary, NC, 2002.

Saxena, K.G. and Ramakrishnan, P.S., Growth and allocation strategies of some perennial weeds of slash and burn agriculture (jhum) in northeastern India, *Can. J. Bot.*, 61, 1300, 1983.

Scott, N., Saggar, S., and McIntosh, P.D., Biogeochemical impact of *Hieracium* invasion in New Zealand's grazed tussock grasslands: Sustainability implications, *Ecol. Appl.*, 11, 1311, 2001.

Shilling, D.G., Beckwick, T.A., Gaffney, J.F., McDonald, S.K., Chase, C.A., and Johnson, E.R.R.L., Ecology, physiology, and management of Cogongrass (*Imperata cylindrica*), Florida Institute of Phosphate Research, Bartow, FL, 1997.

Stock, W.D., Wienard, K.T., and Baker, A.C., Impacts of invading N_2-fixing *Acacia* species on patterns on nutrient cycling in town Cape ecosystems; evidence from soil incubation studies and ^{15}N natural abundance values, *Oecologia*, 101, 375, 1995.

Storer, D.A., A simple high sample volume ashing procedure for determination of soil organic matter, *Commun. Soil. Sci. Plant Anal.*, 15, 759, 1984.

Suding, K.N., LeJeune, K.D., and Seastedt, T.R., Competitive impacts and responses of an invasive weed: Dependencies on nitrogen and phosphorus availability, *Oecologia*, 141, 526, 2004.

Versfeld, D.B. and van Wilgren, B.S., Impact of woody aliens on ecosystem properties, in *The Ecology and Management of Biological Invasions in Southern Africa*, Macdonald, I.A.W., Kruger, F.J., Ferrar, A.A., Eds., Oxford University Press, Cape Town, 239, 1986.

Vitousek, P.M. and Walker, L.R., Biological invasion by *Myrica faya* in Hawaii: Plant demography, nitrogen fixation, ecosystem effects, *Ecol. Monogr.*, 59, 247, 1989.

Willard, T.R., Hall, D.W., Shilling, D.G., Lewis, J.A., and Currey, W.L., Cogongrass (*Imperatacylindrica*) Distribution on Florida Highway Rights-of-Way, *Weed Technology*, 4, 658, 1990.

Windham, L., Effects of an invasive reedgrass, *Phragmites australis*, on nitrogen cycling in brackish tidal marsh on New York and New Jersey, Ph.D. Dissertation, Rutgers University, New Brunswick, NJ, 1999.

Witkowski, E.T.F., Effects of invasive alien acacias on nutrient cycling in the coastal lowlands of the Cape fynbos, *J. Appl. Ecol.*, 28, 1, 1991.

Wolf, J.J., Betty, S.W., and Seastedt, T.R., Soil characteristics of Rocky Mountain National Park grasslands invaded by *Melilotus officinalis* and *M. alba, J. Biogeogr.*, 31, 415, 2004.

Zavaleta, E.S., Hobbs, R.J., and Mooney, H.A.,Viewing invasive species removal in a whole-ecosystem context, *Trends Ecol. Evol.*, 16, 454, 2001.

Section III

Management of Invasive Plants

15 Adaptive Collaborative Restoration: A Key Concept in Invasive Plant Management

James H. Miller and John Schelhas

CONTENTS

15.1 INTRODUCTION

Nonnative invasive species (NNIS) present a severe human dilemma due to their collective threat of replacing and damaging human sustaining ecosystems (U.S. Congress Office of Technology Assessment 1993; Mack et al. 2000; Pimentel 2002). Rapid developments in global trade have caught governments and their regulatory agencies unaware and ill prepared to prevent entries of foreign invasive species across previously insurmountable barriers of oceans, mountains, and desserts (Pierre 1996; Simberloff 1996). New introductions of NNIS have accelerated among and across all continents and have been characterized as bioinvasions of bioterrorists that threaten many countries' biosecurity (Vitousek et al. 1996; Pimentel 2002; Meyerson and Reaser 2003). Of the 20,000 nonnative plant species now free living in the United States, about 4,500 have invasive tendencies, while thousands more reside in our gardens, increasingly in the expanding urban fringe, with unknown consequences to adjoining lands (U.S. Congress Office of Technology Assessment 1993; Pimentel 2002). Deficiencies in policy, deficiencies in consistent research and management funding, and persistent gaps in scientific knowledge have all been identified as root causes of our current invasive dilemma in the United States (Simberloff et al. 2005). We would add that the lack of social organization to counter these invasions is just as obviously a major shortcoming.

Intentional introductions of NNIS for profit, cultural continuity, and support of government programs have been common while unintentional entries by hitchhiking species in ballast water, cargo, and containers go essentially unchecked in the United States and elsewhere (Mack et al. 2000; Pimentel 2002). The U.S. borders, like those of most countries, are relatively porous to plant movement because of the increased volumes of trade, including international internet sales and lack of policies and border surveillance resources (Simberloff et al. 2005). Most plant invaders of wildlands have gained entry to the United States through the plant production industry or by other deliberate introductions, since there is yet little regulation on which species are imported (U.S. Congress Office of Technology Assessment 1993; NRC 2002). If coordinated programs are not immediately institutionalized, future introductions will occur that will markedly and permanently alter forest, agricultural, and conservation lands and waters as NNIS exponentially spread from urban, suburban, and exurban lands and connecting right-of-ways (Liebhold et al. 1995; Simberloff 1996; NRC 2002; Von der Lippe and Kowarik 2006).

In spite of the increasing damage and threats from NNIS, few countries have yet devised effective infrastructures to deal with the alien invasive onslaughts within their borders (Pimentel 2002; Pal 2004; Britton et al. 2004; Schrader 2004). Nonnative invasive plants (NNIP) are currently replacing natural ecosystems in many parts of the globe with currently useless monocultures or stark assemblages of NNIP species (Vitousek et al. 1996; Pimentel 2002; D'Antonio et al. 2004). Ecosystem services that provision human civilizations and regulate and support natural processes and cultural amenities are eroded or drastically altered by encroaching NNIP. At the same time, nonnative and potentially invasive plant species continue to have societal value for soil stabilization, beautification, and restoration, and there are as yet generally few developed native substitutes (Ewel and Putz 2004). Invasive plants thus represent a complex and perplexing societal dilemma, with need for a more comprehensive awareness, management strategies, coordinated programs, and effective laws if we are to avoid bequeathing future generations with degraded ecosystems and ecoservices. It has become clear that a concerted, holistic effort that integrates science with management in new ways will be required for predicting, managing, and mitigating the spread of invasive species (McPherson 2004), and that society needs to develop a new approach to this inconvenient predicament.

15.2 SOCIETY NEEDS A NEW APPROACH TO NNIPs

Natural resource management today in general requires new approaches to deal with the complexity and uncertainty inherent in linked human and natural systems, and management challenges resulting from the multitude of public and private land ownerships that characterize most landscapes. Ecosystems are increasingly recognized as complex and changing, often in response to growing human actions, more rapid climate change, atmospheric pollution, and increasing occupation by alien invasive species. Effective resource management in today's world must work through multiple partners on multiple scales and take uncertainty into account to form systems that integrate efficient social learning.

To address threats to biological diversity, natural resource managers have increasingly incorporated ideas from the new scientific area of adaptive management and the new governance approaches involving collaboration (Lee 1993; Buck et al. 2001; Minteer and Manning 2003; Colfer 2005; Plummer and Armitage 2007). At the same time, ecosystem restoration is rapidly evolving as a science and practice, and both scientists and managers are working to develop the technologies and plant resources to repair invasive-species-impaired ecosystems (Sauer 1998; Taylor and McDaniel 2004). Restoration is complex, and authors differ in their views of its goals. For example, Sauer (1998) envisions restoration to save or reconstruct similar habitat features of the past, while there are no clear *pristine* ecosystems remaining as guides to desired future conditions, only scattered fragmented stands and altered communities (Minteer and Manning 2003). Botkin (2003) at the other extreme would argue that it is only the ecoservices of any human-dependent ecosystem that we must preserve and guarantee, even if invasive plants provide parts of these services. Owing to the degree of current occupation by NNIPs and projected human populations, that compromises in restoration approaches and outcomes will be necessary to maintain sustaining landscapes. However, the degree of compromise required will have much to do with society's awareness, its sense of threat, commitment of resources, and political leadership in forging national and state NNIP plans and initiatives.

Even though major strides have been made within the past 10 years in invasive plant awareness among scientists and professional managers and in invasive plant research, there remains a worrisome lack of strategies and organization for effectively halting introductions, dealing with spreading invasions, and restoring stands and ecosystems. Furthermore, the lack of public awareness and continued widespread sale and planting of invasives suggest that invasive plant presence in future landscapes depends solely on cooperative human action (Colton and Alpert 1998). Thus, we advocate a new, integrated process involving adaptive management cycles carried out through collaborative networks across landscapes for the containment of NNIS and the restoration of impacted stands and ecosystems. We have termed this process *Adaptive collaborative restoration* (Figure 15.1). It is *adaptive*, because we must learn as we go; *collaborative*, because it requires coordinated individual efforts across ownership boundaries and among landowners, managers, and scientists; and *restoration*, because our aim is to restore both sustainable food and fiber production systems and the wildlife habitat and cultural values associated with these. Because the challenges are too great to be shouldered by any single institution, unprecedented collaboration among both pubic and private agencies and entities across multiple levels of organization will be critical. At least in the short term, there are no adequate federal and state agencies with mandates to specifically address this dilemma, and there are few historical precedents for social and political mobilization on the required scale for natural resource management. Immediate collaboration is certainly required for organizing early detection networks that will ensure that widespread epidemics do not continue to occur, or at least that their impacts are mitigated in a timely manner (Meyerson and Reaser 2003). In this chapter, we outline a general

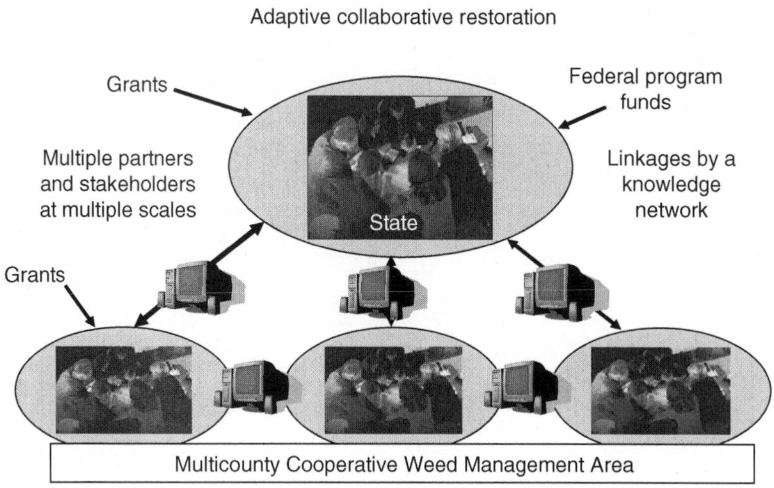

FIGURE 15.1 Horizontal and vertical networks of an ACR program for invasive plants are facilitated by web-based knowledge networks. Funding for programs come from federal, state, and county appropriation and grants. CWMAs are becoming common collaborative networks.

approach to ACR that seeks new synergies in management, science, collaboration, and web-linked technologies to address the historically unique challenges of NNIP invasions.

15.3 WHAT IS ACR?

ACR incorporates elements from three key ecosystem management trends from the 1990s (Sauer 1998; Buck et al. 2001; Plummer and Armitage 2007): adaptive management, collaborative management, and restoration management.

15.3.1 Adaptive Management

Adaptive management generally refers to a process of self-conscious learning-by-doing that incorporates formal processes of goal setting and modeling, monitoring, and rapid incorporation of new knowledge into refined goals and models to create a cyclical process of learning and managing (Walters 1997; Schelhas et al. 2001). Acquiring new information and rapidly incorporating new knowledge and experiences into planning and actions are of the utmost importance with NNIP management due to the number of new species arriving on the scene, evolving perspectives and laws, and the current lack of developed strategies. Instilling adaptive management cycles into an integrated approach can turn reactive management of invasives into a proactive mode (Foxcroft 2004). An example of a simple adaptive management cycle for use at the local level is illustrated in Figure 15.2. For adaptation to work, knowledge networks must play the vital role of providing instant information and connectivity (Jordan et al. 2003). Table 15.1 lists the crucial elements of a knowledge network system for NNIP management, where both real-time information

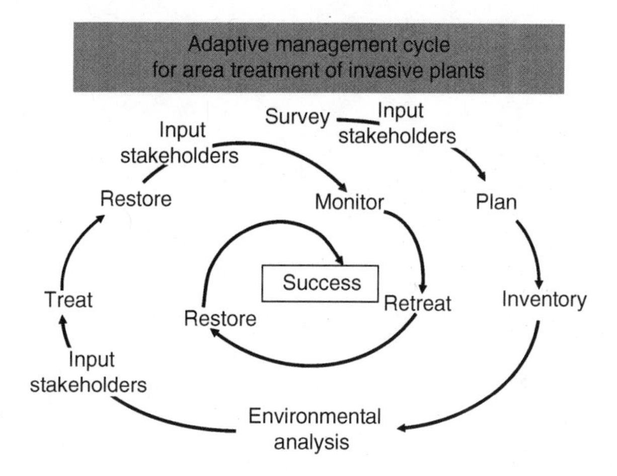

FIGURE 15.2 An adaptive management cycle for an area includes the programmatic steps and repeated treatments required for restoring invasive plant infestations.

and connectivity are subsystems. Table 15.2 enumerates current websites that when linked together could provide knowledge networks hosting formidable information resources. As yet, these websites have little to no connective capabilities, although the linking process is beginning through several national listservs in the United

TABLE 15.1
Web-Accessible Knowledge Network for Invasive Plant Management Must Contain Real-Time Information and Real-Time Connectivity to Facilitate Adaptive Management

Real-Time Information

- Invasive species by
  - Categories of threat
  - Commodity group
  - Land- and water-use categories
- Detailed identification guides
- Occupation maps at expanding scales and spread predictions
- Cost–benefit and risk analyses
- Control, containment, and eradication methods and restoration procedures
- Spread pathways and prevention means
- Comprehensive and multispecies strategies
- Impacts to ecosystem services, safeguards, and mitigation strategies

Real-Time Connectivity

- Decision networks and listservs among collaborative partners (see list in Table 15.3)
- Formal EDRR network
- Directories of service providers for control and restoration
- Directories of native plant sources for restoration using local ecotypes
- A library of pertinent laws, policies, and strategic plans
- Current approved documents such as environmental assessments and environment impact statements

TABLE 15.2
Current Websites That Provide Many of the Functions Needed to Create a Knowledge Network for NNIP Management in the United States

Name	Internet Link
Global	
Global Invasive Species Database	http://www.issg.org/database/welcome/
The Nature Conservancy's Global Invasive Species Team	http://tncweeds.ucdavis.edu/
An International Nonindigenous Species Database Network	http://www.nisbase.org/nisbase/index.jsp
National	
Bugwood Network	http://www.bugwood.org/index.cfm
Institute for Biological Invasion	http://invasions.bio.utk.edu/Default.htm
Invasive and Exotic Species (Invasive.org)	http://www.invasive.org/
National Association of EPPCs	http://www.naeppc.org/
NatureServe	http://www.natureserve.org/explorer/
U.S. Environmental Protection Agency Invasive Species Program	http://www.epa.gov/owow/invasive_species/
Regional	
Invasive Plant Atlas of New England	http://nbii-nin.ciesin.columbia.edu/ipane/index.htm
Mid-Atlantic Exotic Pest Plant Council	http://www.ma-eppc.org/
Mid-West Invasive Plant Network	http://mipn.org/
Nonnative Aquatic Species in the Gulf of Mexico and South Atlantic Regions	http://nis.gsmfc.org/
Southeast EPPC	http://www.se-eppc.org/
Federal Government Agencies	
Alien Plant Invaders of Natural Areas (PCA, National Park Service)	http://www.nps.gov/plants/alien/factmain.htm

Aquatic Nuisance Species Task Force (ANS)	http://www.anstaskforce.gov/default.php
ARS: Exotic & Invasive Weed Research	http://www.ars.usda.gov/main/site_main.htm?modecode=53-25-43-00
CSREES Invasive Species (USDA Cooperative Extension Service)	http://www.csrees.usda.gov/invasivespecies.cfm
FHWA—The Nature of Roadsides and the Tools to Work with it	http://www.invasivespeciesinfo.gov/docs/plants/roadsides/
Forest Inventory and Analysis (FIA)	http://srsfia2.fs.fed.us/nonnative_invasive/southern_nnis.php
National Institute of Invasive Species Science	http://www.niiss.org/cwis438/websites/niiss/About.php?WebSiteID=1
National Invasive Species Information Center	http://www.invasivespeciesinfo.gov/index.shtml
NBII Invasive Species Information Node	http://invasivespecies.nbii.gov/
Nonnative invasive plants of southern forests: A field guide for identification and control	http://www.srs.fs.usda.gov/pubs/gtr/gtr_srs062/index.htm
PLANTS National Database	http://plants.usda.gov/
Sea Grant Nonindigenous Species Site	http://www.sgnis.org/
The U.S. National Arboretum	http://www.usna.usda.gov/Gardens/invasives.html
U.S. Army Environmental Center (USAEC) Pest Management	http://el.erdc.usace.army.mil/pmis/
U.S. Fish and Wildlife Service Invasive Species Program	http://www.fws.gov/invasives/
USDA Animal and Plant Health Inspection Services	http://www.aphis.usda.gov/plant_health/plant_pest_info/weeds/index.shtml
USDA ARS Invaders Database System	http://invader.dbs.umt.edu/Noxious_Weeds/
USDA Forest Service Invasive Species Program	http://www.fs.fed.us/invasivespecies/index.shtml
USDA Forest Service	http://www.fs.fed.us/foresthealth/programs/invasive_species_mgmt.shtml
USGS Nonindigenous Aquatic Species	http://nas.er.usgs.gov/
Weeds Gone Wild: Alien Plant Invaders of Natural Areas	http://www.nps.gov/plants/alien/
State	
National Association of State Department of Agriculture (NASDA)	http://www.nasda.org/
State Laws and Regulations	http://www.invasivespeciesinfo.gov/laws/statelaws.shtml
State Exotic Pest Plant Councils	http://www.naeppc.org/

States that provide unstructured connectivity (e.g., regional exotic pest plant councils, Alien Plant Alliance, and Native Plant Conservation).

15.3.2 Collaborative Management

Collaborative management seeks to develop working linkages among all partners that collectively manage land and water resources across ownerships and jurisdictional boundaries within a defined area. Collaboration for invasive issues has two components. Horizontal connectivity among landowners and managers links people across landscapes, while vertical networks link local, county, state, regional, and national levels (Colfer 2005). Table 15.3 displays the multitude of partners that should be linked within a state at various scales to act in some manner of coordination to enact strategies. Because of federal and state appropriations, most organizational and program formation occurs at the state level, while the *actual work* happens on the ground level. At least 36 states have established some type of interagency invasive species council or working groups to address either selected NNIS or a range of invasive species (Environmental Law Institute 2002). These councils are either nonprofit organizations, governmental entities, or more loose associations of coordinating bodies. The most widely recognized and successful collaborations for invasive plant management in the United States have been Cooperative Weed Management Areas (CWMAs), which are organized at the county, multicounty, or state level (Midwest Invasive Plant Network 2006). A CWMA is a partnership of federal, state, and local government agencies; tribes, individuals, and various interested groups that manage noxious weeds or invasive plants in a defined area (Midwest Invasive Plant Network 2006). Most CWMAs were originally formed in the western United States and now are being organized in the midwestern, northeastern, and southeastern states. While CWMAs are clearly collaborative networks, it is unclear whether they have formalized elements of adaptive management.

15.3.3 Restoration Management

Restoration management is an indispensable part of integrated invasive plant management, providing technology for creating native-based communities and assemblages to replace invasive species infested lands. The restoration or rehabilitation phase requires establishment and/or release of fast-growing native plants that can outcompete and outlast any surviving NNIPs while stabilizing and protecting the soil. It has been learned that reforestation plantings are often necessary to suppress severe invasive grasses and vines, when eradication is not possible, and at the same time can yield a productive tree crop (Otsamo et al. 1995; Harrington et al. 2003). Failures in rangeland NNIP control have been attributed to the absence of a restoration phase that includes controlled recolonization of prairie and prairie-shrub communities resistant to reinvasion (Sheley and Krueger-Mangold 2003). When invasive trees have dramatically changed water courses, perhaps only ecological functions and ecosystem services can be restored when revegetation is enacted along with control measures (Taylor and McDaniel 2004). Natural succession can play a crucial role and be promoted when appropriate NNIP control methods are used that safeguard native species and the soil seed bank (Barnes 2004; Allen et al. 2007).

TABLE 15.3
Potential State Collaborative Partners for an Invasive Plant ACR Program

State level
- Department of agriculture and industries
- Department of conservation and natural resources
- Lands, parks, aquatic fisheries, and wildlife
- Department of transportation
- Forestry commission, department, or service
- Land grant universities and extension service
- Conservation and development districts
- Resource conservation and development districts
- Heritage programs
- Electric power generation and transmission authority
- Department of environmental management or protection
- Port authority, where appropriate

County and city level
- Commissions
- Planning boards
- Roads
- Parks, formal gardens, and lands
- Water providing authorities
- Electric cooperatives
- Land trusts, realtors, and developers
- Citizen groups for natural resource conservation

Federal lands
- Natural Resources Conservation Service
- Farm Services Administration
- U.S. Fish and Wildlife Service
- USDA Forest Service
- U.S. National Park Service
- River authorities, e.g., Tennessee Valley Authority
- Army Corp of Engineers

Industry level
- Commodity producers (livestock, crops, turf, fruit and nuts, aquiculture, etc.)
- Timber producers
- Plant production, wholesale, and retail industry (terrestrial and aquatic)
- Gas and other pipeline companies that manage right-of-ways
- Invasive control consultants
- Restoration consultants
- Herbicide and equipment producers, distributors, and retailers

NGO Partners
- The Nature Conservancy Lands
- Invasive plant councils
- Farmer associations
- Forestry associations
- Crop production associations
- Cattle production associations
- Wildlife, hunting, and fishing associations and federations
- Garden, wildflower, and native plant clubs and associations
- Trail and outdoor recreation associations

Restoration can range from rehabilitation that depends solely on natural succession when invasions are low to more complex reclamation procedures that add plants, structures, and growing medium when the soil and subsoil are highly disturbed or absent (Sauer 1998).

15.4 PROGRAM ELEMENTS

Program elements of an ACR program for NNIP are as follows:

- Cooperative knowledge networks linking stakeholders, land managers and scientists, policy makers, and political representatives at the national, regional, state, multicounty, and county levels, providing real-time information and connectivity. The functioning and power of the network ultimately relies solely on timely contributions and communications of individuals through voluntary, delegated, and assigned responsibilities.
- Collaborative strategies and programs for spread prevention through the following: (1) improved laws, policies, and public education; (2) promotion of new corporate and personal ethics to not sell, buy, or plant invasive plants; (3) sanitization of personnel, equipment, and animals when moving from or among infested sites; and (4) prohibitions regarding the sale and transportation of contaminated products such as extracted native plants, potted plants, fill dirt and rock, and mulch (Bryson and Carter 2004; Evan et al. 2006).
- Effective and efficient early detection and rapid response (EDRR) networks to identify and locate new high-risk introductions, communicate and verify the sites, eradicate the outlier infestations, and restore plant communities resistant to reinvasion (Westbrooks 2004).
- Creation and maintenance of a web-accessible spatially interrelated survey, inventory, and mapping system to corporately track existing and spreading invasions (e.g., Southeast Exotic Pest Plant Council's Early Detection and Distribution Mapping System, http://se-eppc.org/). Such a system with retrievable maps is invaluable for identifying and communicating zones of high infestations, advancing fronts, outliers, and weed-free zones. An example of the value of current survey results is shown in Figure 15.3 for tree-of-heaven (*Ailanthus altissima* (P. Mill.) Swingle) from the Invasive Plant Database of the USDA Forest Service, Southern Research Station's Forest Inventory and Analysis (FIA) Unit in cooperation with state partners. These data and those for 52 other invasive plants in the southern forest region of the United States are posted on a periodically updated website at http://srsfia2.fs.fed.us/nonnative_invasive/southern_nonnative_invasives.htm. Updating cycles for this FIA invasive plant survey depend on the survey activities within individual states, with most updated in part annually.
- Formulation of coordinated control, containment, and eradication programs to include cycles of integrated vegetation management treatments along with monitoring and corporate sharing of successful results and mistakes.

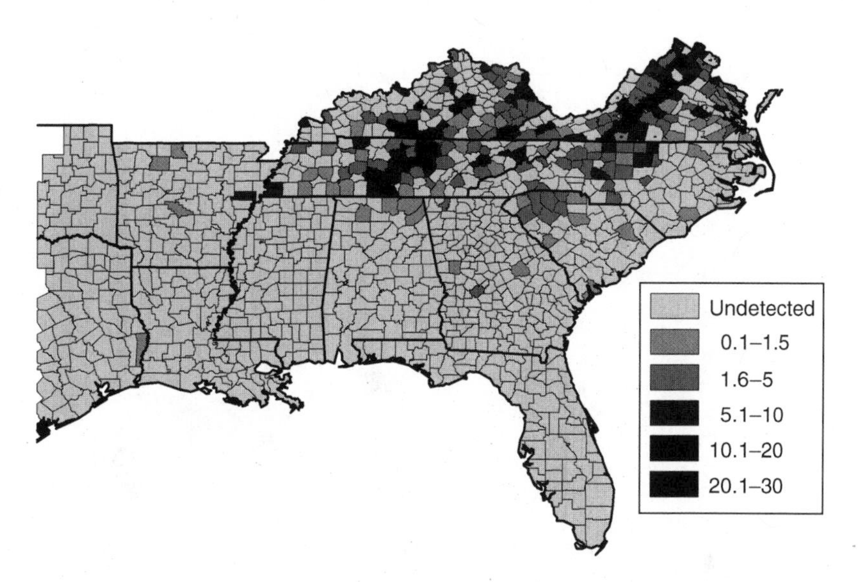

FIGURE 15.3 Percent occurrence of subplots within a county occupied by the invasive tree-of-heaven (*Ailanthus altissima*). (Data from USDA Forest Service, Southern Research Station's FIA Unit, Invasive plant database, Knoxville, TN, http://srsfia2.fs.fed.us/nonnative_invasive/southern_nonnative_invasives.htm. and imported into ArcView GIS.)

As is visible on Figure 15.3 with tree-of-heaven, a coordinated multistate control and eradication program is required to target outliers to stop the spread, contain advancing fronts, and protect special habitats in severely infested zones (Figure 15.4). As part of an integrated effort, prevention programs are crucial as are regional biological control programs for widespread severe NNIPs in certain situations (Simberloff and Stiling 1996; Moran et al. 2005).

- Restoration treatments to dovetail with control and eradication efforts. Adaptive information cycles are especially needed in this rapidly developing field. Restoration will guarantee invasive plant suppression and ensure that ecosystem functions and services are maintained. Continued surveillance and monitoring will be essential for restoration success.
- Focused research and research syntheses are needed with rapid technology transfer through effective networks along with feedback from the field on research needs (McPherson 2004).

The spread of invasives from state to state requires that every state have an Invasive Plant Management Plan with common elements that ensure regional protection. Included in each plan should be working elements and programs for ACR. Adaptive management cycles of learning and sharing advancements in understanding to all stakeholders must be permanently implanted. Regional and state strategies and actions should be nested spatially through collaborative networks across fragmented landscapes with the aim to constrain invasions and restore ecoservices.

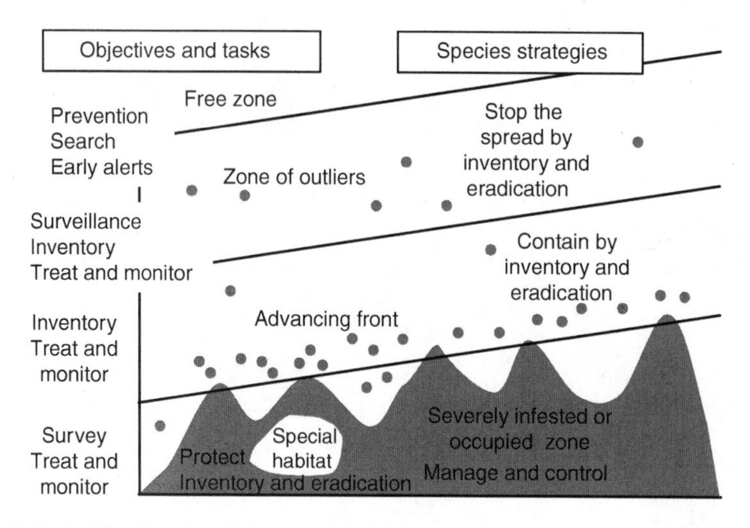

FIGURE 15.4 The degree of infestation shown in the four zones dictates the *objectives and tasks* employed to enact and achieve the *species strategies*. This is a static representation of a dynamic system that could have multiple invasive plants having similar or different zones of occupation.

15.5 IS ACR ACHIEVABLE?

It may seem overly idealistic to think that people can collaborate across institutional and property boundaries and across local to national levels to carry out complex processes of invasive species detection, prevention, and eradication and restore ecosystems in reflective scientific-based, adaptive learning processes. The question might be what would be an alternative approach to achieve the needed objectives? To be sure, there are few fully functioning adaptive collaborative management or restoration processes to serve as real-world models, although many efforts are ongoing (see Buck et al. 2001; Colfer 2005). Yet, natural resource managers worldwide, facing similar management issues, are either adopting adaptive collaborative ideas as a formal approach or drawing on the general principles to improve existing management approaches. Clearly, the ideas of ACR are of great relevance to the common difficulties faced by invasive plant management. New scientific understanding is rapidly accumulating on particular species impacts and means of control, while formal publication of results in scientific journals is too slow and too restrictive. Translating this information into useful technology that is then communicated through collaborative knowledge networks and finally put to use on the ground is the key.

Establishing fully comprehensive ACR processes across logical units of the landscape may seem to be a daunting task, and clearly, it will take time for public awareness and political will to develop to the point that this can happen. With that in mind, it is important to understand that, because many policy makers, managers, and scientists are individually grappling with the same problems, many of the components of ACR are already being put into place including CWMAs, invasive species task forces, and knowledge networks. Guided by the principles of ACR, the concepts and elements presented here can assist in crafting roadmaps for the expansion of interlinked knowledge networks. State and county leaders with their constituents and partners can continue to form cooperative networks that will increasingly carry out collaborative actions and gain funds that move things in the right direction. Individual scientists can create knowledge and syntheses that are available on websites with updating cycles in an adaptive manner, such as current annual state extension weed control recommendations. Agencies and universities can orchestrate linkages among websites and develop intelligent networks that integrate knowledge and site specifics to guide management and restoration prescriptions. ACR must build on existing institutions, issues, and interests at specific places, and will not look the same everywhere. Furthermore, ACR will always be a work in progress—never fully realized and always adapting to a changing world. While our own individual actions may seem insignificant given the magnitude of the NNIS problem, the ACR concept provides a framework with the potential to meld individual actions into a concerted process of effectively stopping new entries, collectively holding lines of defense, and ultimately reversing the current deluge of occupation to restore sustainable ecosystems needed now and tomorrow.

REFERENCES

Allen, S.L., Hepp, G.R., and Miller, J.H., Use of herbicides to control alligator weed and restore native plants in managed marshes, *Wetlands*, 27, 3, 739, 2007.

Barnes, T.G., Strategies to convert exotic grass pastures to tall grass prairie communities, *Weed Technol.*, 18, 1364, 2004.

Botkin, D.B., The naturalness of biological invasions, *West. North Am. Nat.*, 61, 261, 2001.

Britton, K.O., Duerr, D.A., II, and Miller, J.H., Understanding and controlling nonnative forest pests in the South, in *Southern Forest Science: Past, Present, and Future*, Rauscher, H.M. and Johnsen, K., Eds., General Technical Report GTR SRS-75, United States Department of Agriculture, Forest Service, Asheville, NC, 133, Chap. 14, 2004.

Bryson, C.T. and Carter, R., Biological pathways for invasive weeds, *Weed Technol.*, 18, 1216, 2004.

Buck, L.E., Geisler, C.C., Schelhas, J., and Wollenberg, E., Eds., *Biological Diversity: Balancing Interests through Adaptive Collaborative Management*, CRC Press, Boca Raton, FL, 2001.

Colfer, C.J.P., *The Complex Forest: Communities, Uncertainty, and Adaptive Collaborative Management*, Resources for the Future and CIFOR, Washington, DC and Bogor, Indonesia, 2005.

Colton, T.F. and Alpert, P., Lack of public awareness of biological invasions by plants, *Nat. Areas J.*, 18, 262, 1998.

D'Antonio, C.M., Jackson, N.E., Horvitz, C.C., and Hedberg, R., Invasive plants in wildland ecosystems: Merging the study of invasion processes with management needs, *Front. Ecol. Environ.*, 2, 10, 513, 2004.

Environmental Law Institute, *Halting the Invasion: State Tools for Invasive Species Management*, Environmental Law Institute, Washington, DC, 112, 2002.

Evan, C.W., Moorhead, D.J., Bargeron, C.T., and Douce, G.K., *Invasive Plants Responses to Silvicultural Practices in the South*, Bugwood Network, BW-2006-03, The University of Georgia, Athens, GA, 51, 2006.

Ewel, J.J. and Putz, F.E., A place for alien species in ecosystem restoration, *Front. Ecol. Environ.*, 2, 7, 354, 2004.

Foxcroft, L.C., An adaptive management framework for linking science and management of invasive alien plants, *Weed Technol.*, 18, 1275, 2004.

Harrington, T.B., Rader-Dixon, L.T., and Taylor, J.W., Jr., Kudzu (*Pueraria montana*) community responses to herbicides, burning, and high-density loblolly pine, *Weed Sci.*, 51, 965, 2003.

Jordan, N., Becker, R., Gunsolus, J., White, S., and Damme, S., Knowledge networks: An avenue to ecological management of invasive weeds, *Weed Sci.*, 51, 271, 2003.

Lee, K.N., *Compass and Gyroscope: Integrating Science and Politics for the Environment*, Island Press, Washington, DC, 1993.

Liebhold, A.M., McDonald, W.L., Bergdahl, D., and Mastro, V.C., Invasion by exotic forest pests: A threat to forest ecosystems, *For. Sci. Monogr.*, 30, 1, 1995.

Mack, R.N., Simberloff, D., Lonsdale, W.M., Evans, H., Clout, M., and Bazzaz, F., Biotic invasions: Causes, epidemiology, global consequences, and control, *Ecol. Appl.*, 10, 689, 2000.

McPherson, G.R., Linking science and management to mitigate impacts of nonnative plants, *Weed Technol.*, 18, 1185, 2004.

Meyerson, L.A. and Reaser, J.K., Bioinvasions, bioterrorism, and biosecurity, *Front. Ecol. Environ.*, 1, 307, 2003.

Midwest Invasive Plant Network, *CWMA Cookbook: A Recipe for Success*, Midwest Invasive Plant Network, Indianapolis, IN, 22, 2006.

Minteer, B.A. and Manning, R.E., Eds., *Reconstructing Conservation: Finding Common Ground*, Island Press, Washington, DC, 2003.

Moran, V.C., Hoffmann, J.H., and Zimmermann, H.G., Biological control of invasive alien plants in South Africa: Necessity, circumspection, and success, *Front. Ecol. Environ.*, 3, 77, 2005.

NRC (National Research Council), *Predicting Invasions of Nonindigenous Plants and Plant Pests*, National Academy Press, Washington, DC, 2002.

Otsamo, A., Adjers, G., Hadi, T.S., Kuusipalo, J., Tuomela, K., and Vuokko, R., Effects of site preparation and initial fertilization on the establishment and growth of four plantation tree species used in reforestation of *Imperata cylindrica* (L.) Beauv. dominated grasslands, *For. Ecol. Manage.*, 73, 271, 1995.

Pal, R., Invasive plants threaten Segetal weed vegetation in South Hungary, *Weed Technol.*, 18, 1314, 2004.

Pierre, B., A taxonomic, biogeographical and ecological overview of invasive woody plants, *J. Veg. Sci.*, 7, 121, 1996.

Pimentel, D., Ed., *Biological Invasions: Economic and Environmental Costs of Alien Plant, Animal, and Microbe Species*, CRC Press, Washington, DC, 369, 2002.

Plummer, R. and Armitage, D.R., Charting the new territory of adaptive co-management: A Delphi study, *Ecol. Soc.*, 12, 2, 10 [online], Available at http://www.ecologyandsociety.org/vol12/iss2/art10/, 2007.

Sauer, L.J., *The Once and Future Forest: A Guide to Forest Restoration Strategies*, Island Press, Washington, DC, 380, 1998.

Schelhas, J., Buck, L., and Geisler, C., Introduction: The challenge of adaptive collaborative management, in *Biological Diversity: Balancing Interests through Adaptive Collaborative Management*, Buck, L., Geisler, C., Schelhas, J., and Wollenberg, E., Eds., CRC Press, Boca Raton, FL, xix, 2001.

Schrader, G., A new working program on invasive alien species started by the multinational European organization dedicated to protecting plants, *Weed Technol.*, 18, 1342, 2004.

Sheley, R.L. and Krueger-Mangold, J., Principles for restoring invasive plant-infested rangeland, *Weed Sci.*, 51, 260, 2003.

Simberloff, D., Impacts of introduced species in the United States, *Consequences*, 2, 1, 1996.
Simberloff, D., Parker, I.M., and Windle, P.N., Introduced species policy, management, and future research needs, *Front. Ecol. Environ.*, 3, 12, 2005.
Simberloff, D. and Stiling, P., How risky is biological control? *Ecology*, 77, 1965, 1996.
Taylor, J.P. and McDaniel, K.C., Revegetation strategies after saltcedar (*Tamarix* spp.) control in headwater, transitional, and depositional watershed areas, *Weed Technol.*, 18, 1278, 2004.
U.S. Congress, Office of Technology Assessment, *Harmful Non-Indigenous Species in the United States*, OTA-F-565, U.S. Government Printing Office, Washington, DC, 391, 1993.
USDA Forest Service, Southern Research Station's FIA Unit, Invasive plant database, Knoxville, TN, http://srsfia2.fs.fed.us/nonnative_invasive/southern_nonnative_invasives.htm, 2007.
Vitousek, P.M., D'Antonio, C.M., Loope, L.L., and Westbrooks, R., Biological invasions as global environmental change, *Am. Sci.*, 84, 468, 1996.
Von der Lippe, M. and Kowarik, I., Long-distance dispersal of plants by vehicles as a driver of plant invasions, *Conserv. Biol.*, 21, 986, 2006.
Walters, C., Challenges in adaptive management of riparian and coastal ecosystems, *Conserv. Ecol.* [online], 1, 2, 1, Available at http://www.ecologyandsociety.org/vol1/iss2/art1/, 1997.
Westbrooks, R.G., New approaches for early detection and rapid response to invasive plants in the United States, *Weed Technol.*, 18, 1469, 2004.

16 Cogongrass (*Imperata cylindrica*)—A Comprehensive Review of an Invasive Grass*

Gregory E. MacDonald

CONTENTS

16.1 INTRODUCTION

Cogongrass [*I. cylindrica* (L.) Beauv.] is one of the most cosmopolitan grass species throughout the tropical and subtropical regions and is considered to be one of the 10 most troublesome and problematic weedy species in the world. It is considered a weedy pest in over 73 countries with over 100 common names associated with it, including cogongrass, japgrass, speargrass, alang-alang, and bladygrass. Cogongrass

* Adapted and updated from MacDonald, G.E., Cogongrass (*Imperata cylindrica*)—biology, ecology, and management, *Critical Reviews in Plant Science*, 23(5):367–380 (2004). Reproduced from Taylor & Francis Group, LLC. With permission.

is found in every continent except Antarctica and is generally associated with areas disturbed by human activities.

Several natural *I. cylindrica* savannas are commercially and ecologically important but the problems associated with its weediness far outweigh most positive benefits. Cogongrass is a major impediment to reforestation efforts in Southeast Asia, the number one weed in agronomic and vegetable production in many parts of Africa, and is responsible for thousands of hectares of lost native habitat in the southeastern United States. In this chapter, the common name cogongrass is used to describe this species. The objective of this review is to provide a thorough and comprehensive background of this important plant species.

16.2 BIOLOGY

Cogongrass is a warm-season, rhizomatous, perennial grass species that is found throughout the tropical and subtropical regions of the world (Holm et al. 1977). It spreads and dominates in disturbed sites, often in those areas disturbed by human activities (Holm et al. 1977; Brook 1989). Cogongrass invades and persists through several survival strategies including an extensive rhizome system, adaptation to poor soils, drought tolerance, prolific wind-disseminated seed production, fire adaptability, and high genetic plasticity (Hubbard 1944; Holm et al. 1977; Brook 1989; Dozier et al. 1998).

Cogongrass grows in loose to compact tufts, with culms arising from creeping underground rhizomes (Holm et al. 1977; Bryson and Carter 1993) (Figure 16.1). Culms are tufted, with glabrous to pubescent sheaths, and possess membranous ligules. Ayeni (1985) reported that several leaf sheaths are produced tightly rolled together, forming a cylindrical culm. Except for the flowering stalk, cogongrass is virtually stemless. The leaves are slender, flat, and linear-lanceolate, possessing serrated margins and an off-center prominent white midrib (Holm et al. 1977; Hall 1978; Terry et al. 1997). The serrated margins of the leaves accumulate silicates, which deter herbivory (Dozier et al. 1998). The leaves can reach nearly 1.5 m in height under conditions of good moisture and fertility and are amphistomatus (Holm et al. 1977).

Cogongrass rhizomes can comprise over 60% of the total plant biomass, and this low shoot to root–rhizome ratio contributes to its rapid regrowth after burning or cutting (Sajise 1976). Cogongrass rhizomes are white and tough with shortened internodes. Rhizomes are covered with brownish colored cataphylls (scale leaves), which form a protective sheath around the rhizome (Ayeni 1985; English 1998). The rhizomes of cogongrass possess a band of sclerenchymous fibers just below the epidermis, with each vascular bundle surrounded by sclerotic tissue (Holm et al. 1977; English 1998). These anatomical features help in conserving water within the central cylinder and help to resist breakage and disruption when the rhizomes are trampled or disturbed (Holm et al. 1977). Rhizomes are predominantly present in either the top 15 cm of fine textured soils or top 40 cm of coarse textured soils; however, rhizomes have been shown to grow to depths of 120 cm (Holm et al. 1977; Gaffney 1996).

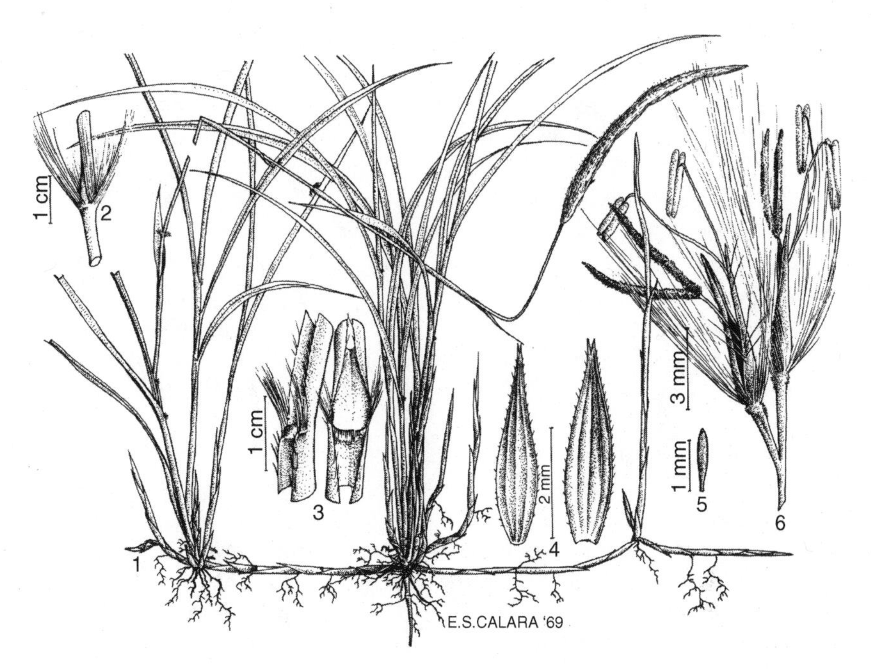

FIGURE 16.1 *I. cylindrica* (L.) Beauv. (1) Plant showing rhizome; (2) node with silky hairs; (3) ligule and portion of sheath and blade, ciliate; (4) lemma and palea; (5) grain; (6) spikelets. (Reprinted from Holm, L.G., Pucknett, D.L., Pancho, J.B., and Herberger, J.P., *The World's Worst Weeds, Distribution and Biology*, University Press of Hawaii, Honolulu, HI, 1977. With permission.)

Ayeni (1985) conducted extensive research on the morphology of rhizome development in cogongrass. He found that rhizomes develop from vegetative fragments at the 3–4 leaf stage, whereas, Tominaga (2003) reported rhizome formation from vegetative pieces occurred within 3 weeks. Rhizome formation in seedling plants is somewhat controversial; Holm et al. (1977) reported that rhizome development occurred 4 weeks after germination, whereas Shilling et al. (1997) reported rhizome formation at 8 weeks after germination. Work by Tominaga (2003) showed the presence of rhizomes after 12 weeks, but tiller formation occurred at 3–4 weeks from seedlings. Moreover, Tominaga (2003) showed that seedlings produce greater numbers of rhizomes and rhizomes of greater length within 1 year of establishment.

Ayeni (1985) showed that initial rhizome growth is plagiotropic (growing at a downward angle to the mother plant). Beginning at the fifth leaf stage, the rhizome grows diageotropic (horizontally) and between the fifth and sixth leaf stage, the rhizome becomes negatively orthogeotropic (turns upward). New shoots arise as the rhizome nears the soil surface. Secondary rhizomes are then formed on the convex side of the upturned rhizome, while secondary shoots are formed on the concave side. Secondary shoot development will precede new rhizomes under stressed conditions, but shoot and rhizome development will occur simultaneously if conditions are ideal. Tominaga (2003) classified cogongrass rhizomes into the following three categories: tillering, secondary colonizing, and pioneer rhizomes. He further suggested (based on earlier work [Grime 1977]) that cogongrass seedlings are defined as R-strategist (ruderal), invading open patches in disturbed habitats, whereas rhizomes from established stands of cogongrass are more defined as a C-strategist (competitor) that can persist in established populations. Lee (1977) measured rhizome density and found over 89 m in length per square meter of soil surface area, providing a tremendous amount of biomass for regeneration after foliar loss.

Ayeni and Duke (1985) also reported on the viability and regenerative capacity of cogongrass rhizomes. They found that regenerative capacity increased in older, more mature rhizomes and that young, newly formed tissue was not able to produce new shoots. This work showed little correlation between rhizome length or weight and regenerative capacity since rhizome fragments weighing as little as 0.1 g could form new plants. Earlier work by Soerjani and Soemarwoto (1969) also showed equal shoot development from 1 to 5 cm rhizome segments. Eussen (1980) reported the production of 350 shoots from rhizomes, covering an area of 4 m^2 in 11 weeks. The production of 168 new rhizomes from rhizome fragments in 87 days was observed by Patterson et al. (1980). Soerjani (1970) performed estimates of cogongrass production and extrapolated that over 4.5 million shoots could be present per hectare. Further estimations reported over 10 metric tons of leaf biomass and 6 tons of rhizome biomass per hectare. More recent research by Terry et al. (1997) suggests fresh weights of rhizomes to be 40 tons/ha.

Shoot regrowth from rhizomes is an area of research that has yielded conflicting results. Wilcut et al. (1988b) reported the lack of axillary bud development in cogongrass due to the lack of shoot emergence from rhizome sections with the apex removed. They also observed no shoot development from rhizomes buried at a depth of 16 cm. However, subsequent research on rhizome development conducted by Gaffney (1996) and English (1998) contradicts the lack of axillary buds in cogongrass. Gaffney (1996) observed shoot development confined to the apical region when the apex was intact, but removal of the apex promoted random shoot development along the length of the rhizome. Gaffney and Shilling (1995) also maintained axillary bud dormancy with applications of indole-3-acetic acid (IAA) to rhizomes with the apex removed. They suggested that auxin-imposed apical dominance maintained dormancy of the axillary buds along the rhizome rather than nitrogen availability (Gaffney and Shilling 1995). Apical dominance was also reported by Sriyani et al. (1996) in cogongrass biotypes from the United States and Indonesia. English (1998) induced axillary bud sprouting of cogongrass rhizomes through exogenous applications of growth-regulating compounds, further supporting the role of auxin-regulated apical dominance. Unlike many rhizomatous grass species, light has a positive effect on sprouting in cogongrass. Holm et al. (1977) report 2–3 times greater sprouting in light compared with dark, and English (1998) showed greater levels of sprouting under red and far-red light regimes compared with dark and green light.

Cogongrass is also a prolific seed producer, with over 3000 seeds per plant. Holm et al. (1977) and Bryson and Carter (1993) provide a detailed botanical description of cogongrass floral structure, and the following descriptions are derived from these works. Cogongrass produces a shortly branched, compacted, and dense seedhead (Figure 16.1). The appearance of the seedhead is cylindrical and spike-like, despite the fluffy, white plumes of the caryopses. The panicle averages 10–20 cm long, occasionally up to 60 cm, and 0.5–2.5 cm wide. Pedicels are unequal and the spikelets oblong to lanceolate and are surrounded by silky hairs. Glumes are membranous and possess long hairs originating from a basal callus. The lower floret is usually infertile, which is typical of the Andropogoneae (Hitchcock 1951). The lemma of the upper floret is similar to the lower floret—oblong, obtuse or toothed, and ciliate. The palea is usually very broad, toothed, and ciliate. The floret contains two stamens, orange to brown in color, whereas the stigmas are purplish to brown. The brownish colored seed (grain) is a caryopsis and single in each spikelet. The caryopsis possesses a plume of long hairs that effects wind dispersal. Pollination occurs as the inflorescence begins to expand, through swelling of lodicules, and this contributes to the fluffy appearance of the panicle (Holm et al. 1977).

Flowering is highly variable depending on region and environment. In the Mediterranean region, cogongrass flowers from May to August, whereas flowering occurs year-round in the Philippines (Holm et al. 1977). In the United States, flowering occurs in the late winter or early spring (Shilling et al. 1997; Willard 1988). Disturbances, including burning, mowing, grazing, frost, or the addition of nitrogen, can also stimulate flowering (Soerjani 1970; Sajise 1972; Holm et al. 1977). The influence of photoperiod in cogongrass flowering has not been researched to date, but Dickens and Moore (1974) and Shilling et al. (1997) suggest that cogongrass in the southeastern United States is not photoperiod dependent. The plumed seeds travel over long distances, but generally, movement is 15 m (Holm et al. 1977). McDonald et al. (1996) reported greater movement from larger clumps of aggregate seeds, and Hubbard (1944) stated that cogongrass seeds could travel up to 24 km over open country.

Although seed production has been reported to be prolific, seed as a major form of spread is questionable, particularly in the United States. Willard et al. (1990) reported that the primary spread in Florida was from rhizome pieces, either through contaminated fill dirt used in construction or intentional plantings for forage. Patterson et al. (1980) and Patterson and McWhorter (1980) also acknowledged rhizomes as a major vector of spread but suggested that the seed was responsible for long-distance infestations in isolated areas of Alabama and Mississippi. Santiago (1965) showed 95% germination from Southeast Asian cogongrass populations within 1 week after harvest and seed viability lasting at least 1 year. Lower germination rates (31%) were observed by Shilling et al. (1997) in Florida, but this was attributed to low spikelet fill. When spikelet fill was adequate, seed germination rates were very high (98%). Seed production from populations in Florida was shown to be self-incompatible: only cross-pollination from geographically isolated, heterogenous populations produced viable seeds (McDonald et al. 1995, 1996). In addition, no dormancy mechanisms were observed in seed from these populations. Shilling et al. (1997) also reported a rapid decline in seed viability over time, with a complete loss of viability after 1 year.

The author of this review speculates that this greater level of viable seed formation in these areas could be attributed to the comingling of the original populations introduced in 1911 from Japan and in 1921 from the Philippines. Previous research has shown that cogongrass is not self-compatible and the original populations more than likely spread via clonally through rhizomes. Once these populations spread to inhabit areas within close proximity, outcrossing was possible and viable seed production occurred. In Florida, cogongrass was thought to have been spread from the original introduction at Brooksville; therefore, the population has remained vastly clonal throughout the entire state. Seedlings tend to emerge in groups, reflecting the dispersion pattern of multiple seeds in a single clump (Shilling et al. 1997). Sajise (1972) showed that cogongrass germination requires light and red to far-red ratios that favor open conditions that were positively correlated to germination rates. Soemarwoto (1959) and Lee (1977) demonstrated that light was necessary to promote cogongrass germination, and Sajise (1976) showed greater seed germination at pH less than 5.0. Burnell et al. (2004) performed research on factors affecting seedling germination in the southeastern United States and found germination to occur from 11°C to 43°C, with an optimum temperature of 30°C.

Shilling and coworkers found that less than 20% of the seedlings survived past the first year and that seedlings are generally poor competitors with other seedling grasses (Willard and Shilling 1990; Shilling et al. 1997; Dozier et al. 1998). Dozier et al. (1998) indicated that seedling establishment would favor areas of limited competition, such as disturbed sites, and further suggested that cogongrass seedlings would be unlikely to establish in areas with >75% sod cover. However, in research on the effects of gap size, King and Grace (2000a) demonstrated the ability of cogongrass seedlings to germinate and establish in nondisturbed wet pine savannas of the United States, suggesting that cogongrass had the ability to invade established native plant communities. They also showed that tillage favored cogongrass establishment by promoting early growth, and burning resulted in greater seedling survival. On the basis of these findings, King and Grace (2000a) postulated that spring burning would favor cogongrass spread from seeds. Jouquet et al. (2004) showed that cogongrass density increased on the mounds of certain termite species in Western Africa, suggesting that the presence and abundance of cogongrass was directly linked to termite activity.

King and Grace (2000b) also showed that high water levels deterred seedling growth, with no germination under flooded conditions. They also observed greater tolerance to flooded conditions as the plants matured. Collins (2005) reported that functional diversity had little impact on cogongrass invasion in studies in northwest Florida and that cogongrass patches had significantly lower pH values compared with the surrounding native vegetation, suggesting that this may be a mechanism of invasion.

16.3 TAXONOMY

Imperata is a genus of the Poaceae or grass family, and within the family, it is in the tribe Andropogoneae, subtribe Saccharine. The genus *Imperata* has nine species worldwide, and is considered to be of major significance primarily due to cogongrass (Gabel 1982). Cogongrass (*I. cylindrica*) is the most varied and cosmopolitan species in the genus, which also includes *I. conferta*, *I. contracta*, *I. brevifolia*, *I. brasiliensis*, *I. tenius*, *I. cheesemanii*, *I. condensata*, and *I. minutiflora*. Gabel (1982) provides an excellent review of the genus and probably the most complete taxonomic study on *Imperata* spp. to date.

I. cylindrica was first described as *Lagurus cylindricus* by Linnaeus in 1759. Since then, over 31 different scientific names have been used to define the species (Gabel 1982). Cirillo, in 1792, was the first to coin the genus *Imperata*, named in honor of Ferante Imperato of Naples. Raeuschel in 1797 was the first to use the combination of *Imperata* and *cylindrica*, but Beauvois in 1812 cited Linnaeus (Laguri) in naming *I. cylindrica*. In current literature, both authorities are used interchangeably, but Gabel (1982) suggests the correct version is *I. cylindrica* (L.) Beauv.

Cogongrass is the most variable species within the genus, and specimens have been found from the western Mediterranean to South Africa, throughout India, Southeast Asia, including the Pacific islands, and Australia (Holm et al. 1977; Gabel 1982; Dozier et al. 1998). This variability has been suggested as the reason for the multitude of scientific names for the species. Hubbard (1944) and Santiago (1980) separated the species into five major varieties based on growth, geographic origin, and morphological characteristics. *I. cylindrica* var. *europa* is found in north Africa, the Mediterranean, and east to Afghanistan and is characterized by a chromosome number of $2n = 40$. *I. cylindrica* var. *major* is indigenous throughout Asia, Australia, the Pacific islands, and east Africa, with a chromosome number of $2n = 20$. *I. cylindrica* var. *africana* is found in west Africa and has a chromosome count of $2n = 60$. *I. cylindrica* var. *latifolia* is native only in northern India, whereas *I. cylindrica* var. *condensata* is found in Chile in the New World. Chromosome number for these varieties has not been reported. Hubbard (1944) and Santiago (1980) also reported differences in floral (spikelet length, enveloping hairs and anther, nervation and texture of glumes, shape of upper lemma) and leaf (hairs present at nodes, length of ligule, texture and leaf blade width) morphology that distinguish the varieties. Most of the research on cogongrass has been conducted with varieties *major* and *africana*, as these are the most widespread, damaging, and variable (Brook 1989). Santiago (1980) and Naiola (1981) further state that there are several ecotypes of *I. cylindrica* var. *major*.

Cogongrass is differentiated from the other species of *Imperata* by the presence of two anthers compared with one for the other species (Hitchcock 1951; Gabel 1982). Gabel (1982) distinguishes *I. condensata* (also possessing two anthers) as a separate species from cogongrass, whereas, Hubbard (1944) describes *condensata* as a variety of *cylindrica*. Gabel (1982) differentiates *I. condensata* by a longer ligule and short pedicels, and plants originating from western South America. Interestingly, Gabel (1982) does not recognize variety distinctions as described by Hubbard (1944), citing that the variability within variety is greater than the variability within the species as a whole. He also cited the lack of a workable key; so identification of a single specimen to a particular variety would be impossible without knowing its geographical origin. Further confusion is evident in Florida, where Hall (1978) does not separate *I. cylindrica* and *I. brasiliensis*, reporting single populations with one or two anthers. In subsequent findings, Hall (1998) suggests these species to be one, with differing variants occurring from the native (New World) genome and the introduced (Old World) foreign genome. The variability observed and the presence of differing anther number within a single population suggest potential hybridization between the species or variants (Hall 1998).

Molecular characterization has been limited for this species, but could greatly aid in addressing some of the questions and controversy surrounding cogongrass. Shilling et al. (1997) performed random amplified polymorphic DNA (RAPD) analysis on 10 populations within Florida. Results from these analyses, coupled with morphological characterization studies, showed that cogongrass was highly outcrossed. However, they also found evidence that some populations were clonal, suggesting that cogongrass spread in Florida was through both seeds and rhizomes. Cheng and Chou (1997) also performed RAPD analyses on cogongrass populations in Taiwan and found evidence of evolved ecotypes. Chou and Tsai (1999) found distinct populations using polymerase chain reaction (PCR)-amplified restriction fragment length polymorphism (RFLP) analyses. These analyses showed that the intergenic spacer (IGS) region of rDNA provides a good genetic marker and a potential tool for studying the microevolutionary process of cogongrass.

16.4 DISTRIBUTION AND HABITAT

Cogongrass is found throughout the world, virtually on every continent. It thrives in areas of human disturbance, and is reported established on over 500 million ha worldwide (Holm et al. 1977; Dozier et al. 1998). In Europe, cogongrass is found in the southern regions along the Mediterranean Sea, and it is also found in northern Africa to the Middle East. It is present in Iran and Afghanistan and throughout India. There are *Imperata* grasslands in northern India that stretch into Nepal. In Asia, cogongrass is most widespread. It occurs throughout Southeast Asia, Indonesia, and the Pacific islands. Asia has been reported to have ~35 million ha (Garrity et al. 1996/1997). Estimates of infestation in Indonesia range from 8.5 million ha (Garrity et al. 1996) to over 64 million ha (Suryatna and McIntosh 1980). These areas possess large, monotypic expanses, often called the mega-grasslands, *Imperata savannas*, or *sheet Imperata*. Many of these expanses are reported to be greater than 10,000 contiguous hectares, and cogongrass represents the climax species in these areas. Large, monotypic expanses also occur in Africa, where cogongrass is prevalent in the western coast, and in the eastern countries of Egypt, Sudan, Ethiopia, and further southward (Holm et al. 1977). It grows as far north as 45° latitude in Japan and as far south as 45° in Australia and New Zealand (Holm et al. 1977). Cogongrass is also found in western South America, and the closely related Brazilian satintail (*I. brasiliensis* Trin.) is found in central South America. Brazilian satintail is also found throughout the Caribbean and south Florida and is often confused with cogongrass.

Cogongrass var. *major* was inadvertently introduced in the United States (Mobile, AL) in 1912 as a packing material in Satsuma oranges from Japan (Tabor 1949, 1952; Dickens 1974). Cogongrass from the Philippines was purposefully introduced in 1921 to Mississippi as potential forage (Hubbard 1944; Dickens and Buchanan 1975; Patterson and McWhorter 1980). Subsequent forage trials were also carried out in Florida, Alabama, and Texas, although the Texas planting died out in the first year (Hubbard 1944; Dickens and Moore 1974). Cattlemen in Florida interested in cogongrass as a forage acquired the grass in 1939, and by 1949, over 1000 acres had been established in central and northwest Florida (Tabor 1952; Hall, 1983; Coile and Shilling 1993). Trials concluded that cogongrass was not an acceptable forage due to the accumulation of silica in the leaf tissue.

Cogongrass tolerates a wide range of soil conditions but appears to grow best in soils with acidic pH, low fertility, and low organic matter. Consequently, cogongrass habitats are quite diverse, ranging from the course sands of shorelines, the fine sands, or sandy loam soils of swamps and river margins to the >80% clay soils of reclaimed phosphate settling ponds. Saxena and Ramakrishnan (1983) report cogongrass to be extremely efficient in nutrient uptake. Brook (1989) also reports association with mycorrhiza, which may help explain its competitiveness on infertile soils. Brewer and Cralle (2003) also suggested that cogongrass is a better competitor for phosphorus than native pine-savanna species in the southern United States, citing that legume species are frequently displaced through this competitive mechanism. Cogongrass is a C_4 grass species (Paul and Elmore 1984), and while it is best adapted to full sun, cogongrass can also thrive under the moderate shade of savannas (Hubbard 1944). Studies conducted by Patterson (1980) suggest that cogongrass can adapt to changes in light levels through changes in specific leaf area and leaf area ratio and could tolerate a 50% reduction in sunlight. Further studies by Gaffney (1996) and Ramsey et al. (2003) showed that cogongrass has a light compensation point of 32–35 μmol m^{-2} s^{-1}, indicating its ability to survive as an understory species. This would explain its ability to rapidly invade deforested areas and persist in plantation crops.

Cogongrass habitats often include subtropical and tropical monsoonal areas. During the dry cycle, the mass of dead leaves often exceeds that of live leaves as leaf production appears to be regulated by soil moisture and balanced by the death of older leaves (Terry et al. 1997). Research has further suggested that this species sacrifices leaves to maintain a healthy rhizome biomass (Terry et al. 1997). This excess leaf biomass provides fuel for fires as cogongrass is a pyrogenic species, relying on fire for survivability and spread (Holm et al. 1977). Cogongrass fires are very intense and hot, with little aboveground vegetation able to survive, thereby preventing natural secondary succession (Eussen and Wirjahardja 1973; Seavoy 1975; Eussen 1980).

Another mechanism by which cogongrass maintains dominance is through allelopathy. Cogongrass has been reported to suppress the growth of crops (Hubbard 1944; Soerjani, 1970). Eussen et al. (1976), Eussen and Soerjani (1975), and Eussen (1979) in a series of experiments showed that cogongrass suppressed the growth of tomato and cucumber and that the factor(s) involved were more active at lower pH (Eussen and Wirjahardja 1973). Studies have also demonstrated potential allelopathy by cogongrass (Eussen 1979; Casini et al. 1998; Koger and Bryson 2003). Interestingly, Suganda and Yulia (1998) showed that crude water extracts of cogongrass rhizomes inhibited the germination of *Fusarium oxysporum* f. sp. *lycopersici* and reduced fusarium wilt of glasshouse grown tomato (*Lycopersicon esculentus* L.) plants. Interference can also be in the form of physical injury; the hard, sharp points of cogongrass rhizomes penetrate the roots, bulbs, and tubers of other plants, leading to infection (Eussen and Soerjani 1975; Boonitee and Ritdhit 1984; Terry et al. 1997). Interestingly, Holly and Ervin (2006) characterized the level of penetration of belowground vegetation by the rhizomes of cogongrass and showed that cogongrass most often penetrated itself.

16.5 IMPACTS

Cogongrass invades and persists in moist tropical areas because of extensive deforestation and fire-based land utilization systems (Holm 1969; Islam et al. 2001). It is considered a primary weedy species in tea (*Camillia sinesis* L.), rubber (*Hevea* spp.), pineapple (*Ananas comosus* Merr.), coconut (*Cocos nucifera* L.), oil palm (*Elaeis* spp.), and other perennial plantation crops in Asia, whereas in Africa it causes the greatest damage in agronomic production (Ivens 1980). In other areas, cogongrass infests natural habitats, destroying many native plant ecosystems in the southeastern United States.

Cogongrass is a major constraint in the establishment of plantation crops such as rubber, pineapple, tea, banana (*Musa* spp.), citrus (*Citrus* spp.), and coconut (Soerianegara 1980; Dela Cruz 1986; Ohta 1990). This weedy species competes directly with the crop for light, nutrients, and water and has been shown to retard the growth of teak trees (*Tectona grandis* L.) by more than 85% in the first year of establishment (Coster 1932, 1939). Cogongrass has also been shown to reduce rubber tree growth by 96% after 5 years (Soedarsan 1980). Holm et al. (1977) reported that when rubber plantations were neglected in Sumatra in the early 1960s due to civil strife, a period of 5–10 years was needed to control the invading cogongrass. In addition to direct competition, cogongrass interferes with crop growth through allelopathy, physical penetration of underground tissues leading to infection and rot, increased fire potential, and soil compaction (Soerianegara 1980; Dela Cruz 1986).

Cogongrass is also a serious weed in slash-and-burn agriculture systems, particularly in Africa (Udensi et al. 1999) and Southeast Asia (Friday et al. 1999). It is reported to be one of the top three weedy pests of cassava (*Manihot* spp.), cotton (*Gossypium* spp.), maize (*Zea mays* L.), peanut (*Arachis hypogaea* L.), upland rice (*Oryza sativa* L.), and sweet potatoes [*Ipomoea batatas* (L.) Poir.] (Holm et al. 1977). Cogongrass is considered the most serious agricultural weed in Benin, Nigeria, and southern Guinea, infesting over 20 crop species (Chikoye et al. 2000). In West Africa, cogongrass has been shown to reduce cassava yields by 62%–80% and yam (*Dioscorea* spp.) production by 78% (Udensi et al. 1999). Chickoye (2001) showed a 50% reduction in maize yield from cogongrass interference, and Akobundu et al. (2000) found that four hand weedings were needed to prevent yield reduction in maize grown in Nigeria. Early reports by the International Institute of Tropical Agriculture (IITA) (1977) showed that 54% of the total crop production budget was for cogongrass weeding. Chickoye et al. (2000) stated that farm size in West Africa was limited by the labor intensiveness of cogongrass, citing reductions in yield and quality, alternative host for pathogens and insects, and physical injury to farm workers through cuts and puncture wounds. Terry et al. (1997) report that vast areas of arable land in west Africa have been abandoned because of the lack of effective cogongrass control, and Darkwa et al. (2001) report that cogongrass is the major factor limiting crop yields in Ghana. This species is also considered to be a major weed in barley (*Hordeum vulgare* L.) in Iran and a weedy pest in vegetables, soybeans (*Glycine max* L.), and potatoes (*Solanum tuberosum* L.) in Japan (Holm et al. 1977).

Cogongrass is occasionally used as forage, but can only be grazed when the plants are very young. Intensive management is needed to maintain the grass in this juvenile vegetative stage, as once the leaves mature they become virtually impalatable. Cogongrass has extremely low nutritive value, with an estimated carrying capacity of less than one-third animal per hectare (Garrity et al. 1993). It possesses a very high carbon to nitrogen ratio (Hartemink and O'Sullivan 2001), and the nutritional value is considered too low for horses. Cattle have been shown to develop scours when forced to graze cogongrass, and there are reports of direct injury through cut muzzles and infected hooves by the sharp tips of the leaves and rhizomes (Holm et al. 1977).

Cogongrass has been shown to harbor locusts, and there is evidence that swards of this grass are a major breeding ground for these pests (Brook 1989). Cogongrass is also host to several polyphagous insects in cereals and an alternative host of the rust *Puccinia refipes* Diet (Vayssiere 1957; Chandrasrikul 1962). Verma et al. (2000) reported the isolation of a 67 kD cross-reactive allergen from cogongrass pollen containing at least three allergen determinants.

In the United States, cogongrass poses the most serious threat to native ecosystems. There are over 500,000 ha with some level of infestation in Florida alone. Several thousand hectares are also infested in the states of Alabama and Mississippi (Bryson and Carter 1993; Matlack 2002). This species can also be found as far west as Louisiana and as far north as Virginia, primarily along the coastal regions. Cogongrass generally invades areas after a disturbance, such as mining or land reclamation, forest operations including clear-cutting and replanting, highway construction, or natural fire or flood.

Once established, cogongrass outcompetes native vegetation, forming large monotypic expanses with extremely low species diversity and richness. Lippencott (2000) found that cogongrass introduced and altered normal fire cycles of southeastern U.S. natural ecosystems (sandhill communities). The fires resulting from swards of cogongrass had higher maximum temperatures at greater heights and increased fire mortality of longleaf pine (*Pinus palustris*), normally a fire-tolerant species. The author further hypothesized that the changes in fire behavior due to cogongrass invasion could shift sandhill ecosystems from a species-diverse pine savanna to a monotypic cogongrass grassland. Cogongrass is also becoming a major constraint in the forestry industry, invading and persisting in newly established pine plantations (Miller 2000; Jose et al. 2002). In addition, cogongrass poses a major fire hazard along state highways and federal interstate highways due to excessive smoke and thus limited visibility. It also encourages wildfires in residential communities developed near wooded areas.

16.6 COMMERCIAL IMPORTANCE

Although cogongrass is generally considered a serious weed, there are many commercial uses for the grass. As with many species, there are conflicting issues regarding cogongrass, especially between government and development agencies and the local people in Southeast Asia (Dove 2004). Cogongrass is used as occasional livestock feed for cattle, goats, camels, and sheep (Holm et al. 1977). Intensive management must be used to maintain the grass as fine-leaved through frequent burning, grazing, or cutting. In the arid and semiarid regions of the former Soviet Union, cogongrass is used for grazing and hay cutting. In Africa and Indonesia, this species is an important alternate forage during periods of drought.

Leaves of cogongrass are considered to be valuable thatching material for roofs for homes, temples, and other buildings. Thatched roofs are common in many parts of Africa and Asia, and in regions of Indonesia, the grass is cultivated for that purpose. The practice of cultivation generally includes burning or cutting, followed by grazing. Once the shoots become unpalatable, the grass is allowed to grow, and the long green, supple leaves are used for thatch. Recently, commercial cogongrass thatch for roofing has been driven by the tourism industry in Indonesia (Potter et al. 2000). Tourists are eager to see the traditional methods of the rural people, and this is so lucrative that cogongrass is being cultivated, especially in Bali and Lombok, and sold as a cash crop for export (Potter 2001). In West Timor, cogongrass is reported to be a scarce commodity for roofing (Potter et al. 2000).

Cogongrass has also been investigated as a source of raw material for papermaking, but research indicated that the grass was too difficult to handle (bulky) and the land would be better utilized with tree crops for paper (Holm et al. 1977). This grass is also utilized as a soil binder, particularly in areas of high rainfall. It is frequently used to stabilize canal and railroad embankments, hold soil around dams and earthen structures, and stabilize sand dunes along coastal areas (Holm et al. 1977; Brook 1989). In the United States, it is often seen along levees constructed for clay settling ponds in Florida (Shilling et al. 1997).

In several areas of the world, cogongrass grasslands are important components of natural ecosystems (Dinerstein, 1979a,b). In Nepal, cogongrass grasslands are threatened by exotic grasses and woody tree invasion (Peet et al. 1999a). These grasslands provide an important interface between forests and open savannas (Peet et al. 1999a). These areas support several endangered species, including the one-horned rhinoceros (*Rhinoceros unicornis*), pygmy hog (*Sus salvalnius*), swamp deer (*Cervus duvauceli*), axis deer (*Axis axis*), hog deer (*A. porcinus*), hispid hare (*Caprolagus hispidus*), and Bengal florican (*Eupodotis benghalensis*) (Moe and Wegge 1997; Peet et al. 1999a). It has also been shown that large deer populations support large tiger (*Panthera tigris*) populations (Peet et al. 1999b). In addition, the grass provides thatch for area communities. In Australia, native cogongrass grasslands are threatened by nonnative legume species such as *Senna obtusifolia* (L.) (Neldner et al. 1997).

Several naturally occurring compounds with medicinal properties have been discovered in cogongrass. Matsunaga et al. (1995) discovered imperanene, a novel phenolic compound that has been shown to inhibit platelet aggregation activity.

Matsunaga et al. (1994c) isolated graminone B, a lignan with vasodilative activity and in a subsequent study, discovered cylindol A, which has 5-lipoxygenase inhibitory activity (Matsunaga et al. 1994b). This group also discovered cylindrene, a sesquiterpeniod that has inhibitory activity on the contractions of vascular smooth muscle (Matsunaga et al. 1994a). More recently, a neuroprotective compound was isolated from cogongrass that shows activity in suppressing glutamate-induced neurotoxicity in rat cortical cells (Yoon et al. 2006).

16.7 MANAGEMENT

Management of cogongrass can be grouped into the following five major areas of control: preventive, cultural, mechanical, biological, and chemical. Although each of these methods provides a certain level of success, an integrated approach using multiple methods is ultimately the most effective way to manage cogongrass.

16.7.1 PREVENTATIVE

The first step in cogongrass management is preventing the weed from entering or establishing within a given area. Cogongrass has been shown to spread via seeds and rhizomes; therefore, measures must target both mechanisms of dispersal. One of the most compelling quotes for preventative or preemptive cogongrass control is from Pendleton (1948) who stated: "Certainly its (cogongrass') hazard as a potential weed for upland crops in the tropical and subtropical portions of the western hemisphere is a very much more serious threat to agriculture than the small amount of benefit it can possibly be as a forage." The level of spread of cogongrass in the southeastern United States has led to the development and testing of satellite imagery to detect cogongrass infestations (Barnett et al. 2003). Most of the researches into mechanisms of spread have been conducted in the United States. Willard et al. (1990) performed a survey on cogongrass infestation in Florida and found the greatest infestation levels near the points of introduction, those being the USDA-ARS station in Brooksville and the University of Florida main campus in Gainesville. Another area of heavy infestation was Marion and Polk Counties, where this plant had been established as forage in the late 1940s (Tabor 1952). Heavy infestations also occurred along major interstates and state routes, suggesting that road construction and/or maintenance had spread contaminated soil. Because of the nature of the infestations (isolated, irregular patches), Willard (1990) further hypothesized that rhizome material was the primary mechanism for these infestations.

However, population studies by Shilling et al. (1997) indicate that seed spread is also a major mechanism in Florida. Patterson et al. (1979) and Wilcut et al. (1988a) observed similar infestations along major southeastern U.S. highways, particularly the I-10 corridor from Mississippi east to Pensacola, FL. The authors suggested that the wind patterns west to east along this corridor created an avenue for seed dispersal along this route. This hypothesis was recently confirmed by Burnell et al. (2004), who showed that cogongrass seed germination from populations in southern Mississippi, United States, was high, with over 95% germination. However, they also reported the lack of seed development in some locations, similar to the earlier research findings in Florida. This suggests that cogongrass has reached a level of expansion in these areas that are now resulting in distinctly different populations, allowing for outcrossing to occur. This pattern of distribution is similar to the observations of Hubbard (1944) in Asia, where inland spread from coastal areas occurs from wind-blown seed. Burnell et al. (2003a) reported that properly timed applications of clethodim, glyphosate, and imazapic reduced seedhead production and seed viability of cogongrass.

Hall (1983) and Patterson et al. (1980) suggested that the spread of cogongrass is unlikely outside the lower coastal plains of the Gulf Coast States due to its reduced competitiveness under cooler environmental conditions and its lack of low-temperature tolerance (Wilcut et al. 1988b). However, the occurrence of cogongrass has increased drastically during the past two decades (Bryson and Carter 1993), and it is currently reported as a weed in Alabama, Florida, Georgia, Louisiana, Mississippi, South Carolina, and Virginia.

The U.S. Department of Agriculture (USDA), Animal and Plant Health Inspection Service (APHIS), Plant Protection Quarantine (PPQ) program has placed cogongrass and Brazilian satintail on the Federal Noxious Weed list. The purpose and charge of this list is to prevent the introduction of parasitic-plant pests and noxious weeds (federally listed or candidates) into the United States. This list also prevents the movement of any species on the Federal Noxious Weed list across a state line within the United States, without permit from USDA. In addition to Federal regulations, many state and local authorities have placed cogongrass under regulatory restrictions (Coile and Shilling 1993).

Another important aspect in the area of prevention is the concern over the sale of cogongrass var. *rubra*, or var. *koenigii*. This variant is widely promoted as an ornamental grass under the names *Rubra*, *Red Baron*, and *Japanese Blood Grass*. These varieties have been reported as nonaggressive, but research by Greenlee (1992) and Bryson 2004 (personal communication) suggested that plants revert to the green, invasive form. The greatest concern, however, is the potential for hybridization between ornamental ecotypes and weedy biotypes found in the southern United States. The ornamental varieties have been shown to survive as far north as Indiana, and this could dramatically extend the host range of this invasive species. Studies by Gabel (1982) and observations by Hall (1998), which suggest a high degree of variability and potential hybridization within the species, further elevate the importance of this issue in preventative cogongrass management.

16.7.2 Cultural

Cultural management, in the author's opinion, is the key to long-term cogongrass management. Cogongrass seedlings have poor seedling vigor. Studies by Shilling et al. (1997) demonstrated that fewer than 20% of cogongrass seedlings survived 1 year after germination. This species has also been shown to be a very poor competitor with other seedling grasses (Willard and Shilling 1990). Studies have shown that 75% or greater sod cover is sufficient to retard cogongrass seedling establishment (Dozier et al. 1998). King and Grace (2000b) also demonstrated that poor seedling growth under high water levels and flooding prevented seed germination.

The addition of fertilizer, particularly nitrogen and phosphorus, has been shown to increase competitiveness of desirable species (primarily legumes) preferentially over cogongrass seedlings. Blair et al. (1978) found nearly a three-fold increase in *Centrosema pubescens* (Benth.) green biomass with the addition of phosphorus and sulfur fertilizers coupled with a 20% decline in cogongrass biomass. Stobbs (1969) also reported suppression of cogongrass by *Stylosanthes gracilis* (Kunth) with phosphorus fertilization. Burnell et al. (2003b) reported a three- to fourfold increase of species diversity and richness where cogongrass was mowed monthly or bimonthly and fertilizer added according to standard recommendations for forage grass hay production in Mississippi. Brewer and Cralle (2003) suggest that short-lived, high levels of phosphorus inputs would reduce the competitive advantage of cogongrass in southern U.S. pine savannas, providing a favorable environment for native legume establishment. They further indicate that ratios of soil phosphorus to nitrogen would provide an indicator of ecosystem resistance to cogongrass invasion.

Cover crops and planting density have been extensively researched for cogongrass suppression, which has been shown to be susceptible to heavy shade (Otsamo et al. 1995a). Traditional methods of control in slash and burn agricultural systems are to abandon the land to natural fallow for 10 years or more (Ahn 1978; Akobundu et al. 1999; Chikoye et al. 1999). Cover crop fallows have been shown to significantly reduce cogongrass populations within 2–5 years (Udensi et al. 1999; Akobundu et al. 2000; Chikoye et al. 2001). Legume-based cover crops are the most favorable due to competitive ability and the addition of nitrogen (Ibewiro et al. 2000). MacDicken et al. (1997) showed that adequate suppression of cogongrass could be obtained with tree fallow systems within 4 years. Cover crops such as velvet bean (*Mucuna pruriens* Bak.) have been shown to suppress cogongrass, allowing for crop production (Versteeg et al. 1998; Udensi et al. 1999; Akobundu et al. 2000). In studies by Chikoye et al. (2002) in West Africa, velvet bean reduced cogongrass rhizome biomass by greater than 95% after 2 years. Kumar and Sood (1998) suggest that improved grass species such as *Setaria anceps* (Stapf ex Massey) and *Panicum maximum* (Jacq.) could be used to eliminate cogongrass.

Menz and Grist (1996) used a bioeconomic modeling approach based on rubber tree density, cogongrass suppression or control, and net returns from latex yield. They found that higher tree densities limited yield because of intertree competition, with maximum yield at a density of 600 trees/ha. However, the control of cogongrass at higher densities offsets the loss in latex yield, and they concluded this approach would be the best option for small landholders. More recently, Kaewkrom et al. (2005) also showed that mixed plantings of fast-growing tree species provided more complete *Imperata* suppression than single species plantations and a more favorable environment for native species establishment and growth.

Tree-based farming systems are successfully implemented in Southeast Asia, especially in areas of marginal soils (Snelder 2001; Goltenboth and Hutter 2004). Anoka et al. (1991) found that *Gliricidia sepium* (Jacq.) Steud. and *Leucaena leucocephala* (Lam.) deWit suppressed cogongrass growth. However, few species survive without a certain level of postplant maintenance (Otsamo et al. 1995b, 1997). In addition, minimum establishment costs for a forest plantation on *Imperata* grassland are US$840/ha. The annual mean increment of a profitable plantation should be at least 25 m^3/ha (Kosonen et al. 1997). The authors also reported that government subsidies or financing was critical to profitability. A recent case study by Suyanto et al. (2005) showed that government involvement alone, including state-owned land, was not adequate to prevent forest degradation and/or cogongrass invasion in Indonesia. They report that local community involvement was the primary factor in the adoption of agroforestry and cogongrass suppression.

Several studies have shown that fast-growing exotic tree species such as *Acacia mangium* (Willd.), *Gmelina arborea* (Roxb.), and *Paraserianthes falcataria* (L.) I. Nielsen provide rapid restoration of forest ecosystems in Indonesia (Tuomela et al. 1996; Turvey 1996; Otsamo et al. 1997; Otsamo 2000a). In subsequent studies, Otsamo (1998a, 2002) studied the effect of four tree species on cogongrass suppression. While high planting density and the type of tree species was critical for cogongrass suppression, ground vegetation development was critical as well. Ground cover played an important role in fire susceptibility, maintenance requirements, and the ultimate promotion of desirable native species (Otsamo 2002). Additional studies by Otsamo (2000b) showed that integration of indigenous species with exotic tree plantings following cogongrass suppression was possible, but further research was needed on silviculture options and economics. Earlier studies showed that removal of *A. mangium* promoted indigenous species growth without reinvasion by cogongrass (Otsamo 1998b). Cummings et al. (2005) suggested the addition of mulch and slashing to deter cogongrass growth and allow native tree establishment. Recent observations in Florida suggest that mulching is effective in controlling cogongrass, but the mulch layer should be 60–100 cm deep (MacDonald 2006, personal communication).

16.7.3 Biological

An extensive review of biological agents associated with cogongrass has recently been conducted, with emphasis on the southeastern United States (Van Loan et al. 2002). Literature records suggest a considerable number of potential natural enemies of cogongrass. These include 80 pathogens, over 90 insects, and several nematodes and mites associated with cogongrass worldwide (Van Loan et al. 2002). Several researchers have studied the gall midge (*Orsioliella javanica* Kieffer), which is reportedly specific to cogongrass (Soerjani 1970, 1986; Mangoendiharjo 1980). This insect destroys the shoot meristem, but only after the grass is cut and the rhizome system has been debilitated. This requirement along with natural enemies of the midge significantly reduces the potential of this control option. In the western hemisphere, Bryson and Carter (1993) reported three native North American skipper butterfly species (*Hesperiidae*) to feed on cogongrass. However, they indicated that the use of these as biological control agents would be unlikely, since these butterfly larvae also attack several commercially important crop species.

Recent research by Yandoc et al. (1999) has shown cogongrass in the United States to be infected with two fungal pathogens, *Bipolaris sacchari* (E. Butler) Shoem. and *Drechslera gigantea* (Heald & F.A. Wolf). Subsequent studies utilizing these pathogens as bioherbicides report good foliage control, but limited activity on rhizomes (MacDonald et al. 2001; Yandoc 2001). However, Yandoc et al. (2004) did demonstrate substantial cogongrass suppression (74% rhizome reduction) when these bioherbicides were used in combination with Bahia grass competition.

Although several organisms have been found and tested, several researchers state that there is little hope of finding a successful biological control for cogongrass (Ivens 1980; Brook 1989). They claim the distribution of this species is so worldwide that the chances of finding a biological control agent in an area where it does not already exist are slight.

16.7.4 Mechanical

Mechanical control of cogongrass is difficult once established, primarily due to regrowth from rhizomes (Hartley 1949). Hand weeding is employed in several areas, with four weedings per season necessary for maize production in Africa (Udensi et al. 1999). Mowing alone does not effectively control cogongrass but has been shown to reduce rhizome and foliage biomass (Willard and Shilling 1990; Willard et al. 1996). Cultivation provides excellent control, with little regrowth occurring under cultivated conditions (Hartley 1949; Peng 1984). A September 1949 research report by USDA Bureau of Animal Industry, Chinsegut Hill Sanctuary, Brooksville, FL, states that "An unwanted 13 acre stand of cogongrass was destroyed in the dry winter of 1948–1949 at this station by four well spaced diskings during a period of protracted winter drought, except close to the pine trees which are scattered through the pasture where the disking could not be deep and thorough" (Anonymous 1949). Johnson et al. (1999) also demonstrated the advantages of integrating tillage for cogongrass control; however, this option is often not feasible or economical in many areas (Van Noordwijk et al. 1996). Lack of deep tillage has been suggested as the major reason for cogongrass problems in Africa (Udensi et al. 1999; Chikoye et al. 2001). Another mechanical method practiced in Indonesia is rolling, where cogongrass foliage is flattened without breaking or removing the leaf tissue (Terry et al. 1997). Crops are then planted into the flattened stand with minimal disturbance. Cogongrass regrowth from this procedure occurs much later than traditional techniques and often provides the addition time needed for crop production.

16.7.5 Chemical

Over the last 30 years, several herbicides have been evaluated for cogongrass control with few successes (Dickens and Buchanan 1975; Sandanam and Jayasinghe 1977; Patterson et al. 1983). Single applications of soil sterilants have been shown to provide acceptable, but expensive control (Dickens and Buchanan 1975; Barnett et al. 2000). Dalapon was shown to have activity on cogongrass, but this material is no longer manufactured (Dickens and Buchanan 1975; Willard et al. 1996). Glufosinate provides good initial control, but regrowth occurs within 6 months (Gaffney 1996; Barnett et al. 2000). Fluazifop-butyl, clethodim, sethoxydim, quizalofop, fenoxaprop, diclofop, and imazapic have provided limited activity in field and greenhouse studies but never satisfactory control (Gaffney 1996; Mask et al. 2000, 2001). However, Avav (2000) demonstrated good control of cogongrass in soybean production in the savanna zone of Nigeria with fluazifop-butyl.

To date, the most effective herbicides for cogongrass management are glyphosate and imazapyr (Willard et al. 1997; Dozier et al. 1998; Udensi et al. 1999; Barnett et al. 2001; MacDonald et al. 2002). These materials are broad-spectrum, systemic herbicides and have been shown to provide good control of cogongrass for 1 year after application (Miller 2000; Ivy et al. 2006). Applications in the fall in the southeastern United States have resulted in greater efficacy for both herbicides (Johnson et al. 1999). This has been attributed to the basipetal flow of photosynthates that occurs at this time of year (Gaffney 1996; Tanner et al. 1992). Generally, imazapyr provides control for a longer period of time because of soil activity, but off-target effects limit its use in certain areas (MacDonald et al. 2002). Studies have shown that removal of the aboveground biomass, especially through burning, will greatly enhance herbicide efficacy (Myers et al. 2006), presumably due to the removal of dead leaf tissue and stimulation of shoot production from rhizomes.

A novel approach of integrating mowing with chemical applications has also been tried for cogongrass, but resulted in poor control compared with conventional application techniques (Marchbanks et al. 2002). Ropewick and other application methods have been tested, but little advantage over conventional techniques is noted (Townson and Butler 1990; Willard et al. 1997). The use of herbicide-tolerant crops has also been evaluated in the United States, whereby crops tolerant to imazapyr herbicide (imidazolinone-tolerant) have been planted directly with cogongrass. The crop is grown, and the cogongrass is treated throughout the growing season with imazapyr (Burns et al. 2006). This approach has little practical benefit in noncropland areas, since control in these studies was not shown to be greater than imazapyr-treated cogongrass in the absence of the crop. However, the use of herbicide-tolerant crops may have merit in other areas, particularly West Africa where cogongrass is a severe problem in agronomic production.

16.7.6 Integrated Control

The use of cover crops such as velvet bean [*M. cochinchinensis* (Lour.) A. Chev] and kudzu [*Pueraria phaseoloides* (Roxb.) Benth] combined with hand weeding and chemical application have been successful in cassava production systems in West Africa (Akobundu and Ekeleme 2000; Chikoye et al. 2001, 2002, 2005). However, when using a similar approach in maize, cover crops caused a grain reduction. Tree crops have been utilized for the control of cogongrass including *Acacia*, *Eucalypta*, and rubber, but intensive mechanical site preparation is often necessary to facilitate establishment (Otsamo et al. 1995b; Bagnall-Oakeley 1996). Further work by Kuusipalo et al. (1995) and Otsamo (1998a) suggested that fast-growing exotic trees were necessary to provide acceptable control of cogongrass before native forest could be reestablished in Indonesia. The strategy is to establish fast-growing exotic tree species, such as *A. mangium*, to eliminate cogongrass, and then allow native species to begin to establish in the understory. Another approach is to use a high proportion of evergreen woody vegetation to minimize the risk of fire, provide grass competition, and enhance secondary succession toward natural forest (Kuusipalo et al. 1995).

In the southeastern United States, pine, oak (*Quercus* spp.), and other hardwood species dominate many natural areas but are poor competitors with cogongrass

(MacDonald et al. 2001). Research by Ramsey et al. (2003) suggests that loblolly pine (*Pinus taeda* L.) in a mid-rotation plantation would not provide adequate cogongrass suppression, regardless of herbicide inputs. Ramsey et al. (2003) suggested that long-term control would be possible with repeated herbicide applications combined with fast-growing evergreen trees that provide continuous heavy shade. Several researchers have recognized the need for revegetation as a component of cogongrass management in the southeastern United States (Johnson et al. 1997, 2000; MacDonald et al. 2002). Faircloth et al. (2003) integrated mowing, cover crops, and chemical applications for cogongrass management along rights-of-way. The authors found that mowing had no impact and that imazapyr herbicide applications followed by revegetation provided excellent control.

In other research by Barron et al. (2003), the combination of imazapyr followed by a planting of Bahia grass (*Paspalum notatum* Fluegge.) provided good cogongrass control. In this study, disking had no impact on cogongrass control, either alone or in combination with herbicides. Additional research has shown that several native southeastern forbs and grasses can be reestablished in areas treated with imazapyr, despite the substantial soil activity of the herbicide (Ketterer et al. 2006).

16.8 CONCLUSIONS

Cogongrass is one of the most troublesome and problematic weedy species throughout the tropical and subtropical regions of the world. Holm et al. (1977) rank it as the seventh worst weed in the world, and report over 100 common names associated with the species. Cogongrass can be considered unique in its weediness, ranging from agronomic production to climax forest communities. The species is extremely well adapted to poor soils, drought conditions, low-light environments, and frequent fire regimes. Cogongrass demonstrates prolific seed and rhizome production, and a high degree of genetic plasticity and variability. Although endangered and commercially useful in certain limited areas, the weediness related to cogongrass far outweighs its value. Cogongrass management must be addressed in a multipronged approach that integrates preventative, cultural, biological, mechanical, and chemical methods of control. Providing a favorable environment that minimizes disturbance for native species will prevent cogongrass infestation. Cover crops, including fast-growing annual legume species, have been beneficial in suppressing or eliminating the grass in agronomic situations. Leguminous tree species have also been shown to eliminate cogongrass in deforested savanna regions, with subsequent introduction of native tree species. Biological control agents are extensive, but research in this area to date is limited. Mechanical control is effective, but the frequency and aggressiveness required often precludes its usefulness. Chemical control is also effective, but several studies have shown the imperativeness of integrating chemical strategies with revegetation strategies for long-term cogongrass management.

For additional information on cogongrass, the author suggests the following references: Dozier et al. (1998), Brook (1989), and Holm et al. (1977). There have also been several workshops and symposia on *Imperata* management in Southeast Asia, most notably (1) *Proceedings of BIOTROP Workshop on Alang-Alang in Bogor*, July 27–29, 1976, BIOTROP Special Publication No. 5, Indonesia and

(2) Garrity, D.P., Guest Editor, 1996/1997, Special Issue: Agroforestry innovations for *Imperata* grassland rehabilitation, *Agroforestry Systems*, 36:1–284. The dissertations of Gabel (1982) and Otsamo (2001) are also particularly useful.

REFERENCES

Ahn, P.M., The optimum length of planned fallow, in *Soils Research in Agroforestry*, Mongi, H.O. and Huxley, P.A., Eds., International Council for Research in Agroforestry (ICRAF), Nairobi, Kenya, 15, 1978.

Akobundu, I.O. and Ekeleme, F.E., Effect of method of *Imperata cylindrica* management on maize grain yield in the derived savanna of south-western Nigeria, *Weed Res. (Oxford)*, 40, 335, 2000.

Akobundu, I.O., Ekeleme, F., and Chikoye, D., The influence of fallow management system and frequency of cropping on weed growth and crop yield, *Weed Res. (Oxford)*, 39, 241, 1999.

Akobundu, I.O., Udensi, U.E., and Chikoye, D., Velvetbean (*Mucuna* spp.) suppresses speargrass (*Imperata cylindrica* (L.) Raeuschel) and increases maize yield, *Int. J. Pest Manage.*, 46, 103, 2000.

Anoka, U.A., Akobundu, I.O., and Okonkwo, S.N.C., Effect of *Gliricidia sepium* (Jacq.) Steud. and *Leucaena leucocephala* (Lam.) deWit on growth and development of *Imperata cylindrica* (L.) Raeuschel, *Agroforest. Syst.*, 16, 1, 1991.

Anonymous, Research report, USDA—Bureau of Animal Industry, Chinsegut Hill Sanctuary, Brooksville, FL, 23, 1949.

Avav, T., Control of speargrass (*Imperata cylindrica* (L.) Raeuschel) with glyphosate and fluazifop-butyl for soybean (*Glycine max* (L.) Merr) production in savanna zone of Nigeria, *J. Sci. Food Agric.*, 80, 193, 2000.

Ayeni, A.O., Observations on the vegetative growth patterns of speargrass [*Imperata cylindrica* (L.) Beauv.], *Agric. Ecosyst. Environ.*, 13, 301, 1985.

Ayeni, A.O. and Duke, W.B., The influence of rhizome features on subsequent regenerative capacity in speargrass (*Imperata cylindrica* (L.) Beauv.), *Agric. Ecosyst. Environ.*, 13, 309, 1985

Bagnall-Oakeley, H., Conroy, C., Faiz, A., Gunawan, A., Gouyon, A., Penot, E., Liangsutthissagon, S., Nguyen, H.D., and Anwar, C., Smallholder *Imperata* management strategies used in rubber-based farming systems, *Agroforest. Syst.*, 36, 83, 1996.

Barnett, J.W., Jr., Byrd, J.D., Jr., and Mask, D.B., Efficacy of herbicides on cogongrass (*Imperata cylindrica*), *Proc. South. Weed Sci. Soc.*, 53, 227, 2000.

Barnett, J.W., Jr., Byrd, J.D., Jr., Bruce, L.M., Li, J., Mask, D.B., Mathur, A., and Burnell, K.D., Cogongrass [*Imperata cylindrica* (L.) Beauv.] can be detected using hyperspectral reflectance data, *Proc. South. Weed Sci. Soc.*, 56, 354, 2003.

Barron, M.C., MacDonald, G.E., Brecke, B.J., and Shilling, D.G., Integrated approaches to cogongrass [*Imperata cylindrica* (L.) Beauv.] management, *Proc. South. Weed Sci. Soc.*, 55, 158, 2003.

Blair, G.J., Pualillin, P., and Samosir, S., Effect of fertilizers on the yield and botanical composition of pastures in South Sulawesi, Indonesia, *Agron. J.*, 70, 559, 1978.

Boonitte, A. and Ritdhit, P., Alleopathic effects of some weeds on mungbean plants (*Vigna radiata*), in *Proc. First Trop. Weed Conf.*, Hat Yai, Songkhia, Thailand, 2, 401, 1984.

Brewer, J.S. and Cralle, S.P., Phosphorus addition reduces invasion of longleaf pine savanna (Southeastern USA) by an non-indigenous grass (*Imperata cylindrica*), *Plant Ecol.*, 167, 237, 2003.

Brook, R.M., Review of literature on *Imperata cylindrica* (L.) Raeuschel with particular reference to South East Asia, *Trop. Pest Manage.*, 35, 12, 1989.

Bryson, C.T. and Carter, R., Cogongrass, *Imperata cylindrica*, in the United States, *Weed Technol.*, 7, 1005, 1993.

Burnell, K.D., Byrd, J.D., Jr., and Meints, P.D., Evaluation of plant growth regulators for cogongrass [*Imperata cylindrica* (L.) Beauv.] seed development and control, *Proc. South. Weed Sci. Soc.*, 56, 342, 2003a.

Burnell, K.D., Byrd, J.D., Jr., Ervin, J.D., Meints, P.D., Barnett, J.W., Jr., and Mask, D.B., Mowing and cultural tactics for cogongrass [*Imperata cylindrica* (L.) Beauv.], *Proc. South. Weed Sci. Soc.*, 56, 352, 2003b.

Burnell, K.D., Byrd, J.D., Jr., Reddy, K.R., and Meints, P.D., Phenological modeling of flower onset in cogongrass [*Imperata cylindrica* (L.) Beauv.], *Proc. South. Weed Sci. Soc.*, 57, 321, 2004.

Burns, B.K., Byrd, J.D., Jr., Chesser, Z.B., Taylor, J.M., and Peyton, B.S., Cogongrass management using a Clearfield cropping system, *Proc. South. Weed Sci. Soc.*, 59, 193, 2006.

Casini, P., Vecchio, V., and Tamantiti, I., Allelopathic interference of itchgrass and cogongrass: Germination and early development of rice, *Trop. Agric.*, 75, 445, 1998.

Chandrasrikul, A., A preliminary host list of plant diseases in Thailand, Tech. Bull. 6, Department of Agriculture, Bangkok, 23, 1962.

Cheng, K.T. and Chou, C., Ecotypic variation of *Imperata cylindrica* populations in Taiwan. I. Morphological and molecular evidences, *Bot. Bull. Acad. Sinica*, 38, 215, 1997.

Chikoye, D., Ekeleme, F., and Ambe, J.T., Survey of distribution and farmers' perceptions of speargrass [*Imperata cylindrica* (L.) Raeuschel] in cassava-based systems in West Africa, *Int. J. Pest Manage.*, 45, 305, 1999.

Chikoye, D., Ekeleme, F., and Udensi, U., Cogongrass suppression by intercropping cover crops in corn/cassava systems, *Weed Sci.*, 49, 658, 2001.

Chikoye, D., Manyong, V.M., Carsky, R.J., Ekeleme, F., Gbehounou, G., and Ahanchede, A., Response of speargrass (*Imperata cylindrica*) to cover crops integrated with handweeding and chemical control in maize and cassava, *Crop Prot.*, 21, 145, 2002.

Chikoye, D., Manyong, V.M., and Ekeleme, F., Characteristics of speargrass (*Imperata cylindrica*) dominated fields in West Africa: Crops, soil properties, farmer perceptions and management strategies, *Crop Prot.*, 19, 481, 2000.

Chikoye, D., Udensi, U.E., and Ogunyemi, S., Integrated management of cogongrass [*Imperata cylindrica* (L.) Rauesch.] in corn using tillage, glyphosate, row spacing, cultivar and cover cropping, *Agron. J.*, 97, 1164, 2005.

Chou, C. and Tsai, C., Genetic variation in the intergenic spacer of ribosomal DNA of *Imperata cylindrica* (L.) Beauv. var. *major* (Cogongrass) populations in Taiwan, *Bot. Bull. Acad. Sinica*, 40, 319, 1999.

Coile, N.C. and Shilling, D.G., Cogongrass, *Imperata cylindrica* (L.) Beauv.: A good grass gone bad! Bot. Circ. No. 28, Florida Department of Agricultural and Consumer Services, Tallahassee, FL, 4, 1993.

Collins, A.R., Implications of plant diversity and soil chemical properties for cogongrass (*Imperata cylindrica*) invasion in Northwest Florida, Ph.D. dissertation, University of Florida, Gainesville, FL, 77, 2005.

Coster, C., Some observations on the growth and control of *Imperata cylindrica*, *Tectona*, 25, 383, 1932.

Coster, C., Grass in teak Taungya plantations, *Indian For.*, 65, 169, 1939.

Cummings, J., Reid, N., Davies, I., and Grants, C., Adaptive restoration of sand-mined areas for biological conservation, *J. Appl. Ecol.*, 42, 160, 2005.

Darkwa, E.O., Oti-Boateng, C., Willcocks, T.J., Terry, P.J., Johnson, B.K., Nyalemegbe, K., and Yangyuoru, M., Weed management of vertisols for small-scale farmers in Ghana, *Int. J. Pest Manage.*, 47, 299, 2001.

Dela Cruz, R.E., Constraints and strategies for the regeneration of *Imperata* grasslands, in *Forest Regeneration in Southeast Asia, Proceedings of the Symposium in Bogor, Indonesia*, Biotropica Special Publ. No. 25, Bogor, Indonesia, 23, 1986.

Dickens, R., Cogongrass in Alabama after sixty years, *Weed Sci.*, 22, 177, 1974.

Dickens, R. and Buchanan, G.A., Control of cogongrass with herbicides, *Weed Sci.*, 23, 194, 1975.

Dickens, R. and Moore, G.M., Effects of light, temperature, KNO_3, and storage on germination of cogongrass, *Agron. J.*, 66, 187, 1974.

Dinerstein, E., An ecological survey of the Royal Karnali-Bardia wildlife reserve, Nepal, Part 1: Vegetation, modifying factors, and successional relationships, *Biol. Conserv.*, 15, 127, 1979a.

Dinerstein, E., An ecological survey of the Royal Karnali-Bardia wildlife reserve, Nepal, Part 2: Habitat/animal interactions, *Biol. Conserv.*, 16, 265, 1979b.

Dove, M.R., Anthropogenic grasslands in Southeast Asia: Sociology of knowledge and implications for agroforestry, *Agrofor. Syst.*, 61, 423, 2004.

Dozier, H., Gaffney, J.F., McDonald, S.K., Johnson, E.R.R.L., and Shilling, D.G., Cogongrass in the United States: History, ecology, impacts, and management, *Weed Technol.*, 12, 737, 1998.

English, R., The regulation of axillary bud development in the rhizomes of cogongrass (*Imperata cylindrica* (L.) Beauv.), Ph.D. dissertation, University of Florida, Gainesville, FL, 123, 1998.

Eussen, J.H.H., Some competition experiments with alang-alang [*Imperata cylindrica* (L.) Beauv.] in replacement series, *Oecologia*, 40, 351, 1979.

Eussen, J.H.H., Biological and ecological aspects of alang-alang [*Imperata cylindrica* (L.) Beauv.], in *Proceedings of BIOTROP Workshop on Alang-Alang in Bogor*, Biotropica Special Publ. No. 5, Bogor, Indonesia, 15, 1980.

Eussen, J.H.H. and Soerjani, M., Problems and control of 'alang-alang' [*Imperata cylindrica* (L.) Beauv.] in Indonesia, *Proc. Fifth Annu. Conf. Asian-Pacific Weed Sci. Soc.*, 5, 58, 1975.

Eussen, J.H.H. and Wirjahardja, S., Studies of an alang-alang, *Imperata cylindrica* (L.) Beauv. vegetation, *Biotrop. Bull.*, 6, 24, 1973.

Eussen, J.H.H., Slamet, S., and Soeroto, D., Competition between alang-alang [*Imperata cylindrica* (L.) Beauv.] and some crop plants, Biotropica Bull. No. 10, SEAMEO Regional Center for Tropical Biology, Bogor, Indonesia, 1976.

Faircloth, W.H., Patterson, M.G., Teem, D.H., and Miller, J.H., Cogongrass (*Imperata cylindrica*): Management tactics on rights-of-way, *Proc. South. Weed Sci. Soc.*, 55, 162, 2003.

Friday, K.S., Drilling, M.E., and Garrity, D.P., *Imperata* grassland rehabilitation using agroforestry and assisted natural regeneration, ICRAF, Southeast Asia Regional Center, Bogor, Indonesia, 1999.

Gabel, M.L., A biosystematic study of the genus *Imperata* (*Gramineae: Andropogoneae*), Ph.D. dissertation, Iowa State University, Ames, IA, 94, 1982.

Gaffney, J.F., Ecophysiological and technical factors influencing the management of cogongrass (*Imperata cylindrica*), Ph.D. dissertation, University of Florida, Gainesville, FL, 111, 1996.

Gaffney, J.F. and Shilling, D.G., Factors influencing the activity of axillary buds in cogongrass (*Imperata cylindrica*) rhizomes, *Proc. South. Weed Sci. Soc.*, 47, 182, 1995.

Garrity, D.P., Ed., Special issue: Agroforestry innovations for *Imperata* grassland rehabilitation, *Agrofor. Syst.*, 36, 1, 1996/1997.

Garrity, D.P., Kummer, D.M., and Guiang, E.S., Country profile: The Philippines, in *Sustainable Agriculture and the Environment in the Humid Tropics*, US National Research Council, National Academy of Science, Washington, DC, 1993.

Garrity, D.P., Soekardi, M., Van Noordwijk, M., De La Cruz, R., Pathak, P.S., Gunasena, H.P.M., Van So, N., Huijun, G., and Majid, N.M., The *Imperata* grasslands of tropical Asia: Area, distribution, and typology, *Agrofor. Syst.*, 36, 3, 1996.

Goltenboth, F. and Hutter, C.P., New options for land rehabilitation and landscape ecology in southeast Asia by "rainforestation farming," *J. Nat. Conserv.*, 12, 181, 2004.

Greenlee, J., *The Encyclopedia of Ornamental Grasses: How to Grow and Use over 250 Beautiful and Versatile Plants*, Michael Friedman, New York, 1992.

Grime, J.P., Evidence for the existence of three primary strategies in plants and its relevance to ecological and evolutionary theory, *Am. Nat.*, 111, 1169, 1977.

Hall, D.W., The grasses of Florida, Ph.D. dissertation, University of Florida, Gainesville, FL, 508, 1978.

Hall, D.W., Weed watch... cogongrass, *Florida Weed Sci. Soc. Newsl.*, 5, 1, 1983.

Hall, D.W., Is cogongrass really an exotic? *Wildland Weeds*, 1, 14, 1998.

Hartemink, A.E. and O'Sullivan, J.N., Leaf litter decomposition of *Piper aduncum*, *Gliricidia sepium* and *Imperata cylindrica* in the humid lowlands of Papua New Guinea, *Plant Soil*, 230, 115, 2001.

Hartley, C.W.S., An experiment on mechanical methods of Lalan eradication, *Malay Agric. J.*, 32, 236, 1949.

Hitchcock, A.S., *Manual of the Grasses of the United States*, 2nd ed., Misc. Publ. No. 200, U.S. Department of Agriculture, Washington, DC, 1051, 1951.

Holly, D.C. and Ervin, G.N., Characterization and quantitative assessment of interspecific and intraspecific penetration of below-ground vegetation by cogongrass [*Imperata cylindrica* (L.) Beauv.] rhizomes, *Weed Biol. Manage.*, 6, 120, 2006.

Holm, L.E., Weed problems in developing countries, *Weed Sci.*, 17, 113, 1969.

Holm, L.G., Pucknett, D.L., Pancho, J.B., and Herberger, J.P., *The World's Worst Weeds, Distribution and Biology*, University Press of Hawaii, Honolulu, HI, 1977.

Hubbard, C.E., *Imperata cylindrica*: Taxonomy, distribution, economic significance, and control, Imp. Agric. Bur. Joint Pub. No. 7, Imperial Bureau Pastures and Forage Crops, Aberystwyth, Wales, Great Britain, 53, 1944.

Ibewiro, B., Sanginga, N., Vanlauwe, B., and Merckx, R., Evaluation of symbiotic dinitrogen inputs of herbaceous legumes into tropical cover-crop systems, *Biol. Fertil. Soils.*, 32, 234, 2000.

International Institute of Tropical Agriculture (IITA), Annual report of 1977, International Institute of Tropical Agriculture, Ibadan, 37, 1977.

Islam, K.R., Ahmed, M.R., Bhuiyan, M.K., and Badruddin, A., Deforestation effects on vegetative regeneration and soil quality in tropical semi-green degraded and protected forests of Bangladesh, *Land Degrad. Dev.*, 12, 45, 2001.

Ivens, G.W., *Imperata cylindrica* (L.) Beauv. in West African agriculture, in *Proceedings of BIOTROP Workshop on Alang-Alang in Bogor*, July 27–29, 1976, Biotropica Special Publ. No. 5, Bogor, Indonesia, 149, 1980.

Ivy, D.N., Byrd, J.D., Jr., Peyton, B.S., Taylor, J.M., and Burnell, K.D., Evaluation of herbicides for activity on cogongrass, *Proc. South. Weed Sci. Soc.*, 59, 196, 2006.

Johnson, E.R.R.L., Gaffney, J.F., and Shilling, D.G., Revegetation as a part of an integrated management approach for the control of cogongrass (*Imperata cylindrica*), *Proc. South. Weed Sci. Soc.*, 50, 141, 1997.

Johnson, E.R.R.L., Gaffney, J.F., and Shilling, D.G., The influence of discing on the efficacy of imazapyr for cogongrass [*Imperata cylindrica* (L.) Beauv.] control, *Proc. South. Weed Sci. Soc.*, 52, 165, 1999.

Johnson, E.R.R.L., Shilling, D.G., MacDonald, G.E., Gaffney, J.F., and Brecke, B.J., Time of year, rate of herbicide application, and revegetation: Factors that influence the control of cogongrass [*Imperata cylindrica* (L.) Beauv.] control, *Proc. South. Weed Sci. Soc.*, 53, 70, 2000.

Jose, S., Cox, J., Miller, D.L., Shilling, D.G., and Merritt, S., Alien plant invasions: The story of cogongrass in southeastern forests, *J. For.*, 100, 41, 2002.

Jouquet, P., Boulain, N., Gignoux, J., and Lepage, M., Association between subterranean termites and grass in a West African savanna: Spatial pattern analysis shows a significant role for *Odontotermes* n. *pauperans*, *Appl. Soil Ecol.*, 27, 99, 2004.

Kaewkrom, P., Gajaseni, J., Jordan, C.F., and Gajaseni, N., Floristic regeneration in five types of teak plantations in Thailand, *For. Ecol. Manage.*, 210, 351, 2005.

Ketterer, E.A., MacDonald, G.E., Ferrell, J.A., Barron, M.C., and Sellers, B.A., Studies to enhance herbicide activity in cogongrass [*Imperata cylindrica* (L.) Beauv.], *Proc. South. Weed Sci. Soc.*, 59, 205, 2006.

King, S.E. and Grace, J.B., The effects of gap size and disturbance type on invasion of wet pine savanna by cogongrass, *Imperata cylindrica* (Poaceae), *Am. J. Bot.*, 87, 1279, 2000a.

King, S.E. and Grace, J.B., The effects of soil flooding on the establishment of cogongrass (*Imperata cylindrica*), a nonindigenous invader of the southeastern United States, *Wetlands*, 20, 300, 2000b.

Koger, C.H. and Bryson, C.T., Effect of cogongrass (*Imperata cylindrica*) residues on bermudagrass (*Cynodon dactylon*) and Italian ryegrass (*Lolium multiflorum*), *Proc. South. Weed Sci. Soc.*, 56, 341, 2003.

Kosonen, M., Otsamo, A., and Kuusipalo, J., Financial, economic and environmental profitability of reforestation of *Imperata* grasslands in Indonesia, *For. Ecol. Manage.*, 99, 247, 1997.

Kumar, P. and Sood, B.R., Renovation of *Imperata cylindrica* dominant natural grassland through the introduction of improved grass species, *Indian J. Agron.*, 43, 183, 1998.

Kuusipalo, J., Adjers, G., Jafarsidik, Y., Otsamo, A., Tuomela, K., and Vuokko, R., Restoration of natural vegetation in degraded *Imperata cylindrica* grassland: Understorey development in forest plantations, *J. Veg. Sci.*, 6, 205, 1995.

Lee, S.A., Germination, rhizome survival, and control of *Imperata cylindrica* (L.) Beauv. on peat, *MARDI Res. Bull.*, 5, 1, 1977.

Lippincott, C.L., Effects of *Imperata cylindrica* (L.) Beauv. (cogongrass) invasion on fire regime in Florida sandhill, *Nat. Areas J.*, 20, 140, 2000.

MacDicken, K.G., Hairiah, K.L., Otsamo, A., Duguma, B., and Majid, N.M., Shade based control of *Imperata cylindrica*: Tree fallows and cover crops, *Agrofor. Syst.*, 36, 131, 1997.

MacDonald, G.E., Shilling, D.G., Meeker, J., Charudattan, R., Minno, M., Van Loan, A., DeValerio, J., Yandoc, C., and Johnson, E.R.R.L., Integrated management of non-native invasive plants in southeastern pine forest ecosystems—Cogongrass as a model system, Final Report, USDA Forest Service, Forest Health Technologies, Washington, DC, 50, 2001.

MacDonald, G.E., Johnson, E.R.R.L., Shilling, D.G., Miller, D.L., and Brecke, B.J., The use of imazapyr and imazapic for cogongrass [*Imperata cylindrica* (L.) Beauv.] control, *Proc. South. Weed Sci. Soc.*, 55, 110, 2002.

Mangoendiharjo, S., Some notes on the natural enemies of alang-alang (*Imperata cylindrica*) in Java, in *Proceedings of BIOTROP Workshop on Alang-Alang in Bogor*, July 27–29, 1976, Biotropica Special Publ. No. 5, Bogor, Indonesia, 47, 1980.

Marchbanks, P.R., Byrd, J.D., Jr., Barnett, J.W., Jr., Mask, D.B., and Burnell, K.D., Comparison of Burch Wet Blade® and conventional boom applications for control of cogongrass (*Imperata cylindrica*), *Proc. South. Weed Sci. Soc.*, 55, 66, 2002.

Mask, D.B., Byrd, J.D., Jr., and Barnett, J.W., Jr., Efficacy of postemergence gramincides on cogongrass (*Imperata cylindrica*), *Proc. South. Weed Sci. Soc.*, 53, 225, 2000.

Mask, D.B., Byrd, J.D., Jr., and Barnett, J.W., Jr., Will postemergent graminicides and mowing control cogongrass (*Imperata cylindrica*)? *Proc. South. Weed Sci. Soc.*, 54, 63, 2001.

Matlack, G.R., Exotic plant species in Mississippi, USA: Critical issues in management and research, *Nat. Areas J.*, 22, 241, 2002.

Matsunaga, K., Ikeda, M., Shibuya, M., and Ohizumi, Y., Cylindol A, a novel biphenylether with 5-lipoxygenase inhibitory activity, and a related compound from *Imperata cylindrica*, *J. Nat. Prod.*, 57, 1290, 1994a.

Matsunaga, K., Shibuya, M., and Ohizumi, Y., Cylindrene, a novel sesquiterpenoid from *Imperata cylindrica* with inhibitory activity on contractions of vascular smooth muscle, *J. Nat. Prod.*, 57, 1290, 1994b.

Matsunaga, K., Shibuya, M., and Ohizumi, Y., Graminone B, a novel lignan with vasodilative activity from *Imperata cylindrica*, *J. Nat. Prod.*, 57, 1734, 1994c.

Matsunaga, K., Shibuya, M., and Ohizumi, Y., Imperanene, a novel phenolic compound with platelet aggregation inhibitory activity from *Imperata cylindrica*, *J. Nat. Prod.*, 58, 138, 1995.

McDonald, S.K., Shilling, D.G., Bewick, T.A., Okoli, C.A.N., and Smith, R., Sexual reproduction by cogongrass, *Imperata cylindrica*, *Proc. South. Weed Sci. Soc.*, 48, 188, 1995.

McDonald, S.K., Shilling, D.G., Okoli, C.A.N., Bewick, T.A., Gordon, D., Hall, D., and Smith, R., Population dynamics of cogongrass, *Proc. South. Weed Sci. Soc.*, 49, 156, 1996.

Menz, K.M. and Grist, P., Increasing rubber planting density to shade *Imperata*: A bioeconomic modelling approach, *Agrofor. Syst.*, 34, 291, 1996.

Miller, J.H., Refining rates and treatment sequences for cogongrass (*Imperata cylindrica*) control with imazapyr and glyphosate, *Proc. South. Weed Sci. Soc.*, 53, 131, 2000.

Moe, S.R. and Wegge, P., The effects of cutting and burning on grass quality and axis deer (*Axis axis*) use of grassland in lowland Nepal, *J. Trop. Ecol.*, 13, 279, 1997.

Myers, M.T., Byrd, J.D., Jr., Peyton, B.S., Burns, B.K., Wright, R.S., and Burnell, K.D., Should aboveground biomass be removed before herbicide applications for control of cogongrass? *Proc. South. Weed Sci. Soc.*, 59, 197, 2006.

Naiola, B.P., Growth variation of some Indonesian alang-alang clones, in *Proc. Eighth Asian-Pacific Weed Sci. Soc. Conf.*, Vol. I, Bangalore, India, 291, 1981.

Neldner, V.J., Fensham, R.J., Clarkson, J.R., and Stanton, J.P., The natural grasslands of Cape York pensinsula, Australia: Description, distribution and conservation status, *Biol. Conserv.*, 81, 121, 1997.

Ohta, S., Influence of deforestation on the soils of the Pantabangan area, Central Luzon, the Philippines, *Soil Sci. Plant Nutr.*, 36, 561, 1990.

Otsamo, A., Effect of nurse tree species on early growth of Anisoptera marginata Korth. (Dipterocarpaceae) on an *Imperata cylindrica* (L.) Beauv. grassland site in South Kalimantan, Indonesia, *For. Ecol. Manage.*, 105, 303, 1998a.

Otsamo, A., Removal of *Acacia mangium* overstorey increased growth of underplanted *Anisoptera marginata* (Dipterocarpaceae) on an *Imperata cylindrica* grassland site in South Kalimantan, Indonesia, *New For.*, 16, 71, 1998b.

Otsamo, A., Secondary forest regeneration under fast-growing forest plantations on degraded *Imperata cylindrica* grasslands, *New For.*, 19, 69, 2000a.

Otsamo, A., Early development of three planted indigenous tree species and natural understorey vegetation in artificial gaps in an *Acacia mangium* stand on an *Imperata cylindrica* grassland site in South Kalimantan, Indonesia, *New For.*, 19, 51, 2000b.

Otsamo, A., Forest plantations on *Imperata* grasslands in Indonesia—Establishment, silviculture and utilization potential, Ph.D. dissertation, University of Helsinki, Helsinki, Finland, 85, 2001. Available at honeybee.helsinki.fi/tropic/Aotsamo.pdf.

Otsamo, A., Early effects of four fast-growing tree species and their planting density on ground vegetation in *Imperata* grasslands, *New For.*, 23, 1, 2002.

Otsamo, A., Adjers, G., Hadi, T.S., Kuusipalo, J., Tuomela, K., and Vuokko, R., Effect of site preparation and initial fertilization on the establishment and growth of four plantation tree species used in reforestation of *Imperata cylindrica* (L.) Beauv. dominated grasslands, *Forest Ecol. Manage.*, 73, 271, 1995a.

Otsamo, A., Hadi, T.S., Adjers, G., Kuusipalo J., and Vuokko, R., Performance and yield of 14 eucalypt species on *Imperata cylindrica* (L.) Beauv. grassland 3 years after planting, *New For.*, 10, 257, 1995b.

Otsamo, A., Adjers, G., Hadi, T.S., Kuusipalo, J., and Vuokko, R., Evaluation of reforestation potential of 83 tree species planted on *Imperata cylindrica* dominated grassland: A case study from South Kalimantan, Indonesia, *New For.*, 14, 127, 1997.

Patterson, D.T., Shading effects on growth and partitioning of plant biomass in cogongrass (*Imperata cylindrica*) from shaded and exposed habitats, *Weed Sci.*, 28, 735, 1980.

Patterson, D.T. and McWhorter, C.G., Distribution and control of cogongrass (*Imperata cylindrica*) in Mississippi, *Proc. South. Weed Sci. Soc.*, 33, 251, 1980.

Patterson, D.T., Flint, E.P., and Dickens, R., Effects of temperature, photoperiod, and population source on the growth of cogongrass (*Imperata cylindrica*), *Weed Sci.*, 28, 505, 1980.

Patterson, D.T., Terrell, E.E., and Dickens, R., Cogongrass in Mississippi, *Miss. Agric. For. Exp. Stn. Res. Rep.*, 46, 1, 1983.

Paul, R. and C.D. Elmore, Weeds and the C_4 syndrome, *Weeds Today*, 15, 3, 1984.

Peet, N.B., Watkinson, A.R., Bell, D.J., and Kattel, B.J., Plant diversity in the threatened sub-tropical grasslands of Nepal, *Biol. Conserv.*, 88, 193, 1999a.

Peet, N.B., Watkinson, A.R., Bell, D.J., and Sharma, U.R., The conservation management of *Imperata cylindrica* grassland in Nepal with fire and cutting: An experimental approach, *J. Appl. Ecol.*, 36, 374, 1999b.

Pendleton, R.L., Cogongrass, *Imperata cylindrica* in the western hemisphere, *Agron. J.*, 40, 1047, 1948.

Peng, S.Y., *The Biology and Control of Weeds in Sugarcane, Developments in Crop Science*, Vol. 4, Elsevier Science, New York, 326, 1984.

Potter, L., Agricultural intensification in Indonesia: Outside pressures and indigenous Strategies, *Asia Pac. Viewp.*, 42, 305, 2001.

Potter, L., Lee, J., and Thorburn, K., Reinventing *Imperata*: Revaluing alang-alang grasslands in Indonesia, *Dev. Change.*, 31, 1037, 2000.

Ramsey, C.L., Jose, S., Miller, D.L., Cox, J., Portier, K.M., Shilling, D.G., and Merritt, S., Cogongrass [*Imperata cylindrica* (L.) Beauv.] response to herbicides and disking on a cutover site in a mid-rotation pine plantation in southern USA, *For. Ecol. Manage.*, 179, 195, 2003.

Sajise, P.E., Evaluation of Cogon (*Imperata cylindrica* L.) Beauv. as a seral stage in Philippine vegetational succession. I. The cogonal seral stage and plant succession. II. Autecological studies on cogon, Ph.D. dissertation, Cornell University, Ithaca, NY, 1972.

Sajise, P.E., Evaluation of cogon [*Imperata cylindrica* (L.) Beauv.] as a serial stage in Philippine vegetational succession. I. The cogonal seral stage and plant succession. II. Autecological studies on cogon, Dissertation Abstracts International B, 3040, 1973 (from *Weed Abstracts* No. 1339, 1976).

Sandanam, S. and Jayasinghe, H.D., Manual and chemical control of *Imperata cylindrica* on tea land in Sri Lanka, *PANS*, 23, 421, 1977.

Santiago, A., Studies on the autecology of *Imperata cylindrica* (L.) Beauv., in *Proceedings of Ninth International Grassland Congress*, San Paulo, Brazil, 499, 1965.

Santiago, A., Gene ecological aspects of the *Imperata* weed and practical implications, in *Proceedings of BIOTROP Workshop on Alang-Alang in Bogor*, July 27–29, 1976, Biotropica Special Publ. No. 5, Bogor, Indonesia, 23, 1980.

Saxena, K.G. and Ramakrishnan, P.S., Growth and allocation strategies of some perennial weeds of slash and burn agriculture (jhum) in northeastern India, *Can. J. Bot.*, 61, 1300, 1983.

Seavoy, R.E., The origin of tropical grasslands in Kalimantan, Indonesia, *J. Trop. Geogr.*, 40, 48, 1975.

Shilling, D.G., Bewick, T.A., Gaffney, J.F., McDonald, S.K., Chase, C.A., and Johnson, E.R.R.L., Ecology, physiology, and management of cogongrass (*Imperata cylindrica*), Final Report, Florida Institute of Phosphate Research, Bartow, FL, 128, 1997.

Snelder, D.J., Soil properties of *Imperata* grasslands and prospects for tree-based farming systems in Northeast Luzon, The Philippines, *Agroforest. Syst.*, 52, 27, 2001.

Soedarsan, A., The effect of alang-alang [*Imperata cylindrica* (L.) Raeuschel] and control techniques on plantation crops, in *Proceedings of BIOTROP Workshop on Alang-Alang in Bogor*, July 27–29, 1976, Biotropica Special Publ. No. 5, Bogor, Indonesia, 71, 1980.

Soemarwoto, O., *The Effect of Light and Potassium Nitrate on the Germination of Alang-Alang [Imperata cylindrica (L.) Beauv.]*, University of Gadjah Mada, Jogjakarta, Indonesia, 13, 1959.

Soenarjo, E., Alang-alang gall midge potential as an alternate host for parasitoids, *Int. Rice Res. Newsl.*, 11, 22, 1986.

Soerianegara, I., The alang-alang (*Imperata cylindrica* (L.) Beauv.) problem in forestry, in *Proceedings of BIOTROP Workshop on Alang-Alang in Bogor*, July 27–29, 1976, Biotropica Special Publ. No. 5, Bogor, Indonesia, 237, 1980.

Soerjani, M., Alang-alang *Imperata cylindrica* (L.) Beav., pattern of growth as related to its problem of control, *Biol. Trop. Bull.*, 1, 88, 1970.

Soerjani, M. and Soemarwoto, O., Some factors affecting germination of alang-alang *Imperata cylindrica* rhizome buds, *PANS*, 15, 376, 1969.

Sriyani, N., Hopen, H.J., Balke, N.E., Tjitrosemito, S., and Soerianegara, I., Rhizome bud kill of alang-alang (*Imperata cylindrica*) as affected by glyphosate absorption, translocation and exudation, *Biotrop. Spec. Publ.*, 58, 93, 1996.

Stobbs, T.H., Animal production from *Hyparthena* grassland oversown with *Stylosanthes gracilis*, *East Afr. Agric. For. J.*, 35, 128, 1969.

Suganda, T. and Yulia, E., Effect of crude water extract of cogongrass (*Imperata cylindrica* Beauv.) rhizome against fusarium wilt disease of tomato, *Int. Pest Control.*, 40, 79, 1998.

Suryatna, E.S. and McIntosh, J.L., Food crops production and control of *Imperata cylindrica* (L.) Beauv. on small farms, 135, in *Proceedings of BIOTROP Workshop on Alang-Alang in Bogor*, July 27–29, 1976, Biotropica Special Publ. No. 5, Bogor, Indonesia, 15, 1980.

Suyanto, S., Permana, R.P., Khususiyah, N., and Joshi, L., Land tenure, agroforestry adoption, and reduction in fire hazard in a forest zone: A case study from Lampung, Sumatra, Indonesia, *Agroforestry Syst.*, 65, 1, 2005.

Tabor, P., Cogongrass, *Imperata cylindrica* (L.) Beauv., in the southeastern United States, *Agron. J.*, 41, 270, 1949.

Tabor, P., Comments on cogon and torpedograsses: A challenge to weed workers, *Weeds*, 1, 374, 1952.

Tanner, G.W., Wood, J.M., and Jones, S.A., Cogongrass (*Imperata cylindrica*) control with glyphosate, *Fla. Sci.*, 55, 112, 1992.

Terry, P.J., Adjers, G., Akobundu, I.O., Anoka, A.U., Drilling, M.E., Tjitrosemito, S., and Utomo, M., Herbicides and mechanical control of *Imperata cylindrica* as a first step in grassland rehabilitation, *Agroforest. Syst.*, 36, 151, 1997.

Tominaga, T., Growth of seedlings and plants from rhizome pieces of cogongrass (*Imperata cylindrica* (L.) Beauv.), *Weed Biol. Manage.*, 3, 193, 2003.

Townson, J.K. and Butler, R., Uptake, translocation and phytotoxicity of imazapyr and glyphosate in *Imperata cylindrica* (L.). Raeuschel: Effect of herbicide concentration, position of deposit and two methods of direct contact application, *Weed Res. (Oxford)*, 30, 235, 1990.

Tuomela, K., Otsamo, A., Kuusipalo, J., Vuokko, R., and Nikles, G., Effect of provenance variation and singling and pruning on early growth of *Acacia mangium* Willd. plantation on *Imperata cylindrica* (L.) Beauv. dominated grassland, *For. Ecol. Manage.*, 84, 241, 1996.

Turvey, N.D., Growth at age 30 months of *Acacia* and *Eucalyptus* species planted in *Imperata* grasslands in Kalimantan Selatan, Indonesia, *For. Ecol. Manage.*, 82, 185, 1996.

Udensi, U.E., Akobundu, I.O., Ayeni, A.O., and Chikoye, D., Management of cogongrass (*Imperata cylindrica*) with velvetbean (*Mucuna pruriens* var. *utilis*) and herbicides, *Weed Technol.*, 13, 201, 1999.

Van Loan, A.N., Meeker, J.R., and Minno, M.C., Cogongrass, in *Biological Control of Invasive Plants in the Eastern United States*, Van Driesche, R., Ed., USDA Forest Service Publication FHTET-2002–04, Washington, DC, 413, 2002.

Van Noordwijk, M., Hairiah, K., Partoharjono, S., Labios, R.V., and Garrity, D.P., Food-crop-based production systems as sustainable alternatives for *Imperata* grasslands? *Agrofor. Syst.*, 36, 55, 1996.

Vayssiere, P., Les mauvaises herbes en Indo-Malasie, *J. Agric. Trop. Bot. Appl.*, 4, 392, 1957.

Verma, J., Singh, B.P., Gangal, S.V., Arora, N., and Sridhara, S., Purification and partial characterization of a 67-kD cross-reactive allergen from *Imperata cylindrica* pollen extract, *Int. Arch. Allergy. Immunol.*, 122, 251, 2000.

Versteeg, M.N., Amadji, F., Eteka, A., Gogan, A., and Koudokpon, V., Farmers adoptability of *Mucuna* fallowing and agroforestry techniques in the coastal savanna of Benin, *Agric. Syst.*, 56, 269, 1998.

Wilcut, J.W., Dute, R.R., Truelove, B., and Davis, D.E., Factors limiting distribution of cogongrass, *Imperata cylindrica*, and torpedograss, *Panicum repens*, *Weed Sci.*, 36, 577, 1988a.

Wilcut, J.W., Truelove, B., Davis, D.E., and Williams, J.C., Temperature factors limiting the spread of cogongrass (*Imperata cylindrica*) and torpedograss (*Panicum repens*), *Weed Sci.*, 36, 49, 1988b.

Willard, T.R., Biology, ecology and management of cogongrass [*Imperata cylindrica* (L.) Beauv.], Ph.D. dissertation, University of Florida, Gainesville, FL, 113, 1988.

Willard, T.R. and Shilling, D.G., The influence of growth stage and mowing on competition between *Paspalum notatum* and *Imperata cylindrica*, *Trop. Grassl.*, 24, 81, 1990.

Willard, T.R., Hall, D.W., Shilling, D.G., Lewis, J.A., and Currey, W.L., Cogongrass (*Imperata cylindrica*) distribution on Florida highway rights-of-way, *Weed Technol.*, 4, 658, 1990.

Willard, T.R., Shilling, D.G., Gaffney, J.F., and Currey, W.L., Mechanical and chemical control of cogongrass (*Imperata cylindrica*), *Weed Technol.*, 10, 722, 1996.

Willard, T.R., Gaffney, J.F., and Shilling, D.G., Influence of herbicide combinations and application technology on cogongrass (*Imperata cylindrica*) control, *Weed Technol.*, 11, 76, 1997.

Yandoc, C., Biological control of cogongrass, *Imperata cylindrica* (L.) Beauv., Ph.D. dissertation, University of Florida, Gainesville, FL, 120, 2001.

Yandoc, C., Charudattan, R., and Shilling, D.G., Suppression of cogongrass (*Imperata cylindrica*) by a bioherbicide fungus and plant competition, *Weed Sci.*, 52, 649, 2004.

Yandoc, C., Charudattan, R., and Shilling, D.G., Enhancement of efficacy of *Bipolaris sacchari* (E.Butler) Shoem., a bioherbicide agent of cogongrass [*Imperata cylindrica* (L.) Beauv.], with adjuvants, *Weed Sci. Soc. Am. Abstr.*, 39, 72, 1999.

Yoon, S.J., Lee, M.K., Sung, S.H., and Kim, Y.C., Neuroprotective 2-(2-phenylethyl)chromones of *Imperata cylindrical*, *J. Nat. Prod.*, 69, 290, 2006.

17 Exotic Plant Species Invasion and Control in Great Smoky Mountains National Park, United States

Michael A. Jenkins and Kristine D. Johnson

CONTENTS

17.1 INTRODUCTION

Invasive species constitute one of the most serious threats to native biodiversity and ecosystem function. According to Wilcove et al. (1998), the spread of alien species is second only to habitat loss as a cause of species endangerment in the United States. Approximately 17,000 species of vascular plants are native to United States, but nearly 5,000 exotic species have escaped cultivation and occur in natural ecosystems (Morse et al. 1995). Many of these have become major management problems by outcompeting native species, homogenizing species composition, and altering ecosystem processes (Levine et al. 2003). In addition to displacing native species, invasion by exotic plants may reduce nutrient availability (Evans et al. 2001; Ehrenfeld 2003), alter hydrology (Loope et al. 1988), degrade wildlife habitat quality (Schmidt and Whelan 1999), and alter disturbance regimes (Gordon 1998). The economic losses and expenditures from damage and control of exotic plants are staggering. For example, exotic aquatic plants cause $10 million per year in damage and require $100 million per year for control (Pimentel et al. 2000). For a single terrestrial species, *Lythrum salicaria* (purple loosestrife), $45 million are spent each year on control efforts (Pimentel et al. 2000).

17.1.1 Exotic Species in National Parks

While national parks cover only 3.2% of the world's land area (Reid and Miller 1989), they serve as cornerstones of regional strategies to protect biodiversity (Margules and Pressey 2000). Ideally, national parks represent the range of biodiversity in an area and protect this diversity from threats to its persistence (Margules and Pressey 2000). However, national parks are not immune to outside threats such as invasive species. Because of land use practices that occur right up to their boundaries, the protection of biodiversity in U.S. national parks is typically analogous to a leaky lifeboat: constant bailing of water is required to stay afloat and any cessation of effort brings dire consequences. Controlling invasive plants is one of the most important management actions undertaken by the National Park Service (NPS) to protect biodiversity.

The NPS was among the first U.S. federal agencies to recognize that the introduction of exotic species was inappropriate in parks set aside to protect natural conditions (Drees 2004). Based upon the findings of reports on wildlife management (Leopold et al. 1963) and research (Robbins et al. 1963), NPS principal biologist Lowell Sumner stated in 1964 that "nonnative species are to be eradicated, or held to a minimum if complete eradication is impossible." This view was formalized in 1968 with publication of the *Administrative Policies for Natural Areas of the National Park Service*, which stated that exotic species may not be introduced into NPS natural areas, and management plans should be developed to control them where they have become established or threaten to invade (NPS 1968; Drees 2004). In 1996, the NPS developed a strategic framework for a national invasive species program. First funded in 2000, the NPS has to date created 16 Exotic Plant Management Teams throughout the country that assist 209 national parks with exotic plant control. Between 1999 and 2004, the NPS controlled exotic plant species on over 76,635 ha. However, ~1 million ha across the National Park System still require control of exotic plants (Drees 2004).

17.1.2 Great Smoky Mountains National Park: A Case Study in Exotic Plant Management

Over 200,000 ha in size, Great Smoky Mountains National Park (GSMNP; Figure 17.1) represents the archetype of a large natural area that plays a vital role in protecting regional biodiversity. First established as a park in 1934, the biological importance and diversity of GSMNP led to its designation as an International Biosphere Reserve in 1976 and World Heritage Site in 1983. The rich diversity of GSMNP results from the wide range of ecological gradients that occur within the Park. Elevations range from 267 to 2025 m and include 16 peaks greater than 1830 m. The rugged topography includes rocky summits, talus slopes, incised drainages, gentle-to-steep side slopes, and level valleys. The geology of the Great Smoky Mountains is extremely complex. While dominated by metamorphosed sandstone, the Park's geology also includes acid-bearing slates, mafic and ultramafic rock, and tectonic windows underlain with limestone (Southworth et al. 2005), resulting in equally variable soils. Annual rainfall varies from 140 cm at low elevations to over 200 cm on some high peaks.

Although GSMNP contains a great diversity of species in many taxa, its vascular flora has received the most study. White et al. (2003) identified 79 unique vegetation communities or associations in GSMNP comprised of over 1300 native species of vascular plants (NPSpecies 2006). This array of species includes 86 species endemic to the southern Appalachians (White 1982), over 80 state-listed species (Bailey 2004; Franklin and Finnegan 2004), and 2 federally listed species (Franklin and Finnegan 2004). Vegetation communities in GSMNP include floodplain forests, species-rich cove forests, mesic–submesic oak–hickory forests, xeric oak and pine forests, northern hardwood forests, mid-high elevation balds dominated by graminoid species or ericaceous shrubs, and high-elevation spruce–fir forests (White et al. 2003). In addition to a diverse array of vegetation communities, GSMNP also contains one of the largest tracts of primary forest in eastern North America; over 20% of the Park (over 40,000 ha) was never cleared of timber and some of these stands are among the most magnificent old-growth forests in the world. Before creation of GSMNP, most parkland was cleared for settlement, logged, broadcast burned, or used for livestock grazing (Pyle 1988).

Despite its large size and protected status, numerous biotic and abiotic factors have altered and continue to threaten biological communities within GSMNP. These factors include invasive exotic plants (Horton and Neufeld 1998; Kuppinger 2007) and exotic forest insects and disease (Smith and Nicholas 1998; Jenkins and White 2002; Vandermast 2005). Among abiotic factors, decades of fire suppression have caused many pine-dominated stands to succeed to hardwood dominance. In addition, inputs of air-borne pollutants including acid deposition (Shaver et al. 1994) and ozone (Chappelka and Samuelson 1998) have likely had detrimental impacts on biotic communities.

Like many natural areas, local and regional population growth has intensified land use and development around GSMNP, which has increased the number and

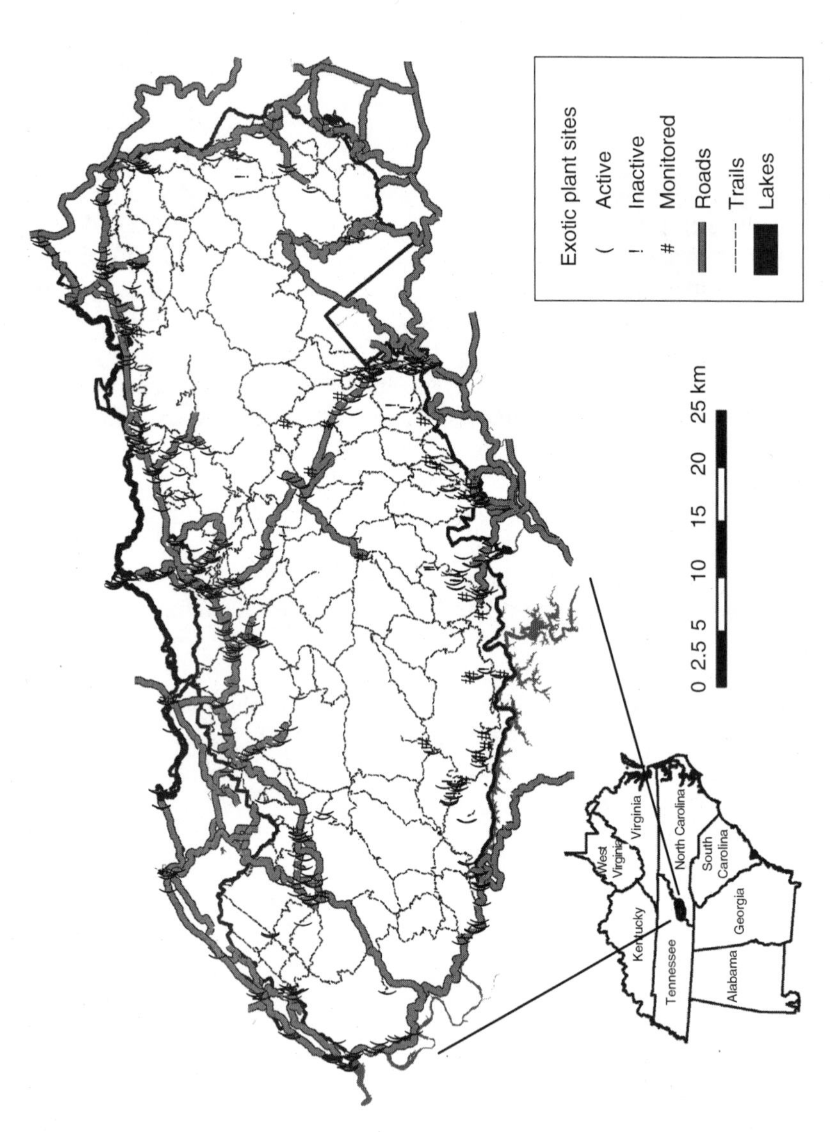

FIGURE 17.1 Exotic plant treatment sites within GSMNP. Active sites were treated during 2006, and monitored sites were inspected for population expansion. Inactive sites remain inactive for only 3 years. After 3 years, they become monitored sites and are inspected for population expansion.

spatial extent of invasions by exotic plants. The human population of the southeastern United States is growing rapidly and increased 30% between 1970 and 1990 (Burkett et al. 2001). Locally, the population of Sevier County, TN, which includes the gateway community of Gatlinburg, is expected to expand by over 60% in the next 20 years (Green et al. 2003). Most of this increase in population is driven by the tourism industry. GSMNP is the most visited of all U.S. national parks with nearly 10 million visitors per year. This increase in population has brought greater development, resulting in more rapid conversion of forests and increased fragmentation of remaining forests and grasslands. These conditions increase habitat for exotic species in areas surrounding the Park. Regionally, these changes result in greater movement of horticultural plants, soil, machinery, and other material and equipment that often result in the intentional or accidental introduction of exotic plants and disease. Consequently, most of the exotic plant species that are management problems in the southern Appalachians are problems in GSMNP as well.

Because of its biological importance, geographic location, and active exotic plant management program, GSMNP is an excellent case study of invasive plant species and their control in a large natural area. Currently, GSMNP actively controls around 50 exotic plant species that are known to displace native species, hybridize with natives, or interfere with cultural landscapes (Table 17.1). Like managers in natural areas across the United States and throughout the world, the vegetation management staff at GSMNP must prioritize threats, focus limited resources, and seek efficient and innovative techniques to control a never-ending onslaught of invasive species. In this chapter, we examine the history of invasive plants in GSMNP, the life-history traits and landscape conditions that facilitate invasion, and the successes and challenges of controlling problem species. Further, we examine ways to minimize the number and impacts of new invasive species through interagency partnerships across agency boundaries and cooperation with the private sector.

17.2 HISTORY OF EXOTIC INVASIVE PLANTS IN GSMNP

17.2.1 Park Establishment and the Onset of Exotics

Some of the earliest invasive plant species introductions in GSMNP were the result of intentional plantings prior to the Park's creation. In the most well-known example, *Pueraria montana* was introduced from China between 1920 and 1950 across the southern United States for erosion control and livestock forage (Mitich 2000). Between 1935 and 1942, the U.S. Soil Conservation Service (SCS) grew 85 million kudzu seedlings and paid farmers to plant them (Simberloff 1996). Likewise, *Rosa multiflora* was introduced to slow erosion and create fence rows. A large suite of species, including *Celastrus orbiculatus*, *Lonicera japonica*, *Ligustrum* spp., and *Ailanthus altissima*, were introduced as ornamental species in the United States as early as the late 1700s to early 1800s (Webster et al. 2006). Still others, including *Alliaria petiolata* and *Tussilago farfara*, were originally introduced as medicinal herbs.

From the beginning, the NPS has limited the introduction of exotic species into the GSMNP. During the first decades after the Park was created, the Civilian

TABLE 17.1
Exotic Species of Concern in GSMNP

Species	Common Name	Habitat	Seed Dispersal	Growth Form
Category-1				
Ailanthus altissima	Tree-of-heaven	Disturbed forests	Wind	Tree
Albizia julibrissin	Mimosa	Disturbed forests, riparian	Gravity, water	Tree
Alliaria petiolata	Garlic mustard	Closed-canopy forests	Mechanical	Biennial herb
Celastrus orbiculatus	Oriental bittersweet	Closed-canopy forests	Birds	Perennial vine
Dioscorea oppositifolia	Chinese yam	Closed-canopy forests, riparian	Aerial tubers, bulbils[a]	Perennial vine
Elaeagnus umbellata	Autumn olive	Old-fields, forest edge, roadsides	Birds	Tree
Euonymus fortunei	Climbing euonymus	Disturbed forests	Birds	Perennial vine
Hedera helix	English ivy	Homesites	Birds	Perennial vine
Lespedeza cuneata	Bush clover	Old-fields	Birds, mammals	Perennial herb
Ligustrum vulgare	Common privet	Closed-canopy forests, riparian	Birds	Perennial vine
Lonicera japonica	Japanese honeysuckle	Homesites, forest edges	Birds, mammals	Perennial vine
Lonicera maackii	Bush honeysuckle	Closed-canopy forests, disturbed forests edges, old-fields	Birds	Shrub
Lythrum salicaria[b]	Purple loosestrife	Wetlands, old-fields	Wind, water	perennial herb
Microstegium vimineum	Japanese grass	Closed-canopy forests	Water, humans	Annual grass
Paulownia tomentosa	Princess tree	Disturbed forests, burns, cliffs	Wind	Tree
Pueraria montana	Kudzu	Roadsides, disturbed forest	Seed production rare[c]	Vine
Sorghum halepense	Johnson grass	Old-fields	Wind	Perennial grass
Spiraea japonica	Japanese spiraea	Old-fields, disturbed forest	Water, humans	Shrub
Rosa multiflora	Multiflora rose	Old-fields, heavily disturbed forests	Birds	Shrub

Category 2				
Berberis thunbergii	Japanese barberry	Closed-canopy forests, disturbed forests	Birds, mammals	Shrub
Carduus nutans	Musk thistle	Old-fields	Wind	Biennial herb
Centaurea biebersteinii	Spotted knapweed	Old-fields, disturbed areas	Gravity, humans	Biennial herb
Cirsium arvense	Canada thistle	Old-fields	Wind	Perennial herb
C. vulgare	Bull thistle	Old-fields	Wind	Biennial herb
Clematis ternifolia	Leatherleaf clematis	Disturbed riparian forest	Wind, gravity, water	Vine
Coronilla varia	Crown vetch	Roadsides, old-fields	Gravity	Perennial herb
E. alata	Burning bush	Closed-canopy forests	Birds	Shrub
Lespedeza bicolor	Bicolor lespedeza	Old-fields	Birds, mammals	Shrub
Lolium arundinaceum	Tall fescue	Old-fields	Mammals, birds	Perennial grass
Melilotus alba	White sweet clover	Old-fields	Wind, gravity, water	Annual or biennial herb
Miscanthus sinensis	Miscanthus	Old-fields, homesites	Wind	Perennial grass
Populus alba	White poplar	Forest edges, old-fields	Wind	Tree
Tussilago farfara	Coltsfoot	Disturbed forests	Wind	Perennial herb
Verbascum thapsus	Common mullein	Old-fields	Mechanical	Perennial herb
Vinca minor	Common periwinkle	Homesites	Gravity, water, wind	Perennial vine
Wisteria floribunda	Wisteria	Forests edges and gaps	Gravity, water	Perennial vine
Category 3				
Barbarea vulgaris	Winter cress	Old-fields, roadsides	Mechanical, animals	Biennial herb
Hieracium aurantiacum	Orange-red hawkweed	Forest edges, old-fields, roadsides	Mammals, humans[c]	Perennial herb
Nasturtium officinale	European watercress	Aquatic	Water[c]	Perennial herb
Picea abies	Norway spruce	Spruce–fir forest, developed areas	Wind	Tree
Phalaris canariensis	Canary grass	Wet fields, water edges	Wind, water[c]	Perennial grass
Pseudosasa japonica	Bamboo	Forest edges and old-fields	Seed production rare[c]	Perennial grass

(continued)

TABLE 17.1 (continued)
Exotic Species of Concern in GSMNP

Species	Common Name	Habitat	Seed Dispersal	Growth Form
Category 4				
Buxus sempervirens	Boxwood	Homesites	Mechanical	Shrub
Forsythia viridissima	Forsythia	Homesites	Gravity, wind	Shrub
Hemerocallis spp.	Day lily	Homesites	Gravity	Perennial herb
Hosta spp.	Plantain lily	Homesites	Gravity, wind	Perennial herb
Iris pseudacorus	Yellow iris	Wet open areas	Water, gravity[c]	Perennial herb
Narcissus spp.	Daffodil	Homesites	Gravity	Perennial herb

Sources: From Clebsch, E.E.C., Clements, R., and Wofford, B.E., Survey and documentation of exotic plants in Great Smoky Mountains National Park, U.S. Department of Interior, National Park Service/University of Tennessee, Knoxville, TN, 1989; Hiebert, R.D. and Stubbendieck, J., in *Handbook for Ranking Exotic Plants for Management and Control*, U.S. Department of Interior, National Park Service, Washington, DC, 1993; Bowen, B., Johnson, K., Franklin, S., Call, G., and Webber, M., *J. Tenn. Acad. Sci.*, 77, 45, 2002.

Note: This list is categorized based on the relative impact of each species and the feasibility of control. All these species are aggressive invaders that expand rapidly from localized communities.

Category 1: Eradication is unlikely once populations are established. Control can be effective in limiting spread and eradicating localized or recently established populations.

Category 2: Eradication of existing populations is possible with considerable effort. Control actions have at least a moderate probability of reducing or eliminating established populations.

Category 3: Invasive but less aggressive, less capable of displacing native species, and are undesirable but not usually noxious. Populations are typically localized with gradual spread. Control actions have a high probability of eradicating or reducing populations.

Category 4: Less invasive, aggressive, and capable of displacing native species. Species are relatively innocuous exotics for which limited control efforts are required.

[a] Sexual reproduction of this species has not been observed in the GSMNP.

[b] Species threatens to invade the GSMNP, but are not yet established. Likely habitats are surveyed annually.

[c] Species spreads locally through vegetative reproduction.

(a)

(b)

FIGURE 17.2 (a) Beds of 1 year old *Cornus florida* (flowering dogwood) seedlings at the plant nursery established by the Civilian Conservation Corps (CCC) at Ravensford, NC, within GSMNP in 1937. The nursery grew native species for restoration efforts during the early years of GSMNP. (b) A CCC nursery worker drying seeds for planting (1937). All seeds used in the nursery were collected within GSMNP. (Photos courtesy of NPS.)

Conservation Corps (CCC) planted hundreds of thousands of trees and shrubs in the Park to reforest areas damaged by agriculture, unregulated logging, and severe wildfire. A large plant nursery was created within the Park to propagate native species for the reforestation effort (Figure 17.2). Native plants were transplanted from areas undergoing road and trail construction for use in reforestation efforts. The wise and fortunate use of native species in these plantings prevented a potential catastrophe of exotic species invasion. If exotic species such as *P. montana* and *R. multiflora* had been introduced during this period, later control efforts would have been considerably more difficult and costly.

According to Brown (2000), soon after its creation GSMNP received a highly unusual request to introduce an exotic species. President Franklin D. Roosevelt, an amateur naturalist, asked the NPS to determine if *Sequoiadendron giganteum* (giant sequoia) and *Sequoia sempervirens* (redwood) trees would grow in the Smoky Mountains. To avoid planting a nonindigenous species in a national park, Secretary of Interior Harold Ickes asked the Bureau of Indian Affairs to try an experimental planting on the Cherokee Reservation that borders GSMNP. Secretary Ickes reported to the President that "The giant sequoia, like the redwood, does not seem to thrive in areas where the summer precipitation and humidity are high."

While the NPS has followed policies to restrict the intentional introduction of exotic species within GSMNP, a few exotic plant populations were the result of deliberate introductions by the NPS. During the 1950s and 1960s *Lolium arundinaceum* was planted in Cades Cove, a historic agricultural area within the Park that was managed as agricultural leases until 1999, to serve as a source of forage and hay for cattle. During the 1960s and early 1970s, *Lespedeza cuneata* was planted in Cades Cove and along some new Park roads. This species, in combination with *Coronilla varia*, was also inadvertently introduced in some areas from contaminated hydroseeding operations. However, the vegetation management program of GSMNP is currently working to restore many of these areas to native grasses and forbs (Price and Weltzin 2003).

17.2.2 Beginning of Control Efforts

For 30 years after the creation of GSMNP, very limited treatments were performed for exotic species. During this period, only two full time employees were dedicated to resource management. Control efforts focused largely on *P. montana* (Figure 17.3a), a species that spread rapidly across the southeastern United States and today blankets 3 million ha in the Southeast, spreading at a rate of 50,000 ha/year (Mitich 2000). These early efforts treated large and conspicuous sites since resources were not available to conduct a Park-wide assessment of *P. montana* coverage. However, these efforts were successful in reducing the coverage of *P. montana* patches that were treated. Although these early managers lacked staffing and site inventory information, they used some very toxic herbicides whose use is no longer legal (Langdon 2006, personal communication). One of these herbicides was 2,4,5-T. During the Vietnam War, 2,4,5-T was combined with the commonly used herbicide 2,4-D to create Agent Orange, a defoliant that produces dioxin as it degrades (Dwernychuk et al. 2002).

As time progressed, managers at GSMNP began to recognize the critical importance of conducting systematic surveys to locate exotic plant populations and prioritize them for treatment. Surveys began in the 1970s when the Uplands Field Research Laboratory of GSMNP conducted several studies of exotic plant

FIGURE 17.3 Exotic species that occur in GSMNP. (a) *Pueraria montana* was the first exotic species to be actively treated in GSMNP. *Pueraria montana* grows rapidly and is able to quickly overtop existing vegetation. Although *P. montana* remains a control problem throughout the southeastern United States, aggressive treatment followed by years of site maintenance have brought the species under control in GSMNP. (b) *Ailanthus altissima* invades disturbed forests through its winged wind-dispersed seeds. Once established, the species forms dense stands through vegetative expansion and suppression of competition through its allelopathic litter. (c) *Microstegium vimineum* is a shade-tolerant invader of forests. Once established, it is nearly impossible to eradicate it without impacting native species. The species is unpalatable to deer, which browse remaining native species from within mats of *M. vimineum* and further promote its dominance. (d) *R. multiflora* produces berries that are widely distributed by birds. This method of seed dispersal allows the species to invade old-fields, pastures, and grasslands where it forms nearly impenetrable thickets that alter the physical structure of vegetation communities. (Photos courtesy of NPS.)

populations in the Park, including *Albizia julibrissin*, *Paulownia tomentosa*, and *A. altissima*. In 1984, the Southeast Regional Office of the NPS began funding exotic species control in GSMNP, and resource managers started documenting several additional species (*P. montana*, *R. multiflora*, *Hedera helix*, *Microstegium vimineum*, and *Wisteria floribunda*), mapping sites, and implementing systematic control. In 1989, a formal exotic plant survey of old homesites and disturbed roadsides was conducted in cooperation with the University of Tennessee (Clebsch et al. 1989). Cooperative research projects were also implemented to examine control techniques for two exotic species, *Paulownia tomentosa* and *M. vimineum*. By the end of the decade, detailed information on exotic plant treatments, including hours worked, herbicide quantities, and area treated were recorded by resource management staff.

This documentation continues today through an efficient database. The documentation of locations, treatments, and results continues to be a valuable tool for tracking the success of the GSMNP vegetation management program.

Control efforts during the 1980s continued to focus on *P. montana*, which was the first species to receive large-scale treatment. In the late 1980s and early 1990s, control efforts were expanded to include *C. orbiculatus*, *Miscanthus sinensis*, *P. tomentosa*, and *R. multiflora*. An agency grant program provided $800,000 in the mid-1990s for exotic plant management over 3 years, funding a crew of 12 workers to focus on exotic plant control throughout the Park. Many new sites were documented, and a comprehensive integrated pest management plan was implemented for 50 target species. Currently, nearly 800 exotic plant sites are documented in the Park's database (Figure 17.1). All treatments and monitoring activities are recorded, and the database is referenced to an Arc Map Geographic Information System. This organized information about past control efforts and population locations is critical for planning future treatments and establishing treatment priorities.

17.2.3 Contemporary Efforts to Control Exotics

The contemporary exotic plant control program follows well-documented protocols to treat exotic species and track overall success. Species are prioritized based on the significance of their impacts and feasibility of control (Clebsch et al. 1989; Hiebert and Stubbendieck 1993; Table 17.1). There are ~380 species of exotic plants that remain from old homesites or have invaded disturbed areas such as roadsides, burns, or construction sites. The NPS manages only about 50 exotic plant species, those known to be invasive enough to displace native plant communities, hybridize with natives, and interfere with cultural landscapes (Table 17.1). Strategies were developed to control each plant based on individual species biology. At present, many of the old homesite infestations are well under control, and the more serious problems are those species invading from outside the Park in gravel, fill dirt, straw, horses, and from windblown and bird-borne seeds. Control methods include hand-pulling, mechanical cutting, and selective use of herbicides (Figure 17.4). The latter are carefully chosen for minimal environmental impact and used as sparingly as possible. Between 1994 and 2005, 278 ha of exotics comprising 50 species have been treated. Control efforts have largely focused on the most invasive species (Table 17.1, exotic species of concern categories 1 and 2). Across the nearly 800 treatment sites, over 98% of exotic species treated since 1994 have belonged to categories 1 and 2 (Figure 17.5). Approximately 25,000 worker hours have been expended over this period, an average of 1915 per year. As shown in Figure 17.6, each year brings a new suite of species requiring treatment. Most exotic species require repeated treatment and subsequent monitoring to maintain control (Figure 17.6).

17.3 PROBLEM INVASIVE SPECIES: MECHANISMS AND TRAITS THAT FAVOR INVASION

The first principle of Integrated Pest Management is to know the biology of a target species (Johnson 1997). By understanding the biology and life-history traits of target

FIGURE 17.4 Treating *Albizia julibrissin*, an invasive tree species. (a) Small stems may be cut with clippers or hand axe. (b) Larger well-established trees require cutting with a chainsaw. (c) Stumps are typically treated with herbicide to avoid resprouting. (d) Foliar spraying is used to treat large populations. (Photos courtesy of NPS.)

species, several important questions may be answered: (1) How does it reproduce and disperse? (2) Is the species perennial, annual, or biennial? (3) Does it have any natural enemies? (4) During what life stage is it most vulnerable to treatment relative to competing native species? With this basic life-history information, the most effective and least toxic form of control may be selected.

Numerous studies have been conducted to determine what life-history traits allow exotic plants to invade communities and outcompete native species (Horton

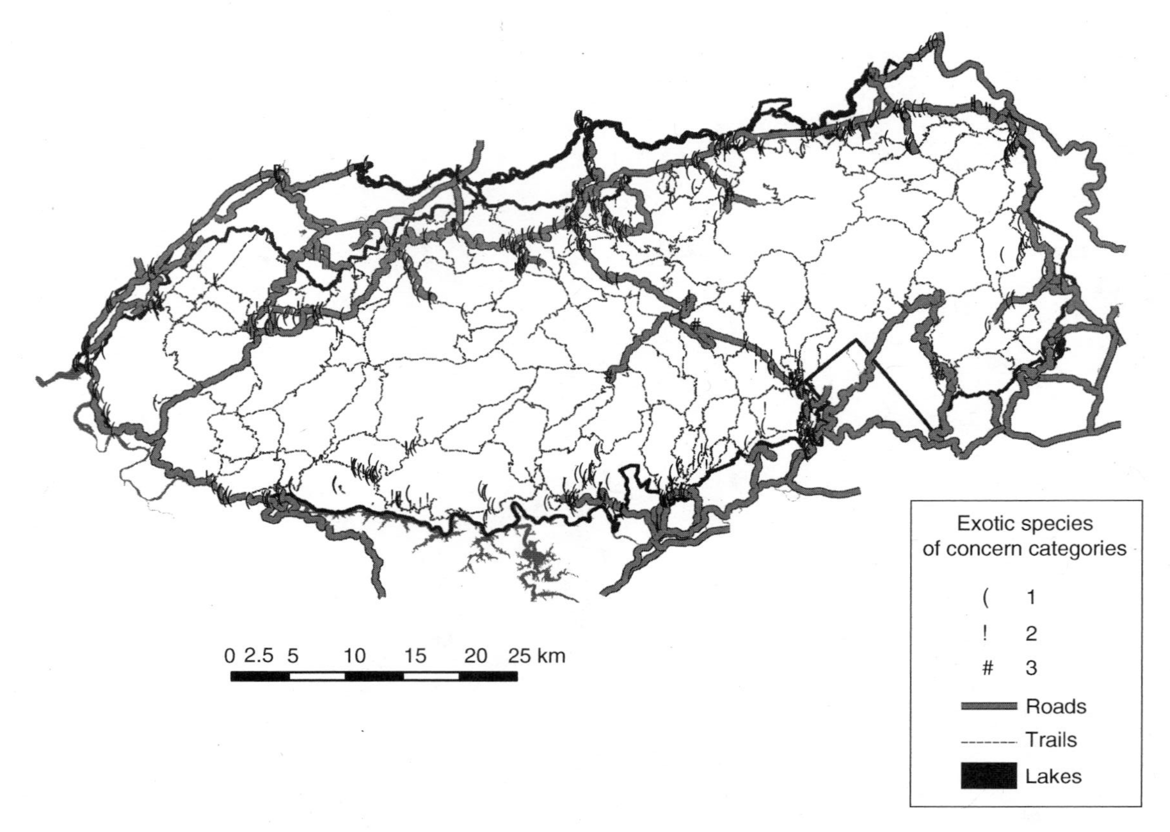

FIGURE 17.5 Exotic plant treatment sites within the GSMNP presented by exotic species of concern category. A full description of categories is provided in Table 17.1. No species in Category 4 were treated.

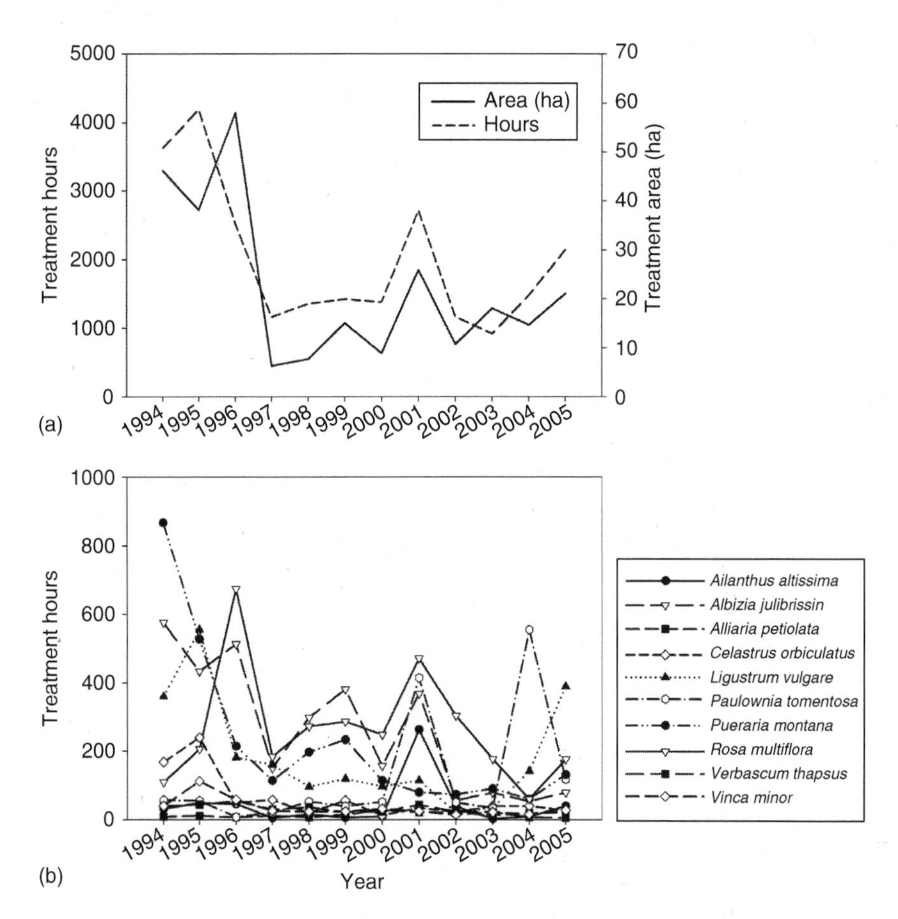

FIGURE 17.6 (a) Total treatment hours and hectares treated by decade (1994–2005) in the GSMNP. (b) Total treatment hours for selected species by decade (1994–2005).

and Neufeld 1998; Claridge and Franklin 2002; Willis et al. 2000; Callaway and Ridenour 2004; Miller and Gorchov 2004). These factors include early maturity, rapid growth, prolific seed production and successful dispersal, rapid vegetative spread, high resource-use efficiency, and the ability to form a persistent seed bank (White 1997; Webster et al. 2006). Additionally, recent studies have shown that invasive plants produce allelopathic or antimicrobial root exudates that inhibit the growth of competing plants or alter plant–soil microbial interactions (Callaway and Aschehoug 2000; Callaway and Ridenour 2004). These exudates may have little effect on natural neighbors because of coevolution, but may greatly inhibit competing species in newly invaded communities. Because there is a tradeoff between growth rate and pest resistance, species that are released from natural enemies may develop more rapid rates of growth and reproduction as a plastic response to a more benign environment (Willis et al. 2000).

While life-history traits play a critical role in determining whether an exotic species will become invasive, extrinsic factors such as land use and disturbance regimes may facilitate rapid colonization by previously noninvasive species (Webster et al. 2006) or increase the rate or spatial extent of on-going invasions. The potential impacts of these factors on invasive plants must be considered in managed ecosystems or ecosystems undergoing frequent or severe disturbance. For example, fire has been shown to facilitate the invasion of early successional exotic plants (Keeley 2006; Kuppinger 2007; Richburg et al. 2000), and the mortality of overstory trees from insects and disease could release exotic plants that employ a sit and wait strategy in mature closed-canopy forest (Greenberg et al. 2001). In addition, altering soil chemical conditions, such as the addition of quarried limestone gravel to trails and roads, may increase the invasion frequency and expansion rate of some species (Hendrickson et al. 2005).

Because of its diversity of community types, stand structural stages, and historic disturbances, GSMNP provides habitat for a wide range of invasive species with a variety of life-history traits that favor their spread. Environments range from shaded to full sunlight, high to low water availability, and from acidic to basic soils. Additionally, disturbances include recovery of former agricultural land, burning by wild and prescribed fires, herbivory by large ungulates, heavy overstory mortality from exotic insects and disease, and tree-fall gaps. Each of these habitat and disturbance scenarios are subject to different pressures from invasive plants, and each require different techniques for detection and treatment of exotics. For this discussion, problematic invasive species in the Park are divided into three broad groups based on their environmental and habitat requirements: (1) disturbance-dependent species of forests that require canopy openings, (2) shade-tolerant species that invade under closed canopies, and (3) species that invade old-fields and pastures. General life-history traits and mechanisms that favor disturbance are discussed, and examples of species from GSMNP are provided.

17.3.1 Invaders of Disturbed Forests

Exotic species that are dependent on canopy disturbance for colonization of new areas comprise a large proportion of the problem invasive species found in GSMNP (Table 17.1). Many of these species (such as *P. tomentosa* and *A. altissima*) exhibit high fecundity and frequently have seeds that are distributed by the wind. Consequently, they are able to rapidly establish large populations when disturbance creates suitable habitat (Figure 17.3b). Some early successional species have long-lived seeds that can persist in the soil until a suitable disturbance occurs. For example, after 5 years, 90% of *A. julibrissin* seeds in the soil seedbank are still viable (Bransby et al. 1992). Once established, many species are able to reproduce vegetatively to further dominate the site. These early successional species require relatively severe canopy disturbance that reduces shading by the overstory, and frequently require the removal or reduction of the forest litter and duff layers.

While invasion is greatest following relatively large-scale canopy disturbance, Knapp and Canham (2000) observed invasion of the shade-intolerant *A. altissima* in tree-fall gaps in a hemlock-dominated old-growth forest in New York. Because conifers display slower lateral growth in canopy gaps than deciduous species (Hibbs 1982), the gaps remained open long enough for *A. altissima* to establish.

Even more troubling, exotic species, including *A. altissima* were observed invading forests in New England following mortality from *Adelges tsugae* (hemlock woolly adelgid; Orwig and Foster 1998). As GSMNP experiences heavy *Tsuga canadensis* (eastern hemlock) mortality from the adelgid, increased monitoring within impacted stands will be necessary to prevent invasion.

In many instances, early successional exotic species are found in road cuts and areas cleared for development, but are unable to establish in closed-canopy forests (Williams 1993; Hendrickson et al. 2005). As in many large natural areas, these types of disturbance are rare within the interior of GSMNP. However, they commonly occur on privately owned lands surrounding the Park and along major roadways adjacent to the Park where there is constant development. Populations of disturbance-dependent species in these areas serve as a constant seed source that can be distributed into GSMNP if an appropriate disturbance occurs. Because these species typically require overstory removal and reduced litter or exposed mineral soil to establish, management activities within natural areas rarely provide the conditions required for colonization. When road work or site clearing does occur in large natural areas, focused efforts can be made to prevent invasion. However, under certain conditions fire can create the early successional habitat these species require.

17.3.1.1 The Role of Fire

Before the early 1900s, fire was a common event in Appalachian forests (Brose et al. 2001). However, following massive wildfires of the late 1800s and early 1900s fire suppression became the norm throughout the eastern United States. The creation of GSMNP in 1934 marked a drastic change in the fire regime of the Park. As in other parks throughout the country, the NPS sought to control fires in GSMNP, increasing the fire return interval from 7 to 13 years before the Park was created to 2000 years after (Harmon 1982). With the exception of one large fire that occurred along the west boundary in 1952, fire suppression was very successful at preventing large fires from occurring in the Park through the 1980s (Figure 17.7). While fire suppression had negative impacts on many vegetation communities, it did reduce the influx of early successional invasive species. However, during the 1980s and 1990s managers in GSMNP began to notice *P. tomentosa* and other exotic species invading after several large wildfires occurred in the Park (Langdon and Johnson 1994).

The frequency and size of both wildfires and prescribed burns have increased greatly over the past 30 years. The total area burned in GSMNP and adjacent Forest Service land increased from 190 ha in the 1980s to 600 ha in the 1990s, to over 6000 ha from 2000 to 2003. Over the past decade, the use of prescribed fire has greatly increased on public lands throughout the eastern United States. In GSMNP, the prescribed burning program has grown steadily from 4 ha of fuel reduction burns in 1996 to nearly 1000 ha burned in 2005 (Figure 17.8). While these burns have been very successful in reducing fuel loads, restoring communities, and creating wildlife habitat, they have also created habitat for invasive exotics. Invasion by exotic species is particularly alarming since, if untreated, it directly undermines the positive ecological effects produced by burning.

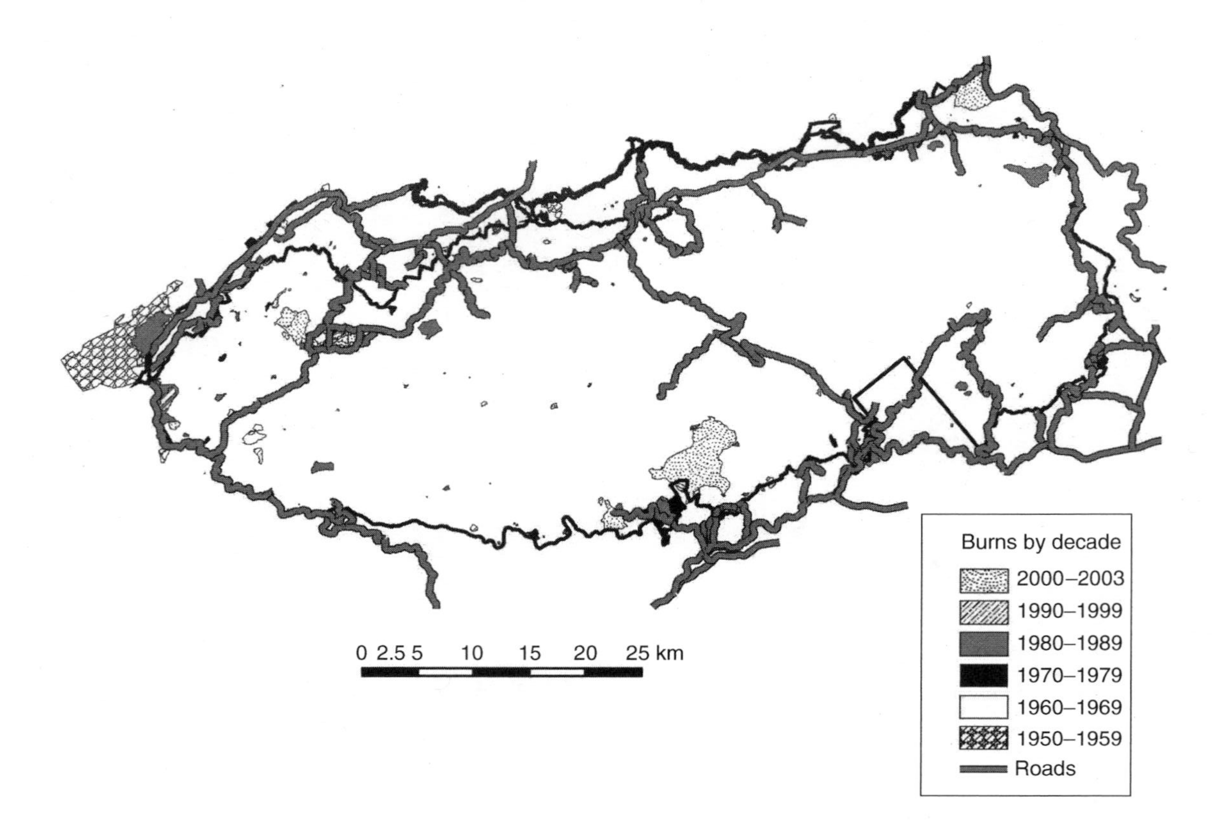

FIGURE 17.7 Past burns in GSMNP by decade (1950–2003). Both wildfires and prescribed burns are included.

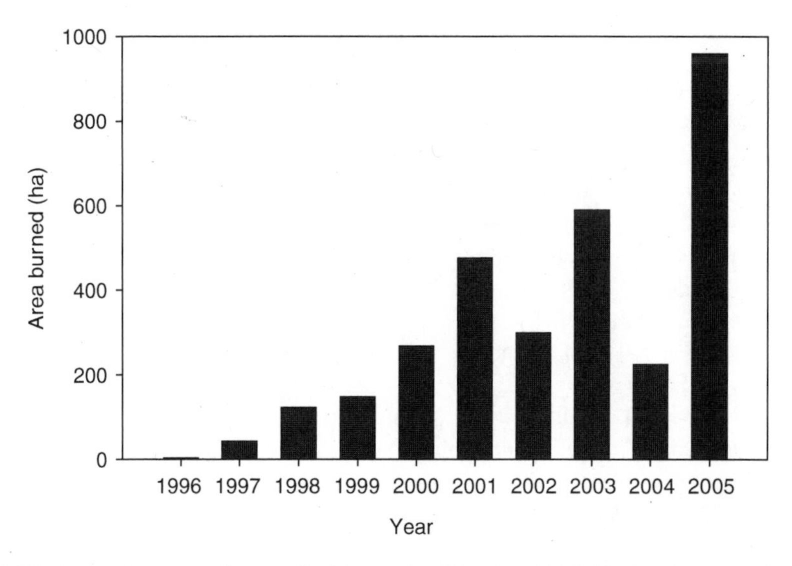

FIGURE 17.8 Hectares of prescribed burns in GSMNP (1996–2005). The size of burns has grown steadily to include landscape-scale burns designed to restore fire-dependent communities. Increased burning has required increased diligence to control the invasion of exotic plants.

17.3.1.2 *Paulownia tomentosa*

Because of the rapid influx of exotic species following burning, treatment is required following both wildfires and prescribed burns. For example, following a 200 ha wildfire that occurred in the summer of 1999, *P. tomentosa* seeds blew in from just outside the Park boundary and quickly became established on the mineral soil left by the fire. Thousands of seedlings were removed; some areas averaged nearly 100 seedlings/ha. With no canopy remaining in many parts of the burned area, *P. tomentosa* and *A. altissima* were in a position to become major components of the postburn species composition. Seedlings are hand-pulled, but if not treated as seedlings, larger stems must be cut and stump treated.

Because of the considerable time and expense devoted to controlling invasive plants following burning, a research project was begun in 2002 to develop predictive models of postburn invasion of exotic species across forested landscapes. Because it is the most common exotic species in recent burns, *P. tomentosa* was selected as the focal point of the study. At the landscape level, Kuppinger (2007) found that, following fire, *P. tomentosa* was most frequently present and occurred in greatest abundance on exposed ridges followed by lower crests and high slopes. Percent slope and aspect were also significantly related to its presence and abundance. All of these variables are associated with site dryness and fire intensity. Factors that influence the ability of *P. tomentosa* seeds to germinate and grow were related to presence and density of the species at the individual stand level. The cover of vegetation above 1 m was the dominant factor at this scale. This variable determines light availability and is highly correlated with fire intensity. In addition, the presence and density of *P. tomentosa* seedlings were positively correlated with the cover of exposed soil and negatively correlated with litter depth. This suggests that seedling density following fire will be greatest on exposed topographic positions where canopy cover and litter depth are reduced. Modeled across the landscape, this allows managers to identify areas within burns that may require intensified postburn control of early successional exotics (Figure 17.9). However, because conditions within burns change rapidly with postfire succession, more information is needed about the persistence of individual seedlings and their ability to compete with native vegetation. Kuppinger (2007) observed that most *P. tomentosa* seedlings died within 4 years after a wildfire at Linville Gorge, NC. The species was only able to persist and compete with native species on the most xeric, exposed, and steep sites that experienced the highest fire intensities during the burn. This suggests that it may be possible to further focus treatment efforts from areas that are likely to be invaded to more extreme xeric areas where the species is mostly likely to persist.

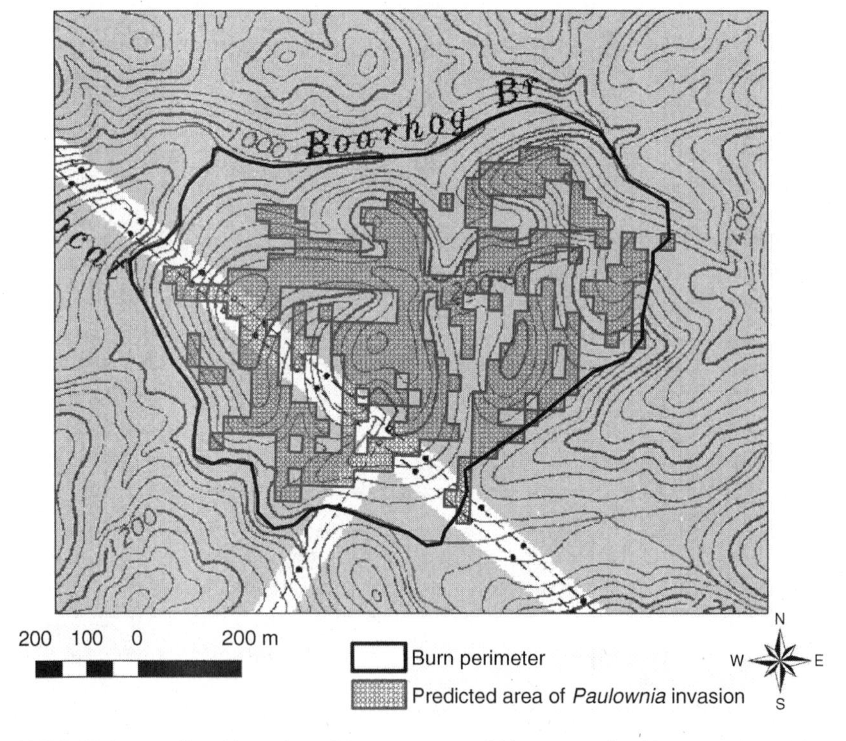

FIGURE 17.9 Predicted invasion of *P. tomentosa* within a prescribed burn conducted in the GSMNP in 2000. Predicted distribution is based on Kuppinger (2007). Spatial models such as this may help managers focus control efforts in areas sensitive to postfire invasion of exotic species. (From Kuppinger, D., Community recovery and invasion by *P. tomentosa* following fire in xeric forests of the southern Appalachians, Ph.D. Dissertation, University of North Carolina, Chapel Hill, NC, 2007.)

17.3.2 Invaders of Closed-Canopy Forests

While the total abundance of exotic species typically decreases with succession and the development of closed-canopy forests (Robertson et al. 1994; Jenkins and Parker 2000), the abundance of shade-tolerant exotic species often increases as forests mature (Meiners et al. 2002). These species often form thick layers in the woody understory and herb layer that impede the establishment and growth of native seedlings and herbaceous cover (Woods 1993; Miller and Gorchov 2004). Because they are able to invade or persist under intact canopies, these species comprise a major threat to the diversity of native plant communities. Often, only a few planted ornamental plants from these species are needed to serve as a seed source for invasions. For example, within GSMNP, two *Berberis thunbergii* sites have produced thousands of seedlings spread over several hectares nearly 4000 seedlings originating from three now-removed mother plants have been hand-pulled over the last 8 years.

Studies have shown that trees and shrubs (both native and exotic) growing along forest edges prevent wind-borne seeds from reaching forest interiors (Brothers and Spingarn 1992; Cadenasso and Pickett 2001). However, many shade-tolerant perennial vines (such as *C. orbiculatus* and *Euonymus fortunei*) and shrubs (such as *Ligustrum vulgare*, *Lonicera maackii*, and *L. japonica*) produce berries which are consumed by birds and mammals, which, in turn, widely distribute seeds into new areas. Consequently, vertebrate-dispersed plants do not necessarily follow an edge to interior pattern of invasion (Bartuszevige et al. 2006), making early detection difficult.

17.3.2.1 *Celastrus orbiculatus*

Celastrus orbiculatus, first introduced from southeast Asia around 1860 as an ornamental vine, is an invader of closed-canopy forests that is a management problem throughout the southern Appalachians. Although an aggressive invader, the species is still planted and maintained as an ornamental and is popular with craft makers because of its brightly colored berries. The species often establishes and persists under undisturbed forest canopies, and following canopy disturbance grows rapidly, overtopping and girdling trees (Greenberg et al. 2001). The seeds of oriental bittersweet are largely distributed by birds that consume the leathery capsule consisting of 3–5 seeds (McNab and Loftis 2002). In addition to its direct impacts on other species, *C. orbiculatus* hybridizes with *Celastrus scandens*, a native species that has become less common in recent years (Pooler et al. 2002). This hybridization degrades the genetic integrity of populations of *C. scandens*.

17.3.2.2 *Alliaria petiolata*

Alliaria petiolata, a shade-tolerant biennial herb in the mustard family, which was first introduced by European settlers as a medicinal herb, is an aggressive invader across much of eastern North America. The species is allelopathic (Prati and Bossdorf 2004) and spreads rapidly in forest understories. Nuzzo (1999) found that patches expanded at an average of 5.4 m/year, with a maximum spread of 36 m.

The seeds of *A. petiolata* are ballistically expelled 1–2 m from the mother plant (Nuzzo 1999), and water and humans serve as long-distance seed vectors (Cavers et al. 1979). The Park has five established *A. petiolata* sites. Four of these sites probably originated during construction projects when seeds were brought into the Park with straw, contaminated seed mix, or soil. The other is in a riparian area where seeds likely washed in from outside the Park. The Park has recently adopted more stringent standards for imported construction materials, since prevention is preferable to subsequent years of control efforts. Over an 8 year period, Park workers removed 139,782 *A. petiolata* plants from the five established populations. The large number of plants removed indicates the potential exponential expansion of *A. petiolata*, which readily outcompetes native spring flora. Because most of the sites are in areas with rich native diversity, very labor-intensive hand-pulling was chosen over herbicides to control *A. petiolata*.

17.3.2.3 *Microstegium vimineum*

Microstegium vimineum (Figure 17.3c), a C_4 annual grass, was first found in the United States in Knoxville, TN in 1919 (Fairbrothers and Gray 1972), only 45 km from GSMNP. Like many invaders of closed-canopy forests, *M. vimineum* is able to adjust to variable light levels (Horton and Neufeld 1998) and nutrient availability (Claridge and Franklin 2002). The species also produces a highly persistent seedbank (Gibson et al. 2002) and forms dense monocultures that inhibit the regeneration and growth of competing species (Barden 1987). In GSMNP, *M. vimineum* became established in many areas before the Park began efforts to control exotic plants. The cover of *M. vimineum* has been reduced in one area of the Park through 15 years of annual cutting with weed eaters prior to seed set in September. However, there are currently no reasonable treatments to control the large and widely distributed populations occurring under forest canopies. In addition, *M. vimineum* is unpalatable to *Odocoileus virginianus* (white-tailed deer), which preferentially browse upon its native competitors. Data collected from deer exclosures in Cades Cove, an area in the GSMNP with historically high deer populations, suggest that browsing by deer is further promoting the dominance of *M. vimineum* (Griggs 2005).

17.3.3 Invaders of Old-Fields and Pastures

Most invaders of pastures and old-fields were intentionally introduced because they thrived within agricultural systems or on highly disturbed lands. Many of the woody species that aggressively invade fields and pastures, such as *R. multiflora* and *Elaeagnus umbellata*, were planted to prevent erosion, provide wildlife habitat, and reclaim disturbed lands (Edgin and Ebinger 2001; Amrine 2002). For example, in the 1930s the U.S. SCS advocated the use of multiflora rose to control soil erosion and serve as a living fence for livestock, and as recently as the late 1960s state conservation agencies in several states provided rooted cuttings to property owners (Schery 1977). By the late 1990s, the species infested over 18 million ha throughout the eastern United States (Underwood et al. 1996).

In state and national parks, invasion of open fields and meadows is a serious problem not only for natural resource management, but for the management of cultural resources as well. Many parks are established to represent a historic event or discrete historical period. These events or periods often occurred before the onset of exotic species and changes in historic disturbance regimes. Consequently, heavy invasion of exotic species can fundamentally alter the historic nature of cultural landscapes. For example, many battles of the American Revolution and Civil War occurred in areas comprising planted fields, meadows, and woodlots. Invasion of these areas by species such as *R. multiflora* and *E. umbellata* changes not only the appearance of the landscape but alters its historical context as well. Consequently, control of exotic species is of critical importance to the successful management of both natural and cultural resources.

17.3.3.1 *Rosa multiflora*

Rosa multiflora (Figure 17.3d) is a serious problem throughout much of the eastern United States where it forms dense impenetrable thickets. A single mature plant can produce up to half a million seeds annually, forming a huge seedbank that can produce seedlings for decades after removal of mature plants (Amrine 2002). Further, severe *R. multiflora* infestations have reduced property values for agriculture, forestry, and recreation (Underwood et al. 1996). In a remote area within the Little Cataloochee section of GSMNP, old homesites and fields were overgrown by *R. multiflora*, which was planted in the 1930s prior to the Park's establishment. By the late 1980s, when control was begun, over 5 contiguous ha of *R. multiflora* grew in thickets of various sizes throughout the area. After successive treatments, many of the former *R. multiflora* sites are growing back in native plants. Treatments are still done in spring to reduce nontarget impacts.

17.3.3.2 *Lolium arundinaceum*

In addition to woody species, herbaceous invasive exotics, including *L. cuneata* and *L. arundinaceum*, were also intentionally introduced into fields and pastures. Other exotics, including *Cirsium arvense* and *Centaurea biebersteinii* were introduced accidentally through contaminated seed and soil (Emery and Gross 2005). *Lolium arundinaceum* was planted to serve as forage for livestock throughout the eastern United States, including Cades Cove in GSMNP. The species has invaded native grasslands, savannas, woodlands, and other high-light natural habitats where it outcompetes and displaces native grasses and forbs. *Lolium arundinaceum* hosts a mutualistic fungal endophyte that makes the species toxic to small mammals and of low palatability to ungulates. Restoration efforts, including burning, mowing, herbicide applications, and planting of native grasses and legumes have been employed in Cades Cove. While these techniques have not reduced the cover of *L. arundinaceum*, they have increased the cover of *Sorghastrum nutans* (Indian grass), a competing native species (Price and Weltzin 2003).

17.4 MOVING FORWARD: SUCCESSES AND CHALLENGES

Overall, the vegetation management program at GSMNP has been very successful in controlling and preventing invasions by exotic species within the Park. The decades of effort expended in the successful control of *P. montana* show that through diligent effort many exotic invasions can be stopped and even reduced to readily managed remnants. However, regardless of effort, existing control techniques cannot control every exotic species. For example, there is no effective technique to control the shade-tolerant grass *M. vimineum* across large expanses of forest. This and other well-established species remain a management problem for many landholders and highlight the need for additional research. Continued research to improve direct treatment techniques is important. However, research beyond herbicide trials is also needed to determine how management techniques, including prescribed fire and silviculture, can be adapted to slow the spread of exotic species.

As is likely true with any exotic plant control program, the greatest challenge faced in GSMNP is maintaining adequate staffing (paid and volunteer) to complete the labor-intensive work of controlling existing exotics while preventing new invasions from sites outside the Park boundaries. Recently, progress has been made in cooperating with the Park maintenance program and outside contractors to prevent introductions through the use of fill dirt, hydroseeding, and other construction activities. Straw is no longer used as mulch in erosion control because of the mix of weed seeds that contaminate even clean straw. In addition, if fill dirt must be brought from outside the Park, the site of origin is inspected first for potential invasives. Several local quarries have been inspected and operators have been advised to control exotics that produce windblown seeds, such as *P. tomentosa* and *A. altissima*. Whenever possible, revegetation of construction sites within GSMNP is accomplished by salvaging plant materials or producing them from seed in the Park greenhouse.

Similar to many others involved in the management of exotics, we have learned that invasive species cannot be controlled by only working within the boundary of GSMNP. Cooperation is needed across agencies and ownerships to effectively slow the spread of exotics. The NPS has created 16 Exotic Plant Management Teams that treat exotic species across multiple parks within a region. This program not only needs to be intensified across the National Park System, but this regional approach should also be expanded across agency boundaries. Much like the interagency structure of federal fire management, the management of exotic species must be better integrated and coordinated at a larger scale.

Improved communication is needed among agencies, land trusts, and private landowners to increase awareness of existing species and identify new threats. The Exotic Plant Councils that have formed across the United States are an important part of this cooperation. Through these councils and other educational programs, the general public and other land managing agencies will hopefully become more aware of the threats exotic plants pose to natural areas, and the risk they pose to real estate values and other management priorities such as wildlife habitat and timber production. The Park has been able to persuade some neighboring landholders to control exotics on their property (especially those that produce seeds distributed by birds or wind), but many are unable or unwilling to cooperate. As with controlling exotic species themselves, convincing others of the dangers of allowing them to spread unchecked is a never ending and difficult endeavor. However, only by making the control of existing species and prevention of new invasions national priorities will we achieve any meaningful success in the long term.

ACKNOWLEDGMENTS

We thank Scott Kichman and Dane Kuppinger for their assistance in preparing figures and Annette Hartigan for her assistance in locating and scanning historical photos. We also thank Janet Rock and Keith Langdon for their helpful reviews of an earlier version of this chapter.

REFERENCES

Amrine, J.W., Multiflora rose, in *Biological Control of Invasive Plants in the Eastern United States*, Van Driesche, R., Lyons, S., Blossey, B., Hoddle, M., and Reardon, R., Eds., Publication FHTET-2002–04, USDA Forest Service, Washington, DC, Chap. 22, 2002.

Bailey, C., Tennessee Natural Heritage Program rare plant list: 2004, Division of Natural Heritage, Tennessee Department of Environment and Conservation, Nashville, TN, 2004.

Barden, L.S., Invasion of *Microstegium vimineum* (Poaceae), an exotic, annual, shade-tolerant, C_4 grass, into a North Carolina floodplain, *Am. Midl. Nat.*, 123, 122, 1987.

Bartuszevige, A.M., Gorchov, D.L., and Raab, L., The relative importance of landscape and community features in the invasion of an exotic shrub in a fragmented landscape, *Ecography*, 29, 213, 2006.

Bowen, B., Johnson, K., Franklin, S., Call, G., and Webber, M., Invasive exotic pest plants in Tennessee, *J. Tenn. Acad. Sci.*, 77, 45, 2002.

Bransby, D.I., Sladden, S.E., and Aiken, G.E., Mimosa as a forage plant: A preliminary evaluation, in *Proc. Forage Grassl.Conf.*, American Forage and Grassland Council, Georgetown, TX, 28, 1992.

Brose, P., Schuler, T.M., Van Lear, D., and Best, J., Bringing fire back: The changing regimes of the Appalachian mixed-oak forests, *J. For.*, 99, 30, 2001.

Brothers, T.S. and Spingarn, A., Forest fragmentation and alien plant invasion of central Indiana old-growth forests, *Conserv. Biol.*, 6, 91, 1992.

Brown, M.L., *The Wild East: A Biography of the Great Smoky Mountains*, University of Florida Press, Gainesville, FL, 137, 2000.

Burkett, V., Ritschard, R., McNulty, S., and O'Brien, J.J., Potential consequences of climate variability and change for the southeastern United States, in *Climate Change Impacts on the United States: The Potential Consequences of Climate Variability and Change: Report of the National Assessment Synthesis Team*, US Climate Change Science Program/US Global Change Research Program, Washington, DC, Chap. 5, 2001.

Cadenasso, M.L. and Pickett, S.T.A., Effect of edge structure on the flux of species into forest interiors, *Conserv. Biol.*, 15, 91, 2001.

Callaway, R.M. and Aschehoug, E.T., Invasive plants versus their new and old neighbors: A mechanism for exotic invasion, *Science*, 290, 521, 2000.

Callaway, R.M. and Ridenour, W.M., Novel weapons: Invasive success and the evolution of increased competitive ability, *Front. Ecol. Environ.*, 2, 436, 2004.

Cavers, P.B., Heagy, M.I., and Kokron, R.F., The biology of Canadian weeds. 35. *Alliaria petiolata* (M. Bieb.) Cavara and Grande, *Can. J. Plant Sci.*, 59, 217, 1979.

Chappelka, A.H. and Samuelson, L.J., Ambient ozone effects on forest trees of the eastern United States: A review, *New Phytol.*, 139, 91, 1998.

Claridge, K. and Franklin, S.B., Compensation and plasticity in an invasive plant species, *Biol. Invasions*, 4, 339, 2002.

Clebsch, E.E.C., Clements, R., and Wofford, B.E., Survey and documentation of exotic plants in Great Smoky Mountains National Park, Cooperative Agreement CA-5460-5-8004, U.S. Department of Interior, National Park Service/University of Tennessee, Knoxville, TN, 1989.

Drees, L., A retrospective on NPS invasive species policy and management, *Park Sci.*, 22, 21, 2004.

Dwernychuk, L.W., Cau, H.D., Hatfield, C.T., Boivin, T.G., Hung, T.M., Dung, P.T., and Thai, N.D., Dioxin reservoirs in southern Viet Nam—A legacy of Agent Orange, *Chemosphere*, 47, 117, 2002.

Edgin, B. and Ebinger, J.E., Control of autumn olive (*Elaeagnus umbellata* Thunb.) at Beall Woods Nature Preserve, Illinois, USA, *Nat. Areas J.*, 21, 386, 2001.

Ehrenfeld, J.G., Effects of exotic plant invasion on soil nutrient cycling processes, *Ecosystems*, 6, 503, 2003.

Emery, S.M. and Gross, K.L., Effects of timing of prescribed fire on the demography of an invasive plant, spotted knapweed *Centaurea maculosa*, *J. Appl. Ecol.*, 42, 60, 2005.

Evans, R.D., Rimer, R., Speny, L., and Belnap, J., Exotic plant invasion alters nitrogen dynamics in an arid grassland, *Ecol. Appl.*, 11, 1301, 2001.

Fairbrothers, D.E. and Gray, J.R., *Microstegium vimineum* (Trin.) A. Camus (Gramineae) in the United States, *Bull. Torrey Bot. Club*, 99, 97, 1972.

Franklin, M.A. and Finnegan, J.T., Natural Heritage Program list of rare plant species of North Carolina: 2004, North Carolina Natural Heritage Program, N.C. Department of Environmental and Natural Resources, Raleigh, NC, 2004.

Gibson, D.J., Spyreas, G., and Benedict, J., Life history of *Microstegium vimineum* (Poaceae), an invasive grass in southern Illinois, *J. Torrey Bot. Soc.*, 129, 207, 2002.

Gordon, D.R., Effects of invasive, non-indigenous plant species on ecosystem processes: Lessons from Florida, *Ecol. Appl.*, 8, 975, 1998.

Green, H.A., Murray, M.N., Marshall, J.L., Couch, S.E., Ransom, A., McLeod, C., Belliveau, K.E., and Gibson, T.A., Population projections for the state of Tennessee: 2005–2025, Tennessee Advisory Committee on Intergovernmental Relations and the University of Tennessee Center for Business and Economic Outreach, Nashville, TN, 2003.

Greenberg, C.H., Smith, L.M., and Levey, D.J., Fruit fate, seed germination, and growth of an invasive vine—An experimental test of "sit and wait" strategy, *Biol. Invasions*, 3, 363, 2001.

Griggs, J.A., Simplified floral diversity and the legacy of a protected deer herd in Cades Cove, Great Smoky Mountains National Park, M.S. Thesis, Michigan Technological University, Houghton, MI, 2005.

Harmon, M.E., Fire history of the westernmost portion of Great Smoky Mountains National Park, *Bull. Torrey Bot. Club*, 109, 74, 1982.

Hendrickson, C., Bell, T., Butler, K., and Hermanutz, L., Disturbance-enabled invasion of *Tussilago farfara* (L.) in Gros Morne National Park, Newfoundland: Management implications, *Nat. Areas J.*, 25, 263, 2005.

Hibbs, D.E., Gap dynamics in a hemlock-hardwood forest, *Can. J. For. Res.*, 12, 522, 1982.

Hiebert, R.D. and Stubbendieck, J., *Handbook for Ranking Exotic Plants for Management and Control*, Natural Resources Report NPS/NRMWRO/NRR-93/08, U.S. Department of Interior, National Park Service, Washington, DC, 1993.

Horton, J.L. and Neufeld, H.S., Photosynthetic responses of *Microstegium vimineum* (Trin.) A. Camus, a shade tolerant, C_4 grass, to variable light environments, *Oecologia*, 114, 11, 1998.

Jenkins, M.A. and Parker, G.R., The response of herbaceous-layer vegetation to anthropogenic disturbance in intermittent stream bottomland forests of southern Indiana, USA, *Plant Ecol.*, 151, 223, 2000.

Jenkins, M.A. and White, P.S., *Cornus florida* mortality and understory composition changes in western Great Smoky Mountains National Park, *J. Torrey Bot. Soc.*, 129, 194, 2002.

Johnson, K.D., IPM—How it works in the Smokies, in *Exotic Pests of Eastern Forests: Conference Proceedings*, Britton, K.O., Ed., USDA Forest Service and Tennessee Exotic Plant Council, Nashville, TN, 289, 1997.

Keeley, J.E., Fire management impacts on invasive plants in the western United States, *Conserv. Biol.*, 20, 375, 2006.

Knapp, L.B. and Canham, C.D., Invasion of an old-growth forest in New York by *Ailanthus altissima*: Sapling growth and recruitment in canopy gaps, *J. Torrey Bot. Soc.*, 127, 307, 2000.

Kuppinger, D., Community recovery and invasion by *Paulownia tomentosa* following fire in xeric forests of the southern Appalachians, Ph.D. Dissertation, University of North Carolina, Chapel Hill, NC, 2007.

Langdon, K.R. and Johnson, K.D., Additional notes on invasiveness of *Paulownia tomentosa* in natural areas, *Nat. Areas J.*, 14, 139, 1994.

Leopold, A.S., Cain, S.A., Cottam, C.M., and Gabrielson, I., Wildlife management in the national parks: Advisory board on wildlife management appointed by Secretary of the Interior, *Trans. North Am. Wildl. Nat. Res. Conf.*, 28, 29, 1963.

Levine, J.M., Vila, M., D'Antonio, C.M., Dukes, J.S., Grigulis, K., and Lavorel, S., Mechanisms underlying the impacts of exotic species invasions, *Proc. R. Soc. Lond.*, 270, 775, 2003.

Loope, L.L., Sanchez, P.G., Tarr, P.W., Loope, W.L., and Anderson, R.L., Biological invasions of arid land reserves, *Biol. Conserv.*, 44, 95, 1988.

Margules, C.R. and Pressey, R.L., Systematic conservation planning, *Nature*, 405, 243, 2000.

McNab, H.W. and Loftis, D.L., Probability of occurrence and habitat features for oriental bittersweet in an oak forest in the southern Appalachian Mountains, USA, *For. Ecol. Manage.*, 155, 45, 2002.

Meiners, S.J., Pickett, S.T.A., and Cadenasso, M.L., Exotic plant invasions over 40 years of old field successions: Community patterns and associations, *Ecography*, 25, 215, 2002.

Miller, K.E. and Gorchov, D.L., The invasive shrub *Lonicera maackii*, reduces growth and fecundity of perennial forest herbs, *Oecologia*, 139, 359, 2004.

Mitich, L.W., 2000, Intriguing world of weeds: Kudzu (*Pueraria lobata* [Wild.] Ohwi.), *Weed Technol.*, 14, 231, 2000.

Morse, L.E., Kartesz, J.T., and Kutner, L.S., 1995, Native vascular plants, in *Our Living Resources: A Report to the Nation on the Distribution, Abundance, and Health of US Plants, Animals, and Ecosystems*, LaRoe, E.T., Farris, G.S., Puckett, C.E., Doran, P.D., and Mac, M.J., Eds., U.S. Department of the Interior, National Biological Service, Washington, DC, 205, 1995.

National Park Service (NPS), Administrative policies for natural areas of the National Park System, Report 16–21, U.S. Government Printing Office, Washington, DC, 1968.

NPSpecies, The National Park Service biodiversity database, Secure online version, https://science1.nature.nps.gov/npspecies, 2006.

Nuzzo, V., Invasion pattern of the herb garlic mustard (*Alliaria petiolata*) in high quality forests, *Biol. Invasions*, 1, 169, 1999.

Orwig, D.A. and Foster, D.R., Forest response to the introduced hemlock woolly adelgid in southern New England, USA, *J. Torrey Bot. Soc.*, 125, 60, 1998.

Pimentel, D., Lach, L., Zuniga, R., and Morrison, D., Environmental and economic costs of nonindigenous species in the United States, *Bioscience*, 50, 53, 2000.

Pooler, M.R., Dix, R.L., and Feely, J., Interspecific hybridizations between the native bittersweet, *Celastrus scandens*, and the introduced invasive species *C. orbiculatus*, *Southeast. Nat.*, 1, 69, 2002.

Prati, D. and Bossdorf, O., Allelopathic inhibition of germination by *Alliaria petiolata* (Brassicaceae), *Am. J. Bot.*, 91, 285, 2004.

Price, C.A. and Weltzin, J.F., Managing non-native plant populations through intensive community restoration in Cades Cove, Great Smoky Mountains National Park, USA, *Restor. Ecol.*, 11, 351, 2003.

Pyle, C., The type and extent of anthropogenic vegetation disturbance in the Great Smoky Mountains before National Park Service acquisition, *Castanea*, 53, 225, 1988.

Reid, W.V. and Miller, K.R., *Keeping Options Alive: The Scientific Basis for Conserving Biodiversity*, World Resources Institute, Washington, DC, 1989.

Richburg, J.A., Dibble, A.C., and Patterson, W.A., Woody invasive species and their role in altering fire regimes of the northeast and mid-Atlantic states, *Tall Timbers Res. Stn. Misc. Publ.*, 11, 104, 2000.

Robbins, W.J., Ackerman, E.A., Bates, M., Cain, S.A., Darling, F.F., Fogg, J.M., Gill, T., Gillson, J.M., Hall, E.R., and Hubbs, C.L., A report by the advisory committee to the National Park Service on research, National Academy of Sciences, National Research Council, Washington, DC, 1963.

Robertson, D.J., Robertson, M.C., and Tague, T., Colonization dynamics of four exotic plants in a northern Piedmont natural area, *Bull. Torrey Bot. Club*, 121, 109, 1994.

Schery, R., The curious double life of *Rosa multiflora*, *Horticulture*, 55, 56, 1977.

Schmidt, K.A. and Whelan, C.J., Effects of exotic *Lonicera* and *Rhamnus* on songbird nest predation, *Conserv. Biol.*, 13, 1502, 1999.

Shaver, C.L., Tonnessen, K.A., and Maniero, T.G., Clearing the air at Great Smoky Mountains National Park, *Ecol. Appl.*, 4, 690, 1994.

Simberloff, D., Impacts of introduced species in the United States, *Consequences*, 2, 1, 1996.

Smith, G.F. and Nicholas, N.S., Patterns of overstory composition in the fir and fir–spruce forests of the Great Smoky Mountains after balsam woolly adelgid infestation, *Am. Midl. Nat.*, 139, 340, 1998.

Southworth, S., Schultz, A., and Denenny, D., Generalized geologic map of bedrock lithologies and surficial deposits in the Great Smoky Mountains National Park Region, Tennessee and North Carolina, Open-File Report 2004-1410, Version 1.0, U.S. Geological Survey, Reston, VA, 2005.

Underwood, J.F., Ioux, M.M., Amrine, J.W., and Bryan, W.B., Multiflora rose control, Ohio State University Extension Bulletin No. 857, Ohio State University, Columbus, OH, 1996.

Vandermast, D.B., Disturbance and long-term vegetation change in the high-elevation deciduous forests of Great Smoky Mountains National Park, Ph.D. Dissertation, University of North Carolina, Chapel Hill, NC, 2005.

Webster, C.R., Jenkins, M.A., and Jose, S., Woody invaders and the challenges they pose to forest ecosystems in the eastern United States, *J. For.*, 104, 366, 2006.

White, P.S., The flora of Great Smoky Mountains National Park: An annotated checklist of the vascular plants and a review of previous floristic work, Research/Resources Management Report SER-55, U.S. Department of the Interior, National Park Service, Atlanta, GA, 1982.

White, P.S., Biodiversity and the exotic species threat, in *Exotic Pests of Eastern Forests: Conference Proceedings*, Britton, K.O., Ed., USDA Forest Service and Tennessee Exotic Plant Council, Nashville, TN, 1, 1997.

White, R.D., Patterson, K.D., Weakley, A., Ulrey, C.J., and Drake, J., Vegetation classification of Great Smoky Mountains National Park: Report submitted to BRD-NPS Vegetation Mapping Program, NatureServe, Durham, NC, 2003.

Wilcove, D.S., Rothstein, D., Dubow, J., Phillips, A., and Losos, F., Quantifying threats to imperiled species in the United States, *Bioscience*, 48, 697, 1998.

Williams, C.E., The exotic empress tree, *Paulownia tomentosa*: An invasive pest of forests? *Nat. Areas J.*, 13, 221, 1993.

Willis, A.J., Memmott, J., and Forrester, M.L., Is there evidence for the post-invasion evolution of increased size among invasive plant species? *Ecol. Lett.*, 3, 275, 2000.

Woods, K.D., Effects of invasion by *Lonicera tartarica* L. on herbs and tree seedlings in four New England forests, *Am. Midl. Nat.*, 130, 62, 1993.

18 Invasive Weeds of Colorado Forests and Rangeland

K. George Beck

CONTENTS

18.1 DESCRIPTION OF COLORADO

Colorado is located in the west central part of the United States. It is nearly rectangular and the highest mountains in the Continental Divide are located within its boundaries (Doesken et al. 2003). The north and south boundaries are the 41° S and 37° N parallels and the east and west boundaries are the 102° E and 109° W meridians. There are about 268,630 km^2 of land in Colorado and about 40% of the

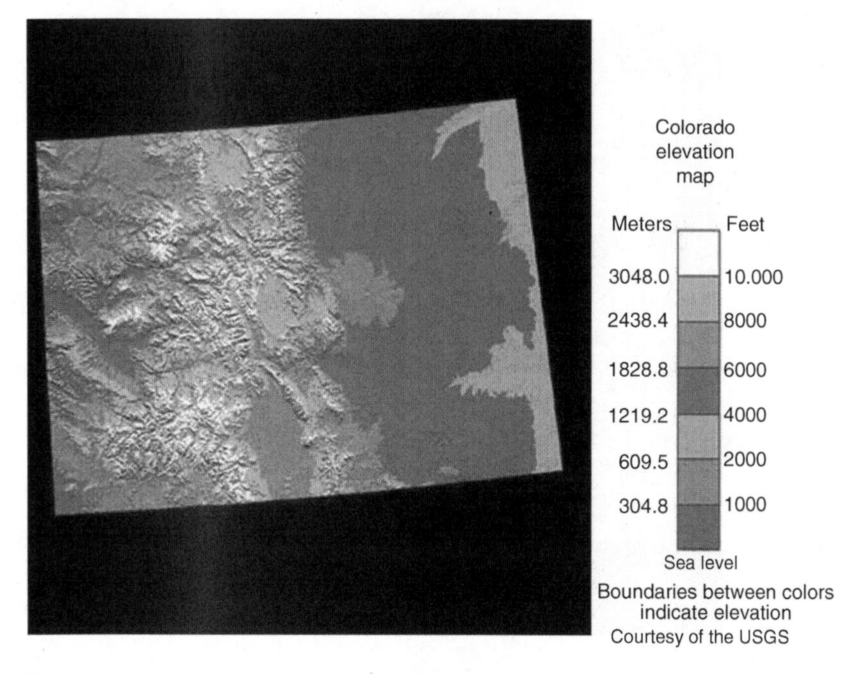

FIGURE 18.1 Relief map of Colorado.

state comprises the eastern High Plains, which are at the western edge of the Great Plains. The average elevation in Colorado is 2,073 m, with the lowest point being about 1,021 m near the town of Holly in southeastern Colorado where the Arkansas River crosses the border into Kansas. The highest peak is Mt. Elbert in the central mountains reaching a height of 4,399 m. Figure 18.1 is a relief map of Colorado that illustrates the tremendous elevational gradient in the state and thus the potential for significant environmental variation. About 7,021,050 ha are considered pasture and rangeland while 8,759,920 ha are forestland (http://csfs.colostate.edu/CO.htm). Federal and state government agencies manage about 6,330,770 ha of forestland and 2,429,150 ha are privately owned and managed.

18.2 LEVEL III ECOREGIONS

Colorado contains parts of six level III ecoregions, and there are 35 corresponding level IV ecoregions, many of which continue into ecologically similar areas in neighboring states (Figure 18.2; Chapman et al. 2006). The Southern Rockies level III ecoregion is comprised of high-elevation, steep, rugged mountains, most of which contains coniferous forests but this is dependent upon elevation. The Southern Rockies region occupies most of Colorado's central and western portions and comprises the majority of the high-elevation forests. Vegetation types at the lowest elevations of the Southern Rockies in Colorado are primarily grasses, forbs, and shrubs, and livestock and wildlife heavily graze vegetation in the lower elevations. Low to middle elevations in the Southern Rockies ecoregion are also grazed readily,

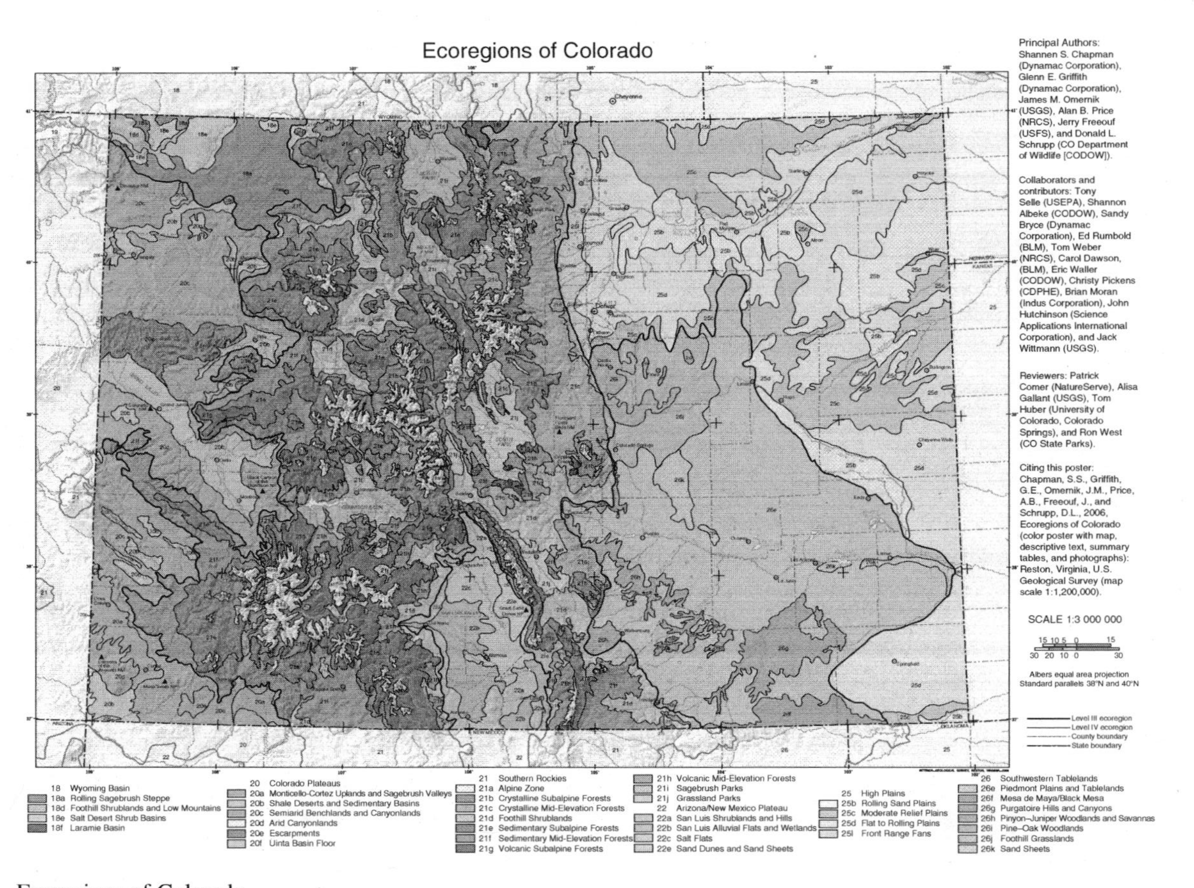

FIGURE 18.2 Ecoregions of Colorado.

but a greater variety of vegetation exists, including Douglas-fir (*Pseudotsuga menziesii*), ponderosa pine (*Pinus ponderosa*), quaking aspen (*Populus tremuloides*), and juniper–oak (*Juniperus* spp.–*Quercus* spp.) woodlands. The mid-to-high elevations of the Southern Rockies are mostly coniferous forests where grazing by livestock is minimal. Mid-elevation forests typically range from 2,134 to 2,743 m. Subalpine forests range in elevation from 2,590 to 3,658 m, but the lower elevational range of subalpine forests in the southern part of Colorado is from 2,473 to 2,985 m. The Alpine Zone starts above timberline, which occurs usually from 3,200 to 3,350 m where deformed and stunted (krummholz vegetation) Engelmann spruce (*Picea engelmannii*), subalpine fir (*Abies lasiocarpa*), and limber pine (*Pinus flexilis*) are found. Sagebrush (*Artemisia* spp.) dominates in Sagebrush Parks ecoregions, whereas Arizona fescue (*Festuca arizonica*), Idaho fescue (*Festuca idahoensis*), mountain muhly (*Muhlenbergia montana*), bluebunch wheatgrass (*Pseduoroegneria spicata*), needleandthread (*Hesperostipa comata*), June grass (*Koeleria macrantha*), and slender wheatgrass (*Elymus trachycaulus*) dominate Grassland Parks ecoregions. North Park and Middle Park are Sagebrush Parks ecoregions and range in elevation from 2,438 to 2,688 m and are located in the north central part of Colorado. South Park is a large Grassland Park ecoregion and its average elevation is 3,048 m and is located approximately in the middle of the state.

The High Plains level III ecoregion is higher in elevation and drier than the Central Great Plains that lies to the east of Colorado (Chapman et al. 2006). The High Plains are smooth to slightly irregular plains that are dominated by the native grasses, blue grama (*Bouteloua gracilis*) and buffalo grass (*Buchloe dactyloides*). Other native vegetation includes sand sage (*Artemisia filifolia*), rabbitbrush (*Chrysothamnus* spp.), yucca (*Yucca* spp.), sideoats grama (*Bouteloua curtipendula*), sand bluestem (*Andropogon hallii*), and Indian ricegrass (*Achnatherum hymenoides*). The High Plains ecoregion occupies the northeastern part of Colorado and some of the southeastern corner as well. There is a considerable amount of cropland on the High Plains and substantial grazing by livestock and wildlife.

The Southwest Tablelands level III ecoregion is the south-central plains of Colorado and lies south and west of the High Plains and just to the east of the Southern Rockies (Chapman et al. 2006). The Southwest Tablelands contains canyons, mesas, badlands, and dissected river breaks. The Colorado portion of this level III ecoregion is considered semiarid rangeland. It is rugged country and far less conducive to agricultural crops than the High Plains. Major natural vegetation includes blue grama, buffalo grass with some juniper–oak–grass savannas on escarpment bluffs. Other native vegetation includes yucca, western wheatgrass (*Pascopyrum smithii*), galleta grass (*Pleuraphis* spp.), alkali sacaton (*Sporobolus airoides*), sand dropseed (*Sporobolus cryptandrus*), sideoats grama, little bluestem (*Schizachyrium scoparium*), big bluestem (*Andropogon gerardii*), and switchgrass (*Panicum virgatum*). The Southwest Tablelands ecoregion in Colorado is grazed by wildlife and livestock, with some locations being heavily grazed by the latter.

The Arizona–New Mexico level III ecoregion is represented by the San Luis Valley, which is in the south-central part of Colorado (Chapman et al. 2006). The San Luis Valley is the upper end of the Rio Grande Valley through which the Rio Grande River flows. There is little to no topographical relief in the San Luis Valley, but it is flanked by the Sangre de Cristo Mountain Range on the east and the San Juan Mountains on the west. The San Luis Valley has a very high water table but very low annual precipitation. Native vegetation includes big sagebrush (*Artemisia tridentata* var. *tridentata*), rabbitbrush, winterfat (*Krascheninnikovia lanata*), shadscale (*Atriplex confertifolia*), saltbush (*Atriplex* spp.), greasewood (*Sarcobatus vermiculatus*), and grasslands with western wheatgrass (*Pascopyrum smithii*), blue grama, green needlegrass (*Nassella viridula*), needleandthread, saltgrass (*Distichilis spicata*), and alkali sacaton. Irrigated agricultural crops abound, and grazing by livestock on rangeland and improved pastures is commonly practiced.

The Colorado Plateaus level III ecoregion comprises the westernmost portion of Colorado. Canyons, mesas, and mountains dominate the landscape where abrupt relief of 305–610 m is fairly common (Chapman et al. 2006). This area of Colorado is hot and dry and many low-lying areas have large plant communities dominated by saltbush and greasewood. The Colorado Plateaus ecoregion has substantial amounts of piñon–juniper (*Pinus edulis–Juniperus* spp.) and Gambel oak (*Quercus gambelii*) woodlands. Other native vegetation includes Wyoming big sagebrush (*Artemisia tridentata* ssp. *wyomingensis*), greasewood, saltbush, shadscale, winterfat, galleta grass, and alkali sacaton. Irrigated and dryland agricultural production are commonly found in low-lying valleys in the Colorado Plateaus ecoregion, and grazing by livestock and wildlife occurs on rangeland and improved pastures throughout the ecoregion.

The Wyoming Basin level III ecoregion juts south into the northwestern portion of Colorado as well as a small portion of Larimer County (north-central Colorado) extending down from Laramie, Wyoming (Chapman et al. 2006). The Wyoming Basin ecoregion as a whole is a large intermontane basin dominated by arid grasslands and shrublands. Native vegetation includes rabbitbrush, fringed sage (*Artemisia frigida*), Wyoming big sagebrush, silver sagebrush (*Artemisia cana*), black sagebrush (*Artemisia nova*), bitterbrush (*Purshia* spp.), prickly pear (*Opuntia* spp.), juniper, mountain mahogany (*Cercocarpus* spp.), western wheatgrass, needleandthread, blue grama, Sandberg bluegrass (*Poa secunda*), bluebunch wheatgrass (*Pseduoroegneria spicata*), and Idaho fescue.

18.3 FOREST TYPES

The distribution of forest types in Colorado is influenced by climate, soils, elevation, aspect, and disturbance history (Rogers et al. 2001). Many of Colorado's forests are characterized as disturbance driven and evolved with naturally occurring disturbances such as wildfires, insect infestations, avalanches, flooding, windstorms, and disease outbreaks. These disturbances served to rejuvenate forests and insured a variety of forest types, densities, and age classes across the state. The lack of disturbance, fire in particular, over the past 100 or more years has created old and dense forests with almost no forests in the 0–20 year old age class.

Major forest types in Colorado include lodgepole pine (*Pinus contorta*), ponderosa pine, aspen, spruce–fir, and piñon–juniper (Colorado State Forest Service 2001). Other forest types include Douglas-fir, southwestern white pine (*Pinus strobiformis*), bristlecone pine (*Pinus aristata*), limber pine, Colorado blue spruce (*Picea pungens*), and cottonwood–willow (*Populus–Salix* spp.).

Lodgepole pine occurs primarily in Colorado's northern Rocky Mountains in the Southern Rockies ecoregion at elevations from 1,829 to 3,353 m (http://csfs.colostate.edu/CO.htm). These level IV ecoregions are the Crystalline and Sedimentary Mid-Elevation Forests and the Crystalline and Sedimentary Subalpine Forests.

Ponderosa pine is widely scattered over the lower elevations of the Southern Rockies ecoregion. It occurs mainly from 1,768 to 2,970 m (http://csfs.colostate.edu/CO.htm) in the Crystalline, Sedimentary, and Volcanic Mid-Elevation Forests of the Southern Rockies (Chapman et al. 2006). Ponderosa pine also is scattered in the Southern Tablelands level III ecoregion Pine–Oak Woodlands and Foothill Grasslands level IV ecoregions between Colorado Springs and Denver on an area known as the Palmer Divide.

Engelmann spruce is found at elevations from 2,438 to 3,353 m and its range overlaps considerably with subalpine fir, which occurs from 2,438 to 3,658 m (http://csfs.colostate.edu/CO.htm). Both species grow in the level IV Crystalline, Sedimentary, and Volcanic Subalpine Forests ecoregions of the Southern Rockies (Chapman et al. 2006).

Quaking aspen are deciduous species that are clonal and often occur in large monocultures but with a lush understory (http://csfs.colostate.edu/CO.htm). Aspen have a broad range and usually are found from 2,103 m to about the timberline, 3,200 m. Pure and mixed stands of aspen occur on about 1,619,400 ha and make them the second-most prevalent forest type in Colorado after spruce–fir (Colorado State Forest Service 2005). Aspen stands are very biologically diverse and are critical habitats for many wildlife species.

Piñon pine and juniper species woodlands are widespread on the lower elevations of western Colorado (Colorado State Forest Service 2001). There are many piñon and juniper species but Colorado piñon pine (*Pinus edulis*) and Utah juniper (*Juniperus osteosperma*) are the dominant species. Mixed stands of piñon and juniper occur from about 1,524 to 2,473 m and are commonly referred to as the "P–J" woodlands, which are readily found on the Colorado Plateaus ecoregions (Chapman et al. 2006).

Cottonwood–willow plant communities are not considered one of the major forest types in Colorado. However, they are extremely important riparian species, and riparian forests in Colorado harbor many invasive weed species, especially at lower elevations. Narrowleaf cottonwood (*Populus angustifolia*) grows at elevations ranging from about 1,524 to 2,440 m, plains cottonwood (*Populus deltoids*) from 1,067 to 1,980 m, and willows (*Salix* spp.) from 1,067 to 2,743 m (http://csfs.colostate.edu/CO.htm). These important riparian species typically occur near water in almost all of the level III ecoregions in Colorado.

18.4 CONDITION OF COLORADO'S FORESTS AND RANGELANDS

18.4.1 Habitat Quality Decline

Upon examining the human history of Colorado, one can discern the factors that contributed to degradation in the quality of natural habitats, which helped to create the invasive weed problem that the citizens share today. Extractive industries such as mining operations, timber, and agricultural development that started in the 1800s created severe disturbances that altered the landscape. Some of these activities continue to this day. Development of recreation industries further stressed the natural environment as skiing and other tourism-related businesses took advantage of Colorado's weather and scenic beauty, but also altered the landscape. The marked urban expansion of the twentieth century continuing into the twenty-first century also has inextricably altered many natural environments in Colorado. All of these factors have, over time, fueled the invasion of weeds by constantly providing for a disturbed environment.

Currently, in Colorado, there are 398 nonnative plant taxa from outside of North America and 105 nonnative plant taxa from elsewhere in the United States (Hartman and Nelson 2001). Thus, there are 498 nonnative plant taxa in Colorado (that we are aware of) that are potentially invasive, which constitutes 14.5% of the flora.

Colorado has about as many different weed species infesting various locations in the state as any other western state in the United States. However, the cumulative size of these infestations for many species tends to be smaller than in other western states. The causes for the discrepancy are largely unknown but differences from time of introduction and climate are attractive hypotheses. The climate of Colorado, in particular, is highly variable in space and time (Doesken et al. 2003). High elevation combined with a mid-latitude continental geography produce cool and dry weather conditions. Drought occurs often in Colorado, particularly on the plains. Highly variable weather from location to location and from year to year, combined with frequent drought, may be at least partially responsible for limited invasive weed infestations compared with some other western states. Even if such a hypothesis was acceptable, the future weed problem in Colorado and elsewhere may worsen in spite of control and management efforts because of atmospheric changes. Increased atmospheric concentration of CO_2 recently was hypothesized as a possible selection pressure for invasion of weedy C_3 forbs in the United States. Under controlled conditions, Ziska (2003) created atmospheric concentrations of CO_2 that were equivalent to those of the beginning of the twentieth century, the current levels, and levels predicted for the end of the twenty-first century, and tested the growth of six invasive weeds under each condition. He found that the increasing CO_2 concentrations that occurred during the twentieth century stimulated the growth of Canada thistle (*Cirsium arvense*), field bindweed (*Convolvulus arvensis*), leafy spurge (*Euphorbia esula*), perennial sow thistle (*Sonchus arvensis*), spotted knapweed (*Centaurea maculosa*), and yellow star thistle (*Centaurea solstitialis*) by an average of 110%. Canada thistle growth was stimulated by 180%. Growth stimulation of weeds from the past through the predicted future CO_2 concentrations averaged 46%. These results suggest that increased amounts of CO_2 in our atmosphere have contributed to the invasion of weeds in the United States and elsewhere and may have been a favorable selection pressure for these weeds as well.

18.4.2 Fire Suppression and Grazing

The history of Colorado again is implicated as a contributing factor to the current invasive weed problem. The lower elevation ponderosa pine forests, for example, were markedly changed because of fire exclusion, logging, and grazing over a century (Arno 2000a). Fire suppression has lead to an accumulation of fuels. Active fire suppression began around 1900 and caused major changes in the fire regimes in Colorado. Historic nonlethal fires have decreased and lethal fires have increased (Arno 2000b). Interruption of frequent burning began in the late 1800s because of relocation of Native Americans who burned these Colorado areas frequently; removal of fine fuels by heavy and extensive livestock grazing; disruption of fuel continuity on the landscape caused by irrigation, cultivation, and development; and adoption of fire exclusion as a governmental management policy. Relatively long fire intervals and mixed burning historically were common along Colorado's Front Range ponderosa pine forests (Laven et al. 1980). Such a fire regime kept these forests in park-like conditions because a mosaic of understory fires and stand replacement fires were thought to occur every 0–34 years along the Front Range.

A century of fire suppression policies have allowed fuels to collect and have stimulated unusually large and severe fires in many dry forest types (Sutherland 2004). Such fires create conditions that favor invasive weeds by exposing mineral soils, eliminating plant competition, and increasing light and nutrient levels (Sutherland 2004). Wildfire can create conditions for the marked expansion of existing weed problems and establishment of new ones. Establishment of invasive weeds after wildfires is a major concern because these weeds impede the reestablishment of native vegetation and associated wildlife habitat (Colorado State Forest Service 2003). Weeds are carried into burned areas by natural means, such as wind, water, and animals, or by humans via clothing, vehicles, or by contaminated seeds used to rehabilitate burned areas, or from contaminated straw used for soil stabilization. Following the major fires of 2002 and 2003 in Colorado, major weed species of concern to federal and state land managers in burned areas include orange hawkweed (*Hieracium aurantiacum*), spotted knapweed, leafy spurge, yellow toadflax (*Linaria vulgaris*), and downy brome (*Bromus tectorum*). Diffuse knapweed (*Centaurea diffusa*), Dalmatian toadflax (*Linaria dalmatica*), Canada thistle, and musk thistle (*Carduus nutans*) may be added to the list of weed species to be concerned about after wildfires throughout Colorado, but particularly in the Crystalline Mid-Elevation Forests and the Foothills Shrublands ecoregions of the Southern Rockies that lie along the Front Range.

Removal of fine fuels by heavy livestock grazing has been implicated in altering the fire regime in Colorado, especially along the Front Range, by eliminating the fuels that allowed nonlethal burns to occur in the understory (Arno 2000b). This process created a significant disturbance and likely favored weed invasion by eliminating competition from desirable grasses and forbs that otherwise might have prevented or at least deterred invasion by downy brome, diffuse knapweed, and Dalmatian toadflax. Downy brome in particular is a problem in the Foothills Shrublands ecoregions of the Southern Rockies along the Front Range and also is a problem in the lower elevation ecoregions of western Colorado. Although many anecdotes circulate as to how livestock grazing has contributed to if not caused the establishment and expansion of invasive weeds—and to be sure, overgrazing has occurred in many Colorado locations—grazing has been shown not to influence the establishment and spread of invasive weeds. Stohlgren et al. (1999b) compared plant diversity and soil characteristics in 26 long-term grazing exclosures with adjacent grazed areas in Colorado, Wyoming, Montana, and North Dakota. Species richness was found to be nearly identical for native and exotic species in grazed and ungrazed plots at the 1000 m^2 scale. These researchers found 31.5 ± 2.5 native species and 3.1 ± 0.5 exotic species in grazing exclosures compared with 31.6 ± 2.9 native species and 3.2 ± 0.6 exotic species in adjacent grazed areas. There were no differences in species diversity, evenness, cover, or various life forms (grasses, forbs, and shrubs), soil texture, or percent soil nitrogen or carbon between grazed and ungrazed sites.

18.5 CURRENT SITUATION IN COLORADO

18.5.1 Public Policy Development

The continued spread of invasive weeds in Colorado has stimulated a series of legal and policy changes in the recent past. Noxious weeds are those weeds that are so troublesome to manipulated and natural environments that laws are passed and regulations promulgated, which mandate their management by all public and private landowners. Most noxious weeds are considered invasive species. Before 1990, there was no statewide noxious weed law in Colorado, but the Colorado Legislature passed a statewide law in 1990 and designated four noxious weeds: diffuse knapweed, spotted knapweed, Russian knapweed (*Acroptilon repens*), and leafy spurge. In 1996, the Legislature amended the 1990 noxious weed law that led to a marked expansion of the noxious weed list to over 80 species and created a statewide noxious weed coordinator position in the Colorado Department of Agriculture. Progress continued and in 1999 the Governor of Colorado issued an executive order (E.O. D 006 99; http://www.ag.state.co.us/CSD/Weeds/weedpublications/order.pdf) concerning noxious and invasive weeds. The gubernatorial executive order instructed the Departments of Agriculture, Natural Resources, and Transportation, and Colorado State University to develop and implement integrated weed management strategies for the land they administer. This executive order has stimulated the affected agencies to communicate more effectively as to how their daily land management activities contribute to the establishment and spread of invasive weeds for themselves as well as the other state agencies involved. At a minimum, invasive weed awareness has increased among these agencies, and educational efforts have been enhanced.

The Colorado Noxious Weed Act was amended again in 2003 (C.R.S. 35.5.5 Colorado Noxious Weed Act; http://www.ag.state.co.us/CSD/Weeds/statutes/weedlaw.pdf) and the most significant change was to the structure of the state noxious weed list. We now have three noxious weed lists in Colorado and the distinction among the lists is largely based on population size and distribution, perceived impact, and difficulty associated with management. The A-list species are those that do not occur yet in Colorado or whose populations are rare or of such limited distribution that eradication is possible and desirable. B-list species are those invasive weeds whose populations are discrete, but often found in many locations throughout the state. In some of these locations, B-list populations are isolated and small enough that eradication is possible and desirable. In other locations, populations of B-list invasive weeds are large and contiguous, or at least potentially so, such that containment and eventual population reduction are more realistic goals. C-list invasive weed species are those whose populations are so great and widespread that suppression and improved management are the statewide goals. Unfortunately, the C-list includes invasive weeds that markedly impact agricultural production, such as field bindweed and downy brome. In addition, because these weeds are so widespread, accurate accounts of their statewide distributions have not been accomplished.

18.5.1.1 A-List Invasive Weeds

18.5.1.1.1 Yellow Starthistle

Examples serve to better explain the differences among the noxious weed lists. A survey in 2004 identified about 32 ha in Colorado infested with yellow star thistle (Figure 18.3). Several very small infestations occur along the Front Range in Larimer and Boulder counties in the High Plains Front Range Fans ecoregion. Other small infestations occur in the western part of Colorado in the lower elevations of Montrose, Ouray, and Mesa counties in the Colorado Plateaus Semiarid Benchlands and Canyonlands ecoregions. Yellow star thistle has been successfully eradicated from eight locations in Colorado.

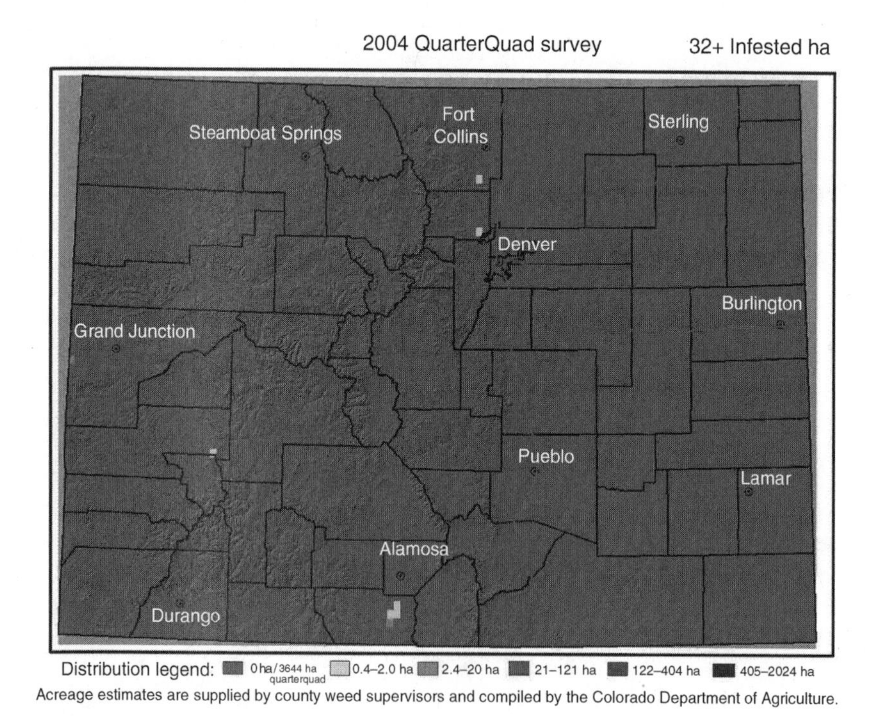

FIGURE 18.3 Statewide distribution of yellow star thistle (*Centaurea solstitialis*) in Colorado.

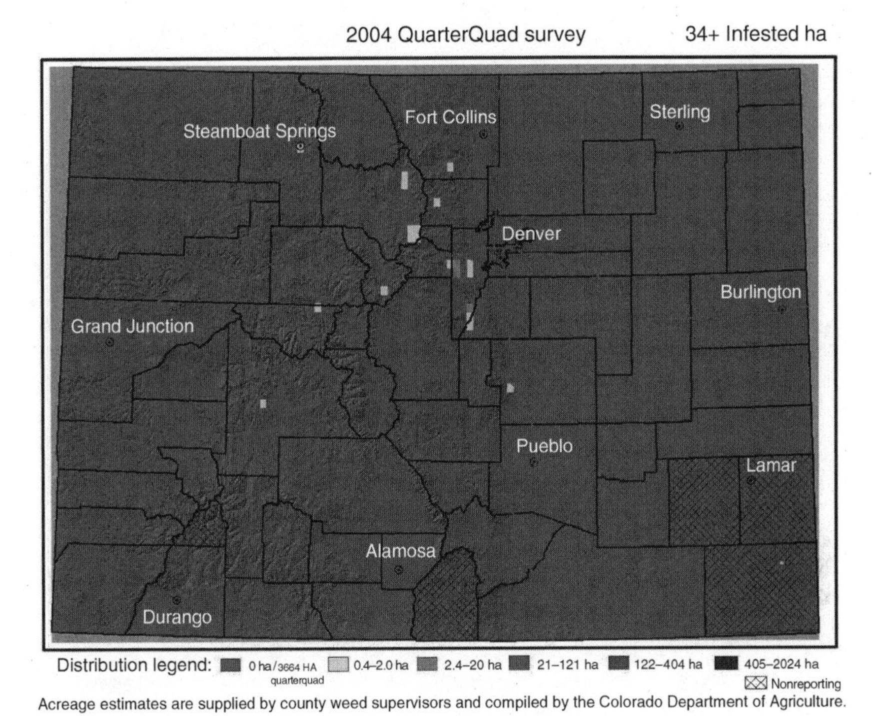

FIGURE 18.4 Statewide distribution of orange hawkweed (*Hieracium aurantiacum*) in Colorado.

18.5.1.1.2 Orange Hawkweed

Orange hawkweed is another A-list species in Colorado and a 2004 survey located about 34 ha infested with this invasive weed (Figure 18.4). Currently, it is found in the Southern Rockies Crystalline Mid-Elevation Forests ecoregion on the eastern side of the Continental Divide in the ponderosa pine forests of Douglas, Jefferson, and Boulder Counties. It is also found in the Southern Rockies Crystalline Mid-Elevation and Subalpine Forests ecoregions west of the Continental Divide in lodgepole pine forests in Grand, Summit, Pitkin, and Routt counties.

18.5.1.2 B-List Invasive Weeds

18.5.1.2.1 Yellow Toadflax

Yellow toadflax (Figure 18.5) is readily found, especially in the Southern Rockies ecoregion and occupies about 39,643 ha in Colorado. It is widespread in the ponderosa pine forests of the Southern Rockies Foothills Shrublands ecoregion along the Front Range into the spruce–fir forests in the Crystalline Mid-Elevation and Subalpine Forests ecoregions. Yellow toadflax can even be found above timberline in the Alpine Zone ecoregion of the Southern Rockies in Rocky Mountain National Park (Connor 2000, personal communication). Other infestations on eastern side of Colorado occur on the Southwestern Tablelands Pine–Oak Woodlands and in

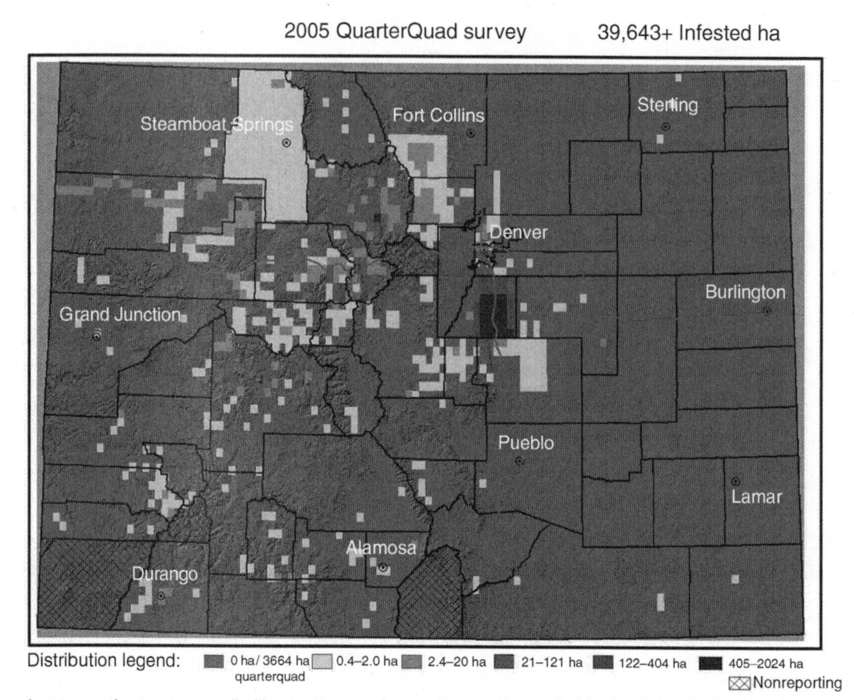

FIGURE 18.5 Statewide distribution of yellow toadflax (*Linaria vulgaris*) in Colorado.

the riparian forests along the South Platte River. Yellow toadflax is also widespread throughout the ponderosa pine and spruce–fir forests of the Southern Rockies Crystalline Mid-Elevation and Subalpine Forests ecoregions, the Sedimentary Mid-Elevation and Subalpine Forests ecoregions, and the Volcanic Mid-Elevation and Subalpine Forests ecoregions on the western side of Colorado. It also extends down into the lower elevations in the Colorado Plateaus ecoregions primarily in or adjacent to riparian areas in the Deserts and Sedimentary Basins ecoregion and the Semiarid Benchlands and Canyonlands ecoregions. Yellow toadflax typically does not occur in the trees of coniferous forests in Colorado except in the ponderosa pine forests of the Foothills Shrublands and Pine–Oak Woodlands ecoregions. In a study conducted in the Flat Tops Wilderness in the Southern Rockies Sedimentary Subalpine Forests ecoregion, Sutton (2003) found that yellow toadflax is most often found in areas that were open-canopy sites (such as meadows and natural parks), along trails, and with higher species diversity. Even though it is difficult to kill yellow toadflax, the small infestations on the northern and southern portions of the High Plains ecoregion will be targeted for eradication.

18.5.1.2.2 Russian Knapweed

There are about 47,911 ha of Colorado rangeland infested with Russian knapweed (Figure 18.6). Infestations occur primarily at lower elevations on rangeland, and it is

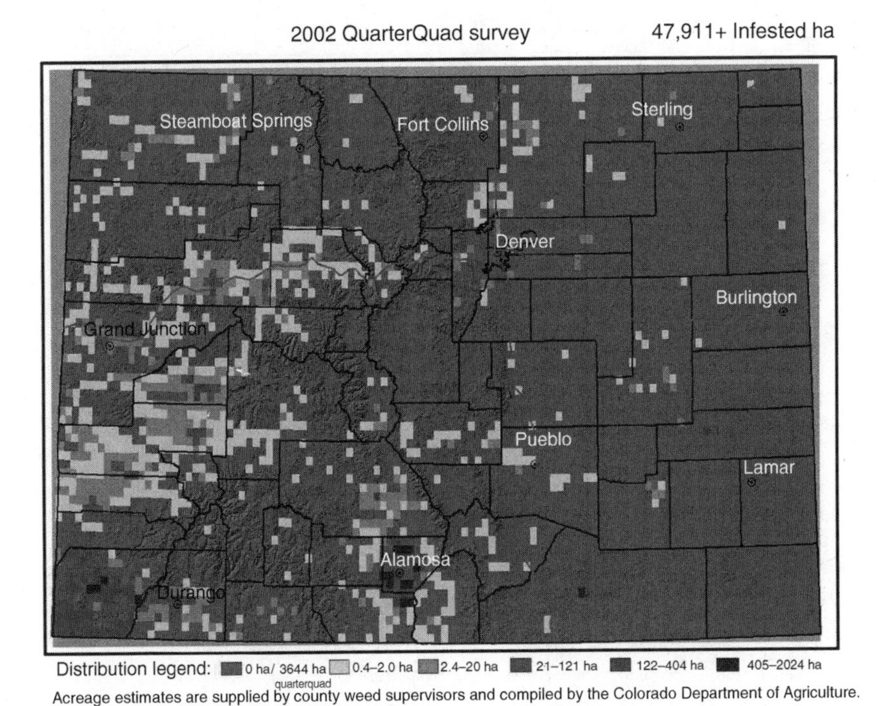

FIGURE 18.6 Statewide distribution of Russian knapweed (*Acroptilon repens*) in Colorado.

more widespread in western Colorado. Russian knapweed usually is not found in forests but occurs at the lower elevations of the Southern Rockies Sedimentary Mid-Elevation Forests ecoregions down into the hot climate of the Colorado Plateaus Monticello-Cortez Uplands and Sagebrush Valleys, Shale Deserts and Sedimentary Basins, and Semiarid Benchlands, and Canyonlands ecoregions. Large and dense populations of Russian knapweed are common in the San Luis Valley, which is part of the Arizona–New Mexico Plateau level III ecoregion in south-central Colorado. Here, Russian knapweed is a problem in the San Luis Shrublands and Hills, San Luis Alluvial Flats and Wetlands, and the Salt Flats ecoregions. Scattered infestations also occur on the eastern plains, particularly the High Plains ecoregions, and large and dense populations are found in the Southwestern Tablelands Purgatoire Hills and Canyons ecoregion. The widely scattered infestations on the High Plains ecoregion would be the only populations targeted for eradication.

18.5.1.2.3 Diffuse Knapweed

Diffuse knapweed infests about 56,050 ha of Colorado and is not as widespread as yellow toadflax or Russian knapweed (Figure 18.7). Most infestations are located near the Denver metropolitan area and surrounding lands where disturbance from development has fueled its spread. Infestations are in open space areas, degraded rangeland and pastures, abandoned lands, new housing construction projects, roadsides, and other disturbed sites. Forest types infested by diffuse knapweed include

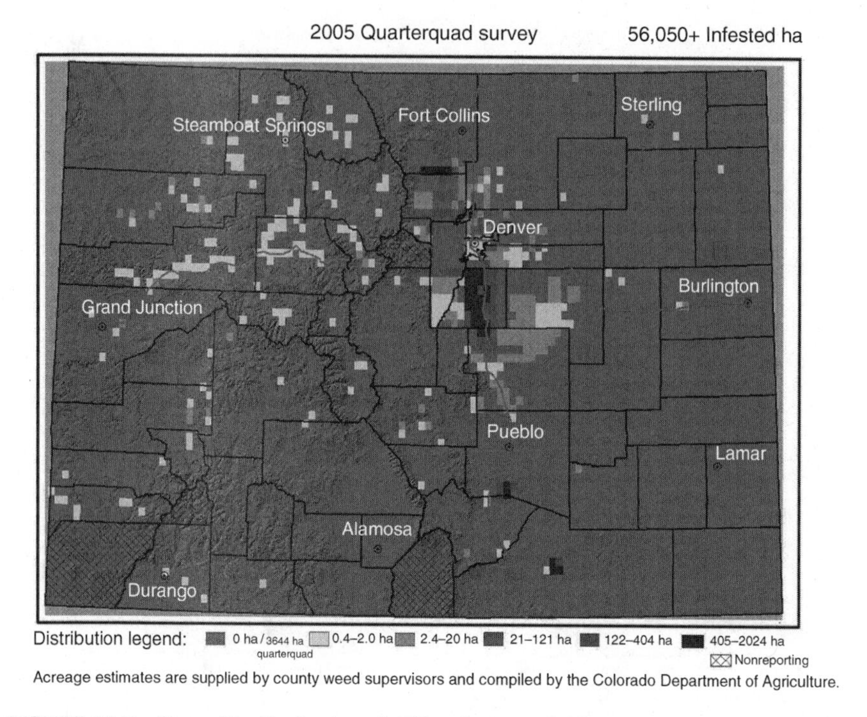

FIGURE 18.7 Statewide distribution of diffuse knapweed (*Centaurea diffusa*) in Colorado.

ponderosa pine, lodgepole pine, spruce–fir, and piñon–juniper but most often it occurs in open-canopy locations exposed to light as opposed to growing under trees. Ecoregions east of the Continental Divide where diffuse knapweed can be found in Colorado include the Southern Rockies Foothills Shrublands and Crystalline and Sedimentary Mid-Elevation Forests ecoregions; the Southwest Tablelands Piedmont Plains and Tablelands, Purgatoire Hills and Canyonlands, Piñon–Juniper Woodlands and Savannas, and Pine–Oak Woodlands, and Foothills Grasslands ecoregions; and the High Plains Front Range Fans, Flat to Rolling Plains, Rolling Sands Plains, and Moderate Relief ecoregions. Diffuse knapweed does not occupy nearly as much as land on the western side of Colorado but is fairly widespread and spreading quickly along roadsides such as Interstate 70, which bisects the state in an east–west direction. Ecoregions on the west side of Colorado where diffuse knapweed currently is found include the Southern Rockies Foothills Shrublands, Crystalline and Sedimentary Mid-Elevation Forests, and the Sagebrush Parks ecoregions. From the higher elevations, diffuse knapweed extends into the lower elevations of the Colorado Plateaus Semiarid Benchlands and Canyonlands, and the Monticello-Cortez Uplands and Sagebrush Valleys ecoregions, and in the Wyoming Basin Rolling Sagebrush Steppe ecoregion. The dense and large stands of diffuse knapweed along the Front Range would be targeted for containment while the outlying populations on the High Plains, Wyoming Basin, Colorado Plateaus, and Southwestern Tablelands ecoregions are scheduled for eradication.

18.5.2 Current State of Weed Management Efforts

The passage of three noxious weed laws in 13 years and other policy actions have stimulated educational efforts throughout Colorado, and people in general are becoming more aware of our invasive weed problem. The 1990 noxious weed law required all counties in the state to have weed districts, and this increased the level of organization and most likely had the greatest effect in raising the level of awareness among Coloradoans. Invasive weed management is better coordinated than it was before 1990, although it is the responsibility of each weed district to locate money to fund management efforts and usually programs are funded from county or municipal general funds. The Colorado Legislature initiated a small Noxious Weed Trust Fund in 1997 to create a competitive environment to fund invasive weed management projects. A fiscal downturn in 2003 caused this source of funding to be decreased to zero but the Trust Fund still receives some financial support from the U.S. Forest Service State and Private Forestry Program. It is our sincere hope that the Colorado Noxious Weed Management Trust Fund will be reinstated and augmented by the Colorado Legislature as the state fiscal situation improves.

When the Colorado Noxious Weed Act was amended in 1996, the Legislature charged the Colorado Department of Agriculture with determining how federal and state agencies were progressing in the battle against invasive weeds. A survey was sent to all Colorado counties in 1997, and 81% responded. The general results showed that while some land management agencies performed better than others, no single agency or landowner had achieved a high degree of compliance with county weed districts (Table 18.1). Although such findings were initially discouraging, optimism returned when overall analysis of the invasive weed situation in Colorado revealed that we are much improved as a state compared with that before 1990 when invasive weed management was largely unheard of and at best, Colorado was a patchwork quilt of uncoordinated weed management efforts. The report of the Colorado Department of Agriculture to the Legislature (Lane 1999) made recommendations that could serve as a template for others to follow:

a. Identify success stories and share the process used to achieve success and the results so that others can learn and benefit.
b. Implementation of invasive weed management models should be a standard objective and means to evaluate overall land management success.
c. Seek out and learn from failures.
d. Cooperate with neighbors to create synergistic weed management efforts to stretch limited resources, particularly budgets.
e. Inventory to determine the extent of the invasive weed problem within a defined area and determine the most appropriate level of management (e.g., eradication, containment, or suppression).

In 2001, Colorado as a state took the next appropriate step to develop better and more coordinated invasive weed management efforts by creating Colorado's

TABLE 18.1
Summary of Adjusted Averaged Scores Describing Degree of Public Agency and Private Landowner Cooperation and Compliance with Colorado Noxious Weed Act as Evaluated by 51 Colorado Counties in 1997–1998

Landowner	n[1]	Awareness[2]	Commitment[2]	Management Plan[2]	Allocation of Resources[2]	Performance[2]	Cooperation[2]	Overall[2]
Soil Conservation Districts/NRCS	43	4.12[a]	3.81[a]	3.48[c]	3.37[a]	3.5[a]	4.13[a]	3.74[a]
Department of Energy	1	4.73	3.4	3.52	4.42	2.44	3.45	3.65
Bureau of Land Management	29	3.75[c]	3.47[b]	3.56[b]	3.23[c]	3.24[c]	3.86[d]	3.48[b]
U.S. Forest Service	38	3.93[b]	3.43[c]	3.26[e]	2.57[abc]	3.15[g]	3.93[b]	3.4[c]
Colorado Division of Wildlife	42	3.56[ae]	3.38[f]	3.28[d]	2.98[e]	3.2[d]	3.92[c]	3.39[d]
National Park Service	7	3.7[d]	3.27[g]	3.82[a]	2.9	3.41[b]	3.44	3.32[g]
CO Department of Transportation	49	3.43[ag]	3.4[e]	2.76[bc]	3.26[b]	3.15[f]	3.71[e]	3.29[ae]
Colorado State Parks	22	3.48[af]	3.42[d]	3.2[f]	2.67	3.16[e]	3.58[f]	3.25[f]
Bureau of Reclamation	10	2.99[ab]	2.97	2.69	3.16[d]	2.94[h]	3.38	2.96[ah]
U.S. Fish and Wildlife Service	4	3.33	2.74	2.79	2.53	2.71	3.37	2.91
Private Landowners	50	2.91[abceg]	2.64[abcdef]	2.41[abcdef]	2.63[abf]	2.66[acdfgi]	3.09[abcde]	2.7[abcdefi]
Bureau of Indian Affairs	3	2.97	2.42	2.55	2.34	1.97[a]	2.17[abcd]	2.4[a]
State Land Board	31	2.38[abcdefg]	2.1[abcdefg]	2.04[abcdefg]	1.95[abcdef]	1.96[abcdefghi]	2.49[abcdef]	2.14[abcdefghi]
Department of Defense	4	2.68[a]	2.33[a]	2.5	1.53[abce]	1.46[abcdefghi]	1.93[abcdef]	2.07[abcdef]

Source: From Lane, 1999.

Note: Averaged scores are presented for all six individual categories and one overall average category. Landowners are sorted in descending order by overall score. 1, Abysmal; 2, Poor; 3, Adequate; 4, Good; 5, Outstanding.

[a] Sample size: number of counties that scored a given landowner.

[b] Within each column, values followed by the same letter are statistically different ($P = .01$). Note that when more than two values are followed by the same letter, the highest average score is statistically greater than the others. For example, the Colorado Division of Wildlife has done a significantly better job of developing and implementing management plans than the State Land Board and private landowners. It is critical to note which scores are statistically significant and which are not. It would be unfair to characterize any one landowner's efforts as superior or inferior to another's if the scores are not statistically significant.

Strategic Plan to Stop the Spread of Noxious Weeds (http://www.ag.state.co.us/CSD/Weeds/strategicplan/StrategicPlan.pdf). This plan was endorsed by 42 very diverse public and private organizations. The updated philosophy and amendments to the 1996 Colorado Noxious Weed Act that were passed in 2003 by the Colorado Legislature flowed from the strategic plan.

Federal land management agencies in Colorado are struggling to enhance their awareness and invasive weed management efforts. State land management agencies are experiencing a similar situation but seem to be making greater improvements relative to education and training. An optimist indeed would see that progress is being made only because, until 1990, very little was being accomplished in any Colorado location to manage invasive weeds. No concerted effort existed except for a few piecemeal efforts by a few county governments largely related to maintaining and improving agricultural production.

Unfortunately, much more progress must be made. People must realize that invasive weed management must be part of an overall land management strategy and combined with other efforts to meet land management goals and objectives. Invasive weed management, however, is both difficult and complex enough that it tends to become a goal in and of itself primarily because the contemporary private or public land manager has received inadequate education concerning pest management, especially invasive weed management. This includes land management leadership at the federal and state government levels. Leadership at the highest levels in federal and state governments must make certain that management of invasive weeds and all other taxa of invasive species are effectively addressed by their agencies. In Colorado, upper* and middle management at the state and federal levels must embrace the importance of invasive weed management and make certain that it is carefully and appropriately integrated into their overall land management goals and objectives. This is especially the case with federal agency middle management where a genuine lack of leadership on invasive species management in general is readily apparent in our state. Until such time that all governmental land management agencies effectively address invasive weed management, private sector land managers will be reluctant participants even though some, such as The Nature Conservancy, choose to lead by example.

One example of a very positive move by a federal agency occurred in 2004 when the Rocky Mountain Region of the U.S. Forest Service announced the requirement to only allow use of certified weed-seed-free feeds for horses being taken into the back country. This policy decision, if enforced, will prevent many new invasive weed infestations from occurring in highly remote and rugged areas in Colorado and represents but one very appropriate and positive step forward.

* Examples of upper and middle management at the state level include state directors of the various departments of natural resources, agriculture, transportation, and commerce and their regional subordinates. Examples of federal middle management include regional foresters, state directors of BLM and similar land management agencies, and national park superintendents.

18.6 INVASIVE WEED RESEARCH IN COLORADO: CAUSES AND CURES

18.6.1 Soil Disturbances

The development of various industries, such as agriculture, mining, timber, transportation, and urban growth, throughout Colorado's history has created a continuous opportunity for invasive weed establishment and expansion because of these disturbances. Activities that disturbed the soil in particular may have increased available soil nitrogen and encouraged weed invasions. Disturbances increase nitrogen mineralization in forest soils (Matson and Vitousek 1981). Changes in nitrogen mineralization and availability following a disturbance may alter species composition of the recovering plant community (Matson and Boone 1984). Many early-seral species are adapted to high nitrogen availability, and some species are stimulated to germinate under conditions of high available soil nitrogen (Bazaaz 1979). Early-seral species, or weeds (especially annual weeds), often dominate previously disturbed landscapes. Increased nitrogen availability due to disturbance in semiarid climates has been shown to delay succession by allowing annuals to establish and maintain dominance of the recovering plant community. McClendon and Redente (1991) conducted an experiment in the Piceance Basin in western Colorado where they disturbed the soil by scraping off the top 5 cm and mixing the next 35 cm by cultivating, which decreased the soil seed reserve by an estimated 90%. They provided annual additions of nitrogen and phosphorous for a period of 5 years and compared the resulting plant communities with similarly disturbed plots where no amendments were made. Annual weeds dominated the nitrogen-treated and control plots for 3 years, but in control plots, during the fourth year, perennial grasses and perennial and biennial forbs increased while annual weeds decreased. This effect was more pronounced during the fifth year. However, annual weeds, such as Russian thistle (*Salsola iberica*), downy brome, and kochia (*Kochia scoparia*), continued to dominate the nitrogen-treated plots 5 years after the experiment was initiated.

Because of the influence that high amounts of available soil nitrogen has on succession by favoring weedy annual species, decreasing available soil nitrogen has been evaluated as a means to stimulate succession in old-fields to a late-seral state. Paschke et al. (2000) examined increasing soil carbon through amendments as a mechanism to decrease available soil nitrogen. Experiments were conducted in early-, mid-, and late-seral plant communities that previously had been cultivated on the shortgrass steppe of eastern Colorado and on one uncultivated site that served as a control. Secondary succession takes about 50 years to occur in this particular region of Colorado (Coffin et al. 1996; Reichardt 1982). This process is typically characterized by a transition from plant communities dominated by annual forbs and annual grasses to herbaceous perennials (Lauenroth and Milchunas 1992). High soil nitrogen conditions were created with annual fertilizer amendments while low nitrogen was created with sucrose additions at 3,788 kg ha^{-1} $year^{-1}$ (1,600 kg C ha^{-1} $year^{-1}$) to immobilize soil nitrogen. The sucrose applications were divided into three equal soil applications each year. Decreasing the available soil nitrogen with carbon additions decreased the relative biomass of downy brome and other weedy annuals and increased the relative biomass of perennial species (grasses, forbs, shrubs, and succulents). Decreasing available soil nitrogen stimulated the rate of succession in early-, mid-, and late-seral plant communities and minimized changes in the uncultivated site. Conversely, increased available soil nitrogen shifted species compositions in the direction of early-seral conditions. The authors concluded that carbon additions may be a useful aid in the rehabilitation of degraded rangeland to late-seral conditions but warned that the cost of using sucrose to do so was prohibitive.

Other research in Colorado could not demonstrate that carbon additions helped to shift plant communities to a later seral state. Reever Morghan and Seastedt (1999) amended soil with sucrose at 1,000 kg ha^{-1} plus sawdust at 625 kg ha^{-1} annually for 3 years. Biomass of diffuse knapweed was decreased by 32%, and biomass of all other species was decreased by 43% by carbon additions, but no species composition changes were detected. The amount of carbon addition from sucrose was about one-third the rate of the additions that others have found successful (McClendon and Redente 1992; Paschke et al. 2000), and perhaps the carbon from sawdust was not sufficiently available in one to three growing seasons in this semiarid ecosystem to effect changes in plant community composition that others have observed. Although their attempts to substitute an expensive source of carbon with one that is much less expensive to stimulate reduction in available soil nitrogen did not yield anticipated results, they appropriately pointed out that ameliorating excess soil nitrogen most likely would prove inadequate to solve invasive weed problems in most situations and indicated that an integrated approach would be necessary to achieve success in altering the plant community to a late-seral state.

18.6.2 Changing Stream Hydrology

In the arid and semiarid ecosystems of the western United States, the plant species that comprise the riparian ecosystems provide critical habitat for many wildlife species, help to maintain water quality and quantity, and stabilize stream banks (Colorado State Forest Service 2001). The conditions of the riparian forests in Colorado have been drastically altered over the past 100 years by water development and flood control, urban development, intensive agriculture, grazing, pollution, and fire. Stands of native cottonwoods and willows have been replaced by tamarisk (*Tamarix* spp.) and Russian olive (*Elaeagnus angustifolia*) because of altered stream flows and other anthropogenic hydrologic alterations. Indeed, exotic species are a major threat to the integrity of riparian areas in the lower montane areas of Colorado (Rocchio 2006).

Natural areas in the western United States with high amounts of resources usually are species rich, particularly with native species. Stohlgren et al. (1999a) indeed found this to be the case but at several scales, particularly at the biome scale, these researchers also found that areas that are rich in native species also are rich in exotic species. Sites with the greatest amount of soil moisture, light, and soil nitrogen had the greatest number of native species, but these same areas also had the greatest number of exotic species; that is, native species richness and exotic species richness were positively correlated. For example, in Rocky Mountain National Park in Colorado, native species richness was greater in wet meadows and yet greater in aspen stands than in dry meadows, ponderosa pine, and lodgepole pine vegetation types. Likewise, exotic species richness also was greater in wet meadows and aspen stands. Wet meadows and aspen stands also had higher levels of soil nitrogen than the dry meadow, ponderosa pine, and lodgepole pine vegetation types. The implications are that species-rich sites such as in aspen stands, wet meadows, or riparian areas have more resources than the native species assemblages use, and these resources therefore are available for other species, including invasive exotic species. Conversely, low species diversity sites, such as the shortgrass steppe on the High Plains in eastern Colorado, are low in available resources, and existing native plant communities successfully dominate use of those limited resources.

Indeed, riparian forests in Colorado are among the most invaded ecosystems in the state. This is likely at least partly due to disturbance and available resources, because the native plant communities do not use all available resources. Examination of one invasive weed species, leafy spurge, demonstrates this effect. The first occurrence of leafy spurge in Colorado is unknown but many agree that it most likely occurred in the 1950s. Regardless of when leafy spurge first established, upon examining the state distribution map (Figure 18.8), one can easily see that this invasive weed has invaded several riparian ecosystems on both sides of the Continental Divide. The Poudre River and South Platte River ecosystems on the High Plains of eastern Colorado are infested almost continuously from the mouth of the Poudre River canyon through Greeley where it flows into the South Platte River and

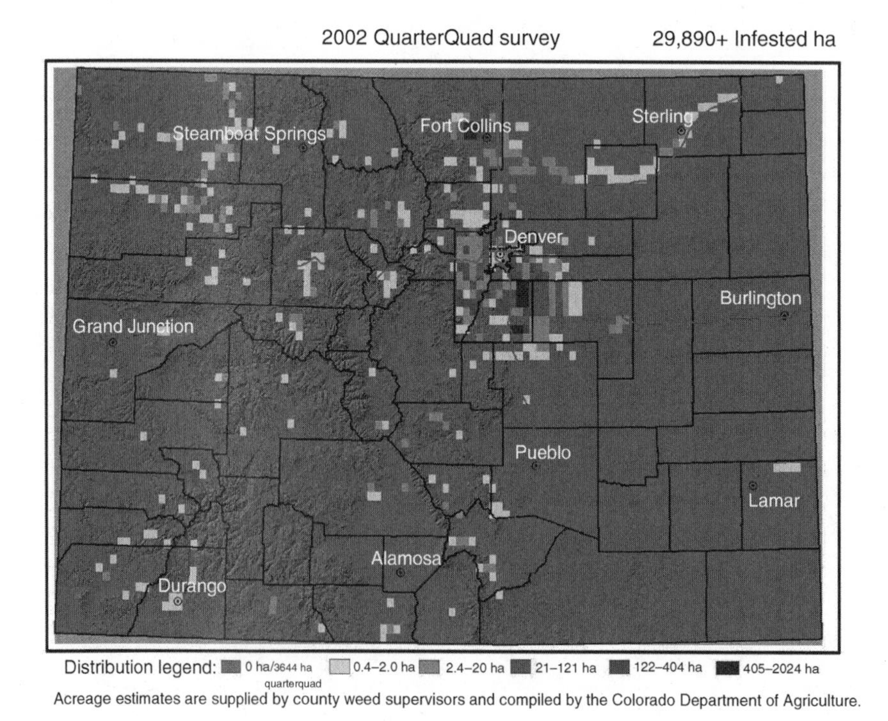

FIGURE 18.8 Statewide distribution of leafy spurge (*Euphorbia esula*) in Colorado.

from there almost to the Nebraska border in the northeast corner of Colorado. Leafy spurge also infests much of the Little Snake River, the Yampa River, and the White River ecosystems in northwest Colorado. Leafy spurge is known to decrease the carrying capacity of rangeland for livestock by 50%–75% (Lym and Messersmith 1985), and we anticipate that leafy-spurge-infested riparian areas also will experience decreased habitat and carrying capacity for wildlife and livestock. Many other invasive weeds infest riparian areas in Colorado including purple loosestrife (*Lythrum salicaria*), Canada thistle, perennial pepperweed (*Lepidium latifolium*), hoary cress (*Cardaria draba*), tamarisk, and Russian olive. Tamarisk and Russian olive are of great concern throughout the western United States, and in Colorado, tamarisk infests 11,450 ha while Russian olive infests 47,911 ha.

18.6.3 Climate and Allelopathy

Many hypotheses have been presented to explain the processes or mechanisms of invasion. The enemy release hypothesis (Mitchell and Power 2003; Torchin et al. 2003) and the evolution of increased competitive ability hypothesis (Blossey and Notzgold 1995) are two such examples. The novel weapons hypothesis or allelopathic advantage over resident species hypothesis (Bais et al. 2003; Callaway and Ridenour 2004) is yet another example and suggests that some invasive weeds can exude biochemicals into their surroundings that often inhibit the growth of neighboring plants and give the invader a competitive advantage.

Allelopathy is very difficult to demonstrate. Stevens (1986), however, conducted rigorous experiments that demonstrated that Russian knapweed exudes allelopathic compounds from its roots. He isolated several polyacetylene compounds from the roots of Russian knapweed and found that one inhibited the growth of lettuce (*Lactuca sativa*), alfalfa (*Medicago sativa*), barnyard grass (*Echinochloa crus-galli*), and red millet (*Panicum miliaceum*) in a laboratory bioassay. When he examined Russian knapweed under field conditions, he found 4–5 ppm of the most active polyacetylene compound in the rhizosphere of the weed during the growing season. Using results from the bioassay, he calculated that 4–5 ppm would cause a 30% root growth reduction in neighboring sensitive plant species.

Field observations in Colorado certainly indicate that in many but not all infestations, Russian knapweed develops into complete monocultures where absolutely no other plants are present. Goslee et al. (2001) used modeling to assess the importance of allelopathy and soil texture for the successful invasion by Russian knapweed on the shortgrass steppe of the High Plains of Colorado. Sensitivity of native species to allelopathy was an important factor in the model. During simulations at moderate sensitivity rates for native species, Russian knapweed became dominant faster and obtained a higher percentage of the total foliar biomass on fine-textured soils compared with coarse-textured soils. This implies that allelopathic effects of Russian knapweed are more pronounced in clay and clay loam soils, which possibly may be related to soil porosity.

In another modeling exercise, Goslee et al. (2003) also demonstrated that success of Russian knapweed invasion in Colorado was dependent upon precipitation, temperatures, and amount of clay-sized particles in the soil. Using county survey information, they identified 528 known locations of Russian knapweed throughout the state except for the northeastern plains and high mountain elevations. Stands often were clustered around reservoirs or in river valleys. Russian knapweed patches were found mostly in low to moderate precipitation zones (18–73 cm annually) and moderate to high mean annual temperatures (1°C–12°C). Many patches were found on fine-textured soils; for example, almost 40% of the stands were found on clay loam or clay soils even though these soil types comprise less than 16% of Colorado soils. Precipitation (mean annual, maximum monthly, minimum monthly, June and December precipitation) and temperature (mean annual, maximum monthly, and June maximum) were identified as important factors to predict the occurrence of Russian knapweed in Colorado. Overall, they found that Russian knapweed in Colorado was more likely to be found on sites with low June precipitation, low elevation, high percentage of clay-sized particles in the soil, and high December maximum temperatures. These environmental and climatic factors that influence Russian knapweed distribution appear to be related to its phenology and physiology. Russian knapweed is a C_3 forb that begins growth in early spring when the soil is no longer frozen (Watson 1980). Locations with warmer winter temperatures may foster Russian knapweed to grow earlier in the year than its competitors. Russian knapweed's deep root system allows it to thrive in dry areas (Selleck et al. 1964; Watson 1980), and this competitive attribute is supported by the survey and modeling results of Goslee et al. (2003). In areas with low spring precipitation, Russian knapweed may survive and outcompete annuals or shallow rooted perennials, and such an advantage may persist for the entire growing season.

The novel weapons hypothesis is quite attractive but difficult to demonstrate. Recently, Bais et al. (2002) examined root exudates of spotted knapweed, isolated with or without catechin from liquid cultures, and reported that spotted knapweed produced 83 $\mu g\ mL^{-1}$ under normal conditions. They also reported catechin to be stable in soil. In separate experiments, Weir et al. (2003) exposed seedlings of several different plant species to varying concentrations of catechin and found that Idaho fescue and prairie junegrass displayed 100% mortality at 50 $\mu g\ mL^{-1}$. Their results strongly suggested that catechin could play a significant role in the invasion success of spotted knapweed.

In other Colorado research, however, the role of catechin was brought into question. Blair et al. (2005) wanted to compare the production of catechin from North American spotted knapweed accessions to those from Europe as a first logical step to test the evolution of increased competitive ability hypothesis (Blossey and Notzgold 1999). They followed the procedures of Bais et al. (2002) and Weir et al. (2003) to extract catechin but were unable to duplicate their results. They then developed a new extraction technique because catechin was insoluble in the solvent that was used previously. The best extraction efficiency by Blair et al. (2005) was 94%. Extraction efficiencies were not reported by Bais et al. (2002) or Weir et al. (2003). Ultimately Blair et al. (2005) found that individual spotted knapweed plants produced 0–2.44 $\mu g\ mL^{-1}$ of catechin with an average of 0.29 $\mu g\ mL^{-1}$ compared with 83 $\mu g\ mL^{-1}$ found by Bais et al. (2002). Blair et al. (2005) found that catechin was especially unstable at pH levels greater than 5 and had a half-life of 0.5 h in water and 13 h in liquid growth media. They also found that root and shoot growth of

Idaho fescue was only slightly inhibited (18% and 27%, respectively) at catechin concentrations 20-fold greater (1000 $\mu g\ mL^{-1}$ compared with 50 $\mu g\ mL^{-1}$) than that previously reported, where the lower concentration caused 100% mortality (Weir et al. 2003).

In a follow-up study, Blair et al. (2006) extracted catechin from field soil collected from spotted knapweed infestations of varying age in Montana at five different times during the growing season. Their previous research allowed them to hypothesize that naturally occurring catechin produced by spotted knapweed would rapidly degrade in wet soils. They found 0–2.9 ppm of catechin in the soil with the greatest amount occurring in dry soils collected on August 1. When these soils were moistened, no catechin was detected 24 h later. They concluded that catechin is very unstable in moist soils and precipitation will never allow catechin to reach toxic levels in soils.

18.6.4 Management

18.6.4.1 Biological Control

Invasive weeds often infest very remote or rugged areas where they are difficult to control simply because they are hard to locate or the distance traveled to invoke control is so great that it is economically prohibitive to do so. Under such circumstances, classical biological control is an attractive alternative to other forms of weed control. Unfortunately, classical biocontrol efforts are not overly successful and historically, only about 30% of the efforts have been successful (McFayden 1998; Meyers 1984). For biological control to be a more useful means of controlling weeds, performance must be enhanced. *Brachypterolus pulicarius* (L.) (Coleoptera: Kateridae) is a univoltine beetle that was inadvertently introduced in North America and has since spread widely by natural and human means (MacKinnon et al. 2005). Adults feed on vegetative shoot tips and flowers of yellow and Dalmatian toadflax while larvae feed on inner floral parts including developing ovaries (Hervey 1927). *B. pulicarius* has decreased seed set of potted yellow toadflax by 90% (McClay 1992) and caused similar seed reductions in potted Dalmatian toadflax (Grubb et al. 2002). Although *B. pulicarius* is virtually ubiquitous on yellow toadflax, low beetle densities, local extinctions, and uncolonized patches of Dalmatian toadflax are commonly found even after redistribution efforts (Nowierski 1995).

Research was conducted in Colorado to determine whether beetle preference for yellow toadflax over Dalmatian toadflax could explain lower densities on the latter, to determine whether host-source influenced preference, and to determine whether *B. pulicarius* aggregated toward conspecifics. In field and laboratory experiments, MacKinnon et al. (2005) found that *B. pulicarius* taken from either Dalmatian or yellow toadflax preferred yellow toadflax. Beetles also preferred yellow toadflax plants where other *B. pulicarius* were already present, suggesting that beetles aggregate toward conspecific pheromones or volatiles given off by the host plant due to beetle activity. These researchers concluded that redistribution efforts onto Dalmatian toadflax currently underway in the United States are not worthwhile because of the beetles' preference for yellow toadflax.

In other Colorado research, experiments were conducted to determine how the process of biological control agent introduction might have influenced genetic variability in the leafy spurge gall midge (*Spurgia capitigena*). Scientists have long recognized that genetic variation within a biological control agent species may influence success or failure of biocontrol introductions (Messenger and van den Bosch 1971; Hoy 1985; Rousch 1990; Roderick and Navajas 2003). However, very few studies have addressed how the process of introduction may influence genetic variation of classical biological control agents. The leafy spurge gall midge was first introduced into North America in 1987; at that time, its scientific name was *Bayeria capitigena* Bremi (Solinas and Pecora 1984). It was renamed by Gagne in 1990 and included a new genus, *Spurgia*, which included two species, *S. esulae* that reportedly occurred exclusively on leafy spurge and *S. capitigena* that occurred only on cypress spurge (*Euphorbia cyrarissias*) (Gagne 1990). Lloyd et al. (2005) examined the variation in mitochondrial sequence and variation in microsatellite loci from introduced North American and native European populations of *S. capitigena* and how the introduction processes many have altered genetic variation. They found evidence to suggest that a mild bottleneck occurred in the introduced North American populations. When they examined the population structure of two European collections from leafy spurge and cypress spurge, they found evidence of local restricted gene flow between the two populations collected on the different weed species, but found no evidence to support the current classification into two distinct fly species. These researchers further addressed the issue of genetic diversity being important for successful control but warned that while additional introductions to increase genetic diversity might increase the fly's ability to adapt to local genotypes, additional introductions and subsequent increase in genetic variability also might cause fitness tradeoffs that could decrease their ability to adapt to host genotypes, which in turn could disrupt an already locally adapted system. Clearly, additional research will be necessary to better understand this situation.

18.6.4.2 Integrated Management and Restoration or Reclamation

Most weed management professionals and scientists agree that management of invasive weeds must incorporate several methods to be most successful (Walker and Buchanan 1982; Sheley et al. 1996; Lym 2005). Several experiments have been conducted in Colorado to assess integrated weed management approaches. Mowing of Canada thistle in combination with fall-applied herbicides was evaluated in subirrigated and upland pastures (Beck and Sebastian 2000). Although two or three mowings before spraying dicamba improved control compared with the herbicide applied alone at the upland site, 37% of the Canada thistle remained uncontrolled. Mowing before spraying did not improve control from picloram, picloram plus 2,4-D, chlorsulfuron, or clopyralid plus 2,4-D, except for the two lowest rates of the latter ($210 + 1{,}120$ and $280 + 1{,}460$ g ha^{-1}) at the subirrigated site. These researchers concluded that the occasional improved control of Canada thistle when mowing preceded herbicide application was inconsistent enough between locations and among herbicide treatments that such an integrated management system should not be commonly recommended.

Classical biological control is a good option for invasive weeds that are widespread or growing in remote and rugged locations. However, classical biological control has experienced limited success worldwide (McFayden 1998; Meyers 1984) and many experiments have been conducted to combine biological control with other weed control techniques to improve the outcome (Lym 2005). Diffuse knapweed is a biennial or short-lived perennial that infests over 1 million ha in the western United States (Roche and Roche 1991). Classical biological control agents have been imported into the United States since 1973 in an effort to decrease impacts caused by diffuse knapweed (Lang et al. 1998). The diffuse knapweed root beetle (*Sphenoptera jugoslavica*) is one such organism and is native to the steppes of Romania, Bulgaria, northern Greece, and northeastern Turkey (Harris and Shorthouse 1996). Larvae of this insect feed within diffuse knapweed roots causing decreased water and nutrient flow to shoots, which may decrease survival of rosettes, delay flowering, and decrease seed production (Powell and Meyers 1988). Under favorable conditions, the beetle can decrease the rate of diffuse knapweed population growth, but beetle populations fluctuate widely and their impact on diffuse knapweed is erratic (Scharer and Schroeder 1993). Control of diffuse knapweed from the root beetle has been best during hot, droughty years (Powell and Meyers 1988). Adult females lay eggs in the juncture of leaves and crowns of diffuse knapweed and after hatching, larvae bore into the crown. Diffuse knapweed must experience a growth arrest when eggs and larvae are present or else the growing weed will crush them. Wilson et al. (2004) established experiments in diffuse knapweed infestations in Colorado where *S. jugoslavica* had been released 4 years earlier to determine whether low rates of herbicides could be used to arrest the growth of diffuse knapweed when eggs and larvae are present and improve control by the beetle. Plots were treated with herbicides in 1998, and results were evaluated in 1999 and 2000. In spring, 1 year after herbicides were applied, picloram at 35, 70, and 140 g ha^{-1} and clopyralid at 35 g ha^{-1} increased the percentage of diffuse knapweed plants infested with the beetle by 25% compared with nonsprayed controls. No effects were observed 2 years after herbicides were applied, indicating that the results were temporary. Results indicate that low rates of picloram or clopyralid can stop growth and mimic the drought-induced growth arrest that diffuse knapweed commonly experiences in the beetle's native range, and possibly make control by the beetle more consistent in space and time.

Russian knapweed can be difficult to manage because of its allelopathic properties and its propensity to form dense monocultures (Watson 1980). Benz et al. (1999) found that suppressing Russian knapweed with herbicides in spring followed by seeding desirable perennial grasses in fall successfully controlled this aggressive invasive weed. Two years after treatments were initiated, clopyralid plus 2,4-D applied at $0.3 + 1.7$ kg ha^{-1} and combined with seeded grasses controlled 66%–92% of Russian knapweed while this same herbicide treatment without seeding controlled only 7% of this highly invasive weed. Mowing also was evaluated as a suppression treatment, and it was ineffective. The best treatment combination 2 years after they were invoked was clopyralid plus 2,4-D applied at $0.3 + 1.7$ kg ha^{-1} plus streambank wheatgrass (*Elymus lanceolatus*) where 92% of Russian knapweed was controlled and yield of streambank wheatgrass at this early stage in its establishment was 242 kg ha^{-1}.

Choice of grass species to sow when reclaiming land that has been infested with Russian knapweed is not an easy decision, again because of its allelopathic properties. Sensitivity of plant species to Russian knapweed allelopathy was identified as an important factor in its invasion success (Goslee et al. 2001) and most likely sensitivity will play a key role in successful reclamation of infested sites. Not all species will be suitable to use during reclamation or restoration efforts because some might be more sensitive to interacting with Russian knapweed than others. Grant et al. (2003) evaluated the germination and establishment of several rangeland grasses native to Colorado as influenced by Russian knapweed allelopathy. Seeds of different native perennial grasses were sown into pots in greenhouse experiments where they encountered roots of Russian knapweed, its litter, or roots plus litter, and compared with the same grass species without interference from Russian knapweed. Similar experiments were conducted under field conditions where seeds were sown either into Russian knapweed stands or in adjacent, noninfested, grass-dominated plant communities. Drip irrigation eliminated competition for moisture in the field experiment and similarly, pots were watered regularly in the greenhouse experiments for the same purpose. Emergence and survival were reduced in the presence of Russian knapweed roots in the greenhouse by 35% for prairie junegrass, 31% for blue grama, and 44% for sand dropseed. Emergence and survival of these grass species in the field experiment were reduced by 57% for prairie junegrass, 32% for blue grama, and 36% for sand dropseed. Western wheatgrass emergence and survival were unaffected by Russian knapweed in field and greenhouse experiments. The latter native grass species represents a good candidate to use when reclaiming Russian knapweed sites because western wheatgrass was not influenced by Russian knapweed interference.

Because reclamation and restoration efforts are difficult, time consuming, and often costly (Benz et al. 1999), a better understanding of the soil processes that aid succession will help to create better systems of invasive weed management. The important role that arbuscular mycorrhizal fungi (AMF) play in successful restoration has long been recognized (Reeves et al. 1979; Allen 1984, 1988; Miller 1987; St. John 1998). Availability and identity of AMF on a disturbed site will influence the relative success of plant species chosen for restoration (Allen 1984, 1988, 1995; Miller 1987). Rowe et al. (2007) examined the role of AMF during restoration of downy brome infested areas of Rocky Mountain National Park in Colorado. Field AMF inoculum soil was collected close to the research site in the park but away from downy brome infestations. This was compared with two commercial sources of inocula. The mycorrhizal infection rate of all inocula and their effects on plant growth were compared for six native plant species and downy brome. Native plants included four late-successional species—fringed sage (*Artemisia frigida*), smooth blue aster (*Aster laevis*), rabbitbrush (*Chrysothamnus viscidiflorus*), and mountain muhly (*Muhlenbergia montana*)—and three early-successional species—flatspine stickseed (*Lappula redowskii*), squirreltail (*Elymus elymoides*), and downy brome. Biomass production of all late-successional species increased with field soil inoculum while biomass of early-successional species was decreased. Plants were poorly colonized by commercial inocula and these did not positively influence plant growth. These researchers concluded that local field soil AMF inoculum would be preferential to use over commercial sources for establishing late-successional species. They also discovered that downy brome was not colonized by any inocula, which is very important to report so that land managers can use AMF treatment to enhance restoration where downy brome is problematic without concern that such treatment will benefit the invasive weed.

18.7 CONCLUDING REMARKS

Colorado has multiple infestations of invasive weed in forests, on rangeland, and in natural areas, but infestations rarely occur among trees in coniferous forests. Progress is being made to engage more public and private land managers and landowners in the battle against invasive weeds, albeit at a much slower rate than what many desire. Insufficient financial resources are always a problem regardless of the environmental issue, and the invasive weed problem in Colorado and elsewhere in the United States is no exception. Although considerable research has been conducted to better understand mechanisms and causes of invasions and how to better manage invasive weeds and restore infested sites in Colorado, much more work is needed. Here too, limited financial resources are a considerable impediment to improving our knowledge base. Only with time and relentless educational efforts will we overcome the financial deficit to better understand and manage invasive weeds and foster complete involvement and cooperation of the people of Colorado in this effort, much less elsewhere in the United States and internationally.

REFERENCES

Allen, E.B., VA mycorrhizae and colonizing annuals: Implications for growth, competition, and succession, in *VA Mycorrhizae and Reclamation of Arid and Semi-Arid Lands*, Williams, S.E. and Allen, M.F., Eds., University of Wyoming Experiment Station, Laramie, WY, 42, 1984.

Allen, E.B., Some trajectories of succession in Wyoming sagebrush grassland: Implications for restoration, in *The Reconstruction of Disturbed Lands: An Ecological Approach*, Allen, E.B., Ed., Westview Press, Boulder, CO, 89, 1988.

Allen, E.B., Mycorrhizal limits to rangeland restoration: Soil phosphorous and soil species composition, in *Rangelands in a Sustainable Biosphere. Proceedings of the V International Rangeland Congress*, Salt Lake City, July 1995, West, N.E., Ed., Society for Range Management, Denver, CO, 57, 1995.

Arno, S.F., Ecosystem-management at lower elevations, in *Proceedings of the Bitterroot Ecosystem Management Research Project: What We Have Learned*, Smith, H.Y., Ed., General Technical Report RMRS-GTR-17, U.S. Department of Agriculture, Forest Service, Rocky Mountain Research Station, Flagstaff, AZ, 17, 2000a.

Arno, S.F., Fire in western forest ecosystems, in *Wildland Fires in Ecosystems, Effects on Flora*, Brown, J.K. and Smith, J.K., Eds., General Technical Report, RMRS-GTR-42-Volume 2, U.S. Department of Agriculture, Forest Service, Rocky Mountain Research Station, Flagstaff, AZ, 97, 2000b.

Bais, H.P., Walker, T.S., Stermitz, F.R., Hufbauer, R.A., and Vivanco, J.M., Enantiomeric-dependent phytotoxic and anti-microbial activity of (±) catechin, a rhizosecreted, racemic mixture from spotted knapweed, *Plant Physiol.*, 128, 1173, 2002.

Bais, H.P., Vepachedu, R., Gilroy, S., Callaway, R.M., and Vivanco, J.M., Allelopathy and exotic plant invasions: From molecules and genes to species interactions, *Science*, 301, 1377, 2003.

Bazaaz, F.A., The physiological ecology of plant succession, *Annu. Rev. Ecol. Syst.*, 10, 351, 1979.

Beck, K.G. and Sebastian, J.R., Combining mowing and fall-applied herbicides to control Canada thistle (*Cirsium arvense*), *Weed Technol.*, 14, 351, 2000.

Benz, L.J., Beck, K.G., Whitson, T.D., and Koch, D.W., Reclaiming Russian knapweed infested rangeland, *J. Range Manage.*, 52, 351, 1999.

Blair, A.C., Hanson, B.D., Brunk, G.R., Marrs, R.A., Westra, P., Nissen, S.J., and Hufbauer, R.A., New techniques and findings in the study of a candidate allelochemical implicated in invasion success, *Ecol. Lett.*, 8, 1039, 2005.

Blair, A.C., Nissen, S.J., Brunk, G.R., and Hufbauer, R.A., Lack of evidence for an ecological role of the putative allelochemical (±)-catechin in spotted knapweed invasion success, *J. Chem. Ecol.*, 32, 2327, 2006.

Blossey, B. and Notzgold, R., Evolution of increased competitive ability in nonindigenous plants: A hypothesis, *J. Ecol.*, 83, 887, 1995.

Callaway, R.M. and Ridenour, W.M., Novel weapons: Invasive success and the evolution of increased competitive ability, *Front. Ecol. Environ.*, 2, 436, 2004.

Chapman, S.S., Griffith, G.E., Omernik, J.M., Price, A.B., Freeouf, J., and Schrupp, D.L., Ecoregions of Colorado (color poster with map, descriptive text, summary tables, and photographs), U.S. Geological Survey (map scale 1:1,200,000), Reston, VA, 2006.

Coffin, D.P., Lauenroth, W.K., and Burke, I.C., Recovery of vegetation in a semi-arid grassland 53 years after disturbance, *Ecol. Appl.*, 6, 538, 1996.

Colorado State Forest Service, Report on the condition of Colorado's forests, http://csfs.colostate.edu/CO.htm, 2001.

Colorado State Forest Service, Report on the health of Colorado's forests, http://csfscolostate.edu/CO.htm, 2003.

Colorado State Forest Service, Report on the health of Colorado's forests, http://csfs.colostate.edu/CO.htm, 2005.

Doesken, N.J., Pielke, R.A., Sr., and Bliss, O.A.P., Climate of Colorado, Climatology of the United States No. 60, Department of Atmospheric Sciences, Colorado State University, Fort Collins, CO, http://ccc.atmos.colostate.edu, 2003.

Gagne, R.J., Gall midge complex (Diptera: Cecidomyiidae) in bud galls of Palearctic *Euphorbia* (Euphorbiaceae), *Ann. Entomol. Soc. Am.*, 83, 335, 1990.

Goslee, S.C., Beck, K.G., and Peters, D.P.C., Distribution of Russian knapweed in Colorado: Climate and environmental factors, *J. Range Manage.*, 56, 206, 2003.

Goslee, S.C., Peters, D.P.C., and Beck, K.G., Modelling invasive weeds in grasslands: The role of allelopathy in *Acroptilon repens* invasion, *Ecol. Model.*, 139, 31, 2001.

Grant, D.W., Peters, D.P.C., Beck, K.G., and Fraleigh, H.D., Influence of an exotic species, *Acroptilon repens*, on seedling emergence and growth of native grasses, *Plant Ecol.*, 166, 157, 2003.

Grubb, R.T., Nowierski, R.M., and Sheley, R.L., Effects of *Brachypterolus pulicarius* (L.) (Coleoptera: Nitidulidae) on growth and seed production of Dalmatian toadflax *Linaria genistifolia* ssp. *dalmatica* (L.) Maire and Petitmengin (Scrophulariaceae), *Biol. Control*, 23, 107, 2002.

Harris, P. and Shorthouse, J.D., Effectiveness of gall inducers in weed biological control, *Can. Entomol.*, 128, 1021, 1996.

Hartman, R.L. and Nelson, B.E., A checklist of the vascular plants of Colorado, Laramie, WY Rocky Mountain Herbarium, Department of Botany, University of Wyoming, Laramie, WY, http://www.rmh.uwyo.edu, 2001.

Hervey, G.R., A European nitidulid, *Brachypterolus pulicarius* L. (Coleoptera: family Nitidulidae), *J. Econ. Entomol.*, 20, 809, 1927.

Hoy, M.A., Improving the establishment of arthropod natural enemies, in *Biological Control in Agricultural IPM Systems*, Hoy, M.A. and Herzog, D.C., Eds., Academic Press, Orlando, FL, 151, 1985.

Lane, E.M., An assessment of federal and state agency weed management efforts in Colorado, a report to the Colorado General Assembly, Colorado Department of the Interior, Littleton, CO, http://www.colorado.gov/cs/satellite/agriculture-main/CDAG/1167928174125, 1999.

Lang, R.F., Piper, G.L., and Coombs, E.M., Establishment and re-distribution of *Sphenoptera jugoslavica* Obenberger (Coleoptera: Buprestidae) for biological control of *Centaurea diffusa* in the midwestern and western United States, *Pan-Pac. Entomol.*, 74, 27, 1998.

Lauenroth, W.K. and Milchunas, D.G., Short-grass steppe, in *Ecosystems of the World 8A: Natural Grasslands—Introduction and Western Hemisphere*, Coupland, R.T., Ed., Elsevier, Amsterdam, 183, 1992.

Laven, R.D., Omni, P.N., Wyant, J.G., and Pinkerton, A.S., Interpretation of fire scar data from a ponderosa pine ecosystem in the central Rocky Mountains Colorado, in *Proceedings of the Fire History Workshop*, Stokes, M.A. and Dieterich, J.H., Eds., General Technical Report RM-81, U.S. Department of Agriculture, Forest Service, Rocky Mountain Forest and Range Experiment Station, 46, 1980.

Lloyd, C.J., Hufbauer, R.A., Jackson, A., Nissen, S.J., and Norton, A.P., Pre- and post-introduction patterns in neutral genetic diversity in the leafy spurge gall midge, *Spurgia capitigena* (Bremi) (Diptera: Cecidomyiidae), *Biol. Control*, 33, 153, 2005.

Lym, R.G., Science and decision making in biological control of weeds: Benefits and risks of biological control, *Biol. Control*, 35, 3, 366, 2005.

Lym, R.G. and Messersmith, C.G., Cost effectiveness of leafy spurge control during a five-year management program, *N.D. Farm Res.*, 43, 1, 7, 1985.

MacKinnon, D.K., Hufbauer, R.A., and Norton, A.P., Host-plant preference of *Brachypterolus pulicarius*, an inadvertently introduced biological control insect of toadflaxes, *Entomol. Exp. Appl.*, 116, 183, 2005.

Matson, P.A. and Boone, R.D., Natural disturbance and nitrogen mineralization: Wave-form dieback of mountain hemlock in the Oregon Cascades, *Ecology*, 65, 1511, 1984.

Matson, P.A. and Vitousek, P.M., Nitrification potentials following clearcutting in the Hoosier national forest in Indiana, *For. Sci.*, 27, 781, 1981.

McClay, A.S., Effects of *Brachypterolus pulicarius* (L.) (Coleoptera: Nitidulidae) on flowering and seed production of common toadflax, *Can. Entomol.*, 124, 631, 1992.

McClendon, T. and Redente, E.F., Nitrogen and phosphorous effects on secondary succession dynamics on a semi-arid sagebrush site, *Ecology*, 72, 6, 2016, 1991.

McClendon, T. and Redente, E.F., Effects of nitrogen limitation on species replacement dynamics during early secondary succession on a semiarid sagebrush site, *Oecologia*, 91, 312, 1992.

McFayden, R.E.C., Biological control of weeds, *Ann. Rev. Entomol.*, 43, 369, 1998.

Messenger, P.S. and van den Bosch, R., The adaptability of introduced biological control agents, in *Biological Control*, Huffaker, C.B., Ed., Plenum, New York, 68, 1971.

Meyers, J.H., How many insect species are necessary for successful biocontrol of weeds? in *Proceedings of VI International Symposium on Biological Control of Weeds*, Delfosse, E., Ed., Agriculture Canada, Ottawa, 77, 1984.

Miller, R.M., Mycorrhizae and succession, in *Restoration Ecology: A Synthetic Approach to Ecological Research*, Jordan, W.R., III, Gilpin, M.E., and Aber, J.D., Eds., Cambridge University Press, New York, NY, 205, 1987.

Mitchell, C.E. and Power, A.G., Release of invasive plants by fungal and viral pathogens, *Nature*, 421, 625, 2003.

Nowierski, R.M., Dalmatian toadflax *Linaria genistifolia* ssp. *dalmatica* L. Maire and Petitmengin (Scrophulariaceae), in *Biological Control in the Western United States Region: Accomplishments and Benefits of Regional Research Project W84 (1964–1989)*, Andres, L.A., Beardsley, J.W., Goeden, R.D., and Jackson, G., Eds., University of California Agricultural and Natural Resource publication No. 3361, University of California Press, Berkeley, CA, 312, 1995.

Paschke, M.W., McClendon, T., and Redente, E.F., Nitrogen availability and old-field succession on a shortgrass steppe, *Ecosystems*, 3, 144, 2000.

Powell, R. and Meyers, J., The effect of *Sphenoptera jugoslavica* on its host plant diffuse knapweed, *J. Appl. Entomol.*, 106, 25, 1988.

Reever Morghan, K.J. and Seastedt, T.R., Effects of soil nitrogen reduction on non-native plants in restored grasslands, *Restor. Ecol.*, 7, 1, 51, 1999.

Reeves, F.B., Wagner, D., Moorman, T., and Kiel, J., The role of endomycorrhizae in revegetation practices in the semi-arid west. A comparison of incidence of mycorrhizae in severely disturbed vs. natural environments, *Am. J. Bot.*, 66, 6, 1979.

Reichardt, K.L., Succession of abandoned fields on the shortgrass prairie, Northeastern Colorado, *Southwest. Nat.*, 27, 299, 1982.

Rocchio, J., Rocky mountain lower montane riparian woodland and shrubland ecological system, ecological integrity assessment, Colorado Natural Heritage Program, Colorado State University, Fort Collins, CO, http://cnhp.colostate.edu/reports.html, 2006.

Roche, B. and Roche, C., Identification, introduction, distribution, ecology, and economics of *Centaurea* species, in *Noxious Rangeland Weeds*, James, L.F., Evans, J.O., Ralphs, M.H., and Child, R.D., Eds., Westview Press, Boulder, CO, 274, 1991.

Roderick, G.K. and Navajas, M., Genes in new environments: Genetics and evolution in biological control, *Nat. Rev. Genet.*, 4, 889, 2003.

Rogers, P., Atkins, D., Frank, M., and Parker, D., Forest health monitoring in the Interior West, General Technical Report RMRS-GTR-75, U.S. Department of Agriculture, Forest Service, Rocky Mountain Research Station, Flagstaff, AZ, 2001.

Rousch, R.T., Genetic considerations in the propagation of entomophagus species, in *New Directions in Biological Control, Alternatives for Suppressing Agricultural Pests and Diseases*, Baker, R.R. and Dunn, P.E., Eds., Alan R. Liss, New York, 373, 1990.

Rowe, H.I., Brown, C.S., and Claassen, V.P., Comparison of mycorrhizal responsiveness with field soil and commercial inoculum for six native montane species and *Bromus tectorum*, *Restor. Ecol.*, 15, 44, 2007.

Scharer, H.M. and Schroeder, D., The biological control of *Centaurea* species in North America: Do insects solve the problem? *Pestic. Sci.*, 37, 343, 1993.

Selleck, G.W., A competition study of *Cardaria* spp. and *Centaurea repens*, *Proc. Seventh Br. Weed Control Conf.*, 7, 569, 1964.

Sheley, R.L., Svejcar, T.J., and Maxwell, B.D., A theoretical framework for developing successional weed management strategies on rangeland, *Weed Technol.*, 10, 766, 1996.

Solinas, M. and Pecora, P., The midge complex (Diptera: Cecidomyiidae) on *Euphorbia* species I., *Entomologica*, 19, 167, 1984.

St. John, T., Mycorrhizal inoculation in habitat restoration, *Land Water*, 42, 17, 1998.

Stevens, K.L., Allelopathic polyacetylenes from *Centaurea repens* (Russian knapweed), *J. Chem. Ecol.*, 12, 6, 1205, 1986.

Stohlgren, T.J., Binkley, D., Chong, G.W., Kalkhan, M.A., Schell, L.D., Bull, K.A., Otsuki, Y., Newman, G., Bashkin, M., and Son, Y., Exotic plant species invade hot spots of native plant diversity, *Ecol. Monogr.*, 69, 1, 25, 1999a.

Stohlgren, T.J., Schell, L.D., and Vanden Heuvel, B., How grazing and soil quality affect native and exotic plant diversity in Rocky Mountain grasslands, *Ecol. Appl.*, 9, 1, 45, 1999b.

Sutherland, S., Fuels planning: Science synthesis and integration; environmental consequences fact sheet 07: Fire and weeds, Research Note RMRS-RN-23-7WWW, U.S. Department of Agriculture, Forest Service, Rocky Mountain Research Station, Flagstaff, AZ, 2004.

Sutton, J.R., Prediction and characterization of yellow toadflax (*Linaria vulgaris* Mill.) infestations at two scales in the Flat Tops Wilderness of Colorado, M.S. Thesis, Colorado State University, Fort Collins, CO, 2003.

Torchin, M.E., Lafferty, K.D., Dobson, A.P., McKenzie, V.J., and Kuris, A.M., Introduced species and their missing parasites, *Nature*, 421, 628, 2003.

Walker, R.H. and Buchanan, G.A., Crop manipulations in integrated weed management systems, *Weed Sci.*, 30, 17, 1982.

Watson, A.K., The biology of Canadian weeds, 43, *Acroptilon (Centaurea) repens* (L.) D.C., *Can. J. Plant Sci.*, 60, 993, 1980.

Weir, T.L., Bais, H.P., and Vivanco, J.M., Intraspecific and interspecific interactions mediated by a phytotoxin (−) catechin secreted by the roots of *Centaurea maculosa* (spotted knapweed), *J. Chem. Ecol.*, 29, 2397, 2003.

Wilson, R., Beck, K.G., and Westra, P., Combined effects of herbicides and *Sphenoptera jugoslavica* on diffuse knapweed (*Centaurea diffusa*) population dynamics, *Weed Sci.*, 52, 418, 2004.

Ziska, L.H., Evaluation of the growth response of six invasive species to past, present, and future atmospheric carbon dioxide, *J. Exp. Bot.*, 54, 381, 395, 2003.

19 Ecology and Management of Tropical Africa's Forest Invaders

Paul P. Bosu, Mary M. Apetorgbor, and Alemayehu Refera

CONTENTS

19.1 INTRODUCTION

The introduction of alien plant species to tropical Africa dates as far back as the fifteenth century when the first Europeans arrived on the shores of the continent. Many of the species that were intentionally or unintentionally introduced today constitute a major proportion of the food, fibre, and wood resources on the continent

(Juhé-Beaulaton 1994; Wild 1968). But not all the introduced species were beneficial, as quite a number of them turned out to be noxious. Indeed, active and passive introductions of new species to the continent of Africa and around the world have continued throughout the centuries, and will continue for many more to come. There is a common belief that the twenty-first century will experience the most introductions of alien species, both plants and animals, to nonnative habitats as a result of globalization and increased international trade. The continent of Africa is both a major donor and recipient of alien genetic resources, which sooner or later become unwanted or invasive in their new environments.

Over the past several decades, many have written on the subject of invasive plant species, from regional or subregional perspectives (Harris 1962; Duffey 1964; Spongberg 1990; Corlett 1992; Strahm 1993). A few reviews or proceedings on a more or less broader perspective have also been published (Duffey 1988; Drake et al. 1989; Ramakrishnan 1991; Whitmore 1991; Pyšek et al. 1995; Binggeli et al. 1998). In spite of these, the literature on forest invasive plant species of Africa is very limited indeed. With the exception of perhaps the Republic of South Africa, most of the countries of Africa lack basic information on the distribution and ecology of some of the invasive species that are found within their borders. A few species with global or continental range have been studied in detail and the information on them can be readily accessed. Otherwise, the presence or absence of any particular invasive or potentially invasive plant in a country is often all the information that one is likely to obtain from many tropical African countries.

19.2 TROPICAL AFRICA REGION

The climate of the continent of Africa is generally stable, but not all countries in Africa are strictly tropical. By definition, tropical Africa includes countries within latitude 10° N (Tropic of Cancer) and latitude 10° S (Tropic of Capricorn) of the equator. This definition therefore eliminates the Republic of South Africa and several of its neighboring countries including Namibia, Bostwana, and Zimbabwe, from the scope of this chapter. Several North African countries including Egypt, Libya, and Morocco are also effectively outside the scope of our discussion. However, a good number of invasive plant species have a broad climatic range and occur in tropical, subtropical, and temperate zones on the continent. This makes it practically impossible to discuss the topic without substantial inputs and reference from African countries outside of the tropical zone.

Areas near the equator and the eastern coast of Madagascar have tropical rainforest climate characterized by heavy rains and high temperatures. Average annual rainfall and temperature are 1780 mm and 26.7°C, respectively. This area is bordered to the north and south by the tropical savannah climate, which is characterized by high temperatures year-round, a wet season during the summer months and dry season during the winter months. Total annual rainfall varies from 550 to 1550 mm. The savannah climate zone grades into the semiarid steppe climate zone, and then into the dry conditions of the arid desert zone. In the highland regions of eastern Africa, rainfall is well distributed throughout the year, and temperatures are quite equable.

19.3 INVASIVE PLANTS OF TROPICAL AFRICA

Many invasive plant species have been recorded in all the major climatic zones of tropical Africa (Table 19.1). The list of species provided in the table is by no means exhaustive. It provides a snapshot of accessible information. *Chromolaena odorata* and *Lantana camara* are by far the most widespread invasive plants in the tropical rainforest zone. These two species are discussed in detail later in the chapter. *Cercropia peltata* is another important invasive plant in the tropical rainforest zone. *C. peltata* was deliberately introduced to the rainforest region of Africa almost a century ago, and is currently one of the most important nonnative invasive tree species in the rainforest of Africa. In Côte d'Ivoire, Cameroon, and the Democratic Republic of Congo, where *C. peltata* was originally introduced, it has been spreading and competing with native species in disturbed forest areas (Binggeli et al. 1998). Another plant of considerable invasive potential, though of relatively recent introduction to the rainforest region, is paper mulberry (*Broussonetia papyrifera*). This species is also discussed in detail later in the chapter. In the savannah and semiarid zones of eastern Africa, *Prosopis juliflora*, *Acacia mearnsii*, *Maesopsis eminii*, and *Senna spectabilis* are among the most troublesome invasives.

A. mearnsii is an important economic species grown for its durable timber used as building material, and for a variety of other uses such as charcoal, pulp, and wood chip production. Bark extracts from the plant are also useful for making resins, thinners, and adhesives. It is native to Australia but now currently distributed the world over. *A. mearnsii* is planted widely in the southern and eastern African regions, where it is now showing invasive characteristics in Tanzania (Chilima et al. 2005). The invasiveness of this species in the Republic of South Africa is already firmly established. Its invasiveness is in part due to its ability to produce large quantities of seeds, formation of large crown, and perceived allelopathic properties (Adair 2002).

S. spectabilis is a medium-to-large tree of tropical American origin, but now widely distributed in many tropical countries. It is moderately invasive in Uganda, Tanzania, and several east African countries. It is believed to have been introduced into East Africa in the colonial era, or by Indian sawmill operators. In Uganda it is regarded as a major problem in the Budongo Forest Reserve (Chilima et al. 2005), where it colonizes vast areas along logging trails and landing sites. The invasive nature of *S. spectabilis* was observed in the Mahale Mountains National Park in western Tanzania sometime in the 1970s, where it suppresses recruitment of native species (Wakibara and Mnaya 2002). Control trials using girdling and felling resulted in a marked decline in plots where control methods were applied, with a marked increase in the native species. A closely related species *Senna* [*Cassia*] *siamea* is widely planted in Ghana for shade and fuelwood, but *S. siamea* is not considered invasive in Ghana.

M. eminii was introduced to East Africa in the early twentieth century (Binggeli et al. 1998) but was not considered as invasive until the late 1970s. It is a fast-growing forest tree and often cultivated in monoculture plantations. Originally introduced to the Amani Botanic Gardens in northeastern Tanzania, *M. eminii* has spread to areas outside of the region, facilitated by widespread planting of the species for reforestation and its natural ability to regenerate in disturbed natural forests (Binggeli and Hamilton

TABLE 19.1
Major Forest Invasive Plants of Tropical Africa

Species	Common Name(s)	Family	Life Form	Country or Region	DI
Acacia hockii De Willd.	Shitim wood	Leguminosae	Small tree	Uganda	1
Acacia mearnsii De Wild.	Black wattle	Mimosaceae	Tree	Tanzania	2
Broussonetia papyrifera (L.) Vent.	Paper mulberry	Moraceae	Tree	Ghana, Uganda	3
Castilla elastica Cerv.	Mexican rubber tree	Moraceae	Small tree	Tanzania	2
Chromolaeana odorata (L.) King & Robinson	Siam weed	Asteraceae	Shrub	West and Central Africa	3
Cedrela odorata L.	Spanish cedar	Meliaceae	Tree	Tanzania, Ghana	1
Cercropia peltata L.	Pumpwood	Cercropiaceae	Tree	Cote d'Ivoire, Cameroon, Zaire	3
Cordia alliodora (Ruiz & Pav.) Oken	Spanish elm	Boraginaceae	Tree	Tanzania	1
Eucalyptus terreticornis Sm.	Forest red gum	Myrtaceae	Tree	Malawi, East Africa	2
Lantana camara L.	Lantana	Verbenaceae	Shrub	Western, Central, and Eastern Africa	3
Leucaena leucocephala (Lam.) De Wit	Leuceaena	Mimosaceae	Small tree	Ghana, Kenya	1
Maesopsis eminii Engl.	Umbrella tree	Rhamnaceae	Tree	Tanzania, East Africa	3
Mimosa pigra L.	Mimosa	Mimosaceae	Small tree	Uganda	2
Pinus patula Schiede & Deppa	Patula pine	Pinaceae	Tree	Malawi, East Africa	2
Prosopis justiflora (Sw.) DC.	Prosopis	Mimosaceae	Small tree	East Africa	3
Rubus ellipticus Smith	Yellow Himalayan raspberry	Rosaceae	Shrub	Tanzania	2
Senna spectabilis DC.	*Cassia*	Caesalpinaceae	Small tree	Uganda, Tanzania, East Africa	2

Note: DI, Degree of invasiveness as used by Binggeli et al. (1998); 1, possibly or potentially invasive; 2, moderately invasive; 3, highly invasive.

1993; Hall 1995). *M. eminii* seedlings are shade tolerant and therefore survive under forest canopy; however, saplings and mature individuals require full sunlight. The species can thus grow to canopy height in large tree-fall gaps (Binggeli and Hamilton 1993). In Tanzania, *M. eminii* invasion and impact on endemic species are affected by heavy human disturbance and climate change in the past several years.

Several other species of considerable invasive potential in other parts of the world, but to a lesser degree in tropical Africa also occur. These include *Mimosa pigra*, a major invasive in northern Australia and listed as one of the world's 100 worst invasive alien species, which now occurs in Uganda (Kiwuso and Otara 2007). *Rubus ellipticus*, an extremely thorny shrub from southern Asia and now widely distributed throughout the world, is recorded in Malawi and may well be present in other east African countries (Chilima 2007). *Castilla elastica* and *Cordia alliodora* are species with invasive characteristics in Tanzania (Maddofe and Mbwambo 2007).

A number of tree species of relatively recent introduction have become moderately or potentially invasive in many parts of tropical Africa. Among these are *Leuceana leucocephala*, *Cedrela odorata*, and *C. neomexicana*. As far as invasive plants of tropical Africa are concerned, *L. leucocephala* occupies an interesting position. Native to Mexico and Central America, this tree has been widely promoted for agroforestry and reforestation throughout world. However, it is now found spreading naturally in monospecific thickets in disturbed forest sites and riparian zones, threatening native species. For this reason, *L. leucocephala* is sometimes described as a "conflict tree" (Neser 1994) and is likely to receive attention as a major forest invasive of tropical Africa in the not too distant future. The species is now present in all the major vegetation zones in Ghana and can be considered noxious in a number of habitats and localities. It competes with crops on farmlands and is difficult to control once established. It has also been observed in unusual proportions along roadsides and other landscapes in Kenya (Mwangi 2007).

C. odorata is a member of the Meliaceae family and a native of the West Indies and Central America. It is planted for its highly valued timber and grows up to 40 m high. It is currently planted as a major plantation species in Ghana, but the species is spreading naturally in some forest reserves, with potential to become a major invasive plant of forest ecosystems in some years to come. Its closely related species, *C. neomexicana*, is reported to be threatening native species regeneration in forest ecosystems in Tanzania (Chilima et al. 2005). In Malawi, *Eucalyptus territicornis* and *Pinus patula* have been reported (Chilima 2007).

19.4 SELECTED HIGHLY INVASIVE SPECIES OF TROPICAL AFRICA

In this section, we discuss in detail four of the most troublesome invasive plants of tropical Africa, namely *Chromolaena odorata*, *Lantana camara*, *Broussonetia payrifera*, and *Prosopis julifolia*.

19.4.1 *Chromolaena odorata* L. King & Robinson Asteraceae

C. odorata (*Eupatorium odoratum* L.), commonly called siam weed is a herbaceous perennial of tropical American origin. *C. odorata* grows in dense stands, reaching

FIGURE 19.1 *Chromolaena odorata* invasion of an abandoned farmland in the moist semi-deciduous forest zone in the southern part of Ghana.

a height of 1.5–2.5 m in many tropical, subtropical, and temperate environments. It is ranked among the most highly invasive species in the tropical regions of Africa and South Asia, and listed among the worst 100 invasive species of the world (Figure 19.1). Young shoots are green and fleshy, grow vigorously, and branch freely. Older stems are brown and woody, especially at the base. The root system is fibrous and shallow and may reach a depth of between 20 and 30 cm in most cases. When growing conditions are right, siam weed can climb nearby vegetation with long rambling branches, reaching a height of between 6 and 10 m (Vanderwoude et al. 2005).

The flowers are white or pale bluish-lilac (Cruttwell McFadyen 1989) and are pollinated by insects. The fruits are small in size, each about 3–5 mm long, ~1 mm wide, and weigh only about 0.2 mg (Vanderwoude et al. 2005). The plant produces copious amount of seeds, which facilitates its dispersal. An estimated 260,000 m^2 of seeds has been reported, with almost 46% viability (Witkowski 2002; Witkowski and Wilson 2001, in Vanderwoude et al. 2005). Fruits are dispersed largely by wind; however, long-distance dispersal to new areas occurs usually by animals or due to anthropogenic causes. Small hooks on the fruits make dispersal by animals both possible and efficient. Percent germination is higher when seeds are buried than on the soil surface, and seeds may remain viable for up to 1 year. *C. odorata* coppices readily from the root crown or stems after fire or death of old stumps, but reproduction by vegetative means does not occur.

19.4.1.1 Habitat and Site Requirements

Like most invasive plants, *C. odorata* thrives in a wide range of soil and vegetation types. The plant grows well in humid tropics, grassland areas, and on a wide range of ecological zones wherever annual rainfall exceeds 1200 mm. Relative humidity of 60%–70% and temperature of 30°C favor good growth of seedlings, but deep canopy shade does not favor growth or seed production (Feleke 2003; Luwum 2002 in Vanderwoude et al. 2005), which explains why the plant is often found in open-canopy gaps, degraded forest sites, abandoned farmlands, open areas, roadsides, and in tree and cash crop plantations. It is highly nutrient demanding (Witkowski 2002) and readily takes advantage of the flush of soil N that becomes available after fire or land clearing for agriculture to flourish (Saxena and Ramakrishnana 1983 in Vanderwoude et al. 2005).

It is believed that *C. odorata* was introduced accidentally to Nigeria, West Africa, in 1947 through seeds of *Gmelina arborea* imported into the country from Sri Lanka (Binggeli et al. 1998). It became a major weed in the country by late 1960s, and gradually spread to the neighboring West African countries of Benin, Togo, and Ghana (Timbilla and Braimah 1994). Though dispersal is by wind, dispersal over long distances occurs through heavy-duty machinery, timber-hauling trucks, light vehicles, etc. Short hooks on the seeds easily cling to clothes, shoes, or gear and are carried to distant locations. The species was, however, introduced deliberately to Cote d'Ivoire, ostensibly to suppress the growth of *Imperata cylindrica* and other obnoxious grasses (Chevalier 1952). The Global Invasive Species database (http://www.issg.org/database/welcome/) lists over 15 African countries where *C. odorata* has been recorded, including other West African countries like Guinea, Sierra Leone, and Liberia. In other parts of Africa, *C. odorata* occurs in Cameroon, The Democratic Republic of Congo, Gabon, Mauritius, and South Africa.

19.4.1.2 Impact of *Chromolaena odorata*

C. odorata forms dense stands in degraded forest gaps, abandoned farmlands, and open areas, and prevents establishment of indigenous species. It competes aggressively with native species due to its allelopathic properties and prevents the regeneration and establishment of native species. It is a major weed in plantations and croplands in tropical Africa, and significantly affects the management and economic returns on plantations of oil palm, citrus, cocoa, coffee, and other trees. A survey in Ghana in 1991 indicated that *C. odorata* covers over two-thirds of the land area of Ghana, being present in all ecological zones except the Sudan savannah (Timbilla and Braimah 1994). The most negatively affected areas include the high rainforest and the semi-deciduous forest zones in the southern half of the country. Apart from its ability to suppress native species regeneration, *C. odorata* is a significant fire hazard during the dry season, and is believed to have greatly facilitated the bushfires experienced in Ghana in the early 1980s (Entsie 2002). In addition, *C. odorata* stands are known to also serve as refugia for the variegated grasshopper (*Zonocerus variegatus*) and other harmful pests, indirectly affecting agricultural production.

In spite of the above negative effects, *C. odorata* also has some positive effects. In West Africa, there are claims that the presence of *C. odorata* impacts positively on the production of food crops such as maize, groundnut (peanuts), and cassava, which can lead to yields sometimes comparable to soils with mineral fertilizers

(Tie Bi 1995). Though such claims are quite popular among farmers in West Africa, evidence is only anecdotal. There are also claims that the introduction of *C. odorata* to West Africa has resulted in a reduction of fallow periods from an average of about 10 years in the past to 3 years presently. This may be true; however, it is also true that croplands today lose their ability to sustain continuous cropping much faster (1–2 years) than in the past when they could sustain optimum yields for up to 5 years or more. Other benefits of *C. odorata* include its medicinal use for curing various ailments such as eye and stomach diseases. It is also used for dressing of fresh wounds because of its inherent ability to stop excessive bleeding.

19.4.1.3 Management of *Chromolaena odorata*

Control strategies using mechanical, chemical, and biological approaches have been pursued. In West Africa, mechanical control by slashing with cutlasses (machetes) is very common, especially during the initial stages of land preparation for farming or plantation establishment. Slashing is followed by burning, usually after a few days to allow the brush to dry. The slashed stumps coppice profusely after the fire and are usually uprooted with hoe or mattock. This form of management does not result in total control or eradication but is only helpful for a while, offering a short-term solution by mitigating its impact on food crops. Frequently, the very large seed bank will germinate to eventually replace the mature stands that have been removed. Many West African farmers use a combination of slashing, burning, and herbicides to suppress the weed. The use of herbicides is, however, of major concern because of the environmental and health hazards posed by the indiscriminate use of these chemicals.

Biological control efforts using the arctid moth *Pareuchaetes pseudoinsulata* (Lepidoptera: Arctiidae) have been pursued in West Africa with partial success. If established, the insects effectively defoliate pure stands, opening the canopy to allow native species to regenerate. For example, regeneration of native shrubs and grasses in a treated site less than 1 year after release of *P. pseudoinsulata* was 68%, compared with 15% in control sites (Timbilla 1996). Biocontrol has been pursued vigorously in Guam, South Africa, and several other countries, but it has generally not been successful.

19.4.2 *Lantana camara* L. Verbanaceae

L. camara is a weed of considerable importance in tropical and subtropical ecosystems of Africa and other parts of the world (Thomas and Ellison 2000). It is a highly variable shrub native to tropical America and known by many different names. It is an artificial hybrid that has undergone extensive hybridization over the course of about 300 years. As a result, it now occurs in many hundreds of forms, and distributed in well over 60 countries and island groups in tropical and subtropical regions. Lantana grows erect in the open, up to a height of about 2 m and scrambling in scrubland, but it can grow to a height of 5–6 m when growing conditions are good (Figure 19.2). It has stout recurved prickles and a very strong odor. It also has a very strong root system, which ensures rapid regeneration of new shoots even after repeated cuttings. The leaves are ovate-oblong, and flowers are small, usually orange but often varying from white to red in various shades. The brightly colored flowers

FIGURE 19.2 *Lantana camara* bushes growing along a roadside near Cape Coast, Ghana.

occur year-round and are pollinated mostly by butterflies (Figure 19.3). Lantana bears small, greenish blue or blackish, and shining fruits, which also occur almost year-round.

19.4.2.1 Habitat and Site Requirements

Lantana thrives quite well in all habitat and soil types, in open and unshaded areas such as forest gaps and edges, abandoned agricultural lands, roadsides, railway

FIGURE 19.3 Close-up view of *L. camara* bushes showing brightly colored flowers.

tracks, wastelands, canals, etc. (Thaman 1974; Winder and Harley 1983; Thakur et al. 1992; Munir 1996 in Day et al. 2003). It tolerates a wide range of temperature and rainfall regimes, commonly occurring in areas of 1000 mm of rainfall per annum, and at altitudes of up to 1000 m. Lantana has been reported from over 10 countries in tropical and subtropical Africa, including Ghana, Liberia, Uganda, Rwanda, Kenya, Tanzania, and Madagascar. Together with *Chromolaena odorata, L. camara* is common in the Republic of South Africa, where it is a major target for control efforts. Though it is not too difficult to distinguish *L. camara* from *C. odorata*, its bushes may be easily confused with that of *C. odorata* bushes when viewed from a distance, especially if it is not in the flowering stage.

19.4.2.2 Impact of *Lantana camara*

Lantana has a competitive advantage over native species as it combines its fast growth rate, allelopathic properties, fire tolerance, and high species variability to overcrowd desirable native species. Being somewhat tolerant to shade, it readily becomes the dominant understory vegetation in disturbed native forests, disrupting natural succession patterns and causing biodiversity loss. It has been observed that the allelopathic qualities of *L. camara* can result in a reduction of vigor and productivity of nearby plant species in orchards, and that the density of Lantana in forest is inversely related to the species richness in the community (Fensham et al. 1994, in Day et al. 2003). Although the distribution and impact of *L. camara* in tropical Africa is less drastic than that of *C. odorata*, Lantana is also quite common along the edges of oil palm, citrus and tree plantations and easily overruns the plantations if not regularly controlled. It encroaches on agricultural land and has potential to reduce the carrying capacity of pastures. In Ghana, the shrub is regarded as a nuisance due essentially to the abrasions that it inflicts on farmers during weeding with cutlass or hoe. In Kenya, *L. camara* is regarded as a threat to native pastures of the sable antelope (Greathead 1971 in Day et al. 2003). In Rwanda, Tanzania, Uganda, and Kenya, *L. camara* is known to provide refugia for tsetse flies that transmit the trypanosome (sleeping sickness) vector that affects humans and livestock (Katabazi 1983; Okoth and Kapaata 1987; Mbulamberi 1990).

The many varieties of lantana in the world today are the result of years of genetic manipulations to improve its ornamental characteristics. However, there is no evidence of the species ever being cultivated or managed deliberately for ornamental purposes in tropical Africa. While it was believed to be a good plant for erosion control and for hedges in some east African countries, dense thickets of lantana have been shown to lead to a reduction in the capacity of the soil to absorb rain water, which actually increases the risk of erosion. In West Africa, it is believed to have the potential to cure a variety of ailments including malaria though it is seldom used. Seeds have been used as lamb feed but there have been reports that the leaves and seeds contain triterpenoids, which cause poisoning or photosensitivity when ingested. There have been reports of death of sheep, goats, and other livestock fed exclusively on lantana bushes (Sharma et al. 1988, in Day et al. 2003).

19.4.2.3 Management of *Lantana camara*

The species has been managed in a variety of ways, including manual, chemical, and biological control methods. According to Day et al. (2003), vigilance is key to effective management, since it is only by repeated suppression of regrowth that success can be achieved. Early detection and elimination of new infestations is critical as this prevents the species from rapidly expanding its range. Manual control by hand-pulling is quite effective especially at the early stages of colonization and in localized populations. However, this is labor intensive and requires repeated follow-up removal of roots and new regenerations over a period of time. In West Africa, this method of control is rarely used because the species usually occurs in dense stands, and the thorny stems do not encourage this control approach. It is frequently managed by manual weeding with cutlass followed by burning after the cut shrubs have been allowed to dry. New regenerations on farms are removed by hoeing. This approach followed by planting of food crops and good farm maintenance often yields good results.

Chemical control is quite effective, but this is not a preferred option because of the cost involved and environmental concerns of excessive chemical application. Nevertheless, chemical control is becoming increasingly popular among local farmers in West Africa because of the availability of cheap and unregulated supply of herbicides on the market.

The use of biological methods to control lantana has been practiced for over a century with some degree of success but altogether has not been too impressive. Over 40 insect biocontrol agents have been tested and released in about as many countries, but none has resulted in total control. Biocontrol agents reported to have resulted in some partial control include *Teleonemia scrupulosa* Stal (Hemiptera), *Octotoma scabripennis* (Coleoptera), *Uroplata giradi* Pic (Coleoptera), and *Ophiomyia lantanae* (Froggatt) (Diptera) (Day et al. 2003). However, biocontrol efforts in Uganda using *T. scrupulosa* Stal were less successful compared with results in Australia and Fiji because the insects also attacked the crop species *Sesamum indicum*. Biological control of lantana is generally considered a failure, and this is attributed to the high genetic diversity of the plant, its high rate of hybridization, and the difficulty in identifying the center of its origin for potential biocontrol agents (Thomas and Ellison 2000).

19.4.3 *Broussonetia papyrifera* (L.) Vent. Moraceae

B. papyrifera (Paper mulberry) is a woody perennial native to Japan and Taiwan. It thrives well in a wide range of habitats including the humid tropics, subtropics, and temperate environments (Whistler and Elevitch 2005). It is believed to have been introduced into the Pacific Island countries in ancient times where it is currently very well established. *B. papyrifera* also occurs in Europe, Asia, and North America. It is a small but fast-growing tree that can attain a height of 3–4 m over a period of 12–18 months, and grow to a total height of 12 m. The leaves are highly variable, ranging from about 8 to 25 cm in length. The small leaves are generally simple with serrate margins but older leaves usually have between 3 and 5 lobes (Figure 19.4). Flowers occur separately on male and female trees. Fruits range in size from 1.0

FIGURE 19.4 Dense mat of *B. papyrifera* root resprouts growing on the understory of a mixed species tree plantation at the Afram Headwaters Forest Reserve at Abofour, Ghana. Note the lobes on the leaves.

to 2.5 cm in diameter and are red to yellow in color. *B. papyrifera* is cultivated for a variety of purposes, as shade tree in home gardens for soil stabilization, for the production of glue from the sap, and more popularly for the production of traditional "tapa" cloth from the bark (Whistler and Elevitch 2005).

Ghana and Uganda are the only two tropical African countries where the species has been reported. It was introduced deliberately to Ghana in 1969 for evaluation of its potential for pulp and paper production. The same reason is also given for its introduction in Uganda, and in both countries, it is considered a serious invasive plant. In Ghana *B. papyrifera* is perhaps the most serious nonindigenous woody invasive plant in the High Forest zone, and the second most important forest invasive after *C. odorata*.

B. papyrifera is known locally in Ghana as "York," nicknamed after the technical officer who worked on the plots during the experimental trials in the 1970s. It is highly concentrated in the two forest reserves where the initial experimental plots were established, namely Pra-Anum and Afram Headwaters Forest Reserves (Figure 19.5). These reserves are located within the moist semi-deciduous and dry semi-deciduous forest zones, respectively. These two forest types are among the most floristically diverse and economically important of all the forest types in Ghana. The high densities of *B. papyrifera* in these two reserves and nearby reserves or forests were facilitated by the high rate of deforestation experienced in recent times. From these two centers, the plant is spreading uncontrollably, and is currently estimated to cover about 81,000 km^2 of the forest zone of Ghana (Apetorgbor and Bosu 2007). Dense stands of *Broussonetia* are very conspicuous in degraded sites and along roadsides in the Pra-Anum and Afram Headwaters Forest Reserves.

FIGURE 19.5 Pole-size trees of *B. papyrifera* along the main Kumasi-Tamale trunk road in Ghana. Trees with yellowing leaves have been treated with chemicals as part of chemical control trials.

Ecological studies on the species were carried out in Ghana during the late 1990s (Agyeman and Kyereh 1997; Agyeman 2000). Evidence indicates that *B. papyrifera* is a fast-growing pioneer species. It has two fruiting seasons, which occur between January and March, and June and September. Fruits are dispersed by bats, birds, and other generalist frugivores. The seeds rarely germinate in dense canopy forests (Agyeman 2000) but rather in large canopy gaps, and regeneration on roadsides and abandoned farmlands is very prolific. It has a fairly high tolerance to fire compared with *Ceiba pentandra*, *Terminalia ivorensis*, and other native tree species with similar ecological characteristics.

19.4.3.1 Invasiveness and Impact

In the Pacific Island countries where the species has long been present, it is not considered invasive because only male clones were introduced. Thus, no seeds are produced and propagation is by vegetative means, using root shoot suckers. However, in Ghana and other places where both fertile male and female plants were introduced, the invasive potential of the plant increases significantly. The invasiveness of *B. papyrifera* is further enhanced by its ability to resprout from the roots. Trees form a dense mat of roots under the surface of the soil, which shoot up randomly at short intervals when growing conditions are favorable. Together with its high growth rate, paper mulberry competitively prevents the regeneration of most indigenous species. In Ghana, the plant poses a considerable threat to farming in parts of the Ashanti and Eastern Regions, and is known to substantially increase the cost of farm maintenance.

19.4.3.2 Management of *Broussonetia papyrifera*

Very little has been done by way of efforts to control the species in Ghana and Uganda. Elsewhere, mechanical and chemical control methods are frequently employed. Local farmers in Ghana use slash and burn methods to control young plants and pole-size trees. However, this method of control is ineffective as the plant coppices vigorously afterward. To prevent regrowth, the new shoots are regularly removed by cutlass or by chemical application. Seedlings and saplings are frequently controlled with herbicides, but chemical control also provides relief for only a few weeks, requiring repeated herbicide application to substantially reduce competition with food crops.

Recognizing the potential dangers of paper mulberry to forestry development in Ghana, the Forest Commission recently tasked the Forestry Research Institute of Ghana (FORIG) to develop strategies to mitigate impact of the plant on forestry in Ghana. FORIG initiated a number of research activities to holistically manage the *Broussonetia* problem. Experimental removal of shrubs (uprooting) as well as cutting of pole-size trees followed by squirting with triclopyr proved to be effective control strategies against the plant. Girdling and squirting with Triclopyr was also effective at killing pole-size trees (Bosu and Apetorgbor 2007).

Preliminary results from studies of the wood properties of *B. papyrifera* indicate that some of its properties compare favorably well with those of some well-known indigenous species such as *Triplochiton scleroxylon*, *Pycnanthus angolensis,* and *Terminalia ivorensis*. There is therefore the potential to use the wood for the manufacture of furniture, moldings, rotary veneer, and plywood core (Ofori et al. 2006, in Foli 2006). The bark of the tree can be processed into a mesh for erosion control and rehabilitation of degraded mine sites. Therefore, continuous extraction of the wood from the forest could minimize the rapid spread of the species.

19.4.4 *Prosopis juliflora* L. Mimosaceae

Species of the genus *Prosopis* are among the most common tree species in the dry tropics. The genus *Prosopis* is represented by about 45 species. They are native to arid and semiarid zones of Africa, Asia, and the Americas, with several American species widely introduced throughout the world over the last 200 years. It is an evergreen, thorny, or sometimes thornless, tree or shrub. It is low branching and has a high coppicing ability. The bark is grayish brown in color. The tree may attain a height of up to 20 m under favorable conditions, while in very dry environments it is reduced to a shrub. Leaves are compound with 1–3 pairs of pinnae and a stalk of 6 cm. The plant has gold yellow flowers which are crowded in spikes. The pods are brittle and sweet with 10–20 hard seeds, which are difficult to extract. Propagation is mainly by seedlings, direct sowing, and coppicing. It starts fruiting at the age of 3–4 years. The pods have a tough pericarp and a cartilaginous endocarp, which does not allow the seeds to escape easily. The seeds germinate readily after passing through the gut of animals without being digested. Germination is around 40%–80%. Natural regeneration takes much longer time. Artificial regeneration is effected either by seeds or vegetatively through root and shoot cuttings (Gichuki et al. 2003).

FIGURE 19.6 *Prosopis* invasion of a rural landscape in Ethiopia.

19.4.4.1 Habitat and Site Requirements

Prosopis spp. can survive on inhospitable sites where little can grow, tolerating some of the hottest temperatures ever recorded, and on poor, even very saline or alkaline soils (Figure 19.6). Because they are nitrogen-fixing trees, *Prosopis* have been noted to improve the fertility and physical characteristics of soils in which they grow. They are deep-rooted, allowing trees to reach water tables, grow, and fruit even in the driest of years, providing an invaluable buffer during droughts. Many species appear to require access to a water table to survive, and some people believe they are responsible for the depletion of ground water reserves (Pasiecznik et al. 2004).

19.4.4.2 Distribution and Impact of *Prosopis*

Available records indicate that exotic *Prosopis* species were first introduced into Senegal, West Africa, in 1822 (Zimmerman 1991; Pasiecznik et al. 2001, as cited by Ngujiri and Choge 2004). In eastern Africa, *Prosopis* spp. were introduced purposely or accidentally as a forestry tree into several Great Horn of Africa countries such as Ethiopia, Sudan, and Somalia (Ngujiri and Choge 2004). In spite of some uses and benefits from some species in the genus, the plant is known for its numerous harmful effects on ecosystems and livelihood of the local people. The benefits of *Prosopis* have been dramatically outweighed by its negative effects.

In Kenya *Prosopis juliflora* and *P. pallida* were included in a tree-planting campaign in the early 1970s for the rehabilitation of quarries near the city of Mombasa, with seed source from Brazil and Hawaii (Johansson 1985, as cited by Ngujiri and Choge 2004). It was not realized at the time that its beneficial attributes to survive extreme arid conditions and produce large quantities of highly nutritious pods would later facilitate its invasive characteristics and devastation of habitats.

FIGURE 19.7 *Prosopis* invasion of forest road in Ethiopia.

Currently, *Prosopis* covers most of the dry parts of Kenya such as the North Eastern, Nyanza, and Rift Valley provinces. The main problems attributed to *Prosopis* are as follows: long powerful and poisonous thorns, invasion and colonization of habitats, elimination of other vegetation, effects on animal health (tooth decay and death from starvation), obstruction or hindrance to communication (blockage of roads, footpaths, etc.,), drying of soil, blockage of water courses and irrigation canals, losses in fishing industry as *Prosopis* takes a lot of water, hosting crop pests, and displacement of people from their settlements and homes (Lenachuru 2004) (Figure 19.7).

In Ethiopia *Prosopis* is now a common sight when traveling north from Arba Minch on the road to Dire Dawa and Harar. Infestations typically originate from the many small villages, extending along the main routes, and are now steadily advancing into the surrounding landscape. This corresponds with movement of animals being driven to markets and nomadic influences. *Prosopis* has also spread to cultivation areas and floodplains along the river Awash, which is of high economic importance to the region (Shiferaw et al. 2004). This is of major concern because the river transects the entire region, thus putting many high potential irrigation areas further downstream at risk. The naturalized extent of invasion is unknown at this stage, but is estimated to be in the order of some 4000 ha.

Invasion of rangelands have caused shortage of grazing land for livestock, which has resulted in drastic reduction of livestock. This is mainly due to reduced land carrying capacity as *Prosopis* trees are displacing desirable grasses that could not withstand its competitive ability. Coppices forming impenetrable thickets prohibit free movement of livestock, as thorns damage the eyes and hooves of camels, donkeys, and cattle with poisons, eventually leading to death of these animals. This is a matter of serious concern for the life of the local people. *Prosopis* is invading potential croplands, forcing local farmers with less capital and machinery to abandon their farmlands. The cost of clearing fallow and land left for reclamation is also increasing for large-scale commercial farmers. Additionally, the presence of *Prosopis* has led to an increase in the incidence of aphids and spider mites by serving as an alternative host.

In recent decades, these "exotic" *Prosopis* have attracted much attention. The wood is an excellent fuel; the use and trade in firewood and charcoal is an important part of the rural economy in many parts of the native range and some countries where it has been introduced. Trees harvested for fuel coppice vigorously after cutting. Straight branches are used for fence posts and poles in construction of shelters and homes. Sawn timber has a pleasant color and grain, and shrinks little on drying. It is used for making furniture and flooring, notably in the United States and Argentina. The wood is also used for tool handles and other household items (Pasiecznik et al. 2004). Honey produced from the tree, which has long and abundant flowering, is of the highest quality. The gum obtained from wounds in the bark is comparable with commercial gum Arabic from *Acacia senegal*. The leaves of most introduced species are rarely browsed by livestock, an advantage in establishment, with only native Indian, African, and a few American species valued for leaf fodder. Leaves are occasionally gathered and used as mulch or compost on agricultural fields, with some noted fungicidal and insecticidal qualities. The bark is a source of tannins, dyes, and fibres, and various plant parts are used in the preparation of medicines for eye, skin, and stomach problems.

19.4.4.3 Management of *Prosopis*

Manual, mechanical, fire, chemical, and biological control approaches have been tried with varying degrees of success. Hand clearing was the first method used to deal with *Prosopis* on pasturelands. This involved felling of trees, and uprooting of stumps and seedlings. Mechanical control of *Prosopis* is not considered economically viable, except on land of high conservation and biodiversity value. Mechanical removal of *Prosopis* from important habitats such as national parks, game reserves, river banks, and irrigation canals may be justified, however. Use of caterpillars and tractors that uproot the trees (by chaining) and deep ploughing the site is common. Chain pulling involves the use of two or more slow moving caterpillar and tractors to fully sever the tree roots, but this is effective only if there are few large trees (Senayit et al. 2004). This method of control is also the most expensive (Jacoby and Ansley 1991, as cited by Choge and Chikamai 2004). In Ethiopia, the control techniques used by commercial farmers include clearing using bulldozer, uprooting using human labor, and burning *Prosopis* trees found on farmlands, roadsides, and irrigation and drainage canals. Commercial farms spend considerably large sums of money to control *Prosopis*. Experience elsewhere shows that clearance with a biomass harvester produces wood chips that can be sold for energy production to offset operational costs (Felker et al. 1990).

Chemical treatments using systemic herbicides on stems or aerial parts have been used to effectively control *Prosopis*. Some of the chemicals that have been used successfully include 2,4,5-T, clopyralid, dicamba, piclorum, triclorpyr, and ammonium sulfamate among others (Jacoby and Ansley 1991, as cited by Choge and

Chikamai 2004). Infested sites often need repeated spraying every 5–7 years to sufficiently suppress regeneration. Cutting and pouring used car oil on the cuttings was practiced in the highly infested areas. However, results of a study by Hailu et al. (2004) revealed that stumping trees at 10 cm below the ground eliminate the chance of resprouting and hence might offer a viable option for controlling and even eliminating the plants.

Fire has been used in conjunction with chemical and mechanical control with some degree of success. Experience with eradication has shown that the use of mechanical and chemical methods (together with fire) has proved more effective than either alone in several cases. The costly and high level of management input limits this strategy. However, total eradication has not been successful because of reinvasion. Methods to prevent reintroduction have not been developed (Choge and Chikamai 2004). The present usage of the plant as fuel wood, besides contributing to the wood fuel requirement of the local community will, to some extent, control and minimize the rapid spread of the plant.

The known biological control agents that feed on *Prosopis* seeds are bruchid beetles, twig girdlers, and psyllids. Many of these species feed only on *Prosopis* and some on a single species (Kingsolver et al. 1977, as cited by Choge and Chikamai 2004). To date, most work on biological control has been done in South Africa where bruchid beetles have been introduced in the Western Cape with considerable success. Further research into the development of a mycoherbicide for *Prosopis* is now in progress. Mycoherbicide is a suspension of fungal spores that can be applied to unwanted trees to cause fungal disease that will kill the treated trees only (Choge and Chikamai 2004). Methods of biological control of *Prosopis* are being tried in other parts of the world, and encouraging results have since been reported.

Observations and experiences in many parts of world have shown that *Prosopis* can be easily controlled under private land tenure system. Ways of privatizing land (as conditions permit) in most of the *Prosopis* affected areas need to be explored as a way of containing *Prosopis* spread. The current trends in the Afar Region, Ethiopia, have shown that trading in *Prosopis* products is a viable undertaking. However, the local districts in Afar Region still restrict *Prosopis* product development to charcoal making and sale of poles.

19.5 CONCLUSIONS

There is no question that managing biotic invasions will constitute one of the major challenges of humankind in the twenty-first century. Scientists, ecologists, conservationists, politicians, and the general public are all alarmed by the spread and impact of exotic species to new ecosystems. Fortunately, current interest in the subject of invasive species among African scientists is relatively high. Perhaps a lot of interest is fuelled by increased international awareness and information flow through television and importantly the internet. Introduction of alien germplasm to the continent of Africa is expected to increase several fold in this century. Increased trade with China, India, and other Asian economies will greatly facilitate these introductions. Of course, such invasions will not be limited to only forest plant resources but also aquatic weeds, animals, fungal and microbial pathogens, insects, and other arthropods, and so on.

While international trade between African countries and the rest of the world will be a major source of alien introductions, recent increases in trade between countries and regional blocks on the continent will promote the spread of introduced species across the continent. Gone are the days when travelers from one region of Africa to another region must of necessity transit to Western Europe. Improvements in transportation and communication infrastructure in many African countries have led to a significant increase in intracontinental travel and trade. In addition, the collapse of the apartheid system in South Africa has promoted trade between various African countries and South Africa. Together with the very porous borders of many African countries, due to lack of capacity and logistics to enforce quarantine regulations, the spread of invasive species in the continent is expected to accelerate in the decades ahead. There are already good examples to show what might be expected on the continent. In the 1980s, the oriental yellow scale insect (*Aonidiella orientalis*) arrived in Africa from Asia, and was first reported from the northern part of Cameroon. In a matter of few years, however, this scale insect had covered and killed neem trees (*Azadirachta indica*) over an estimated 1 million km^2 area in West and Central Africa (Lale 1998). The rapid spread and menace caused by the water hyacinth (*Eichhornia crassipes*) to aquatic resources of many tropical African countries is already a major concern.

The rapid degradation of natural ecosystems as well as increased environmental stresses due to climate change will accelerate the establishment of introduced species into new habitats. More often, invasive plants survive and thrive in disturbed or degraded habitats rather than in natural ecosystems. They are better able to efficiently use the marginal habitat resources than their native competitors do. As more and more habitats become degraded and environmentally stressed, indigenous habitats will become more favorable to exotic invasions.

African scientists have realized the need to enhance the capacity and readiness to combat the spread of forest invasives in the continent. The formation of the Forest Invasive Species Network for Africa (FISNA) is a major step, which seeks to bring all forest health experts on the continent to work toward achieving this common objective. FISNA is currently supported by FAO and has its Secretariat at the Forestry Research Institute of Malawi. The organization has also received considerable support from the U.S. Department of Agriculture Forest Service for its activities and the African Forest Research Network (Chilima et al. 2005). Information exchange and networking are facilitated by the FISNA web site, which is hosted by the FAO (http:/www.fao.org/forestry/site/26951/en/). Regional blocks of FISNA are planned for various parts of the continent to facilitate the organizations' activities. The organization serves to link African forest health scientists and managers to share information, and if possible work together, to manage invasive species problems on the continent. The success or otherwise of managing invasive species in Africa will require the active participation and support of governments, nongovernmental organizations, private sector, and the international community.

REFERENCES

Adair, R., Black wattle: South Africa manages conflict of interest, *CABI Biocontrol News*, 23, 1, 2002.

Agyeman, V.K., Natural regeneration of tropical timber tree species under *Broussonetia papyrifera*: Implications for natural forest management in Ghana, 2000.

Agyeman, V.K. and Kyereh, B., Natural regeneration of tropical tree species under *Broussonetia papyrifera*, *Newsl. Ghana Inst. Prof. For.*, 2, 1, 3, 1997.

Binggeli, P. and Hamilton, A.C., Biological invasion by *Maesopsis eminii* in the East Usambara forests, Tanzania, *Opera Bot.*, 121, 229, 1993.

Binggeli, P., Hall, J.B., and Healey, J.R., *An Overview of Invasive Woody Plants in the Tropics*, School of Agricultural and Forest Science Publication No. 13, University of Wales, Bangor, 1998.

Bosu, P.P. and Apetorgbor, M.M., *Broussonetia papyerifera* in Ghana: Its invasiveness, impact and control attempts, Presented at First Executive Committee Meeting of the Forest Invasive species network for Africa (FISNA), Pietermaritzburg, South Africa, May 16–17, 2007.

Chevalier, A., Deaux Composées permettant de lutter contre l'Imperata et l'empèchant la degradation des sol tropiquaux qu'il faudrait introduire rapidement en Afrique noire, *Rev. Bot. Appl. Agric. Trop.*, 32, 359, 1952.

Chilima, C.Z., An update of forest invasive species in Malawi, Presented at First Executive Committee Meeting of the Forest Invasive species network for Africa (FISNA), Pietermaritzburg, South Africa, May 16–17, 2007.

Chilima, C.Z., Kayambazinthu, D., and Soka, D., *Proceedings of the First Workshop on the Forest Invasive Species Network for Africa (FISNA)*, Sokoine University of Agriculture, Morogoro, Tanzania, August 29, 2005.

Corlett, R.T., The naturalized flora of Hong Kong: A comparison with Singapore, *J. Biogeogr.*, 15, 421, 1992.

Day, M.D., Wiley, C.J., Playford, J., and Zalucki, M.P., *Lantana: Current Management, Status and Future Prospects*, Australian Centre for International Agricultural Research, Canberra, 2003.

Drake, J., di Castri, F., Groves, R., Kruger, F., Rejmánet, M., and Williamson, M., Eds., *Biological Invasions: A Global Perspective*, Wiley, Chichester, UK, 1989.

Duffey, E., Ed., Biological invasions of nature reserves, *Biol. Conserv.*, 44, 1, 1988.

Duffey, E., The terrestrial ecology of Ascension Island, *J. Appl. Ecol.*, 1, 219, 1964.

Entsie, P.-K., Ecologically sustainable *Chromolaena odorata* management in Ghana in the past and present, and the future role of farmers' field schools, in *Proceedings of the Fifth International Workshop on Biological Control and Management of Chromolaena odorata*, Durban, South Africa, Zachariades, C., Muniappan, R., and Strathie, L.W., Eds., ARC-PPRI, 128, 2002.

Fensham, R.J., Fairfax, R.J., and Cannell, R.J., The invasion of *Lantana camara* L. in Forty Mile Scrub National Park, North Queensland, *Aust. J. Ecol.*, 19, 297, 1994.

Foli, E.G., Community forestry management project on farm research services, Report submitted to the Forest Plantation Development Centre, Ministry of Lands, Forestry and Mines, August 2006.

Greathead, D.J., Progress in the biological control of *Lantana camara* in East Africa and discussion of problems raised by the unexpected reaction of some of the more promising insects to Sesamum indicum, in *Proceedings of the IX International Symposium on Biological Control of Weeds*, Moran, V.C. and Hoffman, J.H., Eds., Stellenbosch, South Africa, 261, 1971.

Hall, J.B., *Maesopsis eminii* and its status at Amani, Report to the East Usambara Catchment Forest Project, 37, 1995.

Harris, D.R., Invasion of oceanic islands by alien plants: An example from the Leeward Islands, West Indies, *Inst. Br. Geogr. Trans.*, 31, 67, 1962.

Juhé-Beaulaton, D., Les jardins des forts européens de Ouidah: Premiers jardins d'essai (xviiième siècle), *Cah. du CRA*, 8, 83, 1994.

Katabazi, B.K., A review of biological control of insect pests and noxious weeds in Fiji (1969–1978), *Fiji Agric. J.*, 41, 55, 1983.

Kiwuso, P. and Otara, E., Status of invasive alien species in Uganda, Presented at First Executive Committee Meeting of the Forest Invasive species network for Africa (FISNA), Pietermaritzburg, South Africa, May 16–17, 2007.

Lale, N.E.S., Neem in the conventional Lake Chad Basin area and the threat of oriental yellow scale insects (*Aonidiella orientalis* Newstead) (Homoptera: Psyllidae), *J. Arid Environ.*, 40, 191, 1998.

Lenachuru, C., Impacts of *Prosopis* species in Baringo District, in *Proceedings of the Workshop on Integrated Management of Prosopis in Kenya*, Choge, S.K. and Chikamai, B.N., Eds., Kenyan Forestry Research Institute (KEFRI), Kenya, 53, 2004.

Maddofe, S.S. and Mbwambo, J., Status of forest invasive species in Tanzania, Presented at First Executive Committee Meeting of the Forest Invasive species network for Africa (FISNA), Pietermaritzburg, South Africa, May 16–17, 2007.

Mbulamberi, D.B., Recent outbreaks of human trypanosomiasis in Uganda, *Insect Sci. Appl.*, 11, 289, 1990.

Munir, A.A., A taxonomic review of *Lantana camara* and *L. montevidensis* (Spreng.) Briq. (Verbenaceae) in Australia, *J. Adelaide Bot. Gard.*, 17, 1, 1996.

Mwangi, L., Kenya country report: Invasive tree species and diseases, Presented at First Executive Committee Meeting of the Forest Invasive species network for Africa (FISNA), Pietermaritzburg, South Africa, May 16–17, 2007.

Neser, S., Conflict of interest? The Leucaena controversy, *Plant Prot. News S. Afr.*, 6, 8, 1994.

Ngujiri, F.D. and Choge, S.K., Status and impacts of *Prosopis* species in Kenya, in *Proceedings of the Workshop on Integrated Management of Prosopis in Kenya*, Kenyan Forestry Research Institute (KEFRI), Kenya, 17, 2004.

Okoth, J.O. and Kapaata, R., A study of the resting sites of *Glossina fuscipes* (Newstead) in relation to *Lantana camara* thickets and coffee and banana plantations in the sleeping sickness epidemic focus, Busoga, Uganda, *Insect Sci. Appl.*, 8, 57, 1987.

Pasiecznik, N.M., Harris, P.J.C., and Smith, S.J., *Identifying Tropical Prosopis Species: A Field Guide*, HDRA, Coventry, UK, 2004.

Pyšek, P., Prach, K., Rejmánek, M., and Wade, P.M., Eds., *Plant Invasions, General Aspects and Special Problems*, SPB Academic, Amsterdam, 1995.

Ramakrishnan, P.S., Ed., *Ecology of Biological Invasions in the Tropics*, International Scientific Publications, New Delhi, 1991.

Senayit, R., Agajie, T., Taye, T., Adefires, W., and Getu, E., Invasive alien plant control and prevention in Ethiopia, pilot surveys and control baseline conditions, Report submitted to EARO, Ethiopia and CABI under the PDF B phase of the UNEP GEF project—Removing barriers to invasive plant management in Africa, EARO, Addis Ababa, Ethiopia, 2004.

Sharma, O.P., Makkar, H.P.S., and Dawra, R.K., A review of the noxious plant *Lantana camara*, *Toxicon*, 26, 975, 1988.

Shiferaw, H., Teketay, D., Nemomissa, S., and Assefa, F., Some biological characteristics that foster the invasion of *Prosopis julifora* (Sw.) DC. at Middle Awash Rift Valley Area, north-eastern Ethiopia, *J. Arid Environ.*, 58, 135, 2004.

Spongberg, S.A., *A Reunion of Trees. The Discovery of Exotic Plants and Their Introduction into North American and European Landscapes*, Harvard University Press, Cambridge, MA, 1990.

Strahm, W.A., The conservation of the flora of Mauritius and Rodrigues, PhD Dissertation, University of Reading, Reading, UK, 1993.

Thakur, M.L., Ahmad, M., and Thakur, R.K., Lantana weed (*Lantana camara* var. *aculeate* Linn) and its possible management through natural insect pests in India, *Indian For.*, 118, 466, 1992.

Thaman, R.R., *Lantana camara*: Its introduction, dispersal and impact on islands of the tropical Pacific Ocean, *Micronesia*, 10, 17, 1974.

Thomas, S.E. and Ellison, C.A., A century of classical biological control of *Lantana camara*: Can pathogens make a significant difference? in *Proceedings of the X International Symposium on Biological Control of Weeds*, Spencer, N.R., Ed., Bozeman, MT, 97, 2000.

Tie Bi, T., Contribution a la capacite de fourniture en azote des sols climat tropical humid (Côte d'Ivoire). Application a l'entretien de la productive de terres de culture, Doctrat d'Etat, National University of Côte d'Ivoire, 166, 1995.

Timbilla, J.A., Effect of Biological Control of *Chromolaena odorata* on Biodiversity: A case study in the Ashanti Region of Ghana, in *Proceedings of the Fourth International Workshop on Biological Control and Mangement of Chromolaena odorata*, Bangalore, 1996.

Timbilla, J.A. and Braimah, H., A survey of the introduction, distribution and spread of *Chromolaena odorata* in Ghana, in *Proceedings of the Third International Workshop on Biological Control and Management of C. odorata*, Côte d'Ivoire, 1994.

Vanderwoude, C., Scanlan, J.C., Davis, B., and Funkhouser, S., Plan for national delimiting survey for Siam weed, Natural Resources and Mines Land Protection Services, Queensland Government, Queensland, 2005.

Wakibara, J.V. and Mnaya, B.J., Possible control of *Senna spectabilis (*Caesalpiniaceae), an invasive tree in Mahale Mountains National Park, Tanzania, *Oryx*, 36, 357, 2002.

Whistler, W.A. and Elevitch, C.R., *Broussonetia papyrifera* (paper mulberry) Moraceae (fig family), Species Profiles for Pacific Island Agroforestry, Permanent Agriculture Resources, Holualoa, HI, 13, 2005.

Whitmore, T.C., Invasive woody plants in perhumid tropical climates, in *Ecology of Biological Invasion in the Tropics*, Ramakrishnan, P.S., Ed., International Publications, New Delhi, 35, 1991.

Wild, H., *Weeds and Aliens of Africa: The American Immigrant*, University College of Rhodesia, Salisbury, 1968.

Winder, J.A. and Harley, K.L.S., The phytophagous insects on Lantana in Brazil and their potential for biological control in Australia, *Trop. Pest Manage.*, 29, 346, 1983.

Section IV

Socioeconomic and Policy Aspects

20 Invasive Plants in Namibian Subtropical and Riparian Woodlands

Dave F. Joubert

CONTENTS

20.1 INTRODUCTION

Given that Namibia (Figure 20.1a) is one of the driest African countries south of the Sahara, this chapter may seem out of place in a book on invasive plants in forest ecosystems. However, based on the Food and Agriculture Organisation of the United Nations' (FAO) definition of a forest as "an area covered by at least 10% canopy cover of trees higher than 5 meters and extending for more than half a hectare" (FAO 2000), slightly less than 10% of the country is estimated to be covered by forest (Mendelsohn and el Obeid 2005). This figure excludes numerous riparian forests along perennial and ephemeral rivers.

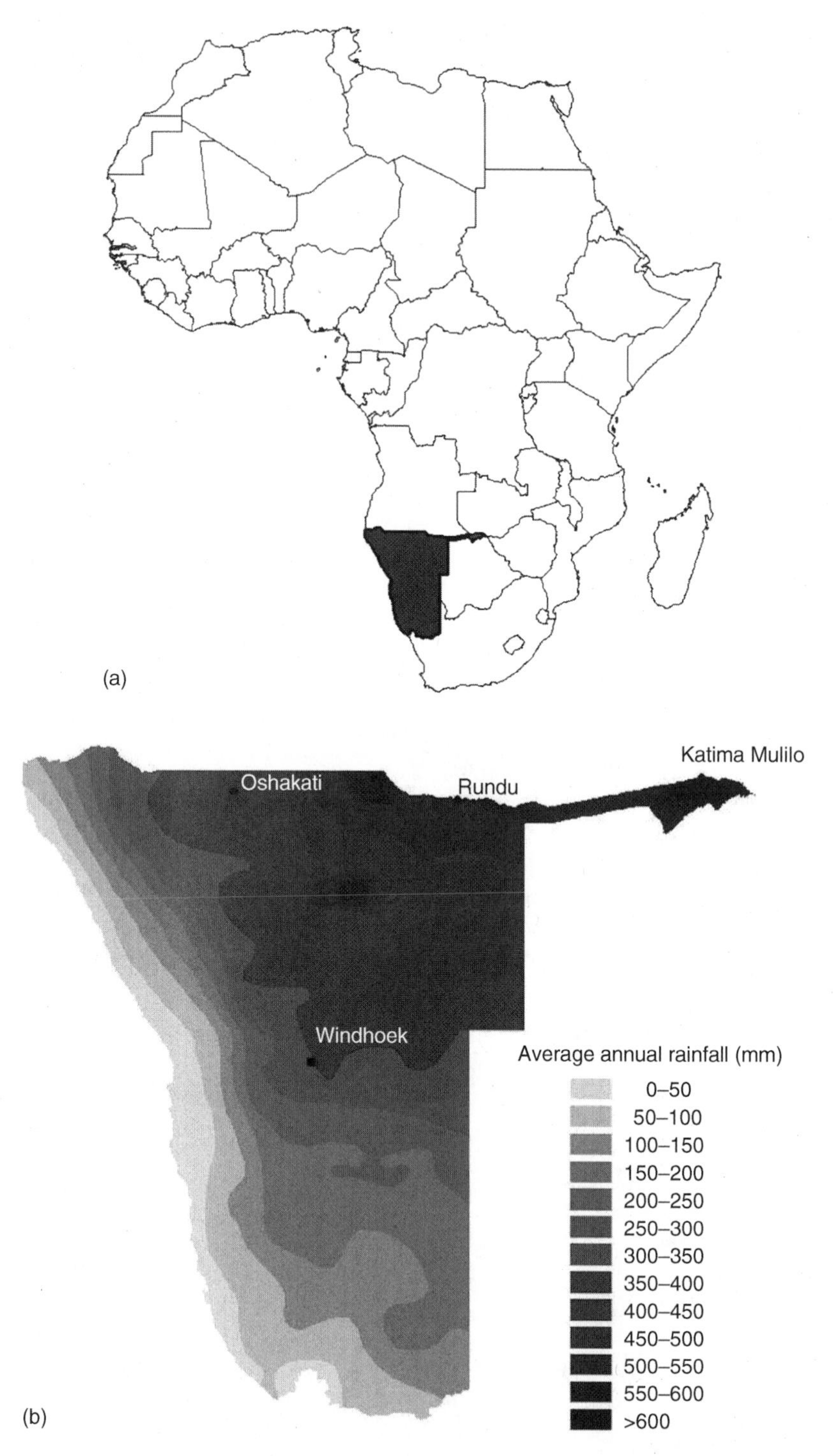

FIGURE 20.1 (a) Namibia in relation to the rest of Africa. (b) Mean annual rainfall in Namibia.

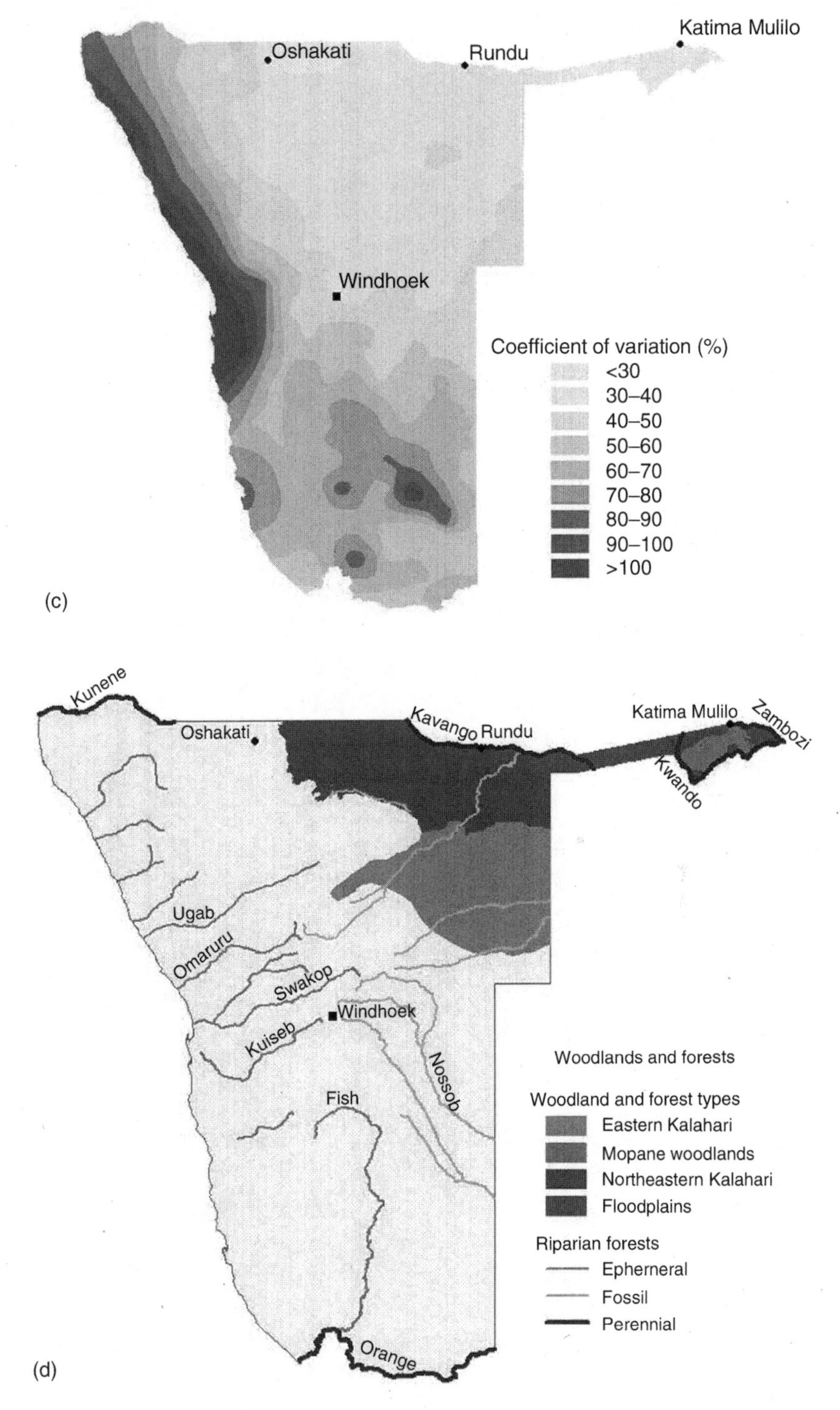

FIGURE 20.1 (continued) (c) Rainfall variability (coefficient of variation) in Namibia (d) Namibia's forest ecosystems.

In many ways, Namibia represents an extreme case with regard to invasive plants and the factors related to them and their control. Table 20.1 summarizes some attributes of Namibia and the impacts these have on alien invasions. Much of Namibia is classified as arid and semiarid and the low rainfall is quite variable, both in time and space (Figure 20.1b and 20.1c). Further, Namibia has one of the lowest population densities in the world, only 2.1 persons per square kilometer, and a population size of just less than 2 million people (Central Bureau of Statistics 2003). Aridity and the associated low population density largely account for Namibia's current relatively modest invasive vegetation problem (see Table 20.1) (Bethune et al. 2004), and

TABLE 20.1
Attributes of Namibia, and Their Effects on Alien Invasion in Forests and in General

Attribute	Effect
Positives	
Low population density	Reduced propagule pressure, reduced land transformation as a pathway for invasion (modest problem). But this causes complacency and lack of capacity to deal with future problems associated with invasions
Low rate of immigration	Reduced propagule pressure, reduced land transformation as a pathway for invasion (modest problem). But this causes complacency and lack of capacity to deal with future problems associated with invasions
Stressful environments (e.g., Kalahari Sand Forests; hyperarid environments elsewhere)	Reduces invasibility of some habitats (e.g., Kalahari Sand Forests with low nutrient status)
Relatively high herbivore densities (both domestic and game species)	May decrease the chances of successful invasion by edible species, but may also increase the spread (e.g., *Leucaena leucocephala* fruits, *Opuntia* fruits, and cladodes)
Enormous amounts of information to draw from globally	If utilized and acted upon, applying the precautionary principle, in many cases, invasive species can be controlled or eliminated cost effectively. Weed risk assessment screening procedures can be implemented
Namibia is signatory to a number of international conventions that oblige it to, among other things, control invasive aliens	Little at present but likely to influence national legislation and policy, and to increase awareness
Neighbor to South Africa, a leading country in the research and management of invasive alien species	Little at present but likely to influence national legislation and policy, and to increase awareness soon, provided Namibia is proactive. Reduced propagule pressure by restricting local nurseries from selling invasive plants normally imported from South Africa
Increased regional co-operation	Little at present, but likely to reduce cross-border flow of invasives and therefore reduce propagule pressure
Negatives	
Modest problem currently induces complacency and lack of capacity to deal with future problems associated with invasions	Little attention given to invasive aliens until problems are more dramatic and control is difficult and costly

TABLE 20.1 (continued)
Attributes of Namibia, and Their Effects on Alien Invasion in Forests and in General

Attribute	Effect
Inadequate, unfocused, and poorly enforced legislation regarding invasive alien control	Little attention given to invasive aliens until problems are more dramatic and control is difficult and costly
Lack of institutional awareness of the potential problems associated with invasive aliens	Little attention given to invasive aliens until problems are more dramatic and control is difficult and costly
Need for development and food security	Pressure to plant *useful* species, which are highly invasive (e.g., *L. leucocephala*), with little concern for environmental impacts, unless these can be shown to have secondary socioeconomic impacts
Lack of trained researchers and personnel involved in conservation	Lack of case studies showing harmful impacts, other than those drawn from other countries. Very little invasive control implemented

natural resource managers are more concerned about the visibly dramatic impacts of deforestation and bush encroachment. Namibia's low average population density is somewhat deceptive, however, since rural population densities reach up to 21 people per square kilometer in the northern region of Ohangwena (Central Bureau of Statistics 2003) and more than 100 people per square kilometer in localized areas. These high-density rural areas coincide with the only perennial rivers Namibia has in the north of the country (and thus riparian forests) and the adjacent Kalahari Sand Forests and Woodlands (Figure 20.1d). The populations in these areas are highly dependent on these associated forests for their livelihoods. These areas currently support a relatively low density of invasive alien plants, largely because of the lack of urban centers in these areas, which often act as nodes of dispersal.

In this chapter, I outline the current situation in Namibia with regard to invasive alien plants, placing the discussion within the context of our forest and woodland ecosystems. I also look toward future scenarios, and recommend the way forward in Namibia.

20.1.1 Forest Ecosystems in Namibia

Based on the FAO definition, over 10% of Namibia is covered by forest (FAO 2000). As shown in Figure 20.1d, most of this is in the wetter, northeastern portion of the country (Mendelsohn and el Obeid 2005). If one includes the Kalahari Woodlands of the northeast (which intergrade with the forests) and the riparian forests found alongside both perennial and ephemeral rivers or supported by their floodplains, the percentage of land area covered is substantially more (though this has not been calculated).

20.1.1.1 Forests and Woodlands of the Northeast

Forests and woodlands represented in these areas are essentially of the same vegetation type, distinguishable only by their amount of canopy cover, which is affected by factors such as browsing pressure, elephant damage, fire damage, and successional state. FAO (2000) refers to "Other Wooded Land" as "land not classified as forest, spanning more than 0.5 hectares; with trees higher than 5 meters and a canopy cover of 5–10% . . ." This then corresponds to Namibia's woodlands. The difference between forest and woodland, according to the definition, is the percent canopy cover. These definitions of forest and woodland exclude the Acacia savannas in the lower rainfall areas, to the south and west, since the dominant canopy cover is below 5 m in height (although scattered larger trees do occur). The forests and woodlands cover the areas with semiarid to subhumid climates, where the mean annual precipitation ranges from ~450 to 700 mm. The extent more or less coincides with the Forest and Woodland Savanna as defined by Giess (1971) or the Northern Kalahari, Northeastern Kalahari Woodland, and Caprivi Mopane Woodland according to a recently refined classification (Mendelsohn et al. 2002) (Figure 20.1d). Most of these forests and woodlands grow on deep, well-drained, acidic, and nutrient-poor Kalahari dune sands (arenosols). They are dominated by large members of the Fabaceae family, including *Baikiaea plurijuga*, *Burkea africana*, *Guibourtia coleosperma* and *Pterocarpus angolensis*, and *Schinziophyton rautanenii* (Euphorbiaceae). These large trees, among others, form the canopy of the forests and woodlands (Figure 20.2a). They are used for a multitude of purposes including timber and wood carving, medicine, and food and beverages (Van Wyk and Gericke 2000; Curtis and Mannheimer 2005). Other forest and woodland types occur within this matrix. For example, the Caprivi Mopane Woodland (dominated by *Colophospermum mopane* [Fabaceae]) occurs on soils with a higher clay percentage (from sandy loam to clay soils) (Mendelsohn and Roberts 1997), but the forests on nutrient-poor sands predominate. From this point, I will refer to them as Kalahari Sand Forests.

20.1.1.2 Riparian Forests alongside Perennial Rivers

In addition to, and generally adjacent to, the dry woodlands of the northeast, denser riparian forests are found alongside the perennial rivers (Figure 20.2b) within this area, on relatively nutrient-rich fluvisols (Mendelsohn et al. 2002). The Zambezi River and its tributaries, the Kwando and Chobe Rivers, form the northeastern borders around the Caprivi Region and drain toward the Indian Ocean, while the endoreic Okavango River borders on Angola and forms the world-famous Okavango Delta in Botswana (Figure 20.1d). These riparian forests are diverse, dense, and luxuriant but some are threatened by habitat destruction and fragmentation (Mendelsohn and el Obeid 2005). The canopy is characterized by species such as *Acacia nigrescens* (Fabaceae), *Albizia versicolor* (Fabaceae), *Diospyros mespiliformis* (Ebenaceae), *Garcinia livingstonei* (Clusiaceae), *Kigelia africana* (Bignoniaceae), *Phoenix reclinata* (Arecaceae), *Philonoptera violaceae* (Fabaceae), *Syzygium cordatum* and *S. guineense* (Myrtaceae), and *Trichilia emetica* (Meliaceae) (Mendelsohn and Roberts 1997; Curtis and Mannheimer 2005), and is generally much more closed than that of the neighboring Kalahari Sand Forests. Some riparian forests also occur along the perennial Kunene and Orange Rivers on the drier northwestern and southern borders of Namibia, respectively (Figure 20.2c). Both rivers flow westward, through a gradient of increasing aridity through the hyperarid Namib Desert, ending as estuaries on the Atlantic coastline. The riparian forests in these arid and hyperarid regions are dependent on the waters of the rivers, which come mainly from better watered catchments upstream in neighboring Angola in the case of the Kunene, and from South Africa and Lesotho in the case of the Orange River. The canopy of these forests along the Kunene River is

FIGURE 20.2 (a) Kalahari Sand Forests in Caprivi Region. *B. africana* is in the foreground at about 8 m tall, with a similar sized *P. angolensis* to the *right*. (Photo courtesy of D. Joubert.) (b) Riparian forests in the Caprivi Region, northeastern Nambia alongside the Zambezi River. (Photo courtesy of D. Joubert.) (c) Riparian forests along the Kunene River at Epupa Falls. *Hyphaene petersiana* (makalani palms) stands are visible through their lighter green foliage. (Photo courtesy of D. Joubert.) (d) Riparian forests along the Kuiseb River, an ephemeral river. The forest here is dissecting dunes and gravelplains in the hyperarid Namib Desert. (Photo courtesy of I. Zimmermann.)

characterized by *Faidherbia albida* (Fabaceae) and almost monostands of *H. petersiana* (Arecaceae) in places. Along the Orange River, the canopy includes *Acacia karroo* (Fabaceae), *Euclea pseudebenus* (Ebenaceae), *Rhus lancea* and *R. pendulina* (Anacardiaceae), *Salix mucronata* (Salicaceae), *Tamarix usneoides* (Tamaricaceae), and *Ziziphus mucronata* (Rhamnaceae). In some parts, the invasive alien *Prosopis* forms monostands. The genus *Prosopis* is represented mostly by three species (*P. chilensis*, *P. glandulosa* var. *torreyana*, and *P. velutina*). I will refer to them as *Prosopis* since there is some confusion regarding the taxonomy of the group and apparent hybridization between species (Smit 2005).

20.1.1.3 Riparian Forests alongside Ephemeral Rivers

Probably the most unusual riparian forests in Namibia are those found alongside the ephemeral rivers in semiarid to hyperarid environments, where they form narrow forest strips commonly referred to as linear oases (Figure 20.2d). The majority of these ephemeral rivers flow from the central highlands, where the mean annual rainfall may be as high as 500 mm (but usually around 300 mm), through a steep gradient of increasing aridity to the Atlantic Ocean where the rainfall is less than 10 mm (Figure 20.1b through 20.1d). A few, such as the Fish River, flow south, also from the inland plateau and drain into or toward the Orange River on Namibia's southern border with South Africa.

These ephemeral rivers typically flow for only a few days a year or not at all in some years, yet many of them support forests, depending upon the amount and reliability of rainfall in their upper catchments, at least along portions of their length, depending upon topography. Some of these, termed fossil rivers (Figure 20.1d), are old river courses that do not truly function as rivers anymore. These, including the Nossob River, occur on extremely flat terrain. Most of these do not support substantial stretches of riparian forest, due to their very erratic flow pattern, but the Nossob River is one of a few exceptions. The trees of these riparian forests rely on groundwater contained in alluvial aquifers that are replenished annually by the floodwaters from rainfall in upper catchments further inland. These floodwaters also bring copious amounts of silt and organic particles that are important for maintaining the relatively lush vegetation there. They also deposit seeds, including those of alien invasive plants. During episodic massive floods, large areas of forest may be destroyed (Jacobson et al. 1995). The canopy of these forests is largely made up of *A. erioloba* and *F. albida* (both Fabaceae) and to a lesser extent *E. pseudebenus* (Ebenaceae), *Ficus sycamorus* (Moraceae), and *Combretum imberbe* (Combretaceae), in varying proportions depending upon the catchment area and the substrate. In the hyperarid western reaches, *T. usneoides* (Tamaricaceae) sometimes forms a more dominant component of the canopy. In some stretches of ephemeral rivers (e.g., in the Swakop River and the Nossob River) *Prosopis* forms dense monostands (Smit 2005). The riparian forests along the westward flowing rivers in the Namib Desert play a crucial role in maintaining production and diversity, and allow this hyperarid desert to support some unlikely fauna, including elephant (*Loxodonta africana*), giraffe (*Giraffa camelopardalis*), black rhinoceros (*Diceros bicornis*), and lion (*Panthera leo*), as well as domestic livestock (Jacobson et al. 1995)

The descriptions of these forest habitats have been necessarily brief. I encourage the reader to read some of the references for a more detailed account.

20.1.2 Awareness of Invasive Species in Namibia

The problem of invasive alien species in Namibia was highlighted 22 years ago at the 1984 annual professional officers meeting of the then Directorate of Nature Conservation and Recreation Resorts. One of the aims of the workshop was to assess the potential for further invasion, in order to implement preventative measures "before the problems get out of hand" (Brown et al. 1985). Workshop sessions at this research meeting resulted in several publications, including a report that determined the distribution of invasive and potentially invasive species in different regions of Namibia (Brown et al. 1985). Most of our information regarding the status of invasive plant species in Namibia prior to independence stems from this and other publications in the 1980s (see Bethune et al. 2004 for a review of these). It is of note that the areas considered most prone to invasion were *river washes*, including forests along ephemeral rivers and perennial rivers (Brown et al. 1985). Even at that time, the riparian forests along perennial rivers in the northeast were considered a future problem due to the envisaged acceleration in development in these previously isolated areas.

Since Namibia achieved independence in 1990, there has been a necessary shift of focus toward economic and social development to address the socioeconomic imbalances of the apartheid colonial past. While this has obviously been welcome and necessary, a collateral effect was that the momentum of the work on invasive aliens initiated in the 1980s was not sustained. This has led to an increase in the introduction of potential invaders, particularly in the field of agroforestry, and a decrease in research and eradication programs. Fortunately, this lapse in awareness has recently begun to be countered through a variety of initiatives, including the development of a strategic action plan that incorporates alien invasive control, alien bashing days, and workshops on invasive alien species, and the development of educational posters, curricula, and research (Bethune et al. 2004). Despite this reawakening, the relatively slow spread of invasive vegetation in Namibia's arid environment sustains the complacency and lack of awareness among the public. It is a common perception (perhaps largely true at this stage), even within Namibia's small scientific community, that populations of invasive plants have a negligible impact on ecosystem functioning, and are therefore not a conservation priority (Bethune et al. 2004). It is well known that prolonged lag times between the time of establishment and the time that species clearly show their invasiveness make it difficult for decision makers to take invasive species seriously, and action is often only taken reactively and expensively (Mack et al. 2000). Ironically, despite the almost complete absence of mention of invasive alien species in Namibia's legislation, Namibia's Vision 2030, a guiding document for development toward a "prosperous and industrialised Namibia," makes mention of enforcing "legislation regarding the illegal.... import and/or propagation of alien invasive species" (Office of the President 2004). It is this inconsistency in awareness that needs to be addressed if Namibia is to successfully control invasive alien species.

20.2 CURRENT STATUS OF INVASIVE PLANTS IN NAMIBIA'S FOREST ECOSYSTEMS

In Namibia, the problem of invasive plants is generally considered to be relatively modest (Barnard et al. 1998) and biodiversity loss and ecosystem degradation are ascribed more to habitat loss and fragmentation. In terms of *forest ecosystems*, there are relatively few problem species currently found in the forests and woodlands of the northeast, although a potentially serious aquatic invasive (*Salvinia molesta*) occurs in the floodplains of the Zambezi system associated with the northeastern forests and woodlands. Currently, *Prosopis* in riparian forests along ephemeral rivers poses by far the greatest problems in terms of ecosystem effects.

Table 20.2 lists the species of plants either known or suspected to be invasive or naturalized in forest ecosystems (based on Bethune et al. 2004). Added to this are species that, although not previously listed as invasive, are now considered to be potentially invasive according to National Botanical Research Institute (NBRI 2002)

TABLE 20.2
Alien Plant Species Known or Suspected to Be Invasive, or with the Suspected Potential to Be Invasive in Namibia's Forest Ecosystems

		Presence		
			Riparian Forests	
Species	Origin[a]	Kalahari Sand Forests	Along Perennial Rivers	Along Ephemeral Rivers
Acanthospermum hispidum (hispid starburr)[b]	South America	S		
Achyranthes aspera (prickly chaff flower)[c]	Southern Asia possibly	L	L	L
Alternanthera pungens (khaki weed)[c]	Neotropics	L	L	L
A. sessilis (sessile joyweed)[c]	Southern Asia	L	L	
Argemone ochroleuca subsp. *ochroleuca* (white flowered Mexican poppy)[b]	Central America	S	S	K
Bidens bipinnata (blackjack)[c]	North America		L	
B. biternata (five leaf blackjack)[b]	America	S	S	
B. pilosa (common blackjack)[b]	Temperate and tropical America	K	K	
Bambusa balcooa (Balcooa bamboo)[b]	India		K	
Caesalpinia gilliesii (bird-of-paradise)[c]	Unknown, Southeast Asia for *C. decapetala*	L	L	
Cardiospermum grandiflorum (balloon vine)[c]	Neotropics	L	L	L
C. halicacabum (heart pea)[c]	North and Tropical America	L	L	L
Chamaesyce hirta (hairy spurge)[c]	Americas	L	L	L
C. prostrata (prostrate sandmat)[c]	Americas	L	L	L
Chenopodium ambrosioides (Mexican tea)[b]	Tropical America			K
Cereus jamacaru (queen of the night)[b]	Northeastern Brazil	S	L	
Chromolaena odorata (triffid weed)[d]	Tropical America	L	L	
Conyza albida (fleabane)[c]	South America	L	L	L
C. bonariensis (hairy fleabane)[c]	South America	L	L	L
Cynodon dactylon (couch grass)[c]	Old World, possibly southern Europe	L	S	
Datura ferox (large thorn apple)[b]	Tropical America	L	L	L
D. innoxia (Downy thorn apple)[b]	North and Central America	S	S	K
D. stramonium (common thorn apple)[b]	Tropical America	S	S	K

TABLE 20.2 (continued)
Alien Plant Species Known or Suspected to Be Invasive, or with the Suspected Potential to Be Invasive in Namibia's Forest Ecosystems

		Presence		
			Riparian Forests	
Species	**Origin**[a]	**Kalahari Sand Forests**	**Along Perennial Rivers**	**Along Ephemeral Rivers**
Dodonaea angustifolia (sand olive)[b]	South Africa	L	K	
Eucalyptus camaldulensis (Red River gum)[c]	Australia		K	
Gomphrena celesioides (Gomphrena weed)[c]	South America	L	L	
Guilleminea densa (small matweed)[c]	North America	L	L	L
Ipomoea purpurea (morning glory)[b]	South America	L	K	
Jatropha curcas (physic nut)[d]	Caribbean	L	L	
Lantana camara (lantana)[b]	Central and South America	L	K	
Leucaena leucocephala (wonder tree)[b]	Tropical America	L	K	
Ligustrum spp. (privet)[c]	Eurasia, Malaysia, tropical Australia	L	L	
Mangifera indica (mango)[b]	Asia		L	
Melia azedarach (Syringa)[b]	East Asia	L	K	
Myoporum tenuifolium (manatoka)[c]	Australia	L	L	
Nicotiana glauca (wild tobacco)[b]	South America	S	S	K
Opuntia ficus-indica (sweet prickly pear and other *Opuntia* spp.)[b]	Central America	S	K	K
Pennisetum clandestinum (kikuyu)[c]	East Africa	L	L	
P. setaceum[b] (fountain grass)	North and East Africa, Middle East	K	L	K
Psidium guajava (guava)[b]	Tropical America	L	L	
Prosopis spp. (mesquite, Prosopis, Honey mesquite, and velvet mesquite)[b]	North and Central America			K
Ricinus communis (castor oil plant)[b]	Tropical East Africa	L	K	K
Pupalia lappacea (creeping cock's comb)[c]	Eurasia	L	L	L
Salsola kali (Russian tumbleweed)[c]	Europe and Asia			L
Schinus molle (pepper tree)[c]	South America	L	L	
S. terebinthifolius (Brazilian pepper tree)[c]	Brazil		L	
Schkuhria pinnata (dwarf marigold)[c]	America	L	L	L
Solanum elaeagnifolium (silver-leaf bitter apple)[c]	America	L	L	

(*continued*)

TABLE 20.2 (continued)
Alien Plant Species Known or Suspected to Be Invasive, or with the Suspected Potential to Be Invasive in Namibia's Forest Ecosystems

		Presence		
			Riparian Forests	
Species	**Origin**[a]	**Kalahari Sand Forests**	**Along Perennial Rivers**	**Along Ephemeral Rivers**
S. mauritianum (bugweed)[b]	Uruguay and southeastern Brazil		L	
S. seaforthianum (potato creeper)[b]	West Indies		L	
Tecoma stans[c]	Mexico and Southern United States	L	L	
Xanthium spinosum (spiny cocklebur)[b]	South America	K	L	
X. strumarium (cocklebur)[c]	Probably New World	L	L	
Total known and suspected species (K and S)		11	16	9
Total species likely to invade (L)		32	33	13
Total species already invading or naturalized or likely to invade	37 from the New World, 10 from Europe and Asia, 3 from Australia (1 shared with Asia), and 4 from Africa	43	49	22

Note: K, Known as invasive or naturalized in habitat; S, Suspected to be invasive or naturalized in habitat; L, Likely to invade habitat potentially, based on preliminary review of literature on climate and habitat data.

This list excludes aquatic invasives.

[a] Information on origins from Henderson (2001), U.S. Forest Service (2006), and Gracela et al. (2004).

[b] Denotes from Bethune et al. (2004).

[c] Denotes this species is from a list of potentially invasive species according to NBRI (2002) and Zimmermann (2002).

[d] Denotes from own literature search after Bethune et al. (2004) report.

and Zimmermann (2002). The latter two are largely based on invasive plant lists in South Africa (Henderson 2001). An additional two species (*Chromolaena odorata* [Asteraceae] and *J. curcas* [Euphorbiaceae]) were added to the list based on a brief review for the purposes of this chapter (there are plans to develop *J. curcas* plantations for biodiesel production in the Kalahari Sand Forests). Some species included might later turn out to be noninvasive. On the other hand, there are potentially invasive species that will have been overlooked. For those plants that are expected to become invasive, broad descriptions of their habitat and climatic conditions, mostly from Henderson (2001) and U.S. Forest Service (2005), were used to tentatively decide whether a species has the potential to invade the three different kinds of forest ecosystems. More detailed climate and habitat matching might be a priority for a dedicated biologist of invasive alien species. As can be seen, the vast majority of species originate from the New World. The riparian forests had the most species currently present, and the most species likely to invade, while the riparian forests along ephemeral rivers had the least. Note that this is not a reflection in any way of the extent of the problem in terms of abundance (see Table 20.3 for comparisons of the invasive problems in the different forest ecosystems).

As pointed out in the national review (Bethune et al. 2004), the current situation is not only a reflection of the *invasiveness* of the various plant species, but also a reflection of their *date(s)* and *rate* of introduction (propagule pressure). Thus, less invasive species, such as popular ornamental garden plants and trees suitable for agroforestry are sometimes more widely distributed, and at higher densities, than potentially more invasive species because of earlier and more frequent reintroductions of higher numbers of individuals. Table 20.3 gives a brief overview of the invasive situation in each of the three forest habitats.

20.2.1 Forests and Woodlands of the Northeast (Kalahari Sand Forests)

The Kalahari Sand Forests in the northeast have been largely spared from invasions until now. Hines et al. (1985) recorded *Argemone ochroleuca* (Papaveraceae), *Bidens pilosa* (Asteraceae), *O. ficus-indica* (Cactaceae), *R. communis* (Euphorbiaceae), and *X. spinosum* (Asteraceae) in Kalahari Sand Forests and Woodlands (Table 20.2), but only around human habitation and in old fields. Since then, no monitoring has been done. However, from personal observations, it would seem that the *status quo* largely remains. This might seem counterintuitive, considering the relatively high rainfall, and in some areas also relatively high human population densities (but this only close to the perennial rivers). Most plantation species, especially of the genus *Eucalyptus* (Myrtaceae), have been rather unsuccessfully introduced, which is largely ascribed to the variable, unpredictable and relatively low rainfall, and low nutrient status of Namibia's sand forest areas (Mendelsohn and el Obeid 2005). Stress, in the form of low nutrient availability and low water availability, is known to decrease the invasibility of a habitat (Alpert et al. 2000). The Kalahari Sand Forests, situated on nutrient-poor, acidic, well-drained sandy soils, match this concept of a stressful environment. Another important reason for this low level of invasion currently is that there have been very few introductions of invasives in the region (low propagule pressure), despite the establishment of exotic plantations since the late 1800s (Erkillä and Siiskonen 1992). This is partially because of the low human population densities away from the perennial rivers, and a lack of urban centers in these areas. Within towns, people tend to have small gardens with a low diversity of plants in them. However, Directorate of Forestry nurseries sell many potential invasive species that are becoming increasingly popular. The presence of large numbers of herbivores may also decrease the invasibility of a habitat (Alpert et al. 2002). Close to human habitation, herbivory by domestic livestock keeps populations of potential invasives in check, while in conservation areas, and away

TABLE 20.3
Current and Future Status of Namibia's Forest Ecosystems and the Reasons Associated with It

	Ecosystem		
Attribute	**Kalahari Sand Forests**	**Riparian Forests along Perennial Rivers**	**Riparian Forests along Ephemeral Rivers**
Invasibility	Low invasibility due to low nutrient sands, long dry season, and fast drainage (does not favor shallow rooted invasives). High browsing pressure may limit invasion of edible species, as may fire (Hines et al. 1985; but see D'Antonio 2000)	Higher invasibility due to high nutrient status and moisture holding ability of soils, transport of seeds, and other propagules. High browsing pressure may limit invasion of edible species, but may promote dispersal of propagules	Higher invasibility due to high nutrient status and moisture holding ability of soils (silts and organic material transported in floods), transport of seeds, and other propagules. High browsing pressure may limit invasion of edible species, but may promote dispersal of propagules
Propagule pressure	Low due to low population densities, almost no crop production and almost no towns	Higher; major towns along rivers, seeds transported; Orange River flows through highly transformed agricultural lands	High; seeds and other forms of propagules transported from catchments situated in commercial farming areas and towns
Current status	Very low infestation levels	Low infestation levels	Moderate to very high infestation levels (e.g., Prosopis infestations in Nossob River). Some species prefer adjoining open sandy river channel habitat. Fewer species invasive and likely to become invasive
Current impacts	Not significant; potential occasional poisoning of herbivores near settlements and waterpoints (e.g., *Datura* sp.)	Unknown, displacement of indigenous species in localized areas, potential poisoning of herbivores; infestations currently not likely to have major ecosystem effects	Displacement of indigenous species; poisoning of herbivores, changes in ecosystem functioning and hydrology; biodiversity losses and changes in other trophic levels in dense infestations of Prosopis otherwise unknown
Invasion pathways	Agroforestry and ornamental species escaping from towns and forestry nurseries. Propagules spread by water and animals	Agroforestry and ornamental species escaping from towns and forestry nurseries. Propagules spread by water and animals	Annual transport of propagules from catchment areas upstream in floodwaters. Alien invasive trees planted on farms and in towns in these catchments
Likely future status (given current trends)	A gradual increase in infestations. Increased propagule pressure (through agroforestry initiatives) and *Green Schemes*, increased disturbance and edges (land clearing and increased fire frequency may increase invasiveness of habitat)	A rapid increase in infestations to very high infestation levels. As for Kalahari Sand Forests, increased propagule pressure	A slow increase to high infestation levels, especially of Prosopis. Propagule pressure not likely to increase substantially. Infestations of some species periodically washed away in main channel

from towns, elephants and other herbivores will do likewise, particularly since propagule pressure is currently too low to counter this high level of herbivory. Obviously, an increase in propagule pressure will increase the chances of establishment, since more individuals find suitable habitat, and can escape the effects of herbivory (Williamson 1996). Namibia has a population growth rate of 2.6% (Central Bureau of Statistics 2003) and a fairly vigorous afforestation campaign led by the Directorate of Forestry. This should result in a dramatic increase in propagule pressure. But is the Kalahari Sand Forest habitat likely to become more invasible? The absence of research in Namibia on invasions and their impacts reduces our ability to predict this. However, we can draw some prediction from insights elsewhere. Land clearing for subsistence farming is increasing along with population growth, and large areas are being cleared for intensive, large scale, irrigated and fertilized crop production in an effort to improve food security through the agricultural sector (referred to as the *Green Scheme*). It is known that edges are more invasible than unfragmented habitat (Alpert et al. 2002). Furthermore, the transformed, disturbed habitat, with its increased soil fertility, is also likely to be more invasible (Alpert et al. 2002). Secondarily, this may increase propagule pressure in the transformed habitat, and these habitats may thus serve as a springboard for more successful invasions into the more pristine sand forests.

Hines et al. (1985) felt that the frequent fires in the area, to which the indigenous vegetation is adapted, reduced the invasibility of the Kalahari Sand Forests. However, fire frequency in these areas has increased markedly with human population growth (Mendelsohn and Roberts 1997), and is probably atypical for the area, if it were left pristine. This increase in frequency has led to a decrease in recruitment of typical canopy species. Indigenous *invasive* shrubs often encroach onto such areas (Mendelsohn and Roberts 1997). D'Antonio (2000) found that fires tend to promote invasive species in most ecosystems. However, there seems to be no evidence to resolve whether fires in these Kalahari Sand Forests, or similar habitats in Africa, should promote or retard invasion by alien plants. Hines et al. (1985) could well be right. There has been a dramatic increase in resource use in general in the area, from increased felling for timber and woodcarving for the burgeoning tourist industry to a similarly increased grass harvesting for an increasingly fashionable roof thatching in the capital, Windhoek. The removal of typically dominant canopy tree species and grass species in these forests and woodlands, such as *Eragrostis pallens*, could promote invasion by woody species such as *Lantana camara* (and other unpalatable and toxic species) as has happened in Australian forests (Duggin and Gentle 1998 in D'Antonio 2000). The potential for invasion in these riparian forests, as reflected by the species likely to become invasive (33 species), is high (Table 20.2).

Until now, no impacts (either biophysical or socioeconomic) have been measured, although it is suspected that occasional poisoning of livestock and other herbivores may occur. Curtis and Mannheimer (2005) have recorded that *J. curcas* can poison children occasionally, but they do not elaborate. It can obviously be inferred that, as invasions increase, indigenous species will be displaced. No trophic, biodiversity, or socioeconomic effects have as yet been investigated, and the significance of these invasions, present and future, is untested.

The neighboring riparian forests along perennial rivers are greater cause for concern, as is explained in the next section. Kalahari Sand Forests may be secondarily invaded by species establishing and spreading from the riparian forests.

20.2.2 Riparian Forests alongside Perennial Rivers

Very little forest remains along the Okavango River, largely as a result of clearing for cultivation as well as elephant damage in overcrowded reserves downstream (Mendelsohn and el Obeid 2005). Further east, the Zambezi River holds few intact forests, partially due to the nature of seasonal floodplains with waterlogged soils that exclude forests. Currently, the riparian forests along perennial rivers have also been surprisingly unaffected by invasive aliens. The most likely reasons for the low invasion rate, considering the high invasibility of riparian forests in general (Alpert et al. 2002), until present are low propagule pressure (with the possible exception of near Katima Mulilo and Rundu, the only major towns on the Zambezi and Okavango River, respectively) as well as browsing pressure on edible invaders. Propagule pressure, particularly of useful agroforestry species, has increased dramatically in recent years. *Leucaena leucocephala* has been sold by Directorate of Forestry nurseries for the past 16 years because of its multitude of purposes, including its value as a fodder tree, nitrogen fixer, and a firewood tree, and also because it grows extremely rapidly. Degraded riparian forests outside Rundu (the major settlement on the Kavango River) are already invaded by young *L. leucocephala*, *Dodonaea angustifolia* (Sapindaceae) (presumably escapees from the nearby local Forestry nursery) (Figure 20.3), and *Eucalyptus camaldulensis* (Myrtaceae)

FIGURE 20.3 Young *L. leucocephala* growing in a clearing in the riparian forest. Three other invasive species are growing just out of picture: *D. angustifolia*, *Eucalyptus camuldulensis*, and *O. ficus-indica*. (Photo courtesy of D. Joubert.)

(escapees from Forestry plantations nearby) as well as *O. ficus-indica* (Cactaceae). No further data are available, other than a few records from a recently completed Tree Atlas program (Curtis and Mannheimer 2005). Between the Kavango and Zambezi Rivers, the Kwando River forests are more intact, although their integrity is threatened by a dramatic increase in elephant populations (Mendelsohn and el Obeid 2005). As far as is known, there are very few infestations of invaders in these forests. Hines et al. (1985) recorded small isolated populations near human habitation along the Kwando River along with small populations of *Bambusa balcooa*, *I. purpurea*, *Solanum mauritianum* (bugweed), and *S. seaforthianum* (potato creeper). Despite the low densities of alien invasive populations in these riparian forests, more species have been recorded than in the other forest habitats, with more species also likely to invade (Table 20.2). Surprisingly, *Lantana camara* has not yet spread extensively from gardens, although an isolated dense infestation occurs in a small isolated forest area in the Waterberg Plateau Park that lies outside the Kalahari Sand Forest and riparian forest areas under consideration (Bethune et al. 2004). Curtis and Mannheimer (2005) note that it has become invasive outside Rundu. *L. camara*'s invasive nature in other forests in tropical and subtropical areas, including South Africa, is well known (e.g., Henderson 2001; Silori and Mishra 2001). Hines et al. (1985) suggested it could become highly invasive in Caprivi and Kavango, the regions where most riparian forests alongside perennial rivers occur. Unfortunately, there is no recent information on its status in these forests. An individual *T. stans* has also recently been observed in a garden abandoned in the 1980s by the South African colonial occupation army in the riparian forest along the Kwando River. It is unknown whether it was planted but escapees of this species have been noted outside towns in areas with much lower rainfall (Cunningham, in press). The most likely reason for this species not being able to invade in the riparian forest to date is the heavy browsing pressure. The individual showed signs of intensive browsing (Figure 20.4), probably by elephant and browsing antelope such impala (*Aepyceros melampus*) and kudu (*Tragelaphus strepsiceros*) all of which occur in very high numbers in the area. This reemphasizes the importance herbivores may have in limiting the successful establishment of invasive species, provided that they are edible, and especially if propagule pressure is low (as in this case). Perhaps the species most likely to be seriously invasive in riparian forests is *Leucaena leucocephala*. However, the Directorate of Forestry feels that its value outweighs its potential threat to the integrity of the forests. This means the points of invasion will increase as rural people take saplings from the towns further into the countryside. Currently, conservation concern in these forests is focused on the primary effects of habitat fragmentation and destruction. These disturbed habitats, ever increasing in surface area, are likely to serve as nodes of invasion by aliens (accidentally and deliberately introduced) into forests. This adds a conservation dimension that needs to be addressed currently, while the relatively modest invasive problem remains just that.

As for Kalahari Sand Forests, no impacts of alien invasives have been studied. *C. odorata* is already a serious invasive problem in South Africa (Henderson 2001). In South Africa, it has also spread to regions of lower annual rainfall (500 mm) than initially predicted, but mainly along watercourses (Zachariades and Goodall 2002).

FIGURE 20.4 Heavily browsed *T. stans* growing, but not spreading in a clearing in the riparian forest along the Kwando River. Browsing pressure may currently be an important limiting factor, while propagule pressure remains low. (Photo courtesy of D. Joubert.)

It then very likely has the potential to invade riparian forests and possibly Kalahari Sand Forests in the higher rainfall areas in the northeast. *C. odorata* is known to suppress tree seedlings by forming a dense canopy that prevents tree seedlings from growing through in Ghana (Honu and Dang 2000). It has a similar effect in South Africa, and also reduces grazing production (Zachariades and Goodall 2000). In KwaZulu-Natal, South Africa, Leslie and Spotila (2001) observed that the reproductive cycle of Nile crocodiles (*Crocodylus niloticus*) was affected. Eggs laid in soils shaded by *Chromolaena odorata* incubated at temperatures likely to be below that needed for producing males, and may have completely arrested embryonic development. Furthermore, the fibrous roots make the soils unsuitable for egg chamber and nest construction. The Nile crocodile is an important predator in the perennial rivers in the north of Namibia, and is currently listed as threatened (Leslie and Spotila 2001). *C. odorata* seeds can be transported by water currents, on animals (they have small spines), hikers' boots, vehicles, and machinery (Wilson 2006). It is therefore conceivable that accidental introduction may occur. The aforementioned is an example of how important it is for Namibia to draw from experiences and research elsewhere.

The Orange and Kunene Rivers, flowing through more arid parts of Namibia on the south- and northwestern borders, respectively, would be expected to have more modest infestations of invasive species than the other riparian forest habitats in the northeast. The Orange River is more under threat than the Kunene River from invasions since it flows through neighboring South Africa, its source being in the highlands of Lesotho, to the southeast. It flows through highly transformed agricultural areas, including extensive vineyards in the hot arid downstream regions. These vineyards and other agricultural fields make use of the river's water for irrigation, and the low humidity of the climate keeps agricultural pests (especially fungal) to a minimum. Stretches of the Orange River are dominated by *Prosopis*, and *J. curcas* has been recorded from one locality.

20.2.3 Riparian Forests alongside Ephemeral Rivers

Although the ephemeral rivers are currently the most invaded habitat in Namibia (although not in terms of number of species—see Table 20.2), they have received little attention since the 1980s, other than work by Smit (2004, 2005) on *Prosopis* in the Nossob River, and recently Parr (unpublished data) in the Kuiseb River. In 1985, Vinejevold et al. (1985) and Tarr and Loutit (1985) recorded only nine species in the Namib Naukluft Park and Skeleton Coast Park in the hyperarid Namib Desert, all from the riparian forest and riverbed habitats. Only three species were considered to be of serious concern at that stage, viz. *Datura innoxia*, *D. stramonium* (Solanaceae), and *Prosopis*. At least in some ephemeral rivers, *Argemone ochroleuca* apparently only established after the 1984 floods (Tarr and Loutit 1985). Despite attempted eradication and control, reinfestation continually occurs since the catchments of these riparian forests are situated on commercial farmlands and townlands. Propagule pressure has been high in these forests for a long time. Riparian forest habitat adjoins open sandy riverbeds, and also opens floodplains in places. It appears that, except for *N. glauca* (Solanaceae) which prefers forest habitat (Parr, unpublished data), and the deeper rooted *Prosopis*, invasive species prefer the sandy main channel and the open floodplain habitats. Presumably, shallow rooted herbaceous and shrubby species are unable to reach the deeper aquifer water in the riparian forests. This means that, in the case of the sandy main channel, infestations are regularly washed away or drowned during annual flooding, only to be replaced by propagules brought downstream in the following floods. *N. glauca* seems to prefer open spaces in the forest (Figure 20.5). Deforestation through logging and lowered water tables (from increased abstraction as well as damming upstream) may therefore increase invasion rates. Boyer and Boyer (1989) found *D. innoxia* and *N. glauca* to be the most abundant invasive plants in a follow-up study. Although infestation levels are high, only *N. glauca* and *Prosopis* are likely to establish relatively stable and long-standing populations. Parr (unpublished data) found that, in the Kuiseb River at least, populations of invasive species have not noticeably increased since the 1980s. In the Kuiseb River, browsing pressure, mainly by livestock, may play a role in this (Parr, unpublished data).

R. communis appears to be negatively affected by browsing pressure (Boyer and Boyer 1989) whereas *D. innoxia* seems to be affected by trampling pressure

FIGURE 20.5 *N. glauca* growing in Kuiseb River floodplain adjacent to the riparian forest in the hyperarid Namib Desert. Long lateral roots allow it to exploit floodwaters that occur for a few days every year. (Photo courtesy of D. Joubert.)

(Vinejevold et al. 1985; Boyer and Boyer 1989). Boyer and Boyer (1989) suggested that despite *D. innoxia*'s high abundance along ephemeral rivers, it has little impact. Farmers report stock losses, however (Parr, unpublished data).

The most notorious invasive plant in Namibia is undoubtedly *Prosopis*. It is particularly problematic and dense in the Nossob River channel (Smit 2005). *P. glandulosa* var. *torreyana* was introduced to southern Africa in the late nineteenth century as a fodder and shade tree, the other two species that make up the hybrid complex following shortly (Smit 2005). Its prolific seed production, ability to produce a persistent seed bank, seed dispersal by animals and water, and resistance to drought have all contributed to this species forming dense thickets in many ephemeral river forests, displacing indigenous species such as *Acacia erioloba*, *A. karroo*, and *Faidherbia albida* and impacting on biodiversity at other trophic levels and ecosystem functioning (Boyer and Boyer 1989; De Klerk 2004; Smit 2005). In areas of the Nossob River it displaces all other plant species, and the copious seed banks can only be reached by small mammals (Smit 2005). *Prosopis* stands in the Kalahari, South Africa, significantly reduced bird species diversity than in indigenous *A. erioloba* woodland in drainage lines (Dean et al. 2002). The raptor guild was missing, and there were fewer frugivores and insectivores. In Namibia, *Prosopis* decreases plant species richness in the Swakop River (Visser 1998). *Prosopis* has other major impacts on these ecosystems. For example, *Prosopis* upstream in the Nossob River may reduce flooding to such an extent that groundwater levels downstream are affected. This might reduce the survival rate of *A. erioloba* trees downstream (Steenkamp, personal communication 2006). Similarly, in the hyperarid reaches of the Kuiseb River, increasing populations of *Prosopis* may be decreasing the level of water in aquifers to the detriment of species such as the endemic Nara melon (*Acanthosicyos horrida* [Cucurbitaceae]) downstream (Parr, unpublished data).

Prosopis is also perceived as useful for fodder and shade by farmers in the south of Namibia (Smit 2005). Smit (2005) reviewed the possibility of incorporating it into the agroforestry industry, owing to its additional value for furniture and charcoal production. *Prosopis* can therefore be seen as a *conflict* tree. In South Africa, two bruchids *Algarobius prosopis* and *Neltumius arizonensis* were introduced in 1987 and 1993, respectively, from Arizona, United States (De Klerk 2004). Today, the beetles occur throughout Namibia where infestations occur. Since mature *Prosopis* trees can produce millions of seeds (according to Smit 2005), even a reported infection of 99% of seeds probably has a negligible effect on the reproductive dynamics of the species. Besides the enormous impact of *Prosopis*, riparian forests along ephemeral rivers appear to be little impacted by alien invasives, despite the high propagule pressure. This assumption has not been properly tested until recently. An MSc student at the University of Namibia (UNAM) is currently investigating the impact of *D. innoxia* on species richness in ephemeral river forests around Windhoek (Namibia's capital city).

20.2.4 State of Invasive Research in Namibia

Although aridity, low population density, and lack of development have played a role in keeping Namibia's alien invasive problem relatively modest and in many cases manageable, they indirectly also pose their own problems, currently and potentially. Namibia's low population density also means that there are very few qualified biologists in relation to the surface area of Namibia, which needs to be conserved (823,680 km^2; Mendelsohn et al. 2002). Researchers involved in invasive alien research are also involved in a number of other disciplines. As a result, this research has focused on the distribution and status of known invasive species in the 1980s (e.g., Brown et al. 1985; MacDonald and Nott 1987) and more recently (e.g., Cunningham et al. 2004), and to a lesser extent the population dynamics of a few species (e.g., Joubert and Cunningham 2002). Impacts on populations, communities, and ecosystem processes have been inferred but not directly measured. Smit (2005) investigated the socioeconomic effects of *Prosopis* invasion, through questionnaires distributed to farmers. Parr (unpublished data) refers mostly to the socioeconomic relationship between farmers and invasive aliens.

The amount and proportion of scientists devoted to conservation topics, and those specifically focused on invasion research, is unlikely to increase in Namibia's midterm future. It is therefore important that research be strongly focused on areas of conservation that can make a difference. Currently, the Ministry of Environment and Tourism does not employ a single researcher devoted to plants or plant ecology. Only a handful of researchers within the Ministry of Agriculture, Water and Forestry (MAWF), the Polytechnic of Namibia (PON), the UNAM, and some nongovernmental organizations such as the Desert Research Foundation of Namibia (DRFN) and the Cheetah Conservation Foundation (CCF) conduct botanical research of any kind. Namibia has no researcher devoted primarily to invasive alien studies. DRFN, CCF, UNAM, and PON have more flexibility, and there has been a recent resurgence in alien invasive research in the country. With the awareness of invasive aliens currently experiencing a revival, particularly at PON and UNAM, there has been an increase in the number of students conducting supervised research on the topic (e.g., Visser 1998; Berry 2000; Gabanakgang 2003; Shapaka et al., in press). However, these studies are also still mostly focused on the distributions of invasive species. It is hoped that the recent introduction of an MSc in Biodiversity at UNAM will generate studies more focused on ecosystem effects as well as the evaluation of different control methods. South Africa, Namibia's neighbor to the south, is at the forefront of invasion biology research, and there are a number of experts in the field. In the southern African 2003 Global Invasive Species Programme (GISP) report (Mac Donald et al. 2003), over 400 references (books, book chapters, journal articles, and theses) on invasive organisms in South Africa are listed. In a random check of five important international journal articles on alien invasives in South Africa (focusing on ecosystem effects), four were excluded from this list, indicating that the list was merely scratching the surface. By contrast, in their exhaustive review of the alien invasive situation in Namibia, Bethune et al. (2004) mustered fewer than 100 references relating to invasive aliens in Namibia. Of these, the overwhelming majority were technical reports and book chapters, many of which only alluded briefly to alien invasives. Not one of the remaining journal articles focused on the determinations of the impacts of alien invasives on ecosystems, beyond distribution and population dynamics. This understandable, but rather alarming, state of affairs can be mitigated by forging much more regional (and international) research co-operation that would help to strengthen research capacity. More importantly, Namibia needs to draw on the wealth of research output from the rest of the world to develop interventions against alien invasions. By way of example, two excellent internet sites provide accounts of species that are invasive in different parts of the world (GISP Species Database; U.S. Forest Service 2005).

20.3 INSTITUTIONAL CAPACITY TO DEAL WITH INVASIVE ALIENS

Being a small country with limited development, Namibia lags well behind its southern neighbor, South Africa, with regard to its capacity to deal with invasive aliens, despite being under its control until 1990. This section briefly outlines the shortcomings and recommendations, as well as recent encouraging developments. For a comprehensive review, see Bethune et al. (2004).

20.3.1 Pertinent Policies and Legislation

A diversity of legislation exists that only indirectly alludes to invasive aliens and their control. However, none of the currently enacted legislation directly refers to *invasive alien* species, although the term is used in some policies. Perhaps the most unambiguous and most actively enforced legislation is the rather outdated Agricultural Pests Act, No. 3 of 1973 (Republic of South Africa 1973). This act makes provision for the control and eradication of, among others, plants at nurseries, plants infected by insects or plant diseases, and the control of plant imports, and also defines the powers of phytosanitary inspectors. Various sections provide for the destruction of, and regulation of the import of, exotic plants as well as allowing the importation of biocontrol agents to eradicate weeds and pests. The Act is essentially aimed at preventing the spread of exotic plants (and others) that may prove detrimental to the agricultural sector (i.e., economic pests). Unfortunately, a species that is not seen as an agricultural pest may still impact natural ecosystems negatively and significantly. Most of the current legislation would require protagonists of invasive alien control to shoulder the burden of proof that targeted species are also weeds or pests of agricultural production, or are implicated in soil erosion, or constitute a health risk to humans. Although the Act refers specifically to agricultural weeds and pests, there is potential to amend it to include invasive alien species in general (Malan, n.d.). The use of the term *precautionary principle* in some legislation is encouraging, and hopefully allows case studies of negative impacts of alien invasives from other countries to be used in lieu of in situ proof. It is disconcerting that there is no mention made in The Forestry Act No. 12 (Ministry of Environment and Tourism 2001) of indigenous versus alien species, nor any guiding principles with which to base decisions regarding invasive alien species. This is despite the fact that most reforestation species are alien plants (Bethune et al. 2004).

Namibia's Environmental Assessment Policy (1995) lists among activities requiring environmental impact assessments the "introduction and/or propagation of invasive alien plant and animal species." Unfortunately, the pending and long-awaited Environmental Management Bill (Ministry of Environment and Tourism 1998) that is based on this policy has still not been tabled in Parliament. This means that Environmental Assessments for significantly impacting projects are still not mandatory.

Another overdue bit of legislation is the draft Parks and Wildlife Management Bill (Ministry of Environment and Tourism 2004). Section 83 of this bill, entitled "Alien/invasive species," has not been formulated, providing an opportunity for those concerned with the issue to make a timely input (Bethune et al. 2004).

Bethune et al. (2004), and especially Malan (n.d.), in reviewing the legislation see opportunities to *tweak* existing legislation in favor of better control of alien invasive organisms in general, by adding clauses pertaining to invasive aliens, by adding regulations listing certain species as prohibited, and simply by means of interpreting legislation. However, it makes sense to introduce new legislation that explicitly refers to invasive alien species. It would be prudent to draw relevant clauses from South Africa's National Environmental Management: Biodiversity Act (Republic of South Africa 2004). In this Act, explicit mention is made to prevent the introduction and spread of invasive alien species, and the management and control of these species to "prevent or minimise harm to the environment and to biodiversity in particular." Hence, the concern is not just with agricultural pests. Namibia urgently needs to develop a national cross-sector policy dealing specifically with invasive alien species (Bethune et al. 2004).

Despite Namibia's rather vague and uncoordinated national legislation, it ratified the Treaty on Biological Diversity (UNCBD 1992) in 1997, and produced a detailed country study the following year (Barnard 1998). Since then, Namibia has actively fulfilled its obligations. Namibia is thus obliged to "as far as possible and as appropriate... prevent the introduction of, control or eradicate their alien species that threaten ecosystems, habitats or species" (UNCBD 1992).

It is encouraging to note new developments regarding regional co-operation over invasive alien control. These may eventually improve legislation and policy within the SADC (Southern African Development Community). For example, Member States are currently drafting guidelines for the control of invasive alien aquatic weeds within the region. The guidelines will encourage coordinated regional efforts. It is hoped that the momentum will be maintained and the same approach extended to invasive alien species in general.

20.3.2 Programs for Controlling Alien Invasives

Since the problem of invasive alien species in Namibia was highlighted in 1984 (Brown et al. 1985), eradication and control programs were implemented in parks and reserves. These programs lost momentum in more recent years. Since the establishment of specialist working groups under the National Biodiversity Task Force about 7 years ago, there has been a revival of interest and awareness in invasive aliens. The Alien Invasive Species Working Group and the Agricultural Biodiversity Working Group developed a poster (Namibia's Nasty Nine—Alien Invasive Species) to raise awareness. One recent effort by the Task Force to coordinate invasive alien activities was the inclusion of alien invasive control as a strategic aim in "Namibia's ten-year strategic plan of action for sustainable development through biodiversity conservation—2001–2010" (Barnard et al., n.d.). Most of these activities, including a review and categorization of information on invasive alien species in Namibia, the establishment of a database and atlas, research on the invasiveness of selected species, the establishment of appropriate legislation and control measures, further promotion of public awareness, and the initiation of more low-impact projects for problematic invasive species, have yet to get off the ground (Bethune et al. 2004). The voluntary members of the Alien Invasive Working Group cite lack of time as the main constraint. Currently, small clearing operations are conducted by nongovernmental organizations such as Scout groups and Greenspace (a group coordinating activities to maintain open spaces in urban areas). Government Agencies are sometimes involved, but the lead ministry (Ministry of Environment and Tourism) does not seem to have co-coordinated efforts to even control invasive aliens. It is likely that an acceleration of these efforts will occur as awareness and interest increases, as it seems to be.

20.4 FUTURE SCENARIOS AND SOLUTIONS IN NAMIBIA

Conflict species in reforestation, such as *L. leucocephala* (Hughes 2006), used for fodder, firewood, building timber, erosion control, *green manures*, shade, and shelter for crops (Richardson 1998) should soon be reaching the explosive phase in their invasion (William 1997). Namibia is under pressure to improve food security and to raise the standard of living of its citizens. With accelerated efforts to achieve this goal, through programs such as the Green Scheme and agroforestry programs, Namibian forests probably now face a significantly greater threat from invasive aliens than in the past. Propagule pressure will continue to increase to a stage where browsing pressure is no longer the limiting factor it appears to be at present (at least for edible species). The surface area of disturbed agricultural lands will increase, which will in turn increase the invasibility of forest ecosystems. The propagation of *conflict* species is likely to continue and increase as new useful species are discovered. No legislation exists regarding restrictions in planting invasive alien species, and there are no formal screening procedures. *J. curcas* (Euphorbiaceae) is a recent candidate for biodiesel fuel production on a commercial scale. Plans are underway to develop *J. curcas* plantations in Namibia for this purpose. The economic incentive to do this is huge, given the current runaway price of oil. The planting of *J. curcas* is also seen as a way to generate carbon credits for Namibia. The initiative can therefore also be promoted as an environmentally sensible one. In other parts of the world (e.g., the Pacific Islands and Western Australia), *J. curcas* has been identified as a noxious weed and invasive plant, being toxic to humans and livestock (U.S. Forest Service 2005; Smith 2006). The species is apparently well adapted to arid and semiarid conditions and occur naturally in a wide range of rainfall conditions, in "well drained soils with good aeration" (Ecoport, U.S. Forest Service 2005). Therefore, the species could well become invasive in the Kalahari Sand Forests. Since environmental assessment is still not compulsory in Namibia, the rather simplistic argument for development is likely to drown out the call for environmental precaution and prudence.

In view of the earlier-mentioned scenarios, Namibia needs to expand on the resurgent interest in invasive aliens (with regional and international influence). There is a window of opportunity to translate this to research, conservation programs, and legislative amendments on invasive aliens, while populations are still controllable. Namibia should focus on preventing entry of likely invasives, eradication of populations that are still in their establishment phase, and maintenance of populations at acceptable levels.

Namibia's southern neighbor, South Africa, is a relative powerhouse regarding invasive alien legislation, and research output and programs, particularly in the field of biocontrol. Namibia needs to strengthen already established relations to draw as much as possible from such countries. Namibia needs to attract research and management partners with which to explore impacts and solutions, and also to increase capacity in these fields. There is sufficient evidence throughout the world that Namibia can draw from to show that a precautionary approach is prudent. The fact that *C. odorata* (see earlier) has not been mentioned in any previously published list of invasives likely to be a problem in Namibia exemplifies this. While published lists are useful, the *C. odorata* example serves as a motivation for Namibia to introduce a weed risk assessment (WRA) system, adapted from the currently highly successful systems applied throughout Australia, Oceania, and New Zealand (e.g., Daehler et al. 2004). Because qualified resource management personnel are stretched thin and focused on other conservation issues, Namibia needs to employ at least one person entirely dedicated to policy formulation and program and research coordination on the control of invasive alien species, preferably under the Ministry of Environment and Tourism. Although the ad hoc efforts of a handful of people involved are commendable, these efforts need to be cemented through the will and commitment of government agencies. Strategies outlined in the Strategic Action Plan need to be implemented.

20.5 CONCLUSIONS

In many ways, Namibia is in an enviable position, with regard to its invasive alien plant problem. The integrity of Namibia's forest ecosystems is currently relatively unscathed by the impacts of invasive aliens (with the exception of ephemeral river forests with severe infestations of *Prosopis*). However, as Table 20.1 shows, past strengths may be tomorrow's constraints and weaknesses, as the invasive alien situation becomes more critical. Namibia represents an example of a country poised at a crossroads. As I have argued, a combination of propagule pressure and increasing invasibility of forest ecosystems in the north suggest a future accelerated rate of invasion. Some species may be about to reach the explosive stage of their growth. To avoid repetition of history, awareness campaigns, eradication and control programs, and entry prohibition programs must be implemented and improved very soon. If implemented and enhanced now, these could result in Namibia's forest ecosystems becoming a positive case study for the rest of the developing world to emulate. Failure to learn from other countries that have already suffered huge economic and ecological losses, and failure to act soon, will result in the need to implement expensive (and perhaps unaffordable and ineffective) control and eradication programs. At that stage, development priorities might preclude such programs from being effective.

ACKNOWLEDGMENTS

The following people are gratefully acknowledged for their contributions: John Mendelsohn provided files to generate maps; Vera De Cauwer created the maps; Shirley Bethune assisted with the layout and was the lead author of a useful report that preceded this chapter; Mike Griffin was coauthor of the mentioned report; Pamela Claassen, Vera De Cauwer, and Elizabeth Jones assisted with the editing of an earlier draft; Rod Randall and Sandy Lloyd from Western Australia were very helpful in providing information on weed risk assessment for the mentioned report.

REFERENCES

Alpert, P., Bone, E., and Holzapfel, C., Invasiveness, invasibility and the role of environmental stress in the spread of non-native plants, *Perspect. Plant Ecol. Evol. Syst.*, 3, 1, 52, 2002.

Barnard, P., Ed., *Biological Diversity in Namibia—A Country Study*, Directorate of Environmental Affairs, Windhoek, 1998.

Barnard, P., Brown, C.J., Jarvis, A.M., Robertson, A., and van Rooyen, L., Extending the Namibian protected area network to safeguard hotspots of endemism and diversity, *Biodivers. Conserv.*, 7, 531, 1998.

Barnard, P., Shikongo, S., and Zeidler, J., Eds., Biodiversity and development in Namibia, Namibia's ten-year strategic plan of action for sustainable development through biodiversity conservation 2001–2010, National Biodiversity Task Force, Ministry of Environment and Tourism, Windhoek, Namibia, n.d.

Berry, M., An investigation into the occurrence, population parameters, and suitable control methods of an alien invasive cactus species, *Cereus jamacaru*, at Waterberg Plateau Park, Namibia, Unpublished B. Tech. thesis, Polytechnic of Namibia, Windhoek, Namibia, 2000.

Bethune, S., Griffin, M., and Joubert, D.F., *National Review of Invasive Alien Species in Namibia. Consultancy to Collect Information on Invasive Alien Species in Namibia*, Directorate of Environmental Affairs discussion document, Directorate of Environmental Affairs, Namibia, 2004.

Boyer, D.C. and Boyer, H.J., The status of alien invasive plants in the major rivers of the Namib Naukluft Park, *Madoqua*, 16, 1, 51, 1989.

Brown, C.J., Macdonald, I.A.W., and Brown, S.E., Invasive alien organisms in South West Africa/Namibia, South African National Scientific Programmes Report 119, CSIR, Pretoria, 1985.

Central Bureau of Statistics, Population and housing census, National report—Basic analysis and highlights, Central Bureau of Statistics, Windhoek, Namibia, 2003.

Cunningham, P.L., *Tecoma stans* as a potential alien invasive in Namibia, *Dinteria*, 30, 33, 2008.

Cunningham, P.L., Joubert, D.F., and Adank, W., *Dodonaea angustifolia* an alien invasive to Namibia? *Dinteria*, 29, 11, 2004.

Curtis, B.A. and Mannheimer, C.A., *Tree Atlas of Namibia*, National Botanical Research Institute, Windhoek, 2005.

Daehler, C.C., Denslow, J.S., Ansari, S., and Kuo, H., A risk-assessment system for screening out invasive pest plants from Hawaii and other Pacific islands, *Conserv. Biol.*, 18, 2, 360, 2004.

D'Antonio, C.M., Fire, plant invasions, and global changes, in *Invasive Species in a Changing World*, Mooney, H.A. and Hobbs, R.J., Eds., Island Press, Washington, DC, 65, 2000.

Dean, W.R.J., Anderson, M.D., Milton, S.J., and Anderson, T.A., Avian assemblages in native *Acacia* and alien *Prosopis* drainage line woodlands in the Kalahari, South Africa, *J. Arid Environ.*, 51, 1, 1, 2002.

De Klerk, J.N., Bush encroachment in Namibia. Report on phase 1 of the Bush encroachment research, monitoring and management project, Ministry of Environment and Tourism, Namibia, 2004.

Duggin, J.A. and Gentle, C.B., Experimental evidence on the importance of disturbance intensity for invasion of *Lantana camara* L. in dry rainforest–open forest ecotones in northeastern NSW, Australia, *For. Ecol. Manage.*, 110, 279, 1998.

Erkillä, A. and Siiskonen, H., *Forestry in Namibia 1850–1990. Silva Carelica*, University of Joensuu, Joensuu, Finland, 1992.

FAO, *On Definitions of Forest and Forest Change*, FRA Working Paper No. 33, Food and Agricultural Organization, Rome, 2000.

Gabanakgang, S.B., Mapping of selected alien plant species along the Klein Windhoek River, Unpublished land management diploma project, Polytechnic of Namibia, Windhoek, Namibia, 2003.

Giess, W., A preliminary vegetation map of South West Africa, *Dinteria*, 4, 1, 1971.

Gracela, N.N., Bussmann, W.R., Gemmill, B., Newton, L.E., and Ngumi, V.W., Utilisation of weed species as sources of traditional medicines in central Kenya, *Lyonia*, 7, 2, 71, 2004.

Henderson, L., *Alien Weeds and Invasive Plants. A Complete Guide to Declared Weeds and Invaders in South Africa*, Plant Protection Research Institute, Agricultural Research Council, South Africa, 2001.

Hines, C.J., Schlettwein, C.H.G., and Kruger, W., Invasive alien plants in Bushmanland, Kavango and Caprivi, in *Invasive Alien Organisms in South West Africa/Namibia*, Brown, C.J., Macdonald, I.A.W. and Brown, S.E., Eds., South African national scientific programmes report no. 119, Pretoria, RSA, CSIR, 6, 1985.

Honu, Y.A.K. and Dang, Q.L., Responses of tree seedlings to the removal of *Chromolaena odorata* in a degraded forest in Ghana, *For. Ecol. Manage.*, 137, 75, 2000.

Hughes, C., *Leucaena leucocephala*, in the Global Invasive Species Database, 2006, available at http://www.issg.org/database/species/ecology.asp?si=23&fr=1&sts=, 2006, accessed July 1, 2006.

Jacobson, P.J., Jacobson, K.M., and Seely, M.K., *Ephemeral Rivers and Their Catchments—Sustaining People and Development in Western Namibia*, DRFN, Windhoek, 1995.

Joubert, D.F. and Cunningham, P.L., The distribution and invasive potential of Fountain Grass, *Pennisetum setaceum*, in Namibia, *Dinteria*, 27, 37, 2002.

Leslie, A.J. and Spotila, J.R., Alien plant threatens Nile crocodile (*Crocodylus niloticus*) breeding in Lake St. Lucia, South Africa, *Biol. Conserv.*, 98, 347, 2001.

MacDonald, I.A.W. and Nott, T.B., Invasive alien organisms in central SWA/Namibia: Results of a reconnaissance survey conducted in November 1984, *Madoqua*, 15, 1, 21, 1987.

MacDonald, I.A.W., Reaser, J.K., Bright, C., Neville, L.K., Howard, G.W., Murphy, S.J., and Preston, G., Eds., *Invasive Alien Species in Southern Africa: National Reports and Directory of Resources*, Global Invasive Species Programme, Cape Town, 2003.

Mack, R.N., Simberloff, D., Lonsdale, W.M., Evans, H., Clout, M., and Bazzaz, F., Biotic invasions: Causes, epidemiology, global consequences and control, *Ecol. Appl.*, 10, 689, 2000.

Malan, J., Discussion document on the regulation of alien invasive species, Unpublished Directorate of Environmental Affairs discussion paper, Directorate of Environmental Affairs, Namibia, n.d.

Mendelsohn, J. and el Obeid, S., *Forests and Woodlands of Namibia*, RAISON, Windhoek, 2005.

Mendelsohn, J. and Roberts, C., *An Environmental Profile and Atlas of Caprivi*, Directorate of Environmental Affairs, Namibia, 1997.

Mendelsohn, J., Jarvis, A., Roberts, C., and Robertson, T., *Atlas of Namibia: A Portrait of the Land and Its People,* David Phillip Publishers, Cape Town, South Africa, 2002.

Ministry of Environment and Tourism, The draft environmental management bill, Ministry of Environment and Tourism, Windhoek, Namibia, 1998.

Ministry of Environment and Tourism, The forestry act no. 12 of 2001, Ministry of Environment and Tourism, Windhoek, Namibia, 2001.

Ministry of Environment and Tourism, The parks and wildlife management bill, draft, Ministry of Environment and Tourism, Windhoek, Namibia, 2004.

NBRI, Unpublished list of specially protected plants in Namibia, National Botanical Research Institute, Windhoek, Namibia, 2002.

Office of the President, Namibia vision 2030. Prosperity, harmony, peace and political stability—Policy framework for long-term national development, Office of the President, Namibia, 2004.

Republic of South Africa, National environmental management biodiversity act (unnumbered) of 2004, Republic of South Africa, South Africa, 2004.

Republic of South Africa, The agricultural pests act 1973 (Act no. 3 of 1973), Republic of South Africa, South Africa, 1973.

Richardson, D.M., Forestry trees as invasive aliens, *Conserv. Biol.*, 12, 1, 18, 1998.

Shapaka, T., Joubert, D.F., and Cunningham, P.L., Alien invasive plants in the Daan Viljoen Game Park, *Dinteria*, 30, 19, 2008.

Silori, C. and Mishra, B., Assessment of livestock grazing pressure in and around the elephant corridors in Mudumalai Wildlife Sanctuary, south India, *Biodivers. Conserv.*, 10, 2181, 2001.

Smit, P., Geo-ecology and environmental change: An applied approach to manage Prosopis-invaded landscapes in Namibia, Ph.D. Thesis, University of Namibia, Windhoek, Namibia, 2005.

Smit, P., *Prosopis*: A review of existing knowledge relevant to Namibia, *J. Namibia Sci. Soc.*, 52, 13, 2004.

Smith, D., Western Australia bans *Jatropha curcas*, Biofuel review, 2006, available online at http://www.biofuelreview.com/content/view/28/2.

Tarr, P.W. and Loutit, R., Invasive alien plants in the Skeleton Coast Park, Western Damaraland and Western Kaokoland, in *Invasive Alien Organisms in South West Africa/Namibia*, Brown, C.J., Macdonald, I.A.W. and Brown, S.E., Eds., South African national scientific program report no. 119, CSIR, RSA, Pretoria, 19, 1985.

United Nations Convention on Biodiversity (UNCBD), Convention on biological diversity, Montreal, Canada, 1992, http://www.cbd.int/default.shtml.

U.S. Forest Service, Pacific Island ecosystems at risk (PIER), *Jatropha curcas*, 2005, available at http: //www.hear.org/Pier/species/Jatropha_curcas, 2005, accessed July 1, 2006.

U.S. Forest Service, Pacific Island ecosystems at risk (PIER), 2006, available at http: //www.hear.org/Pier/species, 2006, accessed July 1, 2006.

Van Wyk, B. and Gericke, N., *People's Plants. A Guide to Useful Plants of Southern Africa*, Briza, Pretoria, 2000.

Vinejevold, R.D., Bridgeford, P., and Yeaton, D., Invasive alien plants in the Namib-Naukluft Park, in *Invasive Alien Organisms in South West Africa/Namibia*, Brown, C.J., Macdonald, I.A.W. and Brown, S.E., Eds., South African national scientific programmes report no. 119, CSIR, RSA, Pretoria, 24, 1985.

Visser, J.N., Effect of *Prosopis glandulosa* on natural vegetation in Lower Swakop River, Unpublished nature conservation diploma project, Polytechnic of Namibia, Namibia, 1998.

William, P.A., *Ecology and Management of Invasive Weeds*, Conservation sciences publication 7, Department of Conservation, Wellington, 1997.

Williamson, M., *Biological Invasions*, Chapman and Hall, London, 1996.

Wilson, C., *Chromolaena odorata*, in the Global invasive species database, 2006, available at http://www.issg.org/database/species/ecology.asp?si=47&fr=1&sts=sss, 2006, accessed July 1, 2006.

Zachariades, C. and Goodall, J.M., Distribution, impact and management of *Chromolaena odorata* in Southern Africa, in *Proceedings of the Fifth International Workshop on Biological Control and Management of Chromolaena odorata*, Durban, South Africa, October 23–25, 2000, Zachariades, C., Muniappan, R. and Strathie, L.W., Eds., ARC-PPRI, Pretoria, 34, 2002.

Zimmermann, H.G., Ed., Country profile report: South Africa, Meeting on prevention and management of invasive alien species: Forging cooperation throughout southern Africa, Lusaka, June 10–12, 2000.

21 The Economics, Law, and Policy of Invasive Species Management in the United States: Responding to a Growing Crisis

Johanna E. Freeman, Rachel Albritton, Shibu Jose, and Janaki R.R. Alavalapati

CONTENTS

21.1 INTRODUCTION

On an almost daily basis, Americans are bombarded with news about threats facing their country: the possibility of another terrorist attack, the proliferation of nuclear weapons, and the threat of global warming. Despite the near-constant attention these topics receive, the American public remains largely unaware of one significant threat to its economy and environment: that of biological invasions. Biological invaders are defined as species nonnative to a region, which cause or are likely to cause economic, environmental, or social harm to that region (Executive Order 13112, United States Government 1999). In recent history, with growth in human population, tourism, and international trade, the transport of biological invaders between continents has escalated at a marked rate. These trends are expected to continue in the coming decades (Le Maitre et al. 2004).

For hundreds of years, plants and animals have been intentionally introduced to the United States for agriculture, horticulture, and sport, with little or no regard for their potential impacts on terrestrial and aquatic ecosystems. Others have entered the country in a flood of unintentional introductions, most often hidden in shipping crates or as stowaways in the ballast water aboard oceangoing ships. Some of these nonnative species, both intentionally and unintentionally introduced, may not pose threats to natural resources or society. However, in many cases, introduced species do become invasive, thereby causing serious financial and ecological harm.

The purpose of this chapter is to examine how economics, law, and policy relate to invasive species management. We will discuss (1) threats posed by invasive species to the natural resources and economy of the United States, (2) historical evolution of U.S. invasive species policy, (3) economics and policy of preventing new invasions and managing existing invasions, and (4) future directions for invasive species management in the United States.

21.2 PROBLEM: THREATS TO THE ECOLOGY AND ECONOMY OF THE UNITED STATES

Invasive exotic species are the second leading cause of biodiversity loss worldwide, due to their ability to replace native species through competition, predation, parasitism, and changes in ecosystem function (Myers 1997). This trend holds true in the United States, where 49% of all imperiled native species are threatened by competition with—or predation by—invasive species (Wilcove et al. 1998). In some extreme examples, native plant communities have been almost entirely replaced by a single invasive species, as with the yellow star thistle (*Centauria solstitalis*) in California grasslands (Pimentel et al. 2005).

Biological invasions have been wholly or partially responsible for at least 3 of the 24 known extinctions of endangered species in the United States (Schmitz and Simberloff 1997), and 400 species listed under the Endangered Species Act (ESA) are being negatively impacted by exotic invaders (Nature Conservancy 1996; Wilcove 1998; USFS 2004). Within the United States, invasive plants cover ~100 million acres and are estimated to spread at a rate of ~3 million acres annually, an area twice the size of Delaware (NISC 2001). Decisions regarding the control and prevention of exotic species invasions therefore have serious implications for biodiversity in the United States.

Ecosystem functions may be altered as a result of biological invasion, minimizing support for native flora and fauna, and impacting ecosystem services such as flood control, erosion control, and nutrient cycling. For example, *Miconia calvescens*, a tree native to South America, has formed monocultures in parts of Hawaii. Its root system, which is much shallower than those of the native species, provides poor anchorage for steep mountain slopes and the result has been numerous landslides in recent years (Kaiser 2006). In aquatic systems, where food webs are the backbone of the ecosystem, invasive species can precipitate the collapse of entire systems that once supported productive fisheries, as is the case with the zebra mussel (*Dresseina polymorpha*) in the Great Lakes.

Estimates of the annual costs to the U.S. economy due to exotic invasive species (including weeds, invertebrates, vertebrates, and pathogens) range between \$1.1 (OTA 1993) and \$120 billion/year (Pimentel et al. 2005). The large discrepancy between these two estimates was primarily due to a greater number of species used by Pimentel et al. (more than 10 times as many). Pimentel et al. (2005) also reported higher numbers for some of the same species used in the OTA estimate. Pimentel et al. (2005) present annual costs to some of the important sectors:

- *Weeds (crops)*: \$27 billion annually.
- The annual crop production loss from exotic weeds is \$24 billion each year. In addition to these losses, ~\$4 billion is spent on herbicides in the United States annually, of which \$3 billion is used for control of exotic weeds.
- *Weeds (pastures)*: \$6 billion annually.
  An estimated \$2 billion in forage losses is caused by weeds in pastures every year. Of these weeds, 45% are exotic, so nearly \$1 billion in forage losses can be attributed to exotic weeds. An additional \$5 billion/year is spent on control of exotic weeds in pastures and rangelands.
- *Vertebrate pests*: \$1.6 billion annually.
  Introduced vertebrates are another source of significant U.S. crop losses. Most notable are European starlings, which cause an estimated \$800 million/year in damage to grain and fruit crops, and feral pigs, which are also believed to cause around \$800 million in yearly damages to grain, peanut, soybean, cotton, hay, and vegetable crops.
- *Insect pests*: \$13.5 billion annually.
  Insects destroy ~13% of potential crop yields annually at a cost of \$33 billion/year. Since 40% of insects are exotic, \$13 billion/year in crop losses can be attributed to exotic insects. Additionally, of the \$1.2 billion spent on pesticides every year, \$500 million can be attributed to exotic insect control.
- *Plant pathogens*: \$21.5 billion annually.
  Plant pathogens cause crop losses of \$33 billion/year. Of this, an estimated \$21 billion/year is attributable to alien plant pathogens. Additionally, \$500 million/year is spent on control of alien plant pathogens.

- *Livestock diseases*: $5 billion annually.
  Microbes and parasites infecting livestock are responsible for ~$5 billion in damages and control costs annually.
- *Forests*: $4.2 billion annually.
  Insects and pathogens together cause $14 billion in losses to the forest products industry. Since 30% of these organisms are exotic, damages of ~$4.2 billion can be attributed to exotic invasives.
- *Aquatic systems*: $7.5 billion annually.
  An estimated total of $145 million is invested annually in the control of invasive exotic aquatic and wetland plants. The estimated loss due to the negative impacts of exotic fish on native fish and their ecosystems is $5.4 billion annually. The annual damage and control cost of zebra and quagga mussels is about $1 billion. The Asian clam also costs an estimated $1 billion in damages to waterways and native species.

21.3 INVASIVE SPECIES LAWS OF THE UNITED STATES

21.3.1 FEDERAL LAWS

Over 30 federal laws governing biological invaders have been enacted in the past century, and many states have enacted similar laws. The timeline in Table 21.1 shows the federal laws of the past 100 years most relevant to invasive species

TABLE 21.1
Timeline of Major U.S. Laws and International Treaties Relating to Invasive Species Management

Date	Title	Authority
1912	Plant Quarantine Act	7 USC §151 et seq.
1931	Animal Damage Control Act	7 USC §426
1944	Organic Act	Organic Act
1952	International Plant Protection Convention	UN Treaty
1957	Federal Plant Pest Act	7 USC §§150aa–150jj
1970	National Environmental Policy Act	Public Law 91–190
1973	Endangered Species Act	16 USC §1531 et seq.
1974	Federal Noxious Weed Act	7 USC §2814
1990	Non-Indigenous Aquatic Nuisance Prevention and Control Act	Public Law 101–646
1995	Agreement on the Application of Sanitary and Phytosanitary Measures	WTO Treaty
1996	National Invasive Species Act	Public Law 104–332
1998	Lacey Act (1900) Amendments	18 USC §42
1999	Executive Order 13112	Executive Order
2000	Plant Protection Act	Public Law 106–224
2004	Noxious Weed Control and Eradication Act	Public Law 108–412

management. The first U.S. law to regulate introduced species was the Plant Quarantine Act of 1912 (7 USC §§151–167), which granted the U.S. Department of Agriculture (USDA) authority over the importation and interstate transfer of plants, and was intended to prevent the spread of harmful pests and diseases. This was followed by the Animal Damage Control Act of 1931 (7 USC §§426–426c), which granted the USDA the authority to control any mammals or birds injurious to crops, forests, horticulture, livestock, and fur-bearing species. The 1944 USDA Organic Act expanded these control responsibilities to insects and diseases, authorizing the Secretary of Agriculture to "detect, eradicate, suppress, control, prevent, or retard the spread of plant pests" in the United States.

For most of the twentieth century, animal and plant inspection and quarantine were carried out by separate branches of the USDA. These functions were consolidated in 1972 with the creation of the USDA Animal and Plant Health Inspection Service (APHIS), which now serves as the lead federal agency for the majority of invasive species issues. APHIS's responsibilities include the following: (1) The prohibition, inspection, treatment, quarantine, and development of mitigation measures prior to allowing plants and animals of any kind to enter the country, (2) the initiation of new quarantine zones within the country to prevent the spread of escaped species that might become invasive, (3) conducting overseas inspections and quarantines of shipments destined for the United States, and (4) providing assistance with overseas eradication and control of pest species.

In the decades following the Plant Quarantine Act, the United States was flooded with countless new plants introduced for horticulture, erosion control, livestock forage, and other purposes. New plant pests and pathogens continued to appear as well, leading to the Federal Plant Pest Act of 1957 (amended in 1994) (7 USC §§150aa–150jj), which prohibited the importation, interstate shipment, or receipt through the mail of "any living stage of insects, mites, nematodes, slugs, snails, protozoa, or other invertebrate animals, bacteria, fungi, other parasitic plants or reproductive parts, viruses, or organisms . . . which can directly or indirectly injure or cause disease or damage in plants. . . ."

In recognition of the growing weed problem, the Federal Noxious Weed Act (7 USC §2814) was passed in 1974, authorizing APHIS to restrict the importation and spread of nonnative weeds, which were defined as

> . . . any living stage . . . of any parasitic or other plant . . . which is of foreign origin, is new to or not widely prevalent in the United States, and can directly or indirectly injure crops, other useful plants, livestock, or poultry or other interests of agriculture, including . . . the fish and wildlife resources of the United States or the public health.

The 1998 amendments to the Lacey Act (18 USC §42) provided APHIS with similar authority over invasive vertebrate and aquatic animal species.

In 2000, the Plant Protection Act (7 USC §§401–441) consolidated and modernized all of the preceding statutes pertaining to plant quarantine and protection. It expanded APHIS's role in addressing weed issues—granting it authority to take both emergency and extraordinary emergency actions to address biological invasions—as well as increasing the maximum civil penalty for violations. Four years later, in 2004, the Plant Protection Act was amended by the Noxious Weed Control and Eradication Act (Public Law 108–412), which authorized the Secretary of Agriculture to establish a grant program (subject to the availability of appropriations) to provide financial assistance for federal, state, local, and private entities seeking to control or eradicate noxious weeds.

One of the most significant federal initiatives in recent years was the creation of the National Invasive Species Council (Executive Order 13112) by President Clinton in 2001. The council was charged with providing national leadership and coordination on invasive species management; furthering early detection, rapid response, and control programs; pursuing international cooperation; developing a national information-sharing system; and increasing education and public awareness about invasive species (NISC 2001).

The Homeland Security Act of 2002 transferred some agricultural quarantine inspections from the USDA to the Department of Homeland Security's (DHS) Division of Customs and Border Patrol (CBP). More than 1800 agriculture specialists were transferred from APHIS to CBP (GAO 2006). In an evaluation of the transfer of inspection authority from APHIS to the DHS, the U.S. General Accounting Office concluded that management and coordination problems created by the change had actually increased the United States' vulnerability to biological invasions. Specifically, the GAO cited the neglect of key pathways, a reduction in funding for certain key ports, and the lack of a risk-based staffing model, which would ensure the placement of adequate numbers of specialists at the areas of greatest vulnerability (GAO 2006). In March of 2007, bills to restore inspection authority back to APHIS were introduced in both the Senate and the House of Representatives (NISIC 2007).

A variety of other federal laws aimed at specific organisms or ecosystems have also been passed over the years. Examples include the 2004 National Plan for the Control and Management of Sudden Oak Death (Public Law 108–488), and the 1992 Hawaii Tropical Forests Recovery Act (Public Law 102–574). New funding has been made available for invasive species control through the 2005 Public Lands Corps Healthy Forests Restoration Act (Public Law 109–154) and §6006 of the 2005 Safe, Accountable, Flexible, Efficient Transportation Equity Act: A Legacy for Users (SAFETEA-LU, Public Law 109–59), which made the control of invasive species in roadway corridors eligible for federal transportation funding. This was an important development, as roadways are known to be significant invasive species vectors.

In addition to the long list of federal laws aimed directly at invasive species control, the National Environmental Policy Act (NEPA) (Public Law 91–190) and the ESA (16 USC §1531 et seq.) are also relevant to invasive species issues. APHIS is required to complete an environmental assessment before approving permits for importing nonnative species (NISIC 2007). Additionally, the ubiquity of invasive species in the ecosystems of the United States makes their management and potential introduction a consideration in many other Environmental Impact Statements issued pursuant to NEPA and state "little NEPAs." Though it has not happened yet, the ESA could potentially be used to compel public or private entities to control invasives, since they are a threat to so many endangered species nationwide.

21.3.2 State Laws

There are numerous state laws governing invasive species, and some states have established centralized weed management departments and programs. The Idaho State Department of Agriculture's Noxious Weeds Management Program (NWMP) is an example of a well-coordinated invasive species program at the state level. The program, which began in 1998, has led to the creation of 32 Cooperative Weed Management Areas (CWMAs), which cover 82% of the state. The CWMAs are locally led, with a mandate to involve all landowners in the CWMA. The CWMAs coordinate and prioritize applications for grants through the Noxious Weed Cost Share Program, which in 2004 was funded jointly by the Idaho state legislature, the Bureau of Land Management, and the Forest Service. In 2004, 29 CWMA statewide projects were funded, at a cost of $200,000. In 2005, competition for funds was greater because of the first year's success, and because CWMA applicants continued to improve their grant-writing skills with experience. In addition to administering the cost-sharing program, the NWMP also serves as a clearinghouse for information (e.g., developing a GIS map of invasives across the state), provides technical support and recommendations to landowners, and makes yearly funding requests and policy recommendations to the governor and state legislature (ISDA 2005).

Many states have their own noxious weed lists, plant quarantine laws, and laws directing the control or eradication of pests and weeds. In Florida, for example, the Florida Aquatic Weed Control Act (Title XXVII, §369.20 of the Florida Statutes) directs the State Department of Environmental Protection to pursue "control, eradication, and regulation of noxious aquatic weeds." Other Florida statutes target specific weeds. For example, Title XXVII, §369.251 of Florida's Statutes states, "A person may not sell, transport, collect, cultivate, or possess any plant, including any part or seed, of the species *Melaleuca quinquenervia*, *Schinus terebinthifolius*, *Casuarina equisetifolia*, *C. glauca*, or *Mimosa pigra*," without a permit from the Department of Environmental Protection, while directing the South Florida Water Management District to "undertake programs to remove such plants from conservation area I, conservation area II, and conservation area III of the district." Florida also maintains its own noxious weed list, which restricts the importation, sale, and use of a large number of exotic plants. California, in addition to having its own set of ballast water discharge laws, a Weed Management Area system, and a noxious weed list, also has a system of Weed Free Areas. These are areas that have not yet been invaded by specific noxious weeds, and the California Department of Agriculture has been directed to ensure that they stay that way (NISIC 2007).

21.4 INVASIVE SPECIES POLICY: PREVENTING NEW INVASIONS

Prevention is the first line of defense in the fight against invasive species, and has consistently been found to be the most cost-effective method (NISC 2001; Wittenberg and Cock 2001). In the case of intentional introductions, many have argued that invasive species law should follow the precautionary principle, according to which introduced species would be presumed guilty until proven innocent, and precautionary measures would be taken even if the cause-and-effect relationships in question had not yet been scientifically established (Wittenberg and Cock 2001; Salmony 2005; McCarraher 2006). Until recently, this was not the principle guiding invasive species management in the United States, but the past decade has seen a shift in this direction, as exemplified by the ballast water initiatives described later.

21.4.1 Ballast Water Regulation

Aquatic systems in the United States are among the ecosystems most severely impacted by exotic invasive species, and the discharge of ballast water is one of the most significant pathways for new invasive species introductions. Large ocean-going ships maintain their stability by using water as ballast, pumping it into onboard tanks at the beginning of a voyage. Suspended plankton, small crabs, shellfish, and other organisms are unintentionally taken in with ballast water through metal grates and periodically released throughout the trip. The final load of ballast water is released when the ship arrives to port, expelling any organisms picked up along the way (Aquino 2006). The EPA (2004) has estimated that ~10,000 species are transferred daily around the globe through the release of ballast water. The infamous zebra mussel (*D. polymorpha*)—which has invaded and irrevocably altered huge portions of the Great Lakes and the Mississippi River ecosystems—is one of the best known examples of a ballast water stowaway.

Two major acts have been passed with the goal of regulating ballast water discharge: the 1990 Non-Indigenous Aquatic Nuisance Prevention and Control Act (Public Law 101–646) and the 1996 National Invasive Species Act (Public Law 104–332). Three acts have been passed pertaining specifically to the Great Lakes: the 1955 Convention of Great Lakes Fisheries between the United States and Canada, the 1999 Water Resources Development Act (Public Law 106–53), and the Great Lakes Fish and Wildlife Restoration Act of 2006 (Public Law 109–326). Additionally, two new bills, the National Aquatic Invasive Species Act and the Great Lakes Collaboration Implementation Act, were introduced into the U.S. Senate and House of Representatives in March of 2007. These bills, which have bipartisan support, are aimed at improving water quality, restoring habitat, and preventing new species introductions.

Given the nature of this vector, it is logical to conclude that it should be regulated as a form of water pollution. The Clean Water Act (CWA) (33 USCA. §1251 et seq.) prohibits the discharge of any pollutant from a point source into navigable waters of the United States without a permit pursuant to the National Pollutant Discharge Elimination System (NPDES) (Aquino 2006). Since the term *pollutant* includes *biological materials*, the term *point source* includes *a vessel or other floating craft*, and the definition of *navigable waters* includes oceanic waters up to 3 miles from shore, it would appear that the regulation of ballast water should fall to the EPA under the CWA. However, in 1973 the EPA created an exemption for discharges "incidental to the normal operation of a vessel," (40 C.F.R. §122.3(a)) and upheld this ballast water exemption for over 30 years (Aquino 2006).

Until recently, the regulation of ballast water fell to the Coast Guard, pursuant to two federal laws mentioned in the previous section—the1990 National Aquatic Nuisance Prevention and Control Act and the 1996 National Invasive Species Act.

The Coast Guard's mandatory regulations required any ballast-water-carrying vessel to file a report 24 hours prior to arrival in U.S. waters. Every vessel had to have a vessel-specific water management plan, which employed one of three practices: (1) ballast water exchange in the open ocean, at least 200 miles from shore, (2) retention of ballast water onboard the vessel, or (3) an alternative method approved by the Coast Guard (Aquino 2006).

In January of 1999, Northwest Environmental Advocates (NEA) petitioned the EPA to repeal its ballast water exemption, arguing that the exemption was inconsistent with the language of the CWA. The EPA did not respond to the petition until the NEA filed suit, eventually leading to a formal denial of the petition by the EPA in 2003. The NEA subsequently filed another lawsuit in the federal district court for the Northern District of California (NEA *v.* EPA 2006), seeking a declaration that the EPA had exceeded its authority under the CWA by exempting ballast water discharges from the NPDES. Six Great Lakes states (Illinois, Michigan, Minnesota, New York, Pennsylvania, and Wisconsin) intervened as plaintiffs (Aquino 2006). In September of 2006, a federal judge upheld the initial ruling and ordered the EPA to regulate the discharge of invasive species and comply with the CWA by September 30, 2008 (EPA 2006).

Aquino (2006) identified some major improvements in ballast water regulation to be expected from the EPA's new jurisdiction. Regulation under the NPDES permitting regime is likely to spur private sector investment in developing new and more effective treatment technologies, as it has with other pollution discharge systems, while the citizen suit provision of the CWA will greatly increase pressure on the federal government to take action on ballast water discharge.

21.4.2 Other Prevention Initiatives

The Agricultural Research Service, a division of the USDA, is making progress in an important aspect of prevention: the systematic classification of the characteristics of known invaders as well as the classification of general characteristics, which may contribute to a species' ability to become invasive (NISC 2001). Some known characteristics often associated with invasiveness are as follows: (1) rapid growth and maturation rate, (2) high level of fertility, and (3) ability to be an ecological generalist (that is, tolerance to a wide range of environmental conditions) (Myers 1997). There are also several ecological and environmental factors, which may be good predictors for the successful establishment of a nonnative species, such as: (1) similarity of the new climate to the species' native climate, (2) whether there are closely related native species in the new ecosystem, (3) disturbance history in the new ecosystem, and (4) intensity and frequency of nonnative species arrival (Kolar and Lodge 2001; Wonham 2006). The development of risk assessment protocols based on species characteristics associated with invasiveness was set as a goal by the 2001 NISC management plan, and by 2005 some progress had been made: Forest Service researchers had adapted the Australian/New Zealand Weed Risk Assessment protocol for use in Hawaii to reduce the sale of potentially invasive species, and NISC had convened several Screening Working Groups to push this initiative forward (NISC 2005).

Another major line of defense, and one that has historically been underfunded, is the identification and monitoring of transportation pathways and ports of entry for unintentional introductions (Wittenburg and Cock 2001). In the 2001 Invasive Species Management Plan, APHIS and FWS set goals to develop new regulations aimed at improving their ability to employ this strategy, and to add human and financial resources in support of these efforts (NISC 2001). As of 2005, both APHIS and FWS had succeeded in securing additional funding for port of entry inspections (bringing APHIS's total annual spending on agricultural quarantine to $144 million), and APHIS was considering several regulatory changes that would strengthen risk evaluation procedures (NISC 2005). ARS had created a list of over 3682 urgent/priority insects and mites, which were intercepted at ports of entry by APHIS (NISC 2005), and the NISC had convened a Pathways Working Group to begin discussions about developing a comprehensive, nationwide system for evaluating invasive species pathways and mitigation strategies.

21.4.3 International Treaties

In 1952, the United States entered into the International Plant Protection Convention, which created a system to control the spread and introduction of plant pests between countries. The system operates via the exchange of phytosanitary certificates between importing and exporting countries' national plant protection offices. This treaty was updated with the establishment of the World Trade Organization treaty in 1995, which contains an Agreement on the Application of Sanitary and Phytosanitary measures (NISIC 2007).

In 2004, the Ballast Water Working Group of the United Nations (UN) Marine Environment Protection Committee adopted the International Convention for the Control and Management of Ships' Ballast Water and Sediments, which will come into effect 1 year after ratification by at least 30 states with a combined merchant fleet of at least 35% of the world's shipping tonnage (McCarraher 2006). As of March 2007, only eight states had signed to the convention, and the United States was not among them.

McCarraher (2006) has proposed that an international treaty addressing all invasive species should also be convened by the UN similar to the Convention on International Trade in Endangered Species (CITES). Although CITES regulates trade in endangered species by identifying and grouping species based on their relative risk of extinction, a global invasive species treaty could categorize species based on their risk of invasiveness. As McCarraher notes, the invasiveness of a species is a function not only of its inherent characteristics, but of the ecosystem into which it is introduced. He therefore proposes that a global invasive species treaty should also incorporate aspects of the Cartagena Protocol on Biosafety, another UN treaty that regulates the movement of genetically modified organisms using a case-by-case risk assessment method.

21.4.4 Economic Modeling

Economics is inextricably linked to the regulation of invasive species. The technology of biological invasion forecasting has improved significantly in recent years

(Leung et al. 2005), and the development of bioeconomic models for species invasions has begun to follow suit.

Bioeconomic models have been actively used for years as a tool for managing renewable biological resources, especially in the regulation of fisheries (Sharov and Liebhold 1998).

These models take human behavior into account (i.e., different levels of harvest intensity in a fishery), as well as external environmental variables and natural population dynamics, in order to optimize profits and ensure population sustainability. Invasion models draw upon this basic framework, but from the opposite perspective—minimizing or eradicating invasive species for the least cost. They also build on concepts developed for ecological and pollution control modeling, the latter particularly in aquatic systems (Eisworth and Johnson 2002). Figure 21.1 provides an example of the conceptual framework for a bioeconomic model.

Bioeconomic analysis can play a key role in managing invasive species as a predictive tool to inform decisions concerning the choice of a management strategy, as a means for evaluating the effectiveness of strategies that have already been implemented, or to improve our understanding and awareness of problem species by assessing damages that have already occurred (Born et al. 2005).

Several ecological–economic invasion models with the potential to be used as decision-support tools have recently been developed for ecosystems of the United States. Sharov and Liebhold (1998) developed a bioeconomic model for slowing the spread of invasive species using the concept of barrier zones, and then applied it to gypsy moth (*Lymantria dispar*) populations. Horan and Lupi (2005) proposed a system of tradable risk permits for commercial shipping in the Great Lakes as a means of preventing future introductions. Settle et al. (2002) demonstrated the importance of accounting for both human behavior and invasive species competitive dynamics in a model predicting the expansion of an exotic lake trout (*Salvelinus nameycush*) population in Yellowstone Park.

A model developed by Leung et al. (2002) to predict the economic impacts of Great Lakes zebra mussels under different management scenarios suggested that greatly increasing near-term investment of public and private funds could markedly undercut future economic losses. Finnoff et al. (2007) further built on the work of Leung et al. (2002), demonstrating that risk aversion (in the form of reluctance to invest in zebra mussel prevention in uninfested areas), which is commonly displayed by those making financial decisions regarding zebra mussels, is the wrong financial strategy.

Buhle et al. (2005) modeled the cost-effectiveness of different control strategies for hypothetical species with different life histories, ranging from short-lived, rapidly reproducing species to species with high survivorship but low fecundity. They applied this model to the real-world case of oyster drills in Willapa Bay, Washington, calculating the relative costs and benefits of egg capsule removal versus adult drill removal for different locations in the bay. Margolis et al. (2005) modeled the effects of trade policies on species invasion rates, focusing on North American Free Trade Agreement. Some ecological–economic modeling has aimed to develop basic principles that can be used in a variety of invasion scenarios. For example, recognizing

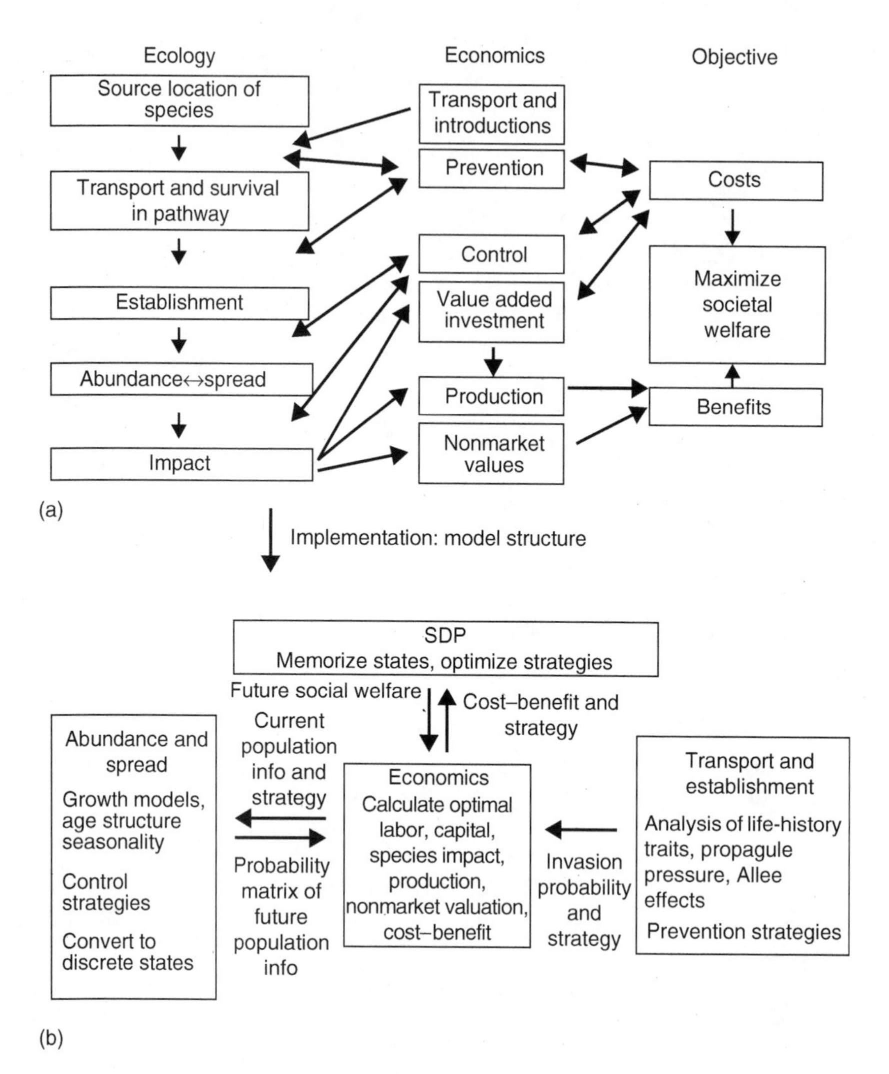

FIGURE 21.1 A generalized bioeconomic model of species invasions. a) Conceptual framework for a model of the invasion process, b) Model implementation. (Reproduced from Leung, B., Lodge, D.M., Finnoff, D., Shogren, J.F., Lewis, M.A., and Lamberti, G., *Proc. Biol. Sci.*, 269, 2407, 2002.)

that many ecological–economic models are too complex for time- and resource-strapped policy makers, Leung et al. (2005) recently developed a set of rules of thumb intended to provide a relatively simple framework to assist in predicting optimal investment strategies.*

* In 2007, a special issue of the *Journal of Agricultural and Applied Economics* (October, 2007) was devoted to bioeconomic models and their applications to manage invasive species.

Although the models described here are complex and the field of bioeconomic invasion modeling is young, simple versions of bioeconomic models are increasingly being used to guide policy. For example, the Maryland Department of Natural Resources commissioned a study to determine potential economic losses associated with uncontrolled nutria populations in the Chesapeake Bay. Nutrias destroy coastal marshes, impacting commercial fisheries, ecosystem services, and biodiversity. The study, which projected direct-use values (such as reduced shellfish harvests) as well as indirect-use biodiversity values (such as lost tourism revenue due to decreased bird habitat) over 50 years, concluded that in 50 years the State of Maryland would face economic losses of $70 million/year more under a no-control scenario than it would if it invested in nutria control immediately (Maryland DNR 2004). The federally coordinated Asian longhorned beetle (ALB) eradication program is another, even larger-scale example of a case in which bioeconomic forecasting was used to guide policy, and is discussed in detail in the following section.

21.5 INVASIVE SPECIES POLICY: EARLY DETECTION AND RAPID RESPONSE

Even the best prevention efforts cannot possibly detect all incoming pests. The next line of defense is early detection and rapid response (EDRR). There is currently no single, comprehensive national system for detecting, responding to, and monitoring invasions. However, the National Invasive Species Council is currently moving toward the development of a national EDRR protocol, in recognition of the fact that piecemeal efforts by public and private entities at the state, national, and international level are less likely to succeed (NISC 2001, 2005).

The Asian longhorned beetle (*Anoplophora glabripennis*) eradication program is an APHIS EDRR success story. The beetle was first discovered in the United States in New York City in 1996, and has since spread to several regions around the New York area. In subsequent years, it was also discovered in Chicago and New Jersey. Arriving by way of wooden shipping crates from China, the beetle is fatal to a large range of tree species in North America, including maples (*Acer* spp.), elms (*Ulmus* spp.), willows (*Salix* spp.), horse chestnuts (*Aesculus* spp.), mulberries (*Morus* spp.), birches (*Betula* spp.), green ash (*Fraxinus pennsylvanica*), sycamore (*Platanus occidentalis*), and London plane tree (*Platanus* × *acerifolia*). Like chestnut blight and Dutch elm disease, the Asian longhorned beetle has the potential to devastate urban tree canopies and permanently change the species composition of northeastern hardwood forests. On the basis of known figures for the beetle's damage in its native, China, a 2001 Forest Service/APHIS study estimated that the value of tree resources lost in New York and Chicago under a scenario of uncontrolled ALB spread would be $2.3 and $1.3 billion, respectively. In a worst-case scenario at the national level, if the beetle were to spread throughout the United States unchecked, the study estimated forest losses of up to $669 billion (USDA 2005).

Since the discovery of the beetle in 1996, APHIS has partnered with multiple state and city agencies in all three regions to stop the spread of the beetle and attempt to eliminate it altogether. The eradication program consists of the destruction of infested trees, the imposition of quarantine zones in which it is illegal to plant or remove host tree species, and regular injections with preventative insecticide. As of December 2005, 8000 infested trees had been removed (USDA 2005). Total expenditures on the national ALB eradication program for the years 1997–2006 (including offices in New York, New Jersey, and Illinois) were $233 million in federal funds and $41 million in state funds. The program employs over 160 federal, state, and city personnel. The eradication and control program is still a work in progress, but results so far are very promising and the number of ALB-infested trees is decreasing (USDA 2005).

It is evident from this case study that advances have been made at the federal level in coordinating early detection/rapid response programs. These advances can be attributed to painful experience (Dutch elm disease, for example), APHIS's expanded emergency authority, the federal government's increased interest in funding invasive species initiatives, and new technology (particularly Global Information Systems). Although an official federal EDRR protocol has yet to be developed, the federal government's response to recent invasions like this one will serve as an excellent prototype.

21.6 FUTURE OF INVASIVE SPECIES MANAGEMENT IN THE UNITED STATES

Continued federal, state, local, and private investment in initiatives like the Asian longhorned beetle eradication program is clearly warranted, but the tools for determining how best to invest those funds are inherently imperfect because both economics and ecology are inexact and complex sciences. This difficulty is compounded by the reality that both elected officials and private landowners are often under pressure to prioritize short-term economic interests over longer-term social and environmental welfare. However, the public investment of over $274 million in the Asian longhorned beetle eradication program is an evidence of an increased willingness on the part of policymakers to embrace bioeconomic modeling and take a precautionary approach to species invasions.

The control of existing, entrenched exotic species infestations was discussed only briefly in this chapter, though it is probably the biggest challenge of all. This is primarily because our chapter focused on invasive species issues at the federal level, whereas the actual combat carried out by resource managers and farmers everyday occurs at the state and local levels. The management of invasives—whether by herbicides, pesticides, biocontrols, or mechanical removal—is costly, site-specific, and species-specific. One of the most urgent needs is for major, multiyear funding allocations to back up the management-oriented grant programs authorized by recent legislation (such as the Noxious Weed Management Act and the National Invasive Species Act).

Securing funding for invasive species control is politically difficult. The existence of most Americans is almost completely removed from the agricultural and natural systems on which we ultimately depend. The urgency of the problem is therefore not apparent to most people, for the same reason that the value of ecosystem services is not apparent to most people. It is hard enough to convince the public that the destruction of natural resources is undesirable when it is visually evident, as in the clearing of land for development. In the case of many species invasions, the ecosystem still looks green, or is still populated by fish, and it can be unclear to the public why one species composition is so much less desirable than another. Furthermore, invasive species control is unglamorous for politicians. There is no ribbon cutting on an ongoing weed control program, and the loss of biodiversity—unless it is directly tied to a charismatic endangered species—can be very difficult to sell as a problem worthy of public investment. Public outreach is therefore an essential part of the equation. There are many species-specific outreach efforts on the part of governments and nonprofit groups, but a national organization devoted solely to advocacy against invasive species does not yet exist.

The 2001 creation of a central coordinating body for long-term invasive species planning and information sharing, the National Invasive Species Council, was a major step in the right direction. In a recent position paper delivered to the NISC, the Ecological Society of America recommended that the next step should be a National Center for Invasive Species Management, which would operate under the NISC and would be modeled after the powerful Centers for Disease Control and Prevention (Courchesne 2006). The new century has seen an auspicious start with regard to invasive species management, and the creation of a National Center for Invasive Species Management would help ensure that the United States continues to face its biological invaders head-on.

Another important future direction for invasive species regulation in the United States is in the retail sector. In spring of 2007, Meijer, a midwestern lawn and garden chain, sold 119 noninvasive trees, shrubs, and perennials carrying the Nature Conservancy's logo and a new *Recommended Noninvasive* tag. The chain also removed two invasives, the Norway maple and the Lombardy poplar, from its inventories. In Florida, the Nature Conservancy and the Exotic Pest Plant councils of Florida and the Southeast have partnered with the Lowe's chain, which has agreed not to sell 45 species of invasive plants in its Florida garden centers (Nature Conservancy 2007). Voluntary initiatives such as these will help combat the spread of invasives, both by curbing the sale of known invasives and by offering consumers an easy alternative they can feel good about buying. Perhaps the next step in this arena should be mandatory labeling regulations, under which all plants sold in the United States would have to bear a label stating whether they are a known invasive, unknown, or certified noninvasive.

Finally, the imposition of penalties and fees related to neglect and bad management practices should be integrated into current policy efforts. Many nonnative species are introduced intentionally for economic reasons, yet those responsible for the introductions are rarely the ones held financially responsible for the harm they cause (McNeely 2001). Many researchers and economists have echoed this sentiment, arguing that the *polluter pays* approach would motivate better practices among those responsible for the transportation of goods and services, as well as generating funds for current and future management needs (OTA 1993; IUCN 2000; Shine et al. 2000; Jenkins 2002). It has been suggested that the United States should implement a fee system similar to the levy fee system currently in place for oil tankers, which could also generate revenue for future rapid response and preventative measures (Jenkins 2002). Such a system would force those responsible for the transportation of goods to follow better standards of practice by increasing awareness of the threat of invasive species.

21.7 CONCLUSION

In many ways, the invasive species problem is most analogous to one of our society's most vexing pollution issues: nonpoint source pollution. Invasives are everywhere, and no single entity is to blame for their introduction. There is no way around the fact that the successful prevention, eradication, and control of species invasions will entail major public investment over many years. However, given the estimated annual cost of invasive species to the U.S. economy, not to mention the intangible costs to native ecosystems, it is clear that such an investment is warranted. In the past 10 years, great strides have been made at all levels of government to organize and fund invasive species management efforts. These efforts must be intensified if the United States is to make a serious stand against its biological invaders.

REFERENCES

Aquino, J.A., Navigating in uncertain waters: 2006 update on the regulation of ballast water discharge in the United States, *Pittsburgh J. Environ. Public Health Law*, 1, 101, 2006.

Born, W., Rauschmayer, F., and Brauer, I., Economic evaluation of biological invasions—A survey, *Ecol. Econ.*, 55, 321, 2005.

Buhle, E.R., Margolis, M., and Ruesink, J.L., Bang for buck: Cost-effective control of invasive species with different life histories, *Ecol. Econ.*, 52, 355, 2005.

Courchesne, C.G., Comprehensive approach to combat invasive species on the horizon, American Bar Association, *ABA Trends*, 37, 6, 1, 2006.

Eisworth, M.E. and Johnson, W.S., Managing nonindigenous invasive species: Insights from dynamic analysis, *Environ. Resour. Econ.*, 23, 319, 2002.

Environmental Protection Agency (EPA), Invasive non-native species, Watershed Academy Web, http://www.epa.gov/OWOW/watershed/wacademy/acad2000/invasive.html, 2004.

Environmental Protection Agency (EPA), September 2006 U.S. District Court Order granting injunctive relief, http://www.epa.gov/owow/invasive_species/091806Remedy_order.pdf, 2006.

Finnoff, D., Shogren, J.F., Leung, B., and Lodge, D., Take a risk: Preferring prevention over control of biological invaders, *Ecol. Econ.*, 62, 2, 16, 2007.

Horan, R.D. and Lupi, F., Tradeable risk permits to prevent future introductions of invasive alien species into the Great Lakes, *Ecol. Econ.*, 52, 289, 2005.

Idaho State Department of Agriculture, Noxious Weeds Program, Overview of Idaho's 2004 noxious weeds management program, 2005.

IUCN (World Conservation Union), IUCN guidelines for the prevention of biodiversity loss caused by alien invasive species, Information paper from the Fifth Meeting of the Conference of the Parties to the Convention on Biological Diversity, Nairobi, Kenya, May 15–26, 2000.

Jenkins, P.T., Paying for protection from invasive species, *Issues Sci. Technol.*, 19, 1, 67, 2002.

Kaiser, B.A., Economic impacts of non-indigenous species: *Miconia* and the Hawaiian economy, *Euphytica*, 148, 1–2, 135, 2006.

Kolar, C.S. and Lodge, M.L., Progress in invasion biology: Predicting invaders, *Trends Ecol. Evol.*, 16, 4, 199, 2001.

Le Maitre, D.C., Richardson, D.M., and Chapman, R.A., Alien plant invasions in South Africa: Driving forces and the human dimension, *S. Afr. J. Sci.*, 100, 103, 2004.

Leung, B., Finnoff, D., Shogren, J.F., and Lodge, D., Managing invasive species: Rules of thumb for rapid assessment, *Ecol. Econ.*, 55, 24, 2005.

Leung, B., Lodge, D.M., Finnoff, D., Shogren, J.F., Lewis, M.A., and Lamberti, G., An ounce of prevention or a pound of cure: Bioeconomic risk analysis of invasive species, *Proc. Biol. Sci.* (Formerly *Proc. R. Soc. Lond. B*), 269, 2407, 2002.

Margolis, M., Shogren, J.F., and Fischer, C., How trade politics affect invasive species control, *Ecol. Econ.*, 52, 305, 2005.

Maryland Department of Natural Resources, Potential economic losses associated with uncontrolled nutria populations in Maryland's portion of the Chesapeake Bay, Maryland Department of Natural Resources, Annapolis, MD, 2004.

McCarraher, A.G., The phantom menace: Invasive species, *N Y Univ. Environ. Law J.*, 14, 73, 2006.

McNeely, J.A., Ed., *The Great Reshuffling: Human Dimensions of Invasive Alien Species*, IUCN, Gland, Switzerland and Cambridge, UK, 2001.

Myers, N., Global biodiversity II: Losses and threats, in *Principles of Conservation Biology*, 2nd ed., Meffe, G.K. and Carroll, C.R., Eds., Sinauer, Sunderland, MA, 123, 1997.

National Invasive Species Council (NISC), Meeting the invasive species challenge: National Invasive Species management plan, Washington, DC, 80, 2001.

National Invasive Species Council (NISC), Progress report on the national invasive species management plan, Washington, DC, 30, 2005.

National Invasive Species Information Center (NISIC), http://www.invasivespeciesinfo.gov/, 2007.

Nature Conservancy, *America's Least Wanted: Alien Species Invasion of U.S. Ecosystems*, The Nature Conservancy, Arlington, VA, 1996.

Nature Conservancy, Seedlings of change, Nature Conservancy Magazine online, http://www.nature.org/magazine/spring2007/misc/index.html, 2007.

Pimentel, D., Zuniga, R., and Morrison, D., Update on the environmental and economic costs associated with alien-invasive species in the United States, *Ecol. Econ.*, 52, 3, 273, 2005.

Salmony, S.E., Invoking the precautionary principle, *Environ. Health Perspect.*, 113, 8, 509A, 2005.

Schmitz, D.C. and Simberloff, D., Biological invasions: A growing threat, *Issues Sci. Technol.*, 13, 4, 33, 1997.

Settle, C., Crocker, T.D., and Shogren, J.F., On the joint determination of biological and economic systems, *Ecol. Econ.*, 42, 301, 2002.

Sharov, A.A. and Liebhold, A.M., Bioeconomics of managing the spread of exotic pest species with barrier zones, *Ecol. Appl.*, 8, 3, 833, 1998.

Shine, C., Williams, N., and Gundling, L., *A Guide to Designing Legal and Institutional Frameworks on Alien Invasive Species*, IUCN, Bonn, 2000.

United States Department of Agriculture (APHIS), Asian Longhorned Beetle Cooperative Eradication Program Strategic Plan, http://www.aphis.usda.gov/ppq/ep/alb/strategic.pdf, 2005.

United States Forest Service (USFS), National strategy and implementation plan for invasive species management, Forest Service FS-805, United States Department of Agriculture, Washington, DC, 2004.

United States General Accounting Office (GAO), Homeland security: Management and coordination problems increase the vulnerability of U.S. agriculture to foreign pests and disease, http://www.gao.gov/new.items/d06644.pdf, 2006.

United States Government, Executive Order 13112 of February 3, 1999, invasive species, *Fed. Regist.*, 64, 25, 6183, 1999.

U.S. Congress, Office of Technology Assessment (OTA), Harmful non-indigenous species in the United States, OTA-F-565, U.S. Government Printing Office, Washington, DC, 1993.

Wilcove, D.S., Rothstein, D., Dubow, J., Phillips, A. and Losos, E., Quantifying threats to imperiled species in the United States, *Bioscience*, 48, 8, 607, 1998.

Wittenberg, R. and Cock, M.J.W., *Invasive Alien Species. How to Address One of the Greatest Threats to Biodiversity: A Toolkit of Best Prevention and Management Practices*, CAB International, Wallingford, Oxon, UK, 2001.

Wonham, M., Species Invasions, in *Principles of Biological Conservation*, 3rd ed., Groom, M.J., Meffe, G.K., and Carroll, R., Eds., Sinauer, Sunderland, MA, 2006.

Index

A

B

C

D

E

F

G

H

I

J

K

L

M

N

Q

R

S

T

U

V

W

Y

Z